Carbon

Ajith Manayil Parambil
Eepsita Priyadarshini • Paulraj Rajamani
Editors

Carbon

Bulk-to-Nano Forms for Detection and Remediation of Environmental Contaminants

Springer

Editors
Ajith Manayil Parambil
Nanotechnology Centre (CNT), Centre for Energy and Environmental Technologies (CEET)
VŠB – Technical University of Ostrava
Ostrava, Czech Republic

Eepsita Priyadarshini
School of Environmental Sciences
Jawaharlal Nehru University
New Delhi, Delhi, India

Paulraj Rajamani
School of Environmental Sciences
Jawaharlal Nehru University
New Delhi, Delhi, India

ISBN 978-3-031-90612-1 ISBN 978-3-031-90613-8 (eBook)
https://doi.org/10.1007/978-3-031-90613-8

This Springer imprint is published by the registered company Springer Nature Switzerland AG
The registered company address is: Gewerbestrasse 11, 6330 Cham, Switzerland

Preface

Environmental contamination poses an ever-growing challenge, demanding innovative, effective, and sustainable solutions. Among the various materials explored for environmental sensing and remediation, *carbon-based materials*—spanning from bulk structures to nanoscale forms—have emerged as versatile and powerful tools. Their remarkable properties, including high surface area, tunable functionalization, and exceptional stability, make them indispensable in the fight against environmental pollution.

This book, *Carbon: Bulk-to-Nano Forms for Detection and Remediation of Environmental Contaminants*, presents a *comprehensive exploration* of carbon-based materials and their diverse environmental sensing and remediation applications. It aims to bridge the gap between fundamental principles and real-world applications, offering insights into how these materials are revolutionizing modern environmental science.

We begin with an *introductory chapter* outlining the significance of *bulk and nanoscale carbon materials* in environmental applications. This sets the stage for understanding their pivotal role in tackling contemporary environmental challenges.

The subsequent chapters delve into *cutting-edge sensing technologies*, including *electrochemical detection, fluorescent carbon nanomaterials, colorimetric sensing*, and *surface-enhanced Raman spectroscopy (SERS)*. Each technique is examined in detail, emphasizing its principles, advantages, and practical applications in detecting environmental pollutants with high sensitivity and accuracy.

Beyond detection, this book explores *remediation strategies* leveraging carbon-based materials. Topics include *adsorption, catalytic degradation, membrane filtration,* and *antibacterial applications*, offering insights into the mechanisms and efficiencies of these approaches. The discussion extends to sustainable methods that minimize environmental impact while maximizing pollutant removal efficacy.

To strengthen the practical relevance of the concepts discussed, this book also includes *case studies and real-world applications*, showcasing the successful implementation of carbon-based sensing and remediation techniques. These examples highlight both the triumphs and challenges encountered in applying these materials to real environmental scenarios.

Designed as a *comprehensive resource*, this book serves as an essential guide for *researchers, environmental scientists, engineers,* and *policymakers* interested in harnessing the potential of carbon-based materials for environmental monitoring and remediation. By illustrating the adaptability, efficiency, and practical significance of carbon-driven solutions, we hope to inspire future innovations in sustainable environmental management.

As global concerns about pollution and resource depletion intensify, *carbon-based materials offer a promising path forward*—one that blends scientific ingenuity with environmental responsibility. We invite readers to explore the possibilities within these pages and contribute to the advancement of cleaner, greener technologies for a sustainable future.

Ostrava, Czech Republic — Ajith Manayil Parambil
New Delhi, Delhi, India — Eepsita Priyadarshini
New Delhi, Delhi, India — Paulraj Rajamani

Description

Carbon: Bulk-to-Nano Forms for Detection and Remediation of Environmental Contaminants offers a *comprehensive and in-depth exploration* of the role of carbon-based materials in addressing environmental pollution. From bulk carbon to nanoscale structures, this book highlights their remarkable potential in *detecting and removing contaminants* from air, water, and soil. It covers advanced sensing technologies, including *electrochemical detection, fluorescence-based techniques, colorimetry, and Raman spectroscopy*, providing insights into their principles, advantages, and applications. Additionally, it delves into cutting-edge *remediation strategies*, such as *adsorption, catalytic degradation, membrane filtration, and antibacterial applications*, showcasing their effectiveness in pollutant removal. Enriched with *real-world case studies and practical applications*, this book serves as a *valuable resource for researchers, environmental scientists, engineers, and policymakers*, offering innovative and sustainable solutions to contemporary environmental challenges.

Acknowledgment

The successful completion of this book, *Carbon: Bulk-to-Nano Forms for Detection and Remediation of Environmental Contaminants*, would not have been possible without the support, dedication, and contributions of numerous individuals and institutions.

First and foremost, we extend our *sincere gratitude to Springer Nature* for providing us with the platform to publish this work and for their unwavering support throughout the publication process. Their commitment to advancing scientific knowledge has been instrumental in bringing this book to fruition. We extend our heartfelt appreciation to the *publishing team at Springer Nature* for their professionalism, guidance, and unwavering support throughout the editorial and production process, ensuring the successful publication of this book.

We are deeply grateful to the *contributing authors* whose expertise and dedication have enriched this book with high-quality content. Their extensive research, insightful perspectives, and commitment to scientific excellence have helped shape a comprehensive resource on carbon-based materials for environmental applications. This book is a reflection of their hard work and invaluable contributions.

We also extend our appreciation to our *colleagues, research collaborators, and institutions* who have provided guidance, encouragement, and valuable insights during the development of this book. Their constructive feedback and scholarly discussions have played a significant role in refining the content and ensuring its relevance to both academic and practical applications.

Finally, we acknowledge our *families and friends*, whose unwavering support and patience have been a source of motivation throughout this journey. Their encouragement has been invaluable in enabling us to dedicate ourselves to this endeavor.

It is our hope that this book will serve as a valuable resource for researchers, environmental scientists, engineers, and policymakers, inspiring further innovations in sustainable environmental management and remediation.

Ajith Manayil Parambil
Eepsita Priyadarshini
Paulraj Rajamani

Contents

About the Editors

Ajith Manayil Parambil, Ph.D. is a dedicated researcher specializing in aqueous contaminant sensing and remediation, with expertise in carbon dots, novel composites, logic gates, device fabrication, and biocomputation. He is a Marie Skłodowska-Curie JAK OP Postdoctoral Fellow at the Nanotechnology Centre, Centre for Energy and Environmental Technology, VSB-Technical University of Ostrava, Czech Republic. He is also a visiting researcher at the School of Chemical Science and Chemical Engineering, Tongji University, Shanghai, China. He earned his Ph.D. from the School of Environmental Sciences, Jawaharlal Nehru University, New Delhi, in 2023, following an academically distinguished journey that includes an M.Phil. from the University of Madras and an M.Sc. from Gandhigram University, where he was a gold medalist and first-rank holder. Dr. Parambil was also a visting researcher at Kaohsiung Medical University, Taiwan.

His contributions to scientific advancement have been recognized through numerous prestigious awards and fellowships, including the Young Scientist Award 2024 from the International Association of Advanced Materials, Sweden, the AWSAR National Award for the Best Science Research Story, the INSPIRE Junior and Senior Research Fellowship from the Department of Science and Technology (DST), Government of India, and the Young Scientist International Travel Grant from the Science and Engineering Research Board (SERB), DST, Government of India. Additionally, he was awarded the Taiwan Experience Education Program

(TEEP) Visiting Researcher Scholarship by the Ministry of Education (MOE), Taiwan, and the Global Talent Internship Scholarship (2021) from the Ministry of Science and Technology (MOST), Taiwan. Dr. Parambil has made significant scholarly contributions, with publications in high-impact journals such as *Small* (Wiley), *Chemical Engineering Journal* (Elsevier), *Journal of Materials Chemistry A* (RSC), and *Journal of Materials Chemistry B* (RSC). His work continues to drive innovation in environmental remediation and advanced material applications.

Eepsita Priyadarshini, Ph.D. is an accomplished researcher specializing in nanomaterial fabrication and its environmental and biomedical applications, with a focus on developing nanocoatings to prevent biofilm infections on medical implants. She holds a Ph.D. in Biological Sciences from CSIR-IMMT, India, and completed her postdoctoral research at Jawaharlal Nehru University, New Delhi. Currently, she serves as an Assistant Professor at Galgotias University, Greater Noida, India. Dr. Priyadarshini was a DST Women Scientist at Jawaharlal Nehru University, New Delhi, where she contributed to advancements in biomedical nanotechnology through cutting-edge in vitro studies. Her academic journey is distinguished by numerous accolades, including the Chancellor's Gold Medal and the University Gold Medal for outstanding performance during her M.Sc. She was also awarded the prestigious Newton-Bhabha Ph.D. Fellowship, which enabled her to conduct research at University College London, United Kingdom.

Beyond her research, Dr. Priyadarshini plays an active role in the scientific community as a reviewer and editor for esteemed journals, such as *ACS Applied Biomaterials*, *Frontiers in Biomaterials*, *Materials Letters*, *Sensors and Actuators*, and the *Chemical Engineering Journal*. Her contributions continue to drive innovation in nanotechnology and biomedical science.

Paulraj Rajamani, Ph.D. is a distinguished researcher and educator in the fields of radiation biology, nanobiotechnology, nanotoxicology, environmental cancer biology, and occupational health. He is a Full Professor at the School of Environmental Sciences, Jawaharlal Nehru University (JNU), New Delhi, India. Dr. Rajamani began his academic journey with postgraduate studies in biology at Madurai Kamaraj University, Tamil Nadu, India. He then pursued M.Phil. and Ph.D. degrees at the School of Environmental Sciences, Jawaharlal Nehru University, where his doctoral research focused on investigating the impact of electromagnetic radiation on the developing brains of young rats. Following his Ph.D., he undertook postdoctoral research at JNU, supported by the Indian Council of Medical Research (ICMR), to further explore the biological effects of environmental and occupational exposures.

With a deep commitment to advancing scientific knowledge, Dr. Rajamani's research encompasses a broad spectrum of interdisciplinary studies, including the biological interactions of nanomaterials, radiation-induced cellular responses, environmental carcinogenesis, and occupational health risks. His work contributes to understanding the potential hazards of emerging technologies while seeking innovative solutions to mitigate their impact on human health and the environment. As a senior academic and researcher, Dr. Rajamani continues to mentor students, collaborate on cutting-edge research projects, and contribute to policy discussions on environmental and occupational health hazards. His expertise and scholarly contributions make him a leading figure in the domain of environmental sciences and biomedical research.

Chapter 1
Introduction to Carbon-Based Materials for Pollution Control

Namal Priyantha and Mishantha Karunathilake

Abstract This chapter addresses the current state of the production and investigation of carbon-based materials, a class of substances in which carbon atoms exhibit different hybridization states, leading to various allotropic structures, toward environmental applications concentrating on pollutant removal. The introduction section of this chapter summarizes the unique properties of carbon, highlighting its versatility, followed by a comprehensive overview of the importance of carbon-derived substances, namely activated carbon, graphite, and graphene, stressing their applications in the broad area of environmental applications, which includes pollution control through various modes. Subsequently, the scope of nano-dimensional carbon-based materials, mainly zero-dimensional fullerenes and quantum dots, nanotubes of 1D single-walled and multi-walled types, and 2D graphene in pollution control is briefed. Moreover, their attractive features, mainly large surface area, very high conductivity for electricity and heat, high stability, and environmental friendliness, will be emphasized with environmental applications. In addition to the fabrication of carbon-based materials, their functionalization is at the basis of several environmental applications, including adsorption, catalysis, and sensing. Novel carbon-based materials, including single-atom catalysis where single atoms are immobilized on the surface of the catalyst support and nanocarbon electrodes, with improved mechanical, electrical, chemical, and optical properties, lead to promising environmental applications. This chapter also addresses the sustainability and challenges associated with carbon-based materials, followed by emerging trends and future directions in developing carbon-based materials for environmental applications.

Keywords Carbon · Environmental applications · Nanomaterials · Pollution mitigation · Remediation

---

N. Priyantha (✉) · M. Karunathilake
Department of Chemistry, University of Peradeniya, Peradeniya, Sri Lanka
e-mail: namalpriyntha@sci.pdn.ac.lk

A. M. Parambil et al. (eds.), *Carbon*,
https://doi.org/10.1007/978-3-031-90613-8_1

Abbreviations

AC	Activated carbon
CDs	Carbon dots
CNDs	Carbon nanodots
CNTs	Carbon nanotubes
CQDs	Carbon quantum dots
GQDs	Graphene quantum dots
IUPAC	International Union of Pure and Applied Chemistry
MWNTs	Multi-walled carbon nanotubes
SACs	Single-atom catalysts
SWNTs	Single-walled carbon nanotubes
WHO	World Health Organization

1.1 Unique Properties of Carbon

1.1.1 Introduction to Carbon

The term "carbon" is derived from the Latin word "carbo," which is referred to charcoal in Roman. Since prehistoric times, carbon has been known in different forms, such as soot or charcoal. In the latter part of the eighteenth century, carbon was first recognized as an element, and several discoveries mark its history (Mostofizadeh et al., 2011). During the same time, Lavoisier demonstrated that diamond is a form of carbon, which is, in fact, a remarkable element essential to the body structure and the biological function of living organisms, and further, for various materials and technologies. Carbon, being an abundant element, secures a great place when its abundance is considered. Carbon exists in many different allotropic forms, expanding its applications (Falcao & Wudl, 2007).

Carbon is an exceptional element that can form a broad spectrum of materials. From sparkling gems to grimy soot, these materials have been the focus of study and utilization for centuries (Ebbesen, 1996). Its unique ability to form stable bonds with a wide range of elements, including itself, leads to its remarkable characteristics. Carbon serves as the basis for organic molecules, resulting in an extraordinary diversity of compounds. Carbon compounds make up more than 95% of all known chemical substances. Even though certain other atoms can bind to one another to create chains and clusters, none of them is comparable to the ability of carbon in forming bonds. It forms compounds with electronegative and electropositive components (Falcao & Wudl, 2007).

Pollution is no doubt a health hazard to all ecosystems and, therefore, to human life as well. The extent of both air pollution and water pollution has been expanded, mainly due to anthropogenic activities (Ajith & James, 2016). Air pollution alone has led to several millions of premature deaths annually. At the same time, it is worse for water pollution, which leads to over nine million deaths according to the

WHO and World Bank statistics, which has become a global challenge to minimize such adverse effects (Gopinath et al., 2020). Activated carbon and carbon nanomaterials support overcoming such challenges by effectively removing many air pollutants and heavy metals, dyes, and pharmaceuticals from water sources.

1.1.2 Allotropes of Carbon

Two centuries ago, carbon was discovered to be a key component of organic molecules and biomolecules. Carbon is also present naturally as amorphous carbon, diamond, and graphite. Although diamond and graphite contain only carbon atoms, their properties differ from each other due to their structural dissimilarities. For instance, diamond, the hardest material known, is an electrical insulator and a 3D transparent material. It comprises carbon atoms with sp^3 hybridization, and hence, each carbon atom is bonded to four carbon atoms in a tetrahedral structure. Conversely, graphite shows extremely high electrical conductivity, and it is a soft, opaque black substance. Among the essential characteristics of carbon that make it unique include the availability of allotropes, catenation (the ability of an element to bond with itself, thereby forming chain-like or ring-like structures), tetravalency, bonding versatility, and hybridization (the mixing of atomic orbitals to create new orbitals of different shapes and different energy, which helps explain molecular shapes and bonding properties). Allotropes of carbon can be categorized in several ways based on the carbon's morphological properties, dimensionality, or hybridization. Carbon allotropes exhibit a wide range of distinct properties (Bhattacharya et al., 2011). Figure 1.1 illustrates different allotropes of carbon.

Fullerene is an allotrope of carbon consisting of carbon atoms connected by single and double bonds, forming a completely or partially closed mesh structure with fused rings of five or six carbon atoms. Fullerene is available in many forms, mainly hollow spheres, ellipsoids, and tubes of different sizes. They are primarily produced by an arc discharge between carbon electrodes in an inert atmosphere. Fullerenes with closed mesh topology are more common, and those having an "n" number of carbon atoms can be represented as C_n. The most famous member of this special family of carbon allotropes is buckminsterfullerene (C_{60}), which was named after Buckminster Fuller, who suggested that the atoms of buckminsterfullerene form a hollow cage. It was the first known example of fullerene. The closed fullerenes, especially C_{60}, are commonly named buckyballs as they resemble a football. Many other types of fullerenes have been found: Nested closed fullerenes (bucky onions), cylindrical fullerenes (carbon nanotubes—CNTs—or buckytubes), and bulk solid form of pure or mixed fullerenes (fullerite) (Kharissova et al., 2019; Bhakta & Barthunia, 2020). Fullerenes were detected only after the accidental synthesis of buckminsterfullerene in 1985, much later than when their existence was predicted. The IUPAC definition of fullerenes is "polyhedral closed cages made up of an n number of carbon atoms, where $n \geq 20$, with pentagonal and hexagonal faces" (Godly & Taylor, 1997). The discovery of fullerenes initiated a new era because of

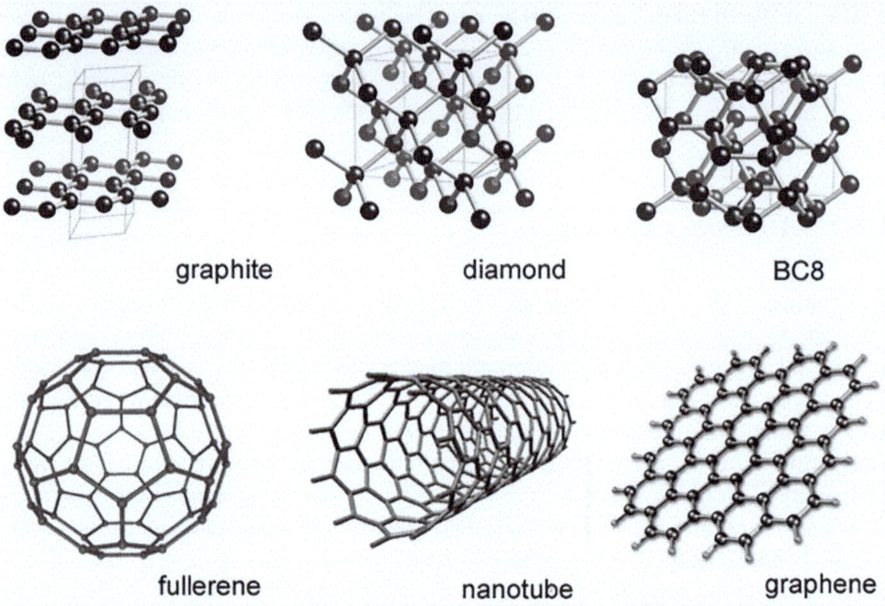

Fig. 1.1 Structural representation of allotropes of carbon. (Reproduced with permission from Oganov et al. Copyright 2013, GeoScienceWorld)

Table 1.1 Properties of common carbon allotropes

	Graphite	Diamond	Fullerenes	Graphene	Carbon nanotubes
Surface area (m^2 g^{-1})	~10–20		80–90	~1500	~1300
Hybridization	sp^2	sp^3	Mostly sp^2	sp^2	Mostly sp^2
Hardness	High	Very hard		Uppermost (for single layer)	High
Conductivity—Electrical (S cm^{-1})	Anisotropic: 2–3 × 10^4	10^{-16}	10^{-10}	~2000	Structure-dependent
Conductivity—Thermal (W m^{-1} K^{-1})	Anisotropic: 1500–2000, 5–10	~2000	0.4	4840–5300	3500

a new type of carbon allotrope and previously known allotropes—graphite, diamond, and amorphous carbon. Consequently, research on fullerenes and their chemical and technological applications became the subject of novel research. Fullerenes are used in many diversified areas, including materials science, electronics, and nanotechnology. A milestone in the discovery and scientific development of fullerenes was the recognition by awarding the Nobel Prize in Chemistry in 1996 to a group of scientists led by Robert F. Curl Jr. (The Royal Swedish Academy of Sciences, 1996). It should be stressed that research and applications of carbon-based materials continue to be investigated in many scientific fields. The properties of common carbon allotropes are given in Table 1.1.

1.1.3 Engineered Types of Carbon

In addition to natural forms of carbon, various engineered carbons have been developed, targeting more applications that could not be easily feasible with natural forms. Engineered carbons include synthetic diamonds, synthetic graphite, glassy carbon, carbon black, carbon for adsorption, coke, graphitic fibers, and diamond-like carbon. Some of their applications are the fabrication of electrodes, gemstones, tires, spacecraft composites, high-performance tennis rackets and cutting wheels, lubricants, shoe polish, catalytic supports, helium gas barriers, toner for photocopying machines, and printing ink. They are also heavily used in gas adsorption applications (Sengupta et al., 2011).

Engineered carbon materials are crucial as catalysts or catalyst supports for degrading environmental pollutants, significantly boosting remediation efficiency. Novel engineering materials, such as covalent organic frameworks, have provided carbon capture solutions, enabling effective absorption and secure storage of atmospheric carbon dioxide. This advancement addresses rising CO_2 levels, highlighting the potential of such material in sustainable environmental technologies (Tiwari et al., 2016).

1.2 Bulk Carbon Materials

1.2.1 Graphite

Graphite is formed through the metamorphism of sedimentary organic matter, which transforms graphite because favorable conditions of the Earth's crust provide the necessary temperature and pressure conditions. Natural graphite resources can be categorized based on their properties, such as grade, crystallinity, grain size, and shape. Accordingly, graphite ores are classified as microcrystalline, flake, and crystalline lump (vein) (Chelgani et al., 2015). Graphite exhibits a layered structure where adjacent layers (sheets) are attracted to each other through weak van der Waals forces, and hence, it shows a soft and greasy texture. These weak forces can easily be overcome when even a small force is applied. Therefore, when graphite pieces are rubbed, marks are left on fingers or paper (Georgakilas et al., 2015). Due to the above reasons, graphite can be used for writing, and the name "graphite" was derived from the Greek word "writing" (Sengupta et al., 2011). Graphite's other important physical and chemical properties include high electrical conductivity, heat conductivity, refractoriness, thermal resistance, and its inertness and solubility in liquid iron. Another desirable characteristic of graphite is that it possesses both metallic and nonmetallic properties, leading to advantages in many industrial applications, such as batteries, bearings, brake linings, conductive coatings, welding rods, refractories, electrical products, crucibles, electrodes, and paints. It is a primary source for producing graphene, which is known for its high quality and strong conductivity when derived from high-grade graphite (Kwiecińska & Petersen, 2004).

At standard room temperature and pressure, elemental carbon exists in its lowest energy state in the form of graphite. Graphite comprises stacked parallel 2D graphene sheets, with carbon atoms arranged in hexagonal rings (More et al., 2013). The graphene sheet in the flat orientation attains its lowest energy state due to the efficient lateral overlap of $2p_z$ orbitals of carbon atoms, which are parallel. As the bonding of carbon atoms within and perpendicular to the plane is different, graphite exhibits anisotropic properties. It exhibits greater strength along its plane compared to diamond due to its higher elastic modulus within the plane than perpendicular to it. The delocalized orbital throughout the graphene sheet also contributes to outstanding electrical and thermal conductivity. The unique structure of graphite consists of stacked graphene sheets arranged three-dimensionally, having a small separation of 0.335 nm (Sengupta et al., 2011).

1.2.2 Activated Carbon (AC)

When carbon is converted to activated carbon, it becomes highly porous, thereby being able to attract foreign molecules. More importantly, chemical functional groups can be introduced on the surface of activated carbon through modification via physical or chemical means. The production of activated carbon is neither complex nor costly, because many natural materials, such as fossil fuels, agricultural byproducts, and lignocellulosic waste material, can be used for its production (Nor et al., 2013). These materials are initially carbonized and then activated to form activated carbon. In the carbonization process, starting materials are heat-treated at temperatures in the range of 400 and 850 °C (Li et al., 2008). During this temperature range, volatile components are removed, leaving char, which contains mainly carbon; however, char produced through thermal treatment has neither high porosity nor high surface area, which are important for applications. Therefore, the activation step is essential to strengthen desirable characteristics. During the activation step of char, not only do previously inaccessible pores become more open and widened, but new pores are also developed. Activation to achieve the desired surface area and porosity involves physical and chemical approaches. Figure 1.2 illustrates different carbonaceous materials developed through synthetic approaches.

At high temperatures, carbonized char is physically activated by treatment with oxidizing gases (e.g., air, CO_2, or steam). It is preferred to use CO_2 due to its purity and controllable activation process, resulting in pores of high uniformity. Conversely, steam is preferred as it produces activated carbon with a high surface area due to steam's efficient diffusion within the char structure. In chemical activation, the starting material is mixed with activators having both oxidizing and dehydrating properties. Common activators are $FeCl_3$, KOH, and NaOH. In contrast to the physical approach, this process allows for simultaneous carbonization and activation at lower temperatures (300–500 °C); hence, initial char production is unnecessary. As a result, pyrolytic decomposition enhances the porous structure and increases the carbon yield. The chemical technique offers a larger surface area, shorter reaction

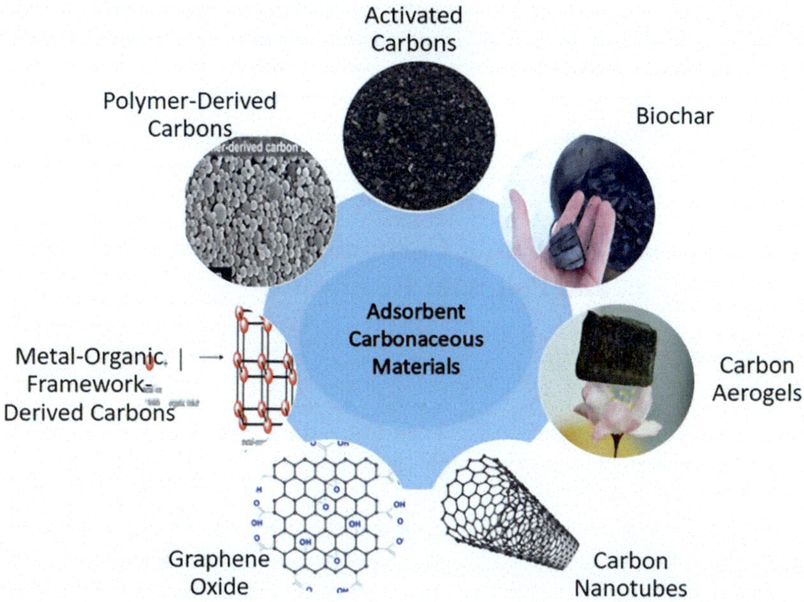

Fig. 1.2 Synthesized carbonaceous materials. (Reproduced with permission from Hoang et al. Copyright 2023, MDBI)

time, extremely microporous structure, and lower temperature requirement than physical methods (Sharma et al., 2022).

1.2.3 Diamond

Due to its unique properties, diamond is suitable for cutting tools, abrasives, and high-performance electronics. Valued for its brilliance, diamonds are very useful in the jewelry industry. Furthermore, diamonds are important in optical devices, laser systems and quantum computing, and further, their nitrogen-vacancy centers are building blocks for quantum information. Diamonds can also be synthesized at high temperatures and pressures on an industrial scale. They are utilized in industrial tools, electronics, and biomedical applications (Zhao et al., 2024; Balmer et al., 2009).

1.2.4 Glassy Carbon

Glassy carbon is another form of carbon with a unique amorphous structure, with atoms not arranged in a crystalline or graphitic form. It is produced by pyrolyzing organic precursors such as resins or polymers at high temperatures. Glassy carbon exhibits excellent thermal stability, electrical conductivity, and chemical resistance. Therefore, it is ideal for electrochemical sensors, electrodes, and high-temperature devices. It is also known for its low porosity and smooth surface, making it suitable for laboratory equipment and advanced materials, including aerospace, automotive, and biomedical applications (Vieira, 2022).

1.3 Traditional Use of Bulk Carbon Materials in Filtration, Adsorption, and Pollutant Removal

1.3.1 Water Filtration

For many years, activated carbon has been utilized as a water filtration medium to purify drinking water. Owing to the high-capacity activated carbon for adsorption of environmental pollutants, resulting from their high surface area and porosity, it is frequently used to remove water pollutants. As the source and processing conditions control the morphology and pore sizes of activated carbon, special attention must be given to selecting the specific type of activated carbon for the intended application. In addition to improving the taste and odor of drinking water, activated carbon would efficiently reduce the biological oxygen demand and the total suspended solids, which are considered important water quality parameters (Yamada et al., 2018). For this reason, most household water filtration systems contain activated carbon as their primary filter material (Isaeva et al., 2021; Sanchez & Garicia, 2023).

1.3.2 Soil Remediation

Biochar is a carbon-rich material created through pyrolysis of organic biomass. It has gained significant interest as a sustainable amendment for soil remediation. As biochar is highly porous and provides high surface area and desirable surface characteristics, it can be used to effectively adsorb both inorganic and organic contaminants (Yang et al., 2019). Among inorganic pollutants, heavy metals (e.g., Pb, Cd, Hg, and As) lead to significant environmental risk due to their hazardous nature and persistence in the environment. Biochar helps immobilize these metals through various mechanisms. One mechanism involves the adsorption of metal cations through electrostatic interaction with functional groups, such as carboxyl, hydroxyl, and carbonyl, enhancing the extent of their removal by biochar (Zhu et al., 2023).

1.4 Nano-Dimensional Carbon Materials

1.4.1 Carbon Dots (CDs)

Purification of single-walled carbon nanotubes led to the accidental discovery of CDs (Xu et al., 2004). CDs have a carbon core with surface functional groups. They are now considered a group of ultrasmall carbon nanostructures with diameters <10 nm. Graphene quantum dots (GQDs), carbon quantum dots (CQDs), carbon nanodots (CNDs), and polymer dots are the common types of CDs. The performance of CDs has been widely investigated, and their applications are being expanded in many disciplines due to their biocompatibility, non-toxicity, and quantum size (Parambil & Rajamani, 2024). The availability of surface functional groups of CDs leads to many characteristic properties, such as photo induced electron transfer, photoluminescence, and resistance to photobleaching. Although no definitive explanation is available, the luminous mechanism of CDs can be attributed to quantum effects, surface states, and electron–hole pair recombination.

CDs have become superior to metal-based quantum dots in many respects due to the above unique characteristics. They are well-suited for environmental applications (e.g., detection of environmental pollutants), as well as in biology (e.g., gene transmission, drug transport, bioimaging agents, biological sensing), chemical analysis (e.g., biosensors, fluorescence sensors), and energy (e.g., energy storage systems) (Song et al., 2017). CDs have the unique ability to accept and transport electrons, demonstrating their ability to be used in photocatalysis (Himaja et al., 2015). The catalytic activity of CDs could be further improved by forming carbon nanomaterials with semiconductor properties. Recent applications of CDs in the new generation batteries have also shown much promise. CDs, considered more advanced than quantum dots, could address undesirable issues, such as the blinking effect and increased size due to bioconjugation, making their delivery into cells more challenging. Moreover, CDs are believed to be less toxic and offer improved optical characteristics (Yao et al., 2019).

CDs are typically produced using two main methods: The "top-down" approach, which utilizes graphite materials as carbon sources, produces CDs through arc discharge, electrochemical synthesis, laser ablation, nanometer etching, or cleavage by hydrothermal or solvothermal oxidation. The "bottom-up" approach uses organic substances as carbon sources. This approach involves combustion, pyrolysis, hydrothermal, solvothermal, or microwave-assisted pyrolysis. Consequently, depending on the approach, CDs can be referred to as graphene nanodots or carbon nanodots, respectively (Xia et al., 2019).

1.4.2 Graphene

Graphene is an extraordinary substance that boasts numerous impressive qualities. It is the thinnest material known to humankind and holds the title of being the strongest material ever measured. Due to its exceptional properties, graphene's single-atom-thick unique 2D honeycomb lattice structure has gathered significant attention from the scientific/engineering community. Since its first production through mechanical exfoliation in 2004, its potential has been expanding for applications in diversified areas, including biology and catalysis, compared to other forms of carbon (Geim, 2009).

It has been revealed that applications of graphene composites have widened, which will further expand with the growing use of nanotechnology. Applications in environmental science include using graphene and its modified forms, including composites, for removing organic, metallic, and radioactive substances from the environment. Although graphene and its derivatives show great promise in biomedicine, its toxicity has become a limitation in such fields, requiring a thorough understanding of their features so that efforts to minimize toxicity can be attempted. Consequently, there has been a rapid increase in nanotoxicological studies on these compounds in recent years. It should be stressed that recent studies focused on assessing the toxicity of graphene have yielded conflicting results, with both toxic and non-toxic effects being observed. Therefore, it would be wise not to make sound conclusions about the safety risks associated with graphene, as these risks could depend on the specific type of material being utilized (Seabra et al., 2014).

1.4.3 Carbon Nanotubes (CNTs)

The history of CNTs goes back to 1991, when a paper on the electronic structure of very small tubes was made available based on theoretical concepts (Tiwari et al., 2016). Initially, the results were considered doubtful for publication because the synthesis of nanotubes at that time was very unlikely. It was later predicted that a sheet of graphite could be rolled into a small tube, and this tubular structure would have a high carrier density similar to that of metals, which is different from that of graphite. In support of the above findings, an investigation of the properties of a sample of carbon soot conducted in Japan using transmission electron microscopy led to the discovery of a tubular structure (Ebbesen, 1996). These nanotubes of carbon (or carbon nanotubes—CNTs) are completely graphitized and capped at both ends with pentagons, similar to fullerene molecules. Thus, they are different from previously known cylindrical carbon tubes and fibers. More importantly, it was observed that the carbon atoms in closed shells of each CNT are arranged, having various degrees of helicity.

This discovery led to many theoretical studies suggesting that CNTs exhibit metallic or semiconductor properties determined by their diameter and helicity.

CNTs were then considered to be the ultimate fiber as far as materials science is concerned, offering an exceptional strength-to-weight ratio. Expectations for CNTs have been expanded since early 1992 (Ebbesen, 1996). Discovery of CNTs is a unique invention in the field of nanotechnology, and it is now known that CNTs contain one or several perfect cylindrical layers of graphite sheets that consist of sp^2 hybrid carbon atoms which form a hexagonal network (Sonkar et al., 2021). This network has no edges and creates a tubular microcrystal of graphite. The exceptional potential of CNTs in various disciplines has promoted the investigation of their characteristics in the recent past. Research on CNTs for future applications requiring lightweight structural components, such as aircraft components, nanocomposites for aircraft coatings, lightweight energy storage systems, space elevators and aerogels, is underway (Ibrahim, 2013; Syduzzaman et al., 2025). Consequently, large amounts of CNTs, mainly substrate-free type, are now produced annually (Tiwari et al., 2016).

CNTs are of two types: single-walled CNTs (SWNTs) and multi-walled CNTs (MWNTs), both with open or closed ends. The former type can be formed by rolling up a single layer of graphite. They are tiny, with diameters in the range of 0.5 and 2.0 nm, which is even smaller than the width of human hair by about 10^5 times. Conversely, MWCNTs consist of many SWNTs as a nested "tube-in-tube" structure. Thus, their dimensions are much larger than those of SWNTs, ranging from 5 to 20 nm in diameter, although they could exceed even 100 nm in some instances. In comparison, the length of CNTs varies from <100 nm to several centimeters, covering a wide range of scales from molecular dimensions up to macroscopic distances (Mazhar, 2023). Exceptional cases of MWCNT are double- and triple-walled CNTs (Dai, 2002).

All carbons are bonded in a hexagonal lattice in an ideal CNT except at their ends. However, mass-produced CNTs often contain defects such as pentagons and heptagons in the sidewalls, altering their desired properties. When the hardness is considered, it is second only to graphene. SWNTs can remarkably tolerate high pressures of up to 24 GPa without altering the geometry. CNTs possess exceptional thermal and electrical properties, which are strong features desired in advancing superconductivity and telecopy technologies. CNTs exhibit ballistic conduction, a unique property that influences their thermal conduction behavior.

1.4.4 Fullerenes

Although the unique structure of fullerene was initially predicted in 1970 by Eiji Osawa, the first reported observation of the possible existence of fullerenes was made in 1985 by Smalley and Kroto. Moreover, the first large-scale production and isolation of fullerenes were completed in 1990 (Withers et al., 1997). Thereafter, different types of chemical transformations of fullerenes have been made effective in synthesizing various derivatives. Fullerenes are now recognized as versatile building blocks in organic chemistry that synthesize compounds with desirable electronic, chemical, geometric, and photophysical properties. The spherical C60, Buckminsterfullerene, is the most extensively studied fullerene containing 30 conjugated double bonds. Unlike classical aromatics such as benzene, fullerenes are three-dimensional and do not contain replaceable hydrogens for substitution reactions (Hirsch, 2010). The reactivity of fullerenes significantly differs from that of classical planar aromatic molecules undergoing mainly addition and redox reactions.

Fullerenes are a diverse group of carbon molecules available in various forms, including single- and multiple-layered tubes and spherical molecules. They have potential applications in diversified areas, such as batteries, catalysis, fuel cells, gas storage, optics, sensors, and separations. The versatility of fullerenes has made it an attractive research area for scientists to be engaged in a broad spectrum of future applications. One valuable application is the ability to inhibit the HIV enzyme by a fullerene derivative (Martinez et al., 2016).

1.4.5 Carbon Nanomaterial Supported Single-Atom Catalysts

Catalysis has become a broad area of science as catalysts can increase the rate of chemical reactions without being consumed because catalysts are regenerated at the end of the reaction. Consequently, catalysts are useful in applications in many areas, such as health, pharmacy, the environment, industry, and nutrition. Chemical reactions facilitated by catalysts enable efficient and high-yielding conversions, resulting in desired products minimizing the formation of undesirable byproducts (Peng et al., 2018).

It is known that metals and metal oxides with variable oxidation states are common in catalytic schemes in industry, including synthetic processes, pharmaceutical preparations, oil and fat production, and clean energy generation. Many heterogeneous catalytic systems use metal catalysts, particularly noble metals, including transition and inner-transition metals (Liu et al., 2017). Nevertheless, metal-based catalysts are associated with many undesirable aspects, mainly economic factors, selectivity issues, poor durability, limited availability, and risk of environmental pollution, and hence, their use has been limited. Solutions to these issues became available with the discovery of carbon-based materials, a new class of readily available and efficient catalysts. The first such finding, reported in 2009, was using carbon-based material, an efficient, low-cost, metal-free alternative to platinum, which had been used as the catalyst for oxygen reduction reactions in fuel cell applications. Thereafter, progress on using metal-free catalysts has been expanded for many catalytic processes (Kakar et al., 2024).

Carbon-based materials offer many advantages in addition to being metal-free catalysts. They are highly porous and have high surface area; they can be prepared with appropriate sizes and shapes; their reactivity can be tuned by incorporating functional groups for targeted applications such as molecular recognitions. Consequently, carbon-based materials are now employed as catalysts or catalytic support materials in energy and biological applications, and their applications will be further expanded in the future. An exciting expansion of carbon-based catalysis, reported recently, is the introduction of single-atom catalysts (SACs), where catalytic properties of metals are used after converting them to individual atoms, the smallest particles of matter. One limitation of individual atoms is that they cannot be kept stable as single atoms, although the efficiency of using metal atoms is maximized. SACs that contain isolated metal atoms are thus made stable by dispersing them on support materials, which is a valuable application in molecular recognition. The selectivity and reactivity of single-atom dispersions can be improved with proper designs to produce well-defined and uniform single-atom dispersion. Among many possible different types of support materials, carbon-supported SACs have been widely investigated due to their extraordinary features, namely tuneable morphology, ordered porosity, and ease of dispersion of many metal atoms, making them highly efficient in catalytic applications (Gawande et al., 2020). Carbon support materials' sustainability and environmentally friendly nature are considered additional advantages. Single-atom catalysts (SACs) are now gaining increasing significance in heterogeneous catalysis. Recent advances in the applications of carbon-based SACs are found in the fields of energy conversion and storage, as well as environmental protection (Su et al., 2024; Zhang et al., 2022).

1.5 Advanced Properties of Carbon Nanomaterials

A desirable feature of carbon nanomaterials is their vast surface area, which could be even greater than 1000 $m^2\ g^{-1}$, thereby making them chemically reactive, leading to applications in adsorption, catalysis, and energy storage devices such as batteries and supercapacitors. Carbon-based nanomaterials are very useful in the industrial sector owing to their many desirable properties.

CNTs, for instance, have an electrical conductivity of up to 5000 S cm^{-1}, and depending on their size and the atomic structure in the walls, they could possess either metallic or semiconducting behavior. Graphene is the basis for all other dimensionalities of carbon compounds, such as 1D nanotubes, 0D buckyballs, and 3D graphite. Due to its many similarities to the structure and properties of CNTs, including large surface area, abundant electrical conditions, and strong mechanical properties, graphene is a highly sought-after material with potential applications in many fields where CNTs have been used. These materials also allow for precise tuning of their electrical characteristics. For example, chemical doping or structural modification can change the bandgap of graphene, a semiconductor with a typical zero-bandgap, making it ideal for advanced transistors and chemical sensors. The graphene/*p*-Si combination formed using chemically vapor-deposited graphene initially yielded a poor % power conversion efficiency of 2.2%, which was later improved with different configurations (Ilya Mazov 2012).

Graphene offers chemical and physical characteristics: transparency of 97.7%, mechanical strength with a Young's modulus of 1 TPa, electrical conductivity of 2.0×10^3 S m^{-1}, thermal conductivity of 5.0×10^3 W m^{-1} K^{-1}, carrier mobility of

2.0×10^5 cm^2 V^{-1} s^{-1} at room temperature, and surface area of 2.6×10^3 m^2 g^{-1}, all of which are exceptional properties. Moreover, graphene is associated with low production cost yet possesses unconventional quantum Hall effect, strong ambipolar electric field effect, presence of relativistic Dirac fermions, outstanding environmental compatibility, and high adsorption capacity toward toxic substances (Geim, 2009; Li et al., 2016).

A comparison of critical physical properties of SWNTs and MWNTs is listed in Table 1.2 (Ilya Mazov 2012).

Carbon nanomaterials, particularly graphene and CNTs, are electrically tuneable and can be applied to various fields. Graphene is a zero-bandgap semiconductor that may be engineered by doping with foreign atoms or structural modifications to be employed in flexible electronics and high-speed transistors. Moreover, the chirality of CNTs governs the electrical conductivity and metallic nature—whether metallic or semiconducting. Because of these tuneable properties—which are critical for applications in nanoelectronics, sensors, and energy storage—carbon nanoparticles are leading the way in developments in electronics and materials research. Since graphene's sensitivity can be altered to detect gases and biomolecules, it is ideal for environmental monitoring and medical diagnostics. Graphene and CNTs are also utilized in thermoelectric devices and solar cells, the efficiency of which is increased by enhanced electrical properties. Additionally, their use in flexible and wearable electronics is expanding, allowing for the creation of tiny, flexible devices with various flexible applications, such as health monitoring and smart textiles (Ajayan & Zhou, 2001).

There are several options for using CNTs in electrical and optoelectronic devices because of their chirality, making them metallic or semiconducting. They are ideal for reinforcing composites in aerospace and automotive applications because of their lightweight and remarkable strength. Carbon nanomaterials, combining high surface area, customizable electrical properties, and extraordinary strength, are highly valuable for developing next-generation materials and technologies in many fields, including materials science, energy, and electronics.

Carbon dots (CDs), another type of zero-dimensional nanomaterial with exceptional optical, electronic, and chemical properties, exhibit strong photoluminescence with tuneable emission wavelengths. Owing to stronger biocompatibility than traditional quantum dots, CDs have been highly recognized in biomedicine. Consequently, CDs are ideal candidates for bioimaging, sensing, and optoelectronics. Further, their excellent electrochemical properties make them

Table 1.2 Some crucial theoretical and experimental data for SWNT and MWNT

Property	SWNT	MWNT
Density (g cm^{-3})	0.8	1.8
Elastic modulus (TPa)	1	0.3–1
Strength (GPa)	50–500	10–60
Resistivity (μΩ cm)	5–50	5–50
Thermal conductivity (W m^{-1} K^{-1})	3000	3000
Thermal stability (°C)	>700 (air), 2800 (vacuum)	>700 °C (air), 2800 (vacuum)
Surface area (m^2 g^{-1})	400–900	200–400

useful in electrochemical devices and energy storage. Moreover, easy functionalization broadens their applications in catalysis and drug delivery (Adam et al., 2022; Su et al., 2020). Moreover, graphene quantum dots (GQDs) are used in electronics, sensors, energy storage, and optoelectronics, including supercapacitors, batteries, and field-effect transistors (Kuntea et al., 2023; Kumar et al., 2020).

1.6 Comparison of Environmental Applications of Nano and Bulk Carbon Materials

1.6.1 Environmental Applications of Bulk Carbon Materials

With their characteristic benefits, bulk and nanocarbon materials are crucial for handling many environmental issues, such as pollution prevention, water filtration, and environmental monitoring. Because of their adsorption power and porous structure, bulk carbon materials—mainly activated carbon—have a long history of being effective in environmental applications. Common types of bulk carbon material include activated carbon in many different forms—granules, pellets, powders, impregnated, and catalytic. Each type of activated carbon should be produced by considering the grade, size, and structural properties specific to the desired application. Therefore, selecting the most suited type of activated carbon for a particular application depends on its mesh size, the type of application, and the chemicals to be removed.

Activated carbon provides the safest and the most efficient technology for removing toxic volatile organic compounds such as benzene, its derivatives, and chlorinated compounds; gaseous pollutants such as H_2S; airborne chemicals; and taste and odor-causing compounds. Impregnating activated carbon with specific reagents could improve the efficiency and selectivity of removing toxic substances (Pui et al., 2019).

Although bulk materials are dependable and economical, their long-term performance may be impacted by potential restrictions on the regeneration and longevity of the adsorbents. Furthermore, bulk carbon-based sensors frequently lack sensitivity and quick response times even though they can monitor air quality and identify higher levels of contaminants.

1.6.2 Characteristic Properties of Nanocarbon Materials for Environmental Applications

The novel qualities of nanocarbon materials and the proven efficiency of bulk carbon materials produce a synergistic approach to improving environmental management techniques. This combination enhances water quality and pollution control and makes a more thorough study of environmental effects.

Compared to bulk carbon materials, nanocarbon materials' remarkably large surface area and distinct reactivity, such as graphene and CNTs, show enhanced strength in adsorbing environmental contaminants. In addition to surface area, high chemical and mechanical stabilities and high electrical conductivity of CNTs support adsorption characteristics (Chen & Wang, 2021). Moreover, porosity, hollow, layered architecture, tiny diameter, and high aspect ratio of CNTs make them popular for adsorbing water pollutants. Charge transfer, hydrophobic interactions, p–p interactions, and electrostatic interactions are the fundamental mechanisms behind pollutant adsorption on CNTs, which has become strong due to the presence of four active adsorption sites: (1) internal sites that make up the hollow tube structure; (2) interstitial channels that provide space between the walls; (3) peripheral nano bundle grooves that provide spaces between them; and (4) exposed active external surface area (Arora & Attri, 2020).

1.6.3 Nanocarbon Materials for Water Purification

The exceptional qualities of nanocarbon materials make them suitable for use in sophisticated filtration systems because such systems can catch contaminants in the air and water more effectively than existing systems. Graphene oxide membranes, for example, are becoming state-of-the-art filtering and desalination technologies having extremely high removal rates of organic molecules and nanoparticles.

Demand for global water supply has been increasing, and the water quality has been decreasing, intensifying water scarcity and limiting access to safe water for drinking and basic hygiene. As only a small percentage of the water available on earth is for human consumption, purifying contaminated water for various uses is highly recommended. In this context, economic, efficient, and environmentally friendly techniques are needed for water purification to maintain water quality and avoid water shortage issues. The most common desalination technology that is used to economically obtain fresh water from saline water is membrane separation. Currently, CNTs are effectively utilized for each step—reverse osmosis, nanofiltration, and membrane distillation—of the entire desalination process (Yang et al., 2019).

1.6.4 Nanocarbon Materials for Pollution Control

Adsorption affinity, in general, is a measure of the number of active functional groups on a surface, which can also be applied to nanomaterials. Having electroactive functional groups on the surface of carbon nanomaterials is beneficial for removing electroactive pollutants. The removal ability is enhanced if the functional groups show catalytic activity. Industrial effluents released from industries such as battery manufacturing, chemical production, mining operations, and metal finishing contain different toxic metal ions. Divalent metal ions present in industrial effluents can effectively be removed through adsorption on CNTs. In this respect, the effect of solution conditions such as the pH and the concentration of the adsorbate (i.e., metal ion) on the extent of removal and adsorption isotherm models has been investigated, demonstrating the strong potential application of CNTs in environmental protection (Kakar et al., 2024). A versatile and highly efficient composite filter based on CNTs to remove heavy metal ions from wastewater with almost 100% efficiency has also been reported (Parham et al., 2013).

Emerging pollutants, comprising endocrine-disrupting chemicals, flame retardants, pesticides, nanomaterials, pharmaceuticals, personal care products, and persistent organic pollutants, are usually present in the environment at very low levels between μg L^{-1} and ng L^{-1}. Nevertheless, they could cause serious harm to the health of ecosystems. Consequently, attention should be focused on removing such pollutants from the environment, especially from industrial pollutants. Importantly, composite membranes based on CNTs have been fabricated in the filtration of the above-emerging pollutants (Wang et al., 2015). Another emerging environmental issue is the production of per- and polyfluoroalkyl compounds, a class of heavily used industrial chemicals. Their applications include water-resistant coatings, firefighting foams, and surfactants. These compounds are persistent in the environment due to their strong carbon–fluorine bonds. MWCNTs have been used efficiently to remove perfluorooctanoic acid and perfluorooctane sulfonate from contaminated water.

1.6.5 Nanocarbon-Based Electrodes for Environmental Monitoring

Electrochemical sensors offer many advantages. Potentiometric sensors can detect ionic pollutants, while amperometric sensors detect electroactive pollutants. Electrochemical sensors can detect a particular metal at different oxidation states. Moreover, they are portable and can be used for field measurements. Some electrochemical sensors do not require tedious sampling procedures and yet can detect metal ions at equilibrium quantitatively. They exhibit reliable and inexpensive detection of environmental pollutants as compared to more sophisticated techniques. However, low sensitivity and limited applicability owing to the overpotential and background issues of conventional electrochemical sensors must be overcome (Sahragard et al., 2023).

With the discovery of nanomaterials, some undesirable issues of traditional electrochemical sensors have been overcome. Electrochemical sensors based on nanomaterials can easily be fabricated by modifying their surfaces. Further, their sensitivity and detection limits are more attractive due to their extremely high surface area, which is available for adsorption and easy modification procedures. Moreover, these sensors can be fabricated by adjusting their size and morphology so that they can be tuned for the intended application. Further, a very low volume of nanomaterials results in the miniaturization of electrochemical devices. Consequently, nanomaterials, in conjunction with hyphenated techniques, are now being used as adsorbents in the remediation of environmental pollutants. Recent attention in this area of research has focused on carbon-based nanomaterials due to their attractive conductivity, stability, and adsorption capacity. Depending on the type of pollutants detected, they can easily be modified to improve selectivity and/or sensitivity through surface modification with functional groups (Lochab et al., 2021).

CNTs, as electrochemical sensing materials, could catalyze electrochemical reactions of environmental pollutants that would otherwise show sluggish kinetics, making them difficult to be detected at low levels (Alvaro et al., 2004). However, for specific detections, surface functionalization of CNTs may be needed. Covalent functionalization involves chemical reactions for attaching functional groups to CNTs, and alcohol, carbonyl, and carboxylic functional groups have been attached for various potential applications (Shen et al., 2017). These functional groups lead to strong interactions with environmental pollutants, making their detection possible. The hydrophilic –OH and –COOH groups, when attached to the surface of CNTs, exhibit excellent adsorption characteristics toward low molar mass and polar analytes, thereby improving detection limits.

Functionalization of the surface of MWCNTs could be even more advantageous because of the greater active surface area as compared to that of CNTs and of stronger ionic interaction with anions as compared to the unmodified MWCNTs (Balasubramanian & Burghard, 2008). The enhanced surface area and molecular-level interactions are important factors for detecting environmental pollutants at very low levels, especially for detecting emerging pollutants. For instance, glassy carbon electrodes modified with HOOC-MWCNTs have exhibited high sensing and adsorption capacity for ionic pollutants. Moreover, voltammetric techniques have simultaneously determined both SO_3^{2-} and NO_2^- at nanomolar levels due to the peak potential difference between the two analytes (Sudha et al., 2018).

1.7 Environmental Impact, Sustainability, and Challenges of Production of Carbon-Based Materials

Manufacturing bulk carbon materials and carbon nanoparticles presents serious sustainability and environmental issues. Carbon nanomaterials, including CNTs and graphene, are usually synthesized by chemical vapor deposition, electrolysis, arc discharge, and hydrothermal and template-directed synthesis. These all require energy, resulting in significant carbon emissions and possible hazards to ecosystems and human health. Furthermore, the production procedures may release hazardous waste, requiring precautions to avoid contaminating the environment. Conversely, mining methods used to get bulk carbon-based materials may cause habitat damage, water pollution, and land degradation. In addition to emitting high levels of CO_2, nanoparticles produced may finally end up in living things, posing a different issue for ecosystems and human health. Furthermore, the synthetic procedures may produce hazardous waste, such as solvents and byproducts. Another challenge to be addressed is to convert carbon nanotubes into low-cost filtration membranes for targeted applications (Kumah et al., 2023).

To minimize the environmental impact and maximize the benefits of carbon nanoparticles and bulk carbon materials, sustainability in their manufacturing and application should be maintained. One method that shows promise for producing carbon nanoparticles is the synthesis using sustainable feedstocks. To improve sustainability, the production of CNTs and graphene from biomass, agricultural waste, and other natural sources is being investigated, promoting the reduction in the use of fossil fuels and lowering carbon footprints. Obtaining activated carbon from waste products not only makes use of easily accessible resources but also encourages waste management, thereby decreasing the environmental damage from using conventional sources.

Improvement of recycling methods of carbon nanomaterial would also contribute to the circular economy by enabling materials to be reused for various purposes. Using effective regeneration techniques would increase the sustainability of bulk carbon materials, especially activated carbon. Reactivation of activated carbon would extend its lifetime, decreasing the amount of the production of new material. Moreover, the environmental impact of producing bulk carbon materials can be reduced by implementing stronger restrictions and implementing sustainable mining techniques. This leads to decreased habitat loss and enhanced water management, ensuring that mining operations are carried out ethically. Both carbon nanoparticles and bulk carbon materials can be used efficiently while reducing their environmental impact by prioritizing these sustainable practices, promoting a more sustainable future (Mishra et al., 2023). Table 1.3 provides the comparison for applying bulk and nanocarbon materials (Cui et al., 2021; Jiwanti et al., 2022).

1.8 Development of Carbon-Based Material for Environmental Applications: Conclusion, Emerging Trends, and Future Directions

Modernization of society, trend of urbanization, industrial development, and advancements in transportation have led to significant environmental pollution issues (Jun et al., 2018). Humans disrupt the natural world by releasing harmful substances, such as antibiotics, pesticides, dyes, heavy metals, greenhouse gases, and volatile organic compounds, into our environment. These pollutants pose serious threats to our health and the environment. While the environmental quality is deteriorating, the demand for comfort is improving. The combination of

Table 1.3 Application of bulk and nanocarbon materials

Broad area of application	Specific application—nanocarbon material	Specific application—bulk carbon materials
Electrochemical	To enhance supercapacitors, batteries, and biosensors	To construct electrodes and sensors
Biomedical	Drug transportation, bioimaging, and biosensing, such as nanotubes and graphene oxide	To perform adsorption on biochar
Environmental	Pollutant removal, photocatalysis	Utilized in filtration (charcoal, activated carbon)
Energy	To develop supercapacitors, lithium-ion batteries, hydrogen storage	To build fuel cells, carbon electrodes, coal-based energy
Industrial use	Advanced materials for lightweight structures and to implement high-performance coatings	To be used in metallurgy, construction, and composites

technological advancements and unique characteristics of carbon has made it possible to overcome some of the above-stated issues, providing necessary service to humankind.

Carbonaceous waste contains constituent substances with carbon–hydrogen, carbon–halogen, or carbon–carbon chemical bonds. It has been explored as a raw material for producing carbon-based functional materials due to its abundance and low cost. Surface-modified forms of carbon-based nanomaterials possess remarkably high specific surface areas along with expanded adsorption sites, forming complexes with heavy metal ions and some organic molecules. Consequently, these nanomaterials are highly effective in efficiently capturing and removing both inorganic and organic pollutants from wastewater, making them an asset in environmental remediation efforts. Additionally, it has been demonstrated that carbon-based nanomaterials are highly effective in removing a variety of atmospheric pollutants, such as NO and H_2S, with high efficiency, which can be further optimized by tuning functionalization together with using nanomaterial with extremely high surface areas, such as graphene oxide, SWCNTs, and MWCNTs. A recent extension of nanocarbon material research has been reported on the adsorption and removal of emerging pollutants. Further development and expansion in this area of research are much needed as there has not been adequate, reliable, and sensitive technology to remove these highly toxic pollutants, some of which could be present in drinking water sources and the environment. It is hoped that technology will be developed to design nanocarbon-based filters for removing emerging pollutants even at low levels.

Another recent advancement in carbon research is the production of carbon-based aerogels through high-temperature pyrolysis of biowaste. These aerogels exhibit remarkable enhancement in oil absorption. While utilizing waste material, this innovative value-addition process offers sustainable solutions for environmental challenges. Furthermore, the emergence of bio-derived carbon materials and hybrid composites marks a progressive shift toward more sustainable solutions, enhancing eco-friendliness and overall cost-effectiveness. The combination of cutting-edge technology and tunability through functionalization, ability of composite formation, and extremely high surface area of carbon-based material would further be expanded, addressing critical challenges in the areas of environmental remediation, industrial effluent treatment, low-level detection of pollutants and removal of emerging substances, as well as energy and power storage using advanced devices such as Li-ion and Na-ion batteries and supercapacitors, contributing to global sustainable development goals (Ma et al., 2017; Gusain et al., 2020).

References

Adam, S., Isaac D., Peter L., Helen M. 2022. "Tax Policies to Reduce Carbon Emissions." Fiscal Studies 43 (3). https://doi.org/10.1111/1475-5890.12308.

Ajayan, P. M., & Zhou, O. Z. (2001). Applications of carbon nanotubes. *Topics in Applied Physics*, 80, 391–425. https://doi.org/10.1007/3-540-39947-x_14

Ajith, M.P. and James, M.K., 2016. Impact of Chamravattam regulator cum bridge on Bharathapuzha river and adjacent areas. *Indian Journal of Economics and Development, 4*(1), pp.2320–9828.

Alvaro, M., Atienzar, P., Delgado, J. L., & Langa, F. (2004). Sidewall functionalization of single-walled carbon nanotubes with nitrile imines. Electron transfer from the substituent to the carbon nanotube. *The Journal of Physical Chemistry B, 108*(34), 12691–12697. https://doi.org/10.1021/jp0480044

Arora, B., & Attri, P. (2020). Carbon nanotubes (CNTs): A potential nanomaterial for water purification. *Journal of Composites Science, 4*(3), 135. https://doi.org/10.3390/jcs4030135

Balasubramanian, K., & Burghard, M. (2008). Electrochemically functionalized carbon nanotubes for device applications. *Journal of Materials Chemistry, 18*(26), 3071. https://doi.org/10.1039/b718262g

Balmer, R. S., Brandon, J. R., Clewes, S. L., Dhillon, H. K., Dodson, J. M., Friel, I., & Inglis, P. N. (2009). Chemical vapour deposition synthetic diamond: Materials, technology and applications. *Journal of Physics: Condensed Matter, 21*(36), 364221. https://doi.org/10.1088/0953-8984/21/36/364221

Bhakta, P., & Barthunia, B. (2020). Fullerene and its applications: A review. *Journal of Indian Academy of Oral Medicine and Radiology, 32*(2), 159. https://doi.org/10.4103/jiaomr.jiaomr_191_19

Bhattacharya, P., Fornari, R., & Kamimura, H. (2011). *Comprehensive Semiconductor Science and Technology. Elsevier Science.*

Chelgani, C., Rudolph, M., Kratzsch, R., Sandmann, D., & Gutzmer, J. (2015). A review of graphite beneficiation techniques. *Mineral Processing and Extractive Metallurgy Review, 37*(1), 58–68. https://doi.org/10.1080/08827508.2015.1115992

Chen, C., & Wang, Y. (2021). The precise engineering of nanostructured carbon materials. *ACS Central Science, 7*(9), 1470–1472. https://doi.org/10.1021/acscentsci.1c00979

Cui, X., Tian, J., Zhang, C., Cai, R., Ma, J., Yang, Z., & Meng, Q. (2021). Comparative study of nanocarbon-based flexible multifunctional composite electrodes. *ACS Omega, 6*(4), 2526–2541. https://doi.org/10.1021/acsomega.0c04313

Dai, H. (2002). Carbon nanotubes: Opportunities and challenges. *Surface Science, 500*(1), 218–241. https://doi.org/10.1016/S0039-6028(01)01558-8

Ebbesen, T.W., Hidefumi H., Margaret E.B., Michael, J. L. Shreeve-K., Robert C.H. 1996. "Decoration of Carbon Nanotubes." Advanced Materials 8 (2): 155–57. https://doi.org/10.1002/adma.19960080212.

Falcao, E. H. L., & Wudl, F. (2007). Carbon allotropes: Beyond graphite and diamond. *Journal of Chemical Technology & Biotechnology, 82*(6), 524–531. https://doi.org/10.1002/jctb.1693

Gawande, M. B., Fornasiero, P., & Zbořil, R. (2020). Carbon-based single-atom catalysts for advanced applications. *ACS Catalysis, 10*(3), 2231–2259. https://doi.org/10.1021/acscatal.9b04217

Geim, A. K. (2009). Graphene: Status and prospects. *Science, 324*(5934), 1530–1534. https://doi.org/10.1126/science.1158877

Georgakilas, V., Perman, J. A., Tucek, J., & Zboril, R. (2015). Broad family of carbon nanoallotropes: Classification, chemistry, and applications of fullerenes, carbon dots, nanotubes, graphene, nanodiamonds, and combined superstructures. *Chemical Reviews, 115*(11), 4744–4822. https://doi.org/10.1021/cr500304f

Godly, E. W., & Taylor, R. (1997). Nomenclature and terminology of fullerenes: A preliminary survey. *Fullerene Science and Technology, 5*(7), 1667–1708. https://doi.org/10.1080/15363839708013349

Gopinath, K. P., Vo, D. V. N., Prakash, D. G., Joseph, A. A., Viswanathan, S., & Arun, J. (2020). Environmental applications of carbon-based materials: A review. *Environmental Chemistry Letters, 19*(1), 557–582. https://doi.org/10.1007/s10311-020-01084-9

Gusain, R., Kumar, N., & Ray, S. S. (2020, February). Recent advances in carbon nanomaterial-based adsorbents for water purification. *Coordination Chemistry Reviews, 405*, 213111. https://doi.org/10.1016/j.ccr.2019.213111

Himaja, A. L., Karthik, P. S., & Singh, S. P. (2015). Carbon dots: The newest member of the carbon nanomaterials family. *The Chemical Record, 15*(3), 595–615. https://doi.org/10.1002/tcr.201402090

Hirsch, A. (2010). ChemInform abstract: Principles of fullerene reactivity. *ChemInform, 30*(18). https://doi.org/10.1002/chin.199918323

Hoang, T. D., Bandh, S. A., Malla, F. A., Qayoom, I., Bashir, S., Peer, S. B., & Halog, A. (2023). Carbon-based synthesized materials for CO_2 adsorption and conversion: Its potential for carbon recycling. *Recycling, 8*(4), 53. https://doi.org/10.3390/recycling8040053

Ibrahim, K. S. (2013). Carbon nanotubes-properties and applications: A review. *Carbon Letters, 14*(3), 131–144. https://doi.org/10.5714/cl.2013.14.3.131

Isaeva, V. I., Vedenyapina, M. D., Kurmysheva, A. Y., Weichgrebe, D., Nair, R. R., Nguyen, N. P. T., & Kustov, L. M. (2021). Modern carbon–based materials for adsorptive removal of organic and inorganic pollutants from water and wastewater. *Molecules, 26*(21), 6628. https://doi.org/10.3390/molecules26216628

Ilya M., Vladimir K., Anatoly R., Valentin S. 2012. "Properties of MWNTContaining Polymer Composite Materials Depending on Their Structure." InTech EBooks, August. https://doi.org/10.5772/48245.

Jiwanti, P. K., Wardhana, B. Y., Sutanto, L. G., Dewi, D. M. M., Putri, I. Z. D., & Savitri, I. (2022). Recent development of nanocarbon material in pharmaceutical application: A review. *Molecules, 27*(21), 7578–7578. https://doi.org/10.3390/molecules27217578

Jun, L. Y., Mubarak, N. M., Yee, M. J., Yon, L. S., Bing, C. H., Khalid, M., & Abdullah, E. C. (2018). An overview of functionalised carbon nanomaterial for organic pollutant removal. *Journal of Industrial and Engineering Chemistry*, 67, 175–186. https://doi.org/10.1016/j.jiec.2018.06.028

Kakar, U. M., Tauanov, Z., Azat, S., Ahmadi, N. A., Hassand, M. H., & Sarwari, A. (2024). Carbon nanotubes as adsorbents for heavy metals: Focus on arsenic and hydrargyrum removal from water. *Journal of Chemistry Studies, 3*(1), 07–20. https://doi.org/10.32996/jcs.2024.3.1.2

Kharissova, O. V., Kharisov, B. I., & Gonzales, C. M. O. (2019). Carbon–carbon allotropic hybrids and composites: Synthesis, properties, and applications. *Industrial & Engineering Chemistry Research, 58*(10), 3921–3948. https://doi.org/10.1021/acs.iecr.8b05857

Kumah, E. A., Fopa, R. D., Harati, S., Boadu, P., Zohoori, F. V., & Pak, T. (2023). Human and environmental impacts of nanoparticles: A scoping review of the current literature. *BMC Public Health, 23*(1), 1059. https://doi.org/10.1186/s12889-023-15958-4

Kumar, Y. R., Deshmukh, K., Sadasivuni, K. K., & Pasha, S. K. K. (2020). Graphene quantum dot based materials for sensing, bio-imaging and energy storage applications: A review. *RSC Advances, 10*(40), 23861–23898. https://doi.org/10.1039/d0ra03938a

Kuntea, M., Chanfon, L. C., Nimbalkara, S., Bunnell, J., Barajasa, E. R., Vazquez, M. E., Trejo-Rodriguez, D., Faucher, C., Smitha, S., & Kassegne, S. (2023). *Insights to molecular and bulk mechanical properties of glassy carbon through molecular dynamics simulation and mechanical tensile testing. ArXiv* (Cornell University). https://doi.org/10.48550/arxiv.2312.02388

Kwiecińska, B., & Petersen, H. I. (2004). Graphite, semi-graphite, natural coke, and natural char classification—ICCP system. *International Journal of Coal Geology, 57*(2), 99–116. https://doi.org/10.1016/j.coal.2003.09.003

Li, W., Yang, K., Peng, J., Zhang, L., Guo, S., & Xia, H. (2008). Effects of carbonization temperatures on characteristics of porosity in coconut shell chars and activated carbons derived from carbonized coconut shell chars. *Industrial Crops and Products, 28*(2), 190–198. https://doi.org/10.1016/j.indcrop.2008.02.012

Li, X., Yu, J., Wageh, S., Al-Ghamdi, A. A., & Xie, J. (2016). Graphene in photocatalysis: A review. *Small, 12*(48), 6640–6696. https://doi.org/10.1002/smll.201600382

Liu, J., Bunes, B. R., Zang, L., & Wang, C. (2017). Supported single-atom catalysts: Synthesis, characterization, properties, and applications. *Environmental Chemistry Letters, 16*(2), 477–505. https://doi.org/10.1007/s10311-017-0679-2

Lochab, A., Sharma, R., Kumar, S., & Saxena, R. (2021). Recent advances in carbon based nanomaterials as electrochemical sensor for toxic metal ions in environmental applications. *Materials Today: Proceedings, 45*, 3741–3753. https://doi.org/10.1016/j.matpr.2021.01.271

Ma, Q., Yu, Y., Sindoro, M., Fane, A. G., Wang, R., & Zhang, H. (2017). Carbon-based functional materials derived from waste for water remediation and energy storage. *Advanced Materials, 29*(13), 1605361. https://doi.org/10.1002/adma.201605361

Martinez, Z. S., Castro, E., Seong, C. S., Cerón, M. R., Echegoyen, L., & Llano, M. (2016). Fullerene derivatives strongly inhibit HIV-1 replication by affecting virus maturation without impairing protease activity. *Antimicrobial Agents and Chemotherapy, 60*(10), 5731–5741. https://doi.org/10.1128/aac.00341-16

Mazhar, M. (2023, June 13). *Present and future commercial applications of carbon nanotubes: A review*. Index.ieomsociety.org. IEOM Society. https://doi.org/10.46254/NA8.20230192

Mishra, R. M., Mentha, S. S., Misra, Y., & Dwivedi, N. (2023, December). Emerging pollutants of severe environmental concern in water and wastewater: A comprehensive review on current developments and future research. *Water-Energy Nexus, 6*, 74–95. https://doi.org/10.1016/j.wen.2023.08.002

More, R., Haubold, A., & Bokros, J. (2013). In B. D. Ratner, A. S. Hoffman, F. J. Schoen, & J. E. Lemons (Eds.), *Biomaterials science: An introduction to materials in medicine* (3rd ed., pp. 209–222). Academic Press.

Mostofizadeh, A., Yanwei L., Bo S., Yudong H. 2011. "Synthesis, Properties, and Applications of Low-Dimensional Carbon-Related Nanomaterials." Journal of Nanomaterials 2011: 1–21. https://doi.org/10.1155/2011/685081

Nor, N. M., Lau, L. C., Lee, K. T., & Mohamed, A. R. (2013). Synthesis of activated carbon from lignocellulosic biomass and its applications in air pollution control—A review. *Journal of Environmental Chemical Engineering, 1*(4), 658–666. https://doi.org/10.1016/j.jece.2013.09.017

Oganov, A. R., Hemley, R. J., Hazen, R. M., & Jones, A. P. (2013). Structure, bonding, and mineralogy of carbon at extreme conditions. *Reviews in Mineralogy and Geochemistry, 75*(1), 47–77. https://doi.org/10.2138/rmg.2013.75.3

Parambil, A.M. and Rajamani, P., 2024. Carbon dots: a promising path towards environmental sustainability. *Environmental Science: Advances, 3*(11), pp.1513–1523

Parham, H., Bates, S., Xia, Y., & Zhu, Y. (2013, April). A highly efficient and versatile carbon nanotube/ceramic composite filter. *Carbon, 54*, 215–223. https://doi.org/10.1016/j.carbon.2012.11.032

Peng, Y., Lu, B., & Chen, S. (2018). Carbon-supported single atom catalysts for electrochemical energy conversion and storage. *Advanced Materials, 30*(48), 1801995. https://doi.org/10.1002/adma.201801995

Pui, W. K., Yusoff, R., & Aroua, M. K. (2019). A review on activated carbon adsorption for volatile organic compounds (VOCs). *Reviews in Chemical Engineering*, 35(5), 649–668. https://doi.org/10.1515/revce-2017-0057

Royal Swedish Academy of Sciences. (1996, October 9). *Press release: The Nobel Prize in chemistry 1996.* https://www.nobelprize.org/prizes/chemistry/1996/press-release/

Sahragard, A., Varanusupakul, P., & Miró, M. (2023, September). Nanomaterial decorated electrodes in flow-through electrochemical sensing of environmental pollutants: A critical review. *Trends in Environmental Analytical Chemistry, 39*, e00208. https://doi.org/10.1016/j.teac.2023.e00208

Sanchez, J. J. G., & Garicia, E. G. (2023, September). *Filtration as gray water treatment using conventional and unconventional adsorbent beds* (pp. 29–46). ECORFAN EBooks. https://doi.org/10.35429/h.2023.5.29.46

Seabra, A. B., Paula, A. J., Lima, R. D., Alves, O. L., & Durán, N. (2014). Nanotoxicity of graphene and graphene oxide. *Chemical Research in Toxicology, 27*(2), 159–168. https://doi.org/10.1021/tx400385x

Sengupta, R., Bhattacharya, M., Bandyopadhyay, S., & Bhowmick, A. K. (2011). A review on the mechanical and electrical properties of graphite and modified graphite reinforced polymer composites. *Progress in Polymer Science, 36*(5), 638–670. https://doi.org/10.1016/j.progpolymsci.2010.11.003

Sharma, G., Sharma, S., Kumar, A., Lai, C. W., Naushad, M., Shehnaz, Iqbal, J., & Stadler, F. J. (2022, March). Activated carbon as superadsorbent and sustainable material for diverse applications. *Adsorption Science & Technology, 2022*, e4184809. https://doi.org/10.1155/2022/4184809

Shen, Y., Rao, D., Sheng, Q., & Zheng, J. (2017). Simultaneous voltammetric determination of hydroquinone and catechol by using a glassy carbon electrode modified with carboxy-functionalized carbon nanotubes in a chitosan matrix and decorated with gold nanoparticles. *Microchimica Acta, 184*(9), 3591–3601. https://doi.org/10.1007/s00604-017-2392-z

Song, P., Lisha Z., Hao L., Meng M., Ting L., Yongmei Y., Rimo X. 2017. "A Multianalyte Fluorescent Carbon Dots Sensing System Constructed Based on Specific Recognition of Fe(Iii) Ions." RSC Advances 7 (46): 28637–46. https://doi.org/10.1039/c7ra04122e.

Sonkar, P. K., Kumar, N., & Gupta, P. K. (2021). Characteristics of carbon nanotubes and their nanocomposites. In *Fundamentals and properties of multifunctional nanomaterials* (pp. 99–118). https://doi.org/10.1016/B978-0-12-822352-9.00011-0

Su, W., Wu, H., Xu, H., Zhang, Y., Li, Y., Li, X., & Fan, L. (2020). Carbon dots: A booming material for biomedical applications. *Materials Chemistry Frontiers, 4*(3), 821–836. https://doi.org/10.1039/C9QM00658C

Su, R., Wang, Z., Liu, Z., Chen, Y., Wang, H., Dai, X., Ge, X., & Luo, Y. (2024, October). Single atoms in environmental catalysis: Breakthroughs in synthesis and application. *Journal of Water Process Engineering, 68*, 106319–106319. https://doi.org/10.1016/j.jwpe.2024.106319

Sudha, V., Murugesan, S., & Thangamuthu, R. (2018, March). Simultaneous electrochemical sensing of sulphite and nitrite on acid-functionalized multi-walled carbon nanotubes modified electrodes. *Journal of Alloys and Compounds, 749*, 990–999. https://doi.org/10.1016/j.jallcom.2018.03.287

Syduzzaman, M., Saad, M. S. I., Piam, M. F., Talukdar, T. A., Shobdo, T. T., & Pritha, N. M. (2025). Carbon nanotubes: Structure, properties and applications in the aerospace industry. *Results in Materials, 25*, 100654.

Tiwari, S. K., Kumar, V., Huczko, A., Oraon, R., Adhikari, A. D., & Nayak, G. C. (2016). Magical allotropes of carbon: Prospects and applications. *Critical Reviews in Solid State and Materials Sciences, 41*(4), 257–317. https://doi.org/10.1080/10408436.2015.1127206

Vieira, L. D. S. (2022, January). A review on the use of glassy carbon in advanced technological applications. *Carbon, 186*, 282–302. https://doi.org/10.1016/j.carbon.2021.10.022

Wang, Y., Zhu, J., Huang, H., & Cho, H. (2015, April). Carbon nanotube composite membranes for microfiltration of pharmaceuticals and personal care products: Capabilities and potential mechanisms. *Journal of Membrane Science, 479*, 165–174. https://doi.org/10.1016/j.memsci.2015.01.034

Withers, J. C., Loutfy, R. O., & Lowe, T. P. (1997). Fullerene commercial vision. *Fullerene Science and Technology, 5*(1), 1–31. https://doi.org/10.1080/15363839708011971

Xia, C., Zhu, S., Feng, T., Yang, M., & Yang, B. (2019). Evolution and synthesis of carbon dots: From carbon dots to carbonized polymer dots. *Advanced Science, 6*(23), 1901316. https://doi.org/10.1002/advs.201901316

Xu, X., Ray, R., Gu, Y., Ploehn, H. J., Gearheart, L., Raker, K., & Scrivens, W. A. (2004). Electrophoretic analysis and purification of fluorescent single-walled carbon nanotube fragments. *Journal of the American Chemical Society, 126*(40), 12736–12737. https://doi.org/10.1021/ja040082h

Yamada, Y., Kawai, M., Yorimitsu, H., Otsuka, S., Takanashi, M., & Sato, S. (2018). Carbon materials with zigzag and armchair edges. *ACS Applied Materials & Interfaces, 10*(47), 40710–40739. https://doi.org/10.1021/acsami.8b11022

Yang, Z., Zhou, Y., Feng, Z., Rui, X., Zhang, T., & Zhang, Z. (2019). A review on reverse osmosis and nanofiltration membranes for water purification. *Polymers, 11*(8), 1252. https://doi.org/10.3390/polym11081252

Yao, B., Huang, H., Liu, Y., & Kang, Z. (2019). Carbon dots: A small conundrum. *Trends in Chemistry, 1*(2), 235–246. https://doi.org/10.1016/j.trechm.2019.02.003

Zhang, H., Liu, W., Cao, D., & Cheng, D. (2022, May). Carbon-based material-supported single-atom catalysts for energy conversion. *IScience, 25*, 104367. https://doi.org/10.1016/j.isci.2022.104367

Zhao, F., He, Y., Huang, B., Zhang, T., & Zhu, H. (2024). A review of diamond materials and applications in power semiconductor devices. *Materials, 17*(14), 3437–3437. https://doi.org/10.3390/ma17143437

Zhu, Z., Duan, W., Chang, Z., Du, W., Chen, F., Li, F., & Oleszczuk, P. (2023). Stability of functionally modified biochar: The role of surface charges and surface homogeneity. *Sustainability, 15*(10), 7745. https://doi.org/10.3390/su15107745

Chapter 2
Diverse Carbon Nanomaterials for Environment: Synthesis Route and Properties

Aswin Kottapurath Vijay and Gifty Sara Rolly

Abstract Manipulation of carbon materials from bulk to nano initiated diverse opportunities in environmental science and technology. Since the discovery of fullerenes in 1985, carbon nanomaterials, comprising carbon nanotubes, graphene, carbon nanofibers, and carbon quantum dots, have gained attention to solve environmental problems due to their unique structural properties. The primary focus of this chapter includes the synthesis methods, unique properties, and environmental significance of diverse carbon nanomaterials. This chapter describes the structural characteristics of carbon nanomaterials. Thereafter, it focuses on standard methods for synthesizing carbon nanomaterials, like hydrothermal, microwave-assisted pyrolysis, laser ablation, arc discharge, chemical vapor deposition, electrochemical, and template-assisted techniques. Each technique is assessed on its efficiency in producing carbon nanomaterials with unique properties. Moreover, a detailed insight into the physiochemical properties of synthesized carbon nanomaterials has been offered that shows their potential to shape a cleaner environment. The environmental significance of carbon nanomaterials is also discussed in the final section. Carbon nanomaterials are remarkable in enabling cleaner technologies as they possess exceptional properties that are beneficial in creating a sustainable environment. They have displayed potential in various fields, such as wastewater filtration, photocatalysis, antibacterial control systems, toxic gas removal, and soil remediation. Therefore, research advancement in carbon nanomaterials aids in addressing environmental challenges for shaping a healthy world.

Keywords Carbon nanoallotropes · Carbon nanomaterials · Environmental significance · Properties · Synthesis methods

A. K. Vijay (✉)
Department of Sanitary Engineering, Faculty of Civil and Environmental Engineering, Gdańsk University of Technology, Gdańsk, Poland
e-mail: aswin.vijay@pg.edu.pl

G. S. Rolly (✉)
Department of Energy Conversion and Storage, Faculty of Chemistry, Gdańsk University of Technology, Gdańsk, Poland
e-mail: gifty.sara.rolly@pg.edu.pl

A. M. Parambil et al. (eds.), *Carbon*,
https://doi.org/10.1007/978-3-031-90613-8_2

Abbreviations

CDs	Carbon nanodots
CNFs	Carbon nanofibers
CNHs	Carbon nanohorns
CNMs	Carbon nanomaterials
CNOs	Carbon nano-onions
CNTs	Carbon nanotubes
CVD	Chemical vapor deposition
DNA	Deoxyribonucleic acid
DWCNTs	Double-walled carbon nanotubes
ECL	Electrochemiluminescence
FRET	Förster resonance energy transfer
GQDs	Graphene quantum dots
MWCNTs	Multi-walled nanotubes
SERS	Surface-enhanced Raman scattering
SWCNTs	Single-walled nanotubes
SWNHs	Single-walled nanohorns
TEM	Transmission electron microscopy

2.1 Introduction

Carbon, a fundamental element in the periodic table, plays a pivotal role in several natural and industrial processes. For decades, carbon has become the focus of extensive research due to its unique ability to form diverse molecular structures (He et al., 2023). Carbon is considered a versatile element as it can form single, double, and triple bonds, resulting in several geometrical shapes and structures that are fundamental to many organic and inorganic compounds. Carbon has different forms, from simple hydrocarbons to complex biomolecules that support life processes. However, over the years, research on carbon has changed from traditional forms to unconventional ones emerging at nanometric scales with unique properties. The rapidly growing interest in carbon-related nanoscience and applications is mainly due to these breakthroughs created by the rise of modern experimental techniques made conceivable through studying materials on an atomic scale (Dobrzański & Dobrzańska-Danikiewicz, 2019). One of the most critical advances in this field is related to processes based on carbon nanomaterials (CNMs), which play a key role in numerous advanced applications, both futuristic technologies and environmental issues (Maduraiveeran & Jin, 2021; Arcudi et al., 2019; Veeman et al., 2021).

CNMs are materials in which carbon atoms are arranged in various configurations, forming allotropes like diamond, graphite, graphene, carbon nanotubes (CNTs), fullerenes, and carbon nanodots (CDs). Nanomaterials possess different physical, chemical, and electronic properties from their bulk counterparts and have several applications (Manimegalai et al., 2023). These materials have more surface area or less resistance to electron transfer at their nano-scale dimensions, lending them greater mechanical strength and chemical stability.

CNMs need to respond effectively and creatively to some environmental issues. The exceptional chemical reactivity and strong adsorption properties make them crucial for pollution governance and resource management (Nasrollahzadeh et al., 2021). CNMs can degrade different kinds of environmental contaminants originating from air, water, and soil by removing heavy metals (Mubarak et al., 2014), organic pollutants (Madima et al., 2020), and pathogens (de Siqueira Brandão Vaz, 2024; Abd-Elsalam, 2020; Ojha, 2020). Their application in filtration systems, membranes, and sensors improves water purification, air filtration, and pollutant detection (Liapun & Motola, 2024) for developing a cleaner environment. Besides, by creating a cleaner environment, CNMs help with renewable energy processes and waste reduction (Ravi & Vadukumpully, 2016). As more research on CNMs is conducted, the application of these materials will get diverse. This will help in environmental remediation, essential for controlling pollution and depleting natural resources.

A classification of CNMs based on their dimensionality can help to understand their manifold properties and applications. In this section, we categorize CNMs based on the dimensions at which the nanomaterial's quantum confinement affects the physical/chemical properties of the material. The dimensional classification of CNMs is organized into four categories, which include zero-dimensional (0D), one-dimensional (1D), two-dimensional (2D), and three-dimensional (3D). The groups include unique carbon nanoallotropes containing different structures suitable for scientific and industrial applications, such as electronics, energy storage, or environmental remediation. According to their dimensions, CNMs are classified in Fig. 2.1.

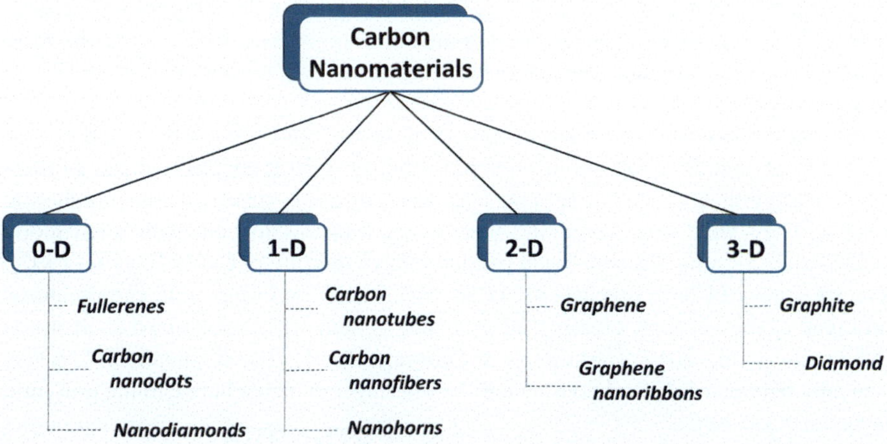

Fig. 2.1 Classification of carbon nanomaterials based on their dimensions: 0D, 1D, 2D, and 3D structures

2.2 Overview of Carbon Nanoallotropes

2.2.1 Zero-Dimensional (0D) Carbon Nanoallotrope

2.2.1.1 Fullerenes and Carbon Nano-Onions

Breakthrough in fullerenes (Kroto et al., 1985) initiated the study of carbon nanostructures. In 1996, Robert F. Curl, Harold W. Kroto, and Richard E. Smalley received the Nobel Prize in Chemistry for this discovery. Fullerenes are characterized as closed, hollow structures with carbon in sp^2-hybridization. These atoms are organized into 12 pentagons, along with a specific number of hexagons. The overall carbon determines the number of hexagons in the molecule. A fullerene composed of $20 + 2n$ carbon atoms will contain n hexagons, while the number of pentagons is 12 in fullerenes with perfectly closed structures. One of the most well-known fullerenes, identified in the early 1980s, is C_{60}, a spherical molecule with a 0.71 nm external diameter, and its chemical properties resemble those of organic compounds. However, it is also considered the smallest carbon nanomaterial and a typical 0D carbon nanoallotrope. Though fullerene soot contains several related structures, such as C_{70}, C_{76}, C_{82}, and C_{84}, showcasing varying sizes and configurations, involving sp^2 and sp^3 hybridized (Zieleniewska et al., 2018) carbon atoms, C_{60} remains the dominant, broadly explored, and heavily utilized fullerene.

C_{60} has a truncated icosahedron structure comprising 12 pentagons and 20 hexagons with sp^2 carbon atoms (Cai et al., 2019). The X-ray diffraction pattern of C_{60} suggests two different kinds of bonds. The first is the 66 bond, with a 1.38 Å bond length, which links carbon atoms to the two neighboring hexagons. The other bond is called the 56 bond, having a 1.45 Å bond length, linking carbon atoms, resulting in a pentagon–hexagon combination. In solid form, C_{60} has a stable face-centered cubic lattice, and the cage is destroyed only when the temperature exceeds 1000 °C (Stetzer et al., 1997). The crystals and aggregates of C_{60} are soluble in several organic solvents, particularly in toluene and carbon disulfide (Ruoff et al., 1993).

Carbon nano-onions (CNOs) are spherical carbon allotropes with closed carbon shells closely similar to an onion's layered structure. CNOs are multi-stacked fullerenes containing layers of graphene organized in quasi-spherical and polyhedral shapes. They were discovered in 1992, using transmission electron microscopy (TEM) after a strong electron beam irradiation in a mixture of CNTs (Ugarte, 1992). The structure of CNOs consists of six and five-membered rings with carbon atoms arranged at the vertices, creating an aromatic system with delocalized π-electrons (Mykhailiv et al., 2017; Bartelmess & Giordani, 2014). The diameter of CNOs is typically between 1.40 and 50 nm, and the interlayer distance between the graphitic layers is 0.335 nm.

2.2.1.2 Carbon Nanodots and Graphene Quantum Dots

Carbon nanodots are nanoparticles with quasi-spherical shapes with 2–10 nm diameters containing a carbonized carbon core with organic functional groups. They possess vast amounts of oxygen and are composed of a combination of turbostratic and graphitic carbon in different volumetric ratios. CDs are usually amorphous and mostly contain sp^3-hybridized carbon. One of the essential features of CDs is their intense photoluminescence, which relies on factors like size, surface functionalization, and excitation wavelength (Wang & Hu, 2014).

Graphene quantum dots (GQDs) are nanomaterials made of sp^2-bonded carbon (Ponomarenko et al., 2008) formed by slicing a graphene monolayer into small fragments of size 2–20 nm. Despite that, the GQDs have a few stacked layers of graphene owing to the nature of the graphenic precursors. The edge effects and quantum confinement of GQDs become significant as the nanosheet size or dimension decreases and gets below 10 nm (Ponomarenko et al., 2008). In contrast to the monolayer and bilayer graphene, the large graphene nanosheet has a zero-width bandgap suitable for electronic and optoelectronic applications.

2.2.1.3 Nanodiamonds

Nanodiamonds are characterized by a specific sp^3-hybridized carbon atom arrangement, making these structures part of the diamond family. The diameters of CNTs are primarily between 1 and 20 nm. Anything larger than about 20 nm is a large nanostructure that behaves much the same as bulk diamonds. Diamond structures of sp^3 carbon less than 1 nm in size also occur and are called diamondoids, which are extracted from petroleum (Dahl et al., 2003). The diamondoids are interesting because they exhibit organic molecule properties, and the carbon atoms on their surface are typically attached to hydrogen and non-carbon atoms. As the sp^3 carbon clusters grow in diameter, a lower fraction of carbons positioned at the surface would result in more ideal diamond-like nanoparticles. Detonation-method-produced nanodiamonds have a mean size of about 4 nm, with carbon atoms nearly 15% occupied at the surface. They bond with a single hydrogen atom and can also combine in different manners to provide an overall nanodiamond structure. Various organic functional groups are present at the nanodiamond surfaces, and their nature depends on the purification methods applied during synthesis.

2.2.2 One-Dimensional (1D) Carbon Nanoallotropes

2.2.2.1 Carbon Nanotubes and Carbon Nanofibers

Carbon nanotubes (CNTs) (Iijima, 1991) are well-ordered cylindrical hollow nanostructures, containing rolled graphene sheets. These nanostructures of carbon atoms are bonded through sp^2 bonds arranged in a hexagonal lattice. CNTs are categorized into single-walled nanotubes (SWCNTs), double-walled (DWCNTs), and multi-walled carbon nanotubes (MWCNTs) according to the number of graphene layers in the walls of the cylindrical structure. Generally, SWCNTs have diameters between 0.4 and 2 nm and lengths of a few micrometers with an empty internal space. The length-to-diameter ratio usually exceeds 10,000; therefore, CNTs are considered the most anisotropic materials ever formed. Based on their chirality, the carbon atoms near the nanotube circumference can be organized as armchair, zig-zag, and chiral models (Kim et al., 2006).

Carbon nanofibers (CNFs) are non-continuous 1D CNMs with cylindrical shapes with layered and curved graphene sheets organized in different ways. They are recognized as sp^2-based linear filaments with diameters in the 5–200 nm range and a high aspect ratio ˃100 (Mallick et al., 2023). Due to the internal arrangement of graphene sheets, several forms of CNFs have been identified: (i) Platelet CNF–graphene layers lying perpendicular to the fiber axis; (ii) herringbone (or fishbone) CNF–graphene sheets tilted to the fiber axis; (iii) ribbon (or tubular) CNF–graphene layers stacked parallel to the fiber axis; (iv) stacked-up and cone stacked CNF-truncated cones arranged to leave a hollow core; and (v) cone-helix CNF–graphite layer creating a helix-spiral structure with an internal hollow core. However, unlike CNTs, CNFs do not have a hollow cavity, a crucial distinguishing feature (Serp, 2003). To differentiate CNTs and CNFs with cones and tilted graphene layers, an angle α is determined, which displays the location of the graphene sheet close to the surface of the sidewall with respect to the axis of the CNFs. When $\alpha = 0$, the carbon nanotube regains its continuous structure. CNFs with cone and tilted graphene layer configurations are characterized by $\alpha > 0$, with the angle playing a crucial role in dictating the physicochemical properties exhibited by the CNFs.

2.2.2.2 Carbon Nanohorns

Single-walled carbon nanohorns (SWCNHs) are cone-shaped structures made from a single layer of graphene (Iijima et al., 1999). Due to van der Waals forces between their open ends, individual SWCNHs assemble into spherical aggregates around 80–100 nm in diameter (Chen et al., 2014). SWCNHs have tip diameters ranging from 1 to 2 nm and base diameters of 4–5 nm. The range between the walls of adjacent single-walled nanohorns (SWNHs) is nearly 0.4 nm. The cone-based structures can be created by slicing a block from a thin single sheet of graphene. Subsequently, the unprotected edges are assembled continuously to form SWNHs.

2.2.3 Two-Dimensional (2D) Carbon Nanoallotropes

2.2.3.1 Graphene

Graphene (Hamwi et al., 1987), a two-dimensional carbon allotrope contains single layers of carbon atoms. The carbon atoms in graphene are arranged in an sp^2-hybridized state, forming σ- and π-bonds in a hexagonal crystal lattice with a 0.142 nm bond length between adjacent carbon atoms. Graphene is a fundamental structural component for other carbon allotropes, including graphite, CNTs, and fullerenes. That is, it can be folded into 0D fullerenes, wrapped into 1D nanotubes, and layered into 3D graphite. Graphene is currently considered the "strongest" material by weight, with a negative thermal expansion coefficient (Bao et al., 2009). Moreover, high electronic conductivity and good thermal stability make graphene ideal for various technological applications in electronics, energetics, mechanics, catalysis, and technology (Geim & Novoselov, 2007).

2.2.4 Three-Dimensional (3D) Carbon Nanoallotropes

2.2.4.1 Graphite and Diamond

Graphite is a naturally occurring crystalline allotrope of carbon with sp^2 hybridized carbon atoms arranged in an infinite array forming distinct layers. Every carbon atom in the hexagonal arrangement is bonded to three neighboring carbon atoms with a 120° C–C–C bond angle. Every carbon atom possesses an unhybridized $2pz$ orbital, in which the fourth electron gets accommodated, forming a delocalized π orbital that stabilizes the bonds within each other. The layers are linked together by weak van der Waals forces with an interlayer distance of 3.354 Å, and the distance between the carbon atoms of each layer is 1.418 Å. Unlike diamond, graphite is soft and can be used as a lubricant as the layers readily cleave (Huai et al., 2021). The planar structure enables electrons to move freely within the planes, allowing graphite to conduct electricity, transfer heat, and absorb light.

Diamond is the hardest natural material, the thermodynamically stable form of carbon at pressures above 60 kbar, making it unique among carbon allotropes. The carbon atoms in diamonds are sp^3 hybridized. Each carbon is bonded covalently to four neighboring carbon atoms at a 1.544 Å distance with 109.5° bond angles. This structure creates a strong, three-dimensional network that gives diamonds exceptional hardness, strength, and durability. The density of diamond (3.514 g/cm^3) is higher than that of graphite, which has a more loosely packed structure and a 2.266 g/cm^3 density. The tetrahedral coordination makes diamonds highly resistant to compression (Field, 2012). Diamonds can be transformed into thermodynamically stable graphite at lower pressures, above 1500 °C, in a vacuum or inert environment. However, under normal conditions, the transformation is negligibly slow.

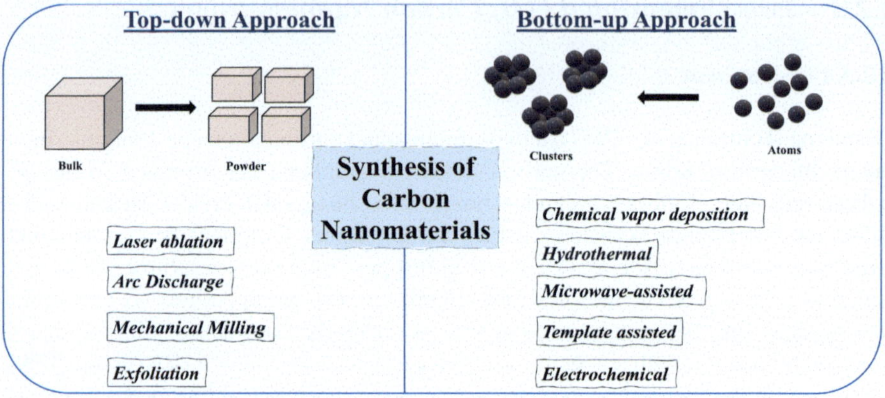

Fig. 2.2 Techniques for the synthesis of carbon nanomaterials

2.3 Common Techniques for Synthesizing Carbon Nanomaterials

Various techniques (see Fig. 2.2) have been developed for CNM synthesis, each having unique advantages and specific strategies for different applications. **Top-down approaches** like arc discharge and laser ablation involve breaking down larger carbon sources into nano-scale materials, allowing accurate regulation over the size and shape of the nanoparticles. Conversely, **bottom-up approaches**, like chemical vapor deposition (CVD) and hydrothermal synthesis, create nanomaterials from molecular precursors, enabling the formation of complex structures with desirable properties. With the increase in demand for CNMs due to their excellent electrical conductivity, good mechanical strength, and high surface area, it is crucial for researchers and industry to know these synthesis methods. These techniques affect the nanoparticle's properties and influence their scalability and ability to be incorporated into various applications. Let us explore some common synthesis methods for CNMs in detail.

2.3.1 Hydrothermal/Solvothermal Methods

The method involves the pyrolysis of organic precursors at high temperatures (150–300 °C) and pressures (1–10 MPa). Here, the precise manipulation of the size and surface functional groups of the carbon nanoparticles occurs (Peixoto et al., 2023). The solvothermal method improves structural and chemical defects in crystal formation by altering the variables that control the crystal growth rate. The crystal size, shape, pore size, and degree of functionalization are improved during the process.

The method is applicable for the economical synthesis of CDs, CNFs, CNTs, and porous carbon due to their low reaction temperature, high yield, and eco-friendliness (Ghodrati et al., 2020; John et al., 2023; Shi et al., 2023). Crystalline materials can be produced in a high-pressure environment by increasing the temperature of a closed reaction chamber above the boiling point of the reactants. The significance of this method is its expertise in managing the process through a liquid phase or multiphase reactions at elevated vapor pressures with the least nanomaterial loss. The technique requires costly instruments and extreme temperatures, which are considered drawbacks.

2.3.2 Microwave-Assisted Pyrolysis

An innovative method for CNM production is microwave-assisted synthesis, which conforms to the principles of green and sustainable chemistry. In this method, bio-based wastes and residues are utilized as carbon precursors (Adeola et al., 2023). These precursors are heated in an oxygen-free environment using microwave energy. During this process, microwave radiation, having a frequency of 2.45 GHz, treats carbon-based materials, such as biomass, polymer, and organic compounds. This process helps in pyrolysis (Allende et al., 2023) and thus forms carbon nanostructure CNTs or nanofibers depending on the used conditions.

The main advantage of microwave-assisted pyrolysis is that it uses energy efficiently, which makes the heat energy process faster, and the temperature can be heated more carefully compared to prevailing techniques (Qiu et al., 2024). This reduces energy consumption as well as processing time. The fast heating rate does not allow the formation of unwanted by-products, thus achieving pure CNMs. This method can also use different starting materials like agricultural waste and polymers, making it sustainable. The CNMs created through this technique are helpful in energy storage, catalysis, cleanup environments, and nanocomposites, considering their tunable structures, large surface area, and excellent electrical conductivity (Dubey & Dube, 2024).

2.3.3 Laser Ablation

A powerful laser is employed to heat and vaporize a carbon-based target to produce carbon nanomaterial (Wazeer et al., 2023). Laser pulses are directed at a solid carbon target in a chamber filled with inert gas. The laser's heat vaporizes the target materials deep beneath the surface (Fig. 2.3a). The CNMs, including CNT, fullerene, and graphene, are generated when the vapor cools down. The CNMs generated by this approach depend on the laser intensity, pulse duration, and gas pressure in the chamber. High-purity CNMs can be produced using the technique. The

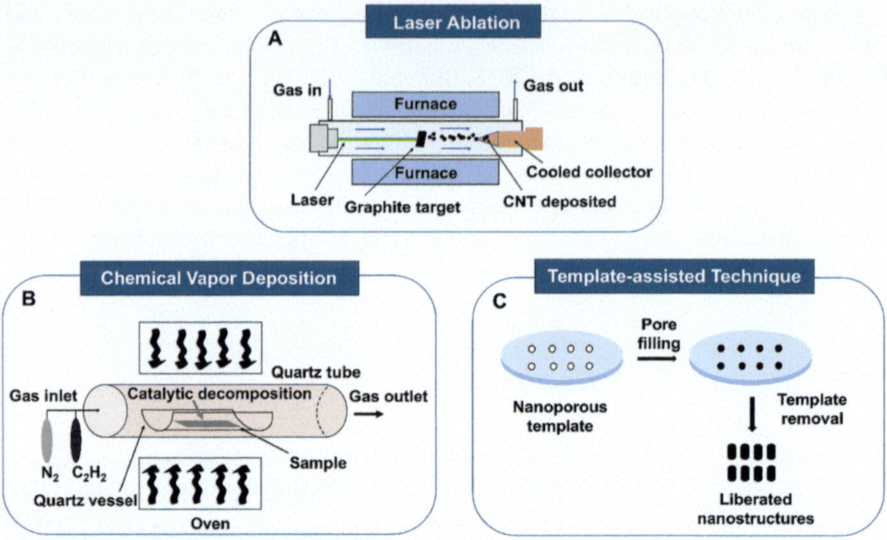

Fig. 2.3 Common synthesis methods for carbon nanomaterials via (**a**) laser ablation (Lu et al., 2019), (**b**) chemical vapor deposition (Ayanda et al., 2024), and (**c**) template-assisted technique (Ayanda et al., 2024)

nanomaterial's size and shape can be adjusted by altering the laser parameters and various target materials (Wang et al., 2018). CNM synthesis using laser ablation does not require stabilizing agents, making it an environmentally friendly approach.

This is an effective method but expensive, and scaling this to larger production is also difficult. The process uses a lot of energy and is costly to operate. Laser ablation is still a common method that continues to be used for research and specific applications, especially for those nanomaterials of high quality required in electronics, energy storage, and advanced materials.

2.3.4 *Arc Discharge Method*

The arc discharge method is commonly used to create CNMs, especially CNTs and fullerenes (Corbella et al., 2022). The process results in high-purity different carbon nanostructures, which occur in various steps. In this process, two graphite electrodes are placed inside the reaction chamber containing inert gas (argon or helium) to avoid oxidation during synthesis. A high-temperature plasma arc is generated when electricity is applied through the electrodes. The heat generated by the arc (>3000 °C) has been used to vaporize the carbon material at the tip of the electrodes, generating plumes of carbon atoms and clusters in the gas phase. As the plasma cools, carbon species undergo nucleation and start growing to produce CNMs. In a cooler environment, carbon atoms prefer to aggregate to form nanostructures like CNTs, fullerenes, or amorphous carbon. Different factors such as arc current, the distance of the electrode, and the pressure of the gas can control the nature and type of nanostructures formed (Shi et al., 1999).

Arc discharge produces CNMs of utmost purity. The inert gas setting stops contaminants from interacting with materials while not changing the material's structure. The method can produce a large amount of CNMs in a short period (Shi et al., 2000). The distance between the electrodes, the arc current, and the inert gas flow rate also regulates the synthesis of required nanomaterials. The technique additionally aids in the production of other nanomaterials like fullerenes (Alekseyev & Dyuzhev, 2003), graphene (Borand et al., 2021), and CNFs (Pacheco-Sotelo et al., 2007). This technique can generate various nanomaterials suitable for electronics, energy storage, and nanocomposites.

2.3.5 Chemical Vapor Deposition

The synthesis of CNMs is dependent on chemical vapor deposition. The vapor-phase precursors undergo chemical reactions, which lead to a thin layer formation on the substrate surface, as shown in Fig. 2.3b. CVD has gained attention considering its low expense, simple equipment, and ease of operation for large-scale management of CNTs (Aminul Islam et al., 2024). In CVD, volatile carbon compounds like methane, acetylene, or ethylene are continuously passed over the catalyst placed in an oven to generate carbon atoms, followed by the creation of CNTs on the surface of the catalyst (nickel, iron, and cobalt) or substrate (Carlsson & Martin, 2010). Due to the heat generated (600–1200 °C), the gas breaks down, depositing carbon atoms on the substrate and developing nanostructures such as CNTs or graphene (Yuge et al., 2010).

The shape of the generated CNMs is influenced by the gas flow, temperature, and the catalyst used in the approach. By altering these conditions, the properties of the formed material can be controlled. High-temperature deposition has several challenges, such as stress creation, increased rate of elemental diffusion from the substrate to the film, size limitations of the chamber, and possible substrate deterioration.

2.3.6 Electrochemical Synthesis

Electrochemical methods can be used to synthesize a variety of CNMs, such as CNTs, graphene, and CNFs. Carbon-rich precursors are subjected to an electric current within an electrolyte solution where the electrodes are made from a conductive material like graphite (Novoselova et al., 2016). Electrochemical reactions are initiated when an electric current passes through the electrochemical cell, oxidizing and reducing the precursor carbon atoms. These carbon atoms get organized on the surface of the electrode and form CNMs. By changing the electrolyte composition, current, or voltage applied, the physical and structural properties of the nanomaterials can be varied.

The electrochemical method has an edge as it is simple and inexpensive. It requires only basic equipment and operates at room temperature and pressure. Hence, it can be easily scaled up for bulk production. The procedure permits accurate control of the nanomaterial's size and form by modifying the reaction conditions. Moreover, the method allows the synthesis to be carried out all in one go without templates. One of the disadvantages is that care must be taken to ensure no side-products are being formed. Also, the process can be slower, and the quality of the products will depend on how they are set up. Despite these challenges, the electrochemical approach is considered sustainable for fabricating CNMs while exhibiting excellence in energy storage, catalysis, and sensors (Maduraiveeran & Jin, 2021).

2.3.7 Template-Assisted Synthesis

This method involves using a template made of silica, polymers, or metals that help in the growth of the required nanomaterial. Various processes, such as CVD or sol-gel, can be employed to carry out the coating of the template with a carbon precursor. After the precursor is deposited, the material is subjected to high temperatures or particular chemical treatment, which causes the precursor to polymerize and/or decompose into the desired carbon nanostructure. After the completion of the synthesis, the template is removed, and a carbon nanomaterial is obtained in the shape of the template. The template-assisted technique is illustrated in Fig. 2.3c.

It is easy to customize the templates to get the desired geometries; thus, they are quite flexible. However, this method encounters some disadvantages, including complicated template removal and contamination during the process. Moreover, the template production cost is high and not scalable for industrial applications. Despite its limitations, template-assisted synthesis is still an excellent method for producing sophisticated CNMs with tremendous potential in the electronics, drug delivery, and energy storage industries.

2.3.8 Biogenic Synthesis

Biogenic synthesis is a green and sustainable method using natural biological processes and organisms to synthesize nanostructured carbon materials (Dai, 2002). The natural ability of plants, microorganisms, algae, and other living organisms has been utilized in the biological approach for producing carbon dots, nanotubes, nano-onion, graphene, etc. (Majumdar et al., 2017). Among the CNMs, CDs have gained much attention as a biological tool in the medical field due to their exceptional biocompatibility (Park et al., 2014).

Biomass-derived CDs can be synthesized using top-down or bottom-up methods. CDs from CNTs, activated carbon, and graphite are frequently produced using top-down methods, such as laser ablation, arc discharge, electrochemical, and ultrasonic treatments (Nagarajan et al., 2023). However, these approaches are considerably more time-consuming than bottom-up procedures and generally require high potential, acidity, and energy. At present, bottom-up techniques are more in practice due to the utilization of non-toxic precursors, cost efficiency, and straightforward synthesis approaches. The hydrothermal carbonization method is most widely preferred for CD synthesis as they are safe and eco-friendly. The precursors are trapped in a hydrothermal reactor containing organic solvents at high temperatures and pressures. Plant extracts containing polyphenols and carbohydrates serve as carbon precursors and reducing agents in synthesizing fluorescent carbon dots (Zhang et al., 2025). Likewise, fungi and bacteria, like microbes, can also be used to synthesize CNMs via microbial metabolism (Cui et al., 2023).

Plants that grow in water, such as algae, make suitable precursors because they have a fast growth rate and high carbon content. This helps with the scalability of biogenic synthesis. The process mainly involves the removal of toxic chemicals and energy-intensive processes generally related to traditional synthesis methods. Also, biogenic synthesis can lead to surface functionalization on the synthesized nanomaterials, which can be used for catalysis, sensing, drug delivery, and environmental remediation (Singh et al., 2024). This sustainable approach follows green chemistry principles and aims to utilize renewable resources to lessen the environmental impacts of nanomaterial production. With scientific knowledge and the optimization of process parameters, biogenic synthesis would be a beneficial and economical process for the large-scale synthesis of carbon nanomaterials in a particular environment. Table 2.1 summarizes the different CNMs produced through these methods and their selected applications.

Table 2.1 Overview of synthesis methods and applications of common carbon nanoallotropes

Carbon nanoallotrope	Synthesis method	Production efficiency	Advantages	Disadvantages	Selected aplications
Fullerenes	Laser ablation Arc discharge CVD	Moderate	High purity, well-defined structure	Expensive, limited scalability	Organic photovoltaics Drug delivery Superconductors
Nano-onions	Hydrothermal Arc discharge	High	Simple synthesis, versatile applications	Limited control over layer uniformity	Lubricants Biomedical imaging Supercapacitor
Nanodots	Electrochemical synthesis Hydrothermal synthesis Laser ablation	Moderate	Excellent fluorescence, biocompatibility	Size distribution inconsistency	Bioimaging Drug delivery Optoelectronics
Nanodiamonds	CVD Laser ablation	Low	Biocompatibility, high hardness	High energy cost, limited scalability	Drug delivery Polishing agents Quantum computing
Nanotubes	Laser ablation Arc discharge CVD	Moderate	Exceptional mechanical and electrical properties	High cost of production, aggregation issues	Electronics Composite materials Drug delivery Sensors
Nanohorns	Arc discharge CVD Laser ablation	Moderate	High surface area	Low scalability	Biosensors Photothermal cancer treatment
Graphene	CVD Mechanical exfoliation	Moderate	Outstanding mechanical and electrical properties	High cost for large-scale production	Flexible electronics Biosensors Energy storage
Graphite	CVD Mining and milling	High	Cost-effective	Limited to bulk applications	Batteries Lubricants

2.4 Properties of Carbon Nanomaterials

2.4.1 *Mechanical Properties*

The unique nanostructures of CNMs and the powerful covalent bonding between carbon atoms give them superior mechanical properties (Eivazzadeh-Keihan et al., 2024). Graphene, with its tensile strength of about 130 GPa, is among the strongest materials, 100 times stronger than steel (Papageorgiou et al., 2020). The configuration of carbons in a two-dimensional hexagonal pattern gives strength to the bonds as they are very strong. Graphene has Young's modulus in the region of 1 TPa, enabling it to be bent and stretched without breaking (Memarian et al., 2015). Graphene can be used primarily on applications requiring flexible components, particularly flexible electronics, composite materials, and advanced structural components and devices.

CNTs can have tensile strengths as high as 100 GPa (Bai et al., 2018), depending on structure (single-walled vs. multi-walled). Like graphenes, CNTs are very stiff materials with Young's modulus near 1 TPa (Zhang et al., 2011). This allows them to retain their shape under deformation, even under very high stress. Combining strength and flexibility, CNTs can be incorporated into nanocomposites and materials that require a high strength-to-weight ratio. Also, CNTs have exceptional toughness against fracture, suggesting that they do not crack easily or fail mechanically. This makes them helpful in strengthening materials in aerospace (Iqbal et al., 2021), automotive, and construction, where strength and lightweight performance are critical (Soni et al., 2020). Overall, the diversity of carbon nanomaterials in many mechanical and structural applications is due to their strength, durability, and flexibility. The physical properties of carbon nanomaterials are compared in Table 2.2.

2.4.2 *Electrical Properties*

CNMs, like graphene, CNTs, and fullerenes, have remarkable electrical properties due to their unique atomic arrangements and bonding patterns. (Georgakilas et al., 2015). The materials are sp^2-hybridized carbon atoms, allowing easy electron flow and conductivity. For instance, graphene constitutes a carbon monolayer in a 2D hexagonal lattice. Due to its structure, electrons can move rapidly with low resistance, providing extremely high electron mobility up to 200,000 $cm^2/V{\cdot}s$ in ideal conditions (Ma, 2024). Graphene is a better conductor because its electrons act like massless particles. Depending on the doping or an electric field applied, the material can display the characteristics of metal and semiconductors. The unique properties of graphene can be used in transistors that operate at high speed, printable, flexible electronic devices, touch screens, and solar cells transparent conductive films (Geng et al., 2017; Shin et al., 2010; Bae et al., 2010; Loh et al., 2016). Additionally, its

Table 2.2 Comparisons of the physical properties among fullerene, carbon nanodiamonds, CNTs, and graphene

Property	Fullerene	Carbon nanodiamonds	Carbon nanotubes	Graphene
Structure	Spherical, 60 carbon atoms organized in a soccer ball-like shape	Spherical or irregular particles	Cylindrical tubes	2D sheet
Bonding (hybridization)	sp^2	sp^3	sp^2	sp^2
Particle size	~0.7 nm (diameter)	~4 to 5 nm (primary particles)	~1 to 2 nm (diameter, single-walled), up to several microns in length	Single layer: 1 atom thick (~0.34 nm)
Density	~1.65 g/cm^3	~3.1 to 3.5 g/cm^3	~1.3 to 1.4 g/cm^3 (single-walled), ~1.5 to 1.8 g/cm^3 (multi-walled)	~2.26 g/cm^3
Surface area	~1000 m^2/g	~300 to 400 m^2/g	~200 to 1300 m^2/g (depending on single-walled or multi-walled)	~2630 m^2/g (single-layer graphene)

exceptional conductivity and high surface area make it suitable for batteries and supercapacitors for energy storage (El-Kady et al., 2016).

Moreover, CNTs show impressive electric properties. The graphene sheet's chirality (or geometric rolling) decides whether a given CNT is metallic or semiconducting. Single-walled carbon nanotubes (SWCNTs) of certain chiralities can display ballistic transport—electrons can travel through the nanotube (and even long distances) without scattering (Liu & Zhang, 2010; Yang et al., 2021). This quality makes them an attractive option for use in nanoelectronics and electrical connections where low resistance is a must. Due to the high conductivity of metallic CNTs, they are used in microchips as a substitute for copper (Wang et al., 2021). They are widely used in many devices like photodetectors and field-effect transistors (FETs) (Liu et al., 2020). Moreover, fullerenes show exceptional conductivity due to their ellipsoidal structure, which is formed by carbon atoms. They can readily accept electrons because of their high electron affinity (Armstrong et al., 2009). They are generally employed in organic light-emitting diodes and organic photovoltaics. They also have electronic properties that can be adjusted to be suitable for modern electronics and optoelectronics.

Other CNMs, including carbon nanofibers and nanodiamonds also find many electrical applications, apart from graphene, CNTs, and fullerenes. Despite being larger than CNTs, CNFs possess good electrical conductivity like CNTs. Due to their large surface areas, the CNFs are ideal for electrode use in capacitor and battery applications (Wu et al., 2007). Nanodiamonds contain mainly sp^3-hybridized carbon atoms. Therefore, they can be different from sp^2-hybridized graphene and CNTs.

However, they can be engineered for electrical applications. These nanodiamonds are incorporated into composites to improve conductivity and mechanical properties (Zhang et al., 2018). Carbon nanomaterials can be modified to achieve different electrical characteristics found in their application in various technologies.

2.4.3 Thermal Properties

The exceptional thermal properties of graphene, CNTs, CNHs, CDs, and CNOs are attributed to their distinctive structures and strong carbon–carbon bonds. Graphene, for example, has a thermal conductivity of 5000 W/m K, which is tremendously greater than copper and other conventional materials (Mohan et al., 2018). Graphene is a good conductor as the lattice vibrations (phonons) conduct heat in its 2D hexagonal structure without much scattering. These traits make graphene products suitable for heat management in gadgets to keep them from overheating. In addition, graphene continues to manage heat flow at a nano-scale thickness, making it useful in miniaturized electronics (Li et al., 2022). But it is getting harder to manage heat as devices shrink.

Single-walled CNTs (SWCNTs) possess great thermal conductivity of nearly 3000 W/m K (Cao & Qu, 2012). Due to the strength of carbon bonds, they are cylindrical and conduct heat along their length. Multi-walled CNTs (MWCNTs) have better thermal conductivity than ordinary thermal materials but not as metals. Due to these properties, CNTs can solve heat problems in electronic devices and work as an additive material in composites. Other CNMs such as CNHs, CDs, and CNOs also play an essential role along with graphene and CNTs but have lower thermal conductivity. Carbon nanohorns offer moderate thermal conductivities of a size that makes them useful in composites that need heat conduction and insulation (Kwon et al., 2023). As they are small carbon particles, CDs are not very thermally conductive and are more often used in bioimaging, where heat management is less of a concern (Kasouni et al., 2019). CNOs, which are layered spherical structures, can be utilized in advanced materials where thermal control is required (Slepičková Kasálková et al., 2021).

These different carbon-based nanomaterials have several applications in thermal management. Heat management is more of an issue as devices get smaller and more powerful in electronics. Graphene and CNTs, which are superior at spreading and dissipating heat, are increasingly used to prevent overheating and enhance the life and performance of electronic components. In energy storage devices like batteries and supercapacitors, CNMs help to regulate the heat released during the charging and discharging cycles (Pomerantseva et al., 2019; Fan et al., 2015). Different types of CNOs and nanohorns with flexible thermal properties are employed in composite materials requiring mechanical strength and effective heat dissipation. Moreover, carbon-based nanomaterials are gaining attention in applications where heat is converted to power. The strong thermal and electrical conductivity of CNMs makes them ideal candidates for improving thermoelectric device efficiency.

2.4.4 *Optical Properties*

The unique optical properties of CNM-based nanocomposites indicate their possible application in photonics, optoelectronics, and sensing. Graphene is a well-known light absorber. It absorbs 2.3% of the incident light that strikes it, irrespective of the wavelength (Li et al., 2019). It is impressive as graphene is just one atom thick. Because of its band structure, it can absorb electromagnetic radiation of wavelengths ranging from ultraviolet (UV) to infrared (IR). In addition, electrons in graphene have high mobility and good light interaction. It indicates applicability in transparent conducting films, photodetectors, and ultrafast laser applications (Chen et al., 2020). Alteration of the chemical structure of the graphene or coating of other substances onto the graphene can alter the optical properties of the graphene further. Optoelectronic devices can also be made from graphene (Tao et al., 2024).

CNTs display appealing optical properties, notably photoluminescence and strong interactions with light. Specific types of chirality associated with CNTs alter their electronic band gaps, allowing them to emit and absorb near-infrared light. Due to this exceptional property, they are used in phototherapy and biomedical imaging (Acharya et al., 2024). In addition, semiconducting CNTs could have strong excitonic effects in enhancing light emission, making such CNTs suitable for light-emitting devices and lasers (Han et al., 2024). CNHs exhibit strong light scattering power and high surface-enhanced Raman scattering (SERS) with respect to their conical nano-shape for sensing applications (Na et al., 2023). CDs refer to the photoluminescent carbon nanoparticles with high quantum yield. Thus, it makes an ideal candidate for bioimaging and sensing agents (Chatzimitakos et al., 2023). The optical properties can be varied by altering the size, surface modification, and synthesis of the CDs.

2.4.5 *Biocompatibility*

CNMs can contribute significantly to biological research for future scientific and technological developments. The importance of CNMs in biology as channels for bioimaging, biosensing, and innovative biomimetic materials has been recognized in the past. CNTs have been utilized in several biological applications due to their tunable surface chemistry, which enables the functionalization of recognition elements and biomolecules in the nanotubes. CNTs are employed in enzyme-based biosensors as electron transport channels, e.g., conductive nanowires (Patolsky et al., 2004). The glucose sensors are common enzyme-based biosensors used for monitoring the blood glucose levels of diabetic patients (Vaddiraju et al., 2010). CNTs coupled with deoxyribonucleic acid (DNA) are efficient in sensing applications. It was found that DNA readily gets adsorbed onto the CNT surfaces, forming supramolecular complexes that are prominent chemical entities in hydrated solution as they possess robust non-covalent interactions between CNTs and DNA (Zheng et al., 2003; Zheng & Semke, 2007). DNA-CNT composites are utilized as

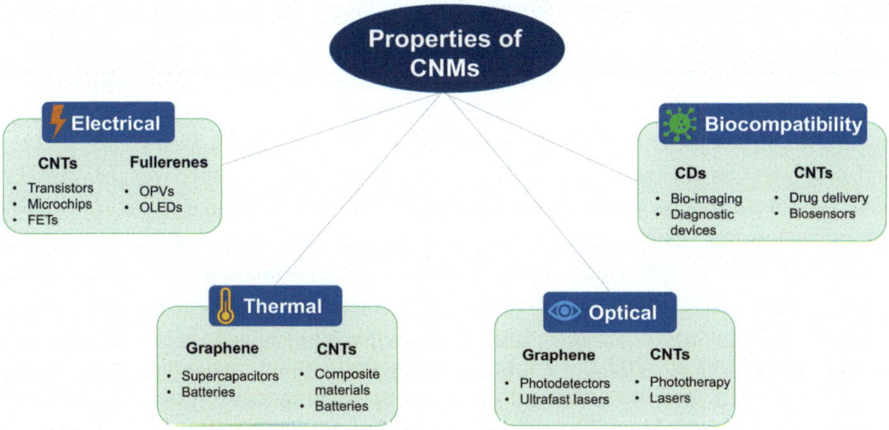

Fig. 2.4 Illustration of important properties of CNMs in various fields

biosensors for detecting different analytes, such as peroxide (Liang et al., 2007), proteins (Wu et al., 2009), and dopamine (Hu et al., 2005). Cancer detection using CNT-based materials was reported through electrochemical impedimetric immunosensing using CNTs modified with gold nanoparticles for the detection of cancer biomarkers (Arkan et al., 2015).

GQDs and CDs have also been used as platforms for glucose detection (Zhang et al., 2014). Numerous immunosensors based on these CNMs have been advanced for disease diagnosis. COOH-functionalized CDs coupled with aptamers detect the carcinoembryonic antigen, a cancer biomarker (Miao et al., 2016). Electrochemiluminescence (ECL) sensors employing GQDs or CDs are designed to detect cancer biomarkers. Although these CNMs generate poor ECL signals, coupling them with metal or semiconductor nanoparticles enhances their ECL response by increasing the concentration of CDs (Li et al., 2014). Interaction of modified CDs with oligonucleotides enhances fluorescence, making them suitable conduits for DNA sensing. Förster resonance energy transfer (FRET) between C-dots and gold nanoparticles aids in ultrasensitively sensing DNA sequences linked to HIV (Qaddare & Salimi, 2017). The fluorescence properties, particularly excitation-dependent emissions, make CDs applicable in biological imaging (Nandi et al., 2014). An illustration of the significant properties of carbon nanomaterials and their selected applications is provided in Fig. 2.4.

2.5 Environmental Significance of Carbon-Based Nanomaterials

Environmental pollution is a major global challenge as various pollutants contaminate urban and agricultural regions. The introduction of innovative approaches to pollution remediation strategies is essential for achieving a sustainable environment. In this context, CNMs can considerably contribute as they possess excellent adsorption potential due to their large surface area (Gao et al., 2019).

CNMs are utilized to improve wastewater filtration systems where adsorption is crucial. It has been reported that CNTs have shown notable adsorption capacity toward lead (Yang et al., 2019), copper (III) (Bystrzejewski & Pyrzyńska, 2011), and microcystins (cyanobacterial toxins) (Wei et al., 2017). Moreover, MWCNTs have been used for the sorption of herbicides, antibiotics, or nitrogen and phosphorus in wastewater (Deng et al., 2012; Zhao et al., 2016; Jiang et al., 2021). As the desorption of the contaminants is effortless, the filters based on CNMs can be easily recycled. CNTs and fullerenes exhibit a mobilization potential for several organic pollutants like persistent polychlorinated biphenyls and lindane (agricultural insecticide) (Majeed et al., 2020). Graphene, particularly functionalized graphene and graphene oxides, has also attained attention in treating wastewater streams and removing heavy metal ions and organic compounds.

Photocatalytic oxidation/reduction reactions integrated with adsorption effectively remove various pollutants from wastewater. The photocatalysis process degrades organic pollutants into less harmful compounds, which can be eliminated by adsorption. CNTs can function as suitable photocatalyst carriers as they possess a large surface area, excellent chemical stability, unique electronic structure, exceptional absorbability, and nano-scale hollow tube characteristics. Photocatalysis studies involving CNT–titanium dioxide catalysts have been found effective in the removal of dyes, pharmaceuticals, and other chemicals. Graphene and its composites are also considered effective in the photocatalytic degradation of pollutants (Wang et al., 2017). Graphene has multiple benefits compared to other carbon compounds, as it possesses greater transmittance for UV-visible light, exceptional electrical and thermal conductivities, and corrosion resistance.

CNMs also possess antimicrobial properties, and some studies suggest the applications of CNTs for disinfection or antimicrobial surface coatings. Silver-coated CNTs have shown antimicrobial activity, making them suitable for biomedical devices and antibacterial control systems (Nie et al., 2017). CNMs can be used in developing novel, efficient biochemical sensors for detecting considerably low amounts of chemical compounds in different environments, making them applicable in agriculture. A combination of metal–organic frameworks with CNMs such as CNTs and graphene oxide has shown the ability to improve the physiochemical properties of the original material, thereby making them efficient in applications, including the removal of toxic gases and volatile organic compounds present in the environment. These composites have been evaluated for their capacity to remove gases like ammonia, nitrogen dioxide, and hydrogen disulfide (Feng et al., 2021). The environmental applications of CNMs are illustrated in Fig. 2.5.

In short, CNMs have numerous applications, including water and air purification, soil remediation, and the enhancement of energy sources to address environmental challenges. Through research advancement, these materials will help control pollution and utilize resources for a healthy world.

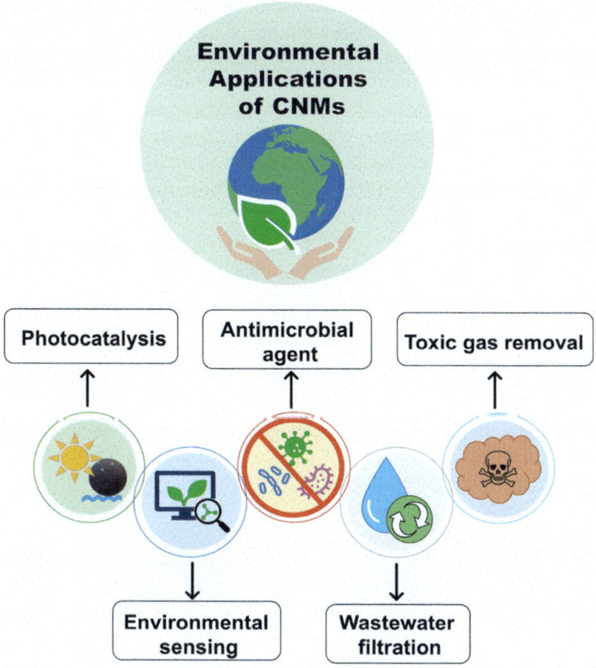

Fig. 2.5 Environmental applications of carbon nanomaterials

2.6 Conclusion

Carbon nanomaterial's unique characteristics and versatile synthesis techniques make them promising for solving environmental challenges. This chapter details the synthesis methods and physicochemical properties, followed by the reactivity and efficiency of the CNMs used for environmental applications. The properties of CNMs disclose their potential for creating a sustainable environment. These nanomaterials can be employed for future developments in controlling pollution and water purification, among other ecological challenges. This chapter provides a broad outlook on the current significance of carbon nanomaterials. Besides, this chapter aims for more advancements in designing CNMs to encourage their incorporation for shaping an eco-friendly environment.

References

Abd-Elsalam, K. A. (2020). Carbon nanomaterials: 30 years of research in agroecosystems. In *Carbon nanomaterials for agri-food and environmental applications* (pp. 1–18). Elsevier. https://doi.org/10.1016/B978-0-12-819786-8.00001-3

Acharya, R., Patil, T. V., Dutta, S. D., Lee, J., Ganguly, K., Kim, H., Randhawa, A., & Lim, K.-T. (2024). Single-walled carbon nanotube-based optical nano/biosensors for biomedical applications: Role in bioimaging, disease diagnosis, and biomarkers detection. *Advanced Materials Technologies, 9*(20). https://doi.org/10.1002/admt.202400279

Adeola, A. O., Duarte, M. P., & Naccache, R. (2023, July). Microwave-assisted synthesis of carbon-based nanomaterials from biobased resources for water treatment applications: Emerging trends and prospects. *Frontiers in Carbon, 2*. https://doi.org/10.3389/frcrb.2023.1220021

Alekseyev, N. I., & Dyuzhev, G. A. (2003). Fullerene formation in an arc discharge. *Carbon, 41*(7), 1343–1348. https://doi.org/10.1016/S0008-6223(03)00058-7

Allende, S., Brodie, G., & Jacob, M. V. (2023, June). Breakdown of biomass for energy applications using microwave pyrolysis: A technological review. *Environmental Research, 226*, 115619. https://doi.org/10.1016/j.envres.2023.115619

Aminul Islam, M., Hasan, M., Rahman, M., Mobarak, M. H., Mimona, M. A., & Hossain, N. (2024, November). Advances and significances of carbon nanotube applications: A comprehensive review. *European Polymer Journal, 220*, 113443. https://doi.org/10.1016/j.eurpolymj.2024.113443

Arcudi, F., Đorđević, L., & Prato, M. (2019). Design, synthesis, and functionalization strategies of tailored carbon nanodots. *Accounts of Chemical Research, 52*(8), 2070–2079. https://doi.org/10.1021/acs.accounts.9b00249

Arkan, E., Saber, R., Karimi, Z., & Shamsipur, M. (2015). A novel antibody–antigen based impedimetric immunosensor for low level detection of HER2 in serum samples of breast cancer patients via modification of a gold nanoparticles decorated multiwall carbon nanotube-ionic liquid electrode. *Analytica Chimica Acta, 874*(May), 66–74. https://doi.org/10.1016/j.aca.2015.03.022

Armstrong, N. R., Wang, W., Alloway, D. M., Placencia, D., Ratcliff, E., & Brumbach, M. (2009). Organic/organic′ heterojunctions: Organic light emitting diodes and organic photovoltaic devices. *Macromolecular Rapid Communications, 30*(9–10), 717–731. https://doi.org/10.1002/marc.200900075

Ayanda, O. S., Mmuoegbulam, A. O., Okezie, O., Iya, N. I. D., Mohammed, S.'a. E., James, P. H., Muhammad, A. B., et al. (2024). Recent progress in carbon-based nanomaterials: Critical review. *Journal of Nanoparticle Research, 26*(5), 106. https://doi.org/10.1007/s11051-024-06006-2

Bae, S., Kim, H., Lee, Y., Xu, X., Park, J.-S., Zheng, Y., Balakrishnan, J., et al. (2010). Roll-to-roll production of 30-inch graphene films for transparent electrodes. *Nature Nanotechnology, 5*(8), 574–578. https://doi.org/10.1038/nnano.2010.132

Bai, Y., Zhang, R., Ye, X., Zhu, Z., Xie, H., Shen, B., Cai, D., et al. (2018). Carbon nanotube bundles with tensile strength over 80 GPa. *Nature Nanotechnology, 13*(7), 589–595. https://doi.org/10.1038/s41565-018-0141-z

Bao, W., Miao, F., Chen, Z., Zhang, H., Jang, W., Dames, C., & Lau, C. N. (2009). Controlled ripple texturing of suspended graphene and ultrathin graphite membranes. *Nature Nanotechnology, 4*(9), 562–566. https://doi.org/10.1038/nnano.2009.191

Bartelmess, J., & Giordani, S. (2014, November). Carbon nano-onions (multi-layer fullerenes): Chemistry and applications. *Beilstein Journal of Nanotechnology, 5*, 1980–1998. https://doi.org/10.3762/bjnano.5.207

Borand, G., Akçamlı, N., & Uzunsoy, D. (2021). Structural characterization of graphene nanostructures produced via arc discharge method. *Ceramics International, 47*(6), 8044–8052. https://doi.org/10.1016/j.ceramint.2020.11.158

Bystrzejewski, M., & Pyrzyńska, K. (2011). Kinetics of copper ions sorption onto activated carbon, carbon nanotubes and carbon-encapsulated magnetic nanoparticles. *Colloids and Surfaces A: Physicochemical and Engineering Aspects, 377*(1–3), 402–408. https://doi.org/10.1016/j.colsurfa.2011.01.041

Cai, W., Chen, C.-H., Chen, N., & Echegoyen, L. (2019). Fullerenes as nanocontainers that stabilize unique actinide species inside: Structures, formation, and reactivity. *Accounts of Chemical Research, 52*(7), 1824–1833. https://doi.org/10.1021/acs.accounts.9b00229

Cao, A., & Qu, J. (2012). Size dependent thermal conductivity of single-walled carbon nanotubes. *Journal of Applied Physics, 112*(1). https://doi.org/10.1063/1.4730908

Carlsson, J.-O., & Martin, P. M. (2010). Chemical vapor deposition. In *Handbook of deposition technologies for films and coatings* (pp. 314–363). Elsevier. https://doi.org/10.1016/B978-0-8155-2031-3.00007-7

Chatzimitakos, T. G., Kasouni, A. I., & Stalikas, C. D. (2023, November). Multi-purpose nitrogen and phosphorous codoped carbon nanodots for fluorescence-based sensing and bioimaging. *Chemical Engineering Journal, 476*, 146686. https://doi.org/10.1016/j.cej.2023.146686

Chen, D., Wang, C., Nie, X., Li, S., Li, R., Guan, M., Liu, Z., et al. (2014). Photoacoustic imaging guided near-infrared photothermal therapy using highly water-dispersible single-walled carbon nanohorns as theranostic agents. *Advanced Functional Materials, 24*(42), 6621–6628. https://doi.org/10.1002/adfm.201401560

Chen, X., Shehzad, K., Gao, L., Long, M., Guo, H., Qin, S., Wang, X., et al. (2020). Graphene hybrid structures for integrated and flexible optoelectronics. *Advanced Materials, 32*(27), e1902039. https://doi.org/10.1002/adma.201902039

Corbella, C., Portal, S., Kundrapu, M. N., & Keidar, M. (2022). Nanosynthesis by atmospheric arc discharges excited with pulsed-DC power: A review. *Nanotechnology, 33*(34), 342001. https://doi.org/10.1088/1361-6528/ac6bad

Cui, X., Wang, X., Chang, X., Bao, L., Wu, J., Tan, Z., Chen, J., et al. (2023). A new capacity of gut microbiota: Fermentation of engineered inorganic carbon nanomaterials into endogenous organic metabolites. *Proceedings of the National Academy of Sciences of the United States of America, 120*(20), e2218739120. https://doi.org/10.1073/pnas.2218739120

Dahl, J. E., Liu, S. G., & Carlson, R. M. K. (2003). Isolation and structure of higher diamondoids, nanometer-sized diamond molecules. *Science, 299*(5603), 96–99. https://doi.org/10.1126/science.1078239

Dai, H. (2002). Carbon nanotubes: Synthesis, integration, and properties. *Accounts of Chemical Research, 35*(12), 1035–1044. https://doi.org/10.1021/ar0101640

de Siqueira Brandão Vaz, L. E. V. (2024). Nanomaterials for air and water purification. In *Nanomaterials for air- and water purification* (pp. 55–75). Wiley. https://doi.org/10.1002/9783527838059.ch3

Deng, J., Shao, Y., Gao, N., Deng, Y., Tan, C., Zhou, S., & Hu, X. (2012, June). Multiwalled carbon nanotubes as adsorbents for removal of herbicide diuron from aqueous solution. *Chemical Engineering Journal, 193–194*, 339–347. https://doi.org/10.1016/j.cej.2012.04.051

Dobrzański, L. A., & Dobrzańska-Danikiewicz, A. D. (2019). Why are carbon-based materials important in civilization progress and especially in the Industry 4.0 stage of the industrial revolution. *Materials Performance and Characterization, 8*(3), 337–370. https://doi.org/10.1520/MPC20190145

Dubey, A., & Dube, C. L. (2024, May). Microwave processing of carbon-based materials: A review. *Nano-Structures & Nano-Objects, 38*, 101136. https://doi.org/10.1016/j.nanoso.2024.101136

Eivazzadeh-Keihan, R., Sadat, Z., Lalebeigi, F., Naderi, N., Panahi, L., Ganjali, F., Mahdian, S., et al. (2024). Effects of mechanical properties of carbon-based nanocomposites on scaffolds for tissue engineering applications: A comprehensive review. *Nanoscale Advances, 6*(2), 337–366. https://doi.org/10.1039/D3NA00554B

El-Kady, M. F., Shao, Y., & Kaner, R. B. (2016). Graphene for batteries, supercapacitors and beyond. *Nature Reviews Materials, 1*(7), 16033. https://doi.org/10.1038/natrevmats.2016.33

Fan, L., Li, B., Zhang, N., & Sun, K. (2015). Carbon nanohorns carried iron fluoride nanocomposite with ultrahigh rate lithium ion storage properties. *Scientific Reports, 5*(1), 12154. https://doi.org/10.1038/srep12154

Feng, Q., Huang, B., & Li, X. (2021). Graphene-based heterostructure composite sensing materials for detection of nitrogen-containing harmful gases. *Advanced Functional Materials, 31*(41). https://doi.org/10.1002/adfm.202104058

Field, J. E. (2012). The mechanical and strength properties of diamond. *Reports on Progress in Physics, 75*(12), 126505. https://doi.org/10.1088/0034-4885/75/12/126505

Gao, R., Dai, Q., Du, F., Yan, D., & Dai, L. (2019). C60-adsorbed single-walled carbon nanotubes as metal-free, PH-universal, and multifunctional catalysts for oxygen reduction, oxygen evolution, and hydrogen evolution. *Journal of the American Chemical Society, 141*(29), 11658–11666. https://doi.org/10.1021/jacs.9b05006

Geim, A. K., & Novoselov, K. S. (2007). The rise of graphene. *Nature Materials, 6*(3), 183–191. https://doi.org/10.1038/nmat1849

Geng, H., Yao, B., Zhou, J., Liu, K., Bai, G., Li, W., Song, Y., Shi, G., Doi, M., & Wang, J. (2017). Size fractionation of graphene oxide nanosheets via controlled directional freezing. *Journal of the American Chemical Society, 139*(36), 12517–12523. https://doi.org/10.1021/jacs.7b05490

Georgakilas, V., Perman, J. A., Tucek, J., & Zboril, R. (2015). Broad family of carbon nanoallotropes: Classification, chemistry, and applications of fullerenes, carbon dots, nanotubes, graphene, nanodiamonds, and combined superstructures. *Chemical Reviews, 115*(11), 4744–4822. https://doi.org/10.1021/cr500304f

Ghodrati, M., Mousavi-Kamazani, M., & Zinatloo-Ajabshir, S. (2020). $Zn_3V_3O_8$ nanostructures: Facile hydrothermal/solvothermal synthesis, characterization, and electrochemical hydrogen storage. *Ceramics International, 46*(18), 28894–28902. https://doi.org/10.1016/j.ceramint.2020.08.057

Hamwi, A., Mouras, S., Djurado, D., & Cousseins, J. C. (1987). New synthesis of first stage graphite intercalation compounds with fluorides. *Journal of Fluorine Chemistry, 35*(1), 151. https://doi.org/10.1016/0022-1139(87)95120-7

Han, B., Li, Y., Wu, W., Cai, X., Qiu, S., He, X., & Wang, S. (2024, January). Infrared light-emitting diodes based on chirality-sorted carbon nanotube films. *ACS Applied Materials & Interfaces*. https://doi.org/10.1021/acsami.3c11990

He, H., Zhang, R., Zhang, P., Wang, P., Chen, N., Qian, B., Zhang, L., Yu, J., & Dai, B. (2023). Functional carbon from nature: Biomass-derived carbon materials and the recent progress of their applications. *Advanced Science, 10*(16). https://doi.org/10.1002/advs.202205557

Hu, C., Zhang, Y., Bao, G., Zhang, Y., Liu, M., & Wang, Z. L. (2005). DNA functionalized single-walled carbon nanotubes for electrochemical detection. *The Journal of Physical Chemistry B, 109*(43), 20072–20076. https://doi.org/10.1021/jp0550457

Huai, W., Zhang, C., & Wen, S. (2021). Graphite-based solid lubricant for high-temperature lubrication. *Friction, 9*(6), 1660–1672. https://doi.org/10.1007/s40544-020-0456-2

Iijima, S. (1991). Helical microtubules of graphitic carbon. *Nature, 354*(6348), 56–58. https://doi.org/10.1038/354056a0

Iijima, S., Yudasaka, M., Yamada, R., Bandow, S., Suenaga, K., Kokai, F., & Takahashi, K. (1999). Nano-aggregates of single-walled graphitic carbon nano-horns. *Chemical Physics Letters, 309*(3–4), 165–170. https://doi.org/10.1016/S0009-2614(99)00642-9

Iqbal, A., Saeed, A., & Ul-Hamid, A. (2021). A review featuring the fundamentals and advancements of polymer/CNT nanocomposite application in aerospace industry. *Polymer Bulletin, 78*(1), 539–557. https://doi.org/10.1007/s00289-019-03096-0

Jiang, S., Wang, J., Qiao, S., & Zhou, J. (2021, November). Phosphate recovery from aqueous solution through adsorption by magnesium modified multi-walled carbon nanotubes. *Science of the Total Environment, 796*, 148907. https://doi.org/10.1016/j.scitotenv.2021.148907

John, B. K., Mathew, S., John, N., Mathew, J., & Mathew, B. (2023, March). Hydrothermal synthesis of N,S-doped carbon quantum dots as a dual mode sensor for azo dye tartrazine and fluorescent ink applications. *Journal of Photochemistry and Photobiology A: Chemistry, 436*, 114386. https://doi.org/10.1016/j.jphotochem.2022.114386

Kasouni, A., Chatzimitakos, T., & Stalikas, C. (2019). Bioimaging applications of carbon nanodots: A review. *C, 5*(2), 19. https://doi.org/10.3390/c5020019

Kim, Y. A., Muramatsu, H., Hayashi, T., Endo, M., Terrones, M., & Dresselhaus, M. S. (2006). Fabrication of high-purity, double-walled carbon nanotube buckypaper. *Chemical Vapor Deposition, 12*(6), 327–330. https://doi.org/10.1002/cvde.200504217

Kroto, H. W., Heath, J. R., O'Brien, S. C., Curl, R. F., & Smalley, R. E. (1985). C60: Buckminsterfullerene. *Nature, 318*(6042), 162–163. https://doi.org/10.1038/318162a0

Kwon, T., Lee, S. H., Kim, J. H., Hong, S.-K., Kim, M., Kim, M., Kim, D.-K., et al. (2023). Polypropylene nanocomposites doped with carbon nanohorns for high-voltage power cable insulation applications. *Advanced Composites and Hybrid Materials, 6*(5), 167. https://doi.org/10.1007/s42114-023-00746-w

Li, S., Luo, J., Yang, X., Wan, Y., & Liu, C. (2014, July). A novel immunosensor for squamous cell carcinoma antigen determination based on CdTe@carbon dots nanocomposite electrochemiluminescence resonance energy transfer. *Sensors and Actuators B: Chemical, 197*, 43–49. https://doi.org/10.1016/j.snb.2014.02.066

Li, Q., Lu, J., Gupta, P., & Qiu, M. (2019). Engineering optical absorption in graphene and other 2D materials: Advances and applications. *Advanced Optical Materials, 7*(20). https://doi.org/10.1002/adom.201900595

Li, X., Wang, Y., Zhao, Y., Zhang, J., & Qu, L. (2022). Graphene materials for miniaturized energy harvest and storage devices. *Small Structures, 3*(1). https://doi.org/10.1002/sstr.202100124

Liang, Z., Lao, R., Wang, J., Liu, Y., Wang, L., Huang, Q., Song, S., Li, G., & Fan, C. (2007). Solubilization of single-walled carbon nanotubes with single-stranded DNA generated from asymmetric PCR. *International Journal of Molecular Sciences, 8*(7), 705–713. https://doi.org/10.3390/i8070705

Liapun, V., & Motola, M. (2024). Carbon nanostructures functionalization for air filtration and purification. In *Handbook of functionalized carbon nanostructures* (pp. 1821–1858). Springer. https://doi.org/10.1007/978-3-031-32150-4_53

Liu, C.-H., & Zhang, H.-L. (2010). Chemical approaches towards single-species single-walled carbon nanotubes. *Nanoscale, 2*(10), 1901. https://doi.org/10.1039/c0nr00306a

Liu, P., Yang, S. E., Chen, Y., Ma, Y., Liu, S., Fang, X., Fan, F., & Han, J. (2020). Carbon nanotube-based heterostructures for high-performance photodetectors: Recent progress and future prospects. *Ceramics International, 46*(12), 19655–19663. https://doi.org/10.1016/j.ceramint.2020.05.067

Loh, K. P., Tong, S. W., & Wu, J. (2016). Graphene and graphene-like molecules: Prospects in solar cells. *Journal of the American Chemical Society, 138*(4), 1095–1102. https://doi.org/10.1021/jacs.5b10917

Lu, Z., Raad, R., Safaei, F., Xi, J., Liu, Z., & Foroughi, J. (2019, June). Carbon nanotube based fiber supercapacitor as wearable energy storage. *Frontiers in Materials, 6*. https://doi.org/10.3389/fmats.2019.00138

Ma, Z. (2024). Advances in graphene-assisted flexible substrate sensors for human motion monitoring. *International Journal of Electrochemical Science, 19*(9), 100760. https://doi.org/10.1016/j.ijoes.2024.100760

Madima, N., Mishra, S. B., Inamuddin, I., & Mishra, A. K. (2020). Carbon-based nanomaterials for remediation of organic and inorganic pollutants from wastewater. A review. *Environmental Chemistry Letters, 18*(4), 1169–1191. https://doi.org/10.1007/s10311-020-01001-0

Maduraiveeran, G., & Jin, W. (2021, October). Carbon nanomaterials: Synthesis, properties and applications in electrochemical sensors and energy conversion systems. *Materials Science and Engineering: B, 272*, 115341. https://doi.org/10.1016/j.mseb.2021.115341

Majeed, N., Panigrahi, K. C. S., Sukla, L. B., John, R., & Panigrahy, M. (2020). Application of carbon nanomaterials in plant biotechnology. *Materials Today: Proceedings, 30*, 340–345. https://doi.org/10.1016/j.matpr.2020.01.618

Majumdar, B., Mandani, S., Bhattacharya, T., Sarma, D., & Sarma, T. K. (2017). Probing carbocatalytic activity of carbon nanodots for the synthesis of biologically active dihydro/spiro/glyco quinazolinones and aza-Michael adducts. *The Journal of Organic Chemistry, 82*(4), 2097–2106. https://doi.org/10.1021/acs.joc.6b02914

Mallick, P., Gadtya, A. S., Tripathy, D., Satpathy, S. K., & Moharana, S. (2023). Carbon–metal hybrid nanomaterials for high technologies. In *Nanoparticles reinforced metal nanocomposites* (pp. 273–295). Springer. https://doi.org/10.1007/978-981-19-9729-7_9

Manimegalai, S., Vickram, S., Deena, S. R., Rohini, K., Thanigaivel, S., Manikandan, S., Subbaiya, R., Karmegam, N., Kim, W., & Govarthanan, M. (2023, January). Carbon-based nanomate-

rial intervention and efficient removal of various contaminants from effluents—A review. *Chemosphere, 312*, 137319. https://doi.org/10.1016/j.chemosphere.2022.137319
Memarian, F., Fereidoon, A., & Darvish Ganji, M. (2015, September). Graphene Young's modulus: Molecular mechanics and DFT treatments. *Superlattices and Microstructures, 85*, 348–356. https://doi.org/10.1016/j.spmi.2015.06.001
Miao, H., Wang, L., Zhuo, Y., Zhou, Z., & Yang, X. (2016, December). Label-free fluorimetric detection of CEA using carbon dots derived from tomato juice. *Biosensors and Bioelectronics, 86*, 83–89. https://doi.org/10.1016/j.bios.2016.06.043
Mohan, V. B., Lau, K.-t., Hui, D., & Bhattacharyya, D. (2018). Graphene-based materials and their composites: A review on production, applications and product limitations. *Composites Part B: Engineering, 142*(June), 200–220. https://doi.org/10.1016/j.compositesb.2018.01.013
Mubarak, N. M., Sahu, J. N., Abdullah, E. C., & Jayakumar, N. S. (2014). Removal of heavy metals from wastewater using carbon nanotubes. *Separation & Purification Reviews, 43*(4), 311–338. https://doi.org/10.1080/15422119.2013.821996
Mykhailiv, O., Zubyk, H., & Plonska-Brzezinska, M. E. (2017, November). Carbon nano-onions: Unique carbon nanostructures with fascinating properties and their potential applications. *Inorganica Chimica Acta, 468*, 49–66. https://doi.org/10.1016/j.ica.2017.07.021
Na, Y., Zhang, J., Zhang, S., Liang, N., & Zhao, L. (2023, October). Fluorescence sensor for zearalenone detection based on oxidized single-walled carbon nanohorns/N-doped carbon quantum dots-aptamer. *Journal of Fluorescence, 34*, 2557. https://doi.org/10.1007/s10895-023-03466-y
Nagarajan, D., Gangadharan, D., & Venkatanarasimhan, S. (2023). Synthetic strategies toward developing carbon dots via top-down approach. In *Carbon dots in analytical chemistry* (pp. 1–13). Elsevier. https://doi.org/10.1016/B978-0-323-98350-1.00016-5
Nandi, S., Malishev, R., Kootery, K. P., Mirsky, Y., Kolusheva, S., & Jelinek, R. (2014). Membrane analysis with amphiphilic carbon dots. *Chemical Communications, 50*(71), 10299–10302. https://doi.org/10.1039/C4CC03504F
Nasrollahzadeh, M., Sajjadi, M., Iravani, S., & Varma, R. S. (2021, January). Carbon-based sustainable nanomaterials for water treatment: State-of-art and future perspectives. *Chemosphere, 263*, 128005. https://doi.org/10.1016/j.chemosphere.2020.128005
Nie, C., Yang, Y., Cheng, C., Ma, L., Deng, J., Wang, L., & Zhao, C. (2017, March). Bioinspired and biocompatible carbon nanotube-Ag nanohybrid coatings for robust antibacterial applications. *Acta Biomaterialia, 51*, 479–494. https://doi.org/10.1016/j.actbio.2017.01.027
Novoselova, I. A., Kuleshov, S. V., Volkov, S. V., & Bykov, V. N. (2016, September). Electrochemical synthesis, morphological and structural characteristics of carbon nanomaterials produced in molten salts. *Electrochimica Acta, 211*, 343–355. https://doi.org/10.1016/j.electacta.2016.05.160
Ojha, A. (2020). Nanomaterials for removal of waterborne pathogens. In *Waterborne pathogens* (pp. 385–432). Elsevier. https://doi.org/10.1016/B978-0-12-818783-8.00019-0
Pacheco-Sotelo, J. O., Pacheco, M. P., Barrientos, R. V., de Lourdes, M., Lopez, J., Flores, J. F. R., Torres, C. E., Fabela, J. L. T., & Azocar, A. C. (2007). Carbon nanofibers synthesized by glow-arc high-frequency discharge. *IEEE Transactions on Plasma Science, 35*(2), 460–466. https://doi.org/10.1109/TPS.2007.891621
Papageorgiou, D. G., Li, Z., Liu, M., Kinloch, I. A., & Young, R. J. (2020). Mechanisms of mechanical reinforcement by graphene and carbon nanotubes in polymer nanocomposites. *Nanoscale, 12*(4), 2228–2267. https://doi.org/10.1039/C9NR06952F
Park, S. Y., Lee, H. U., Park, E. S., Lee, S. C., Lee, J.-W., Jeong, S. W., Kim, C. H., Lee, Y.-C., Huh, Y. S., & Lee, J. (2014). Photoluminescent green carbon nanodots from food-waste-derived sources: Large-scale synthesis, properties, and biomedical applications. *ACS Applied Materials & Interfaces, 6*(5), 3365–3370. https://doi.org/10.1021/am500159p
Patolsky, F., Weizmann, Y., & Willner, I. (2004). Long-range electrical contacting of redox enzymes by SWCNT connectors. *Angewandte Chemie International Edition, 43*(16), 2113–2117. https://doi.org/10.1002/anie.200353275

Peixoto, D. A., Silva, S. C., Borges, P. H. S., Lima, R. C., & Nossol, E. (2023). Hydrothermal synthesis as a versatile tool for the preparation of metal hexacyanoferrates: A review. *Journal of Materials Science, 58*(7), 2993–3024. https://doi.org/10.1007/s10853-023-08190-3

Pomerantseva, E., Bonaccorso, F., Feng, X., Cui, Y., & Gogotsi, Y. (2019). Energy storage: The future enabled by nanomaterials. *Science, 366*(6468). https://doi.org/10.1126/science.aan8285

Ponomarenko, L. A., Schedin, F., Katsnelson, M. I., Yang, R., Hill, E. W., Novoselov, K. S., & Geim, A. K. (2008). Chaotic dirac billiard in graphene quantum dots. *Science, 320*(5874), 356–358. https://doi.org/10.1126/science.1154663

Qaddare, S. H., & Salimi, A. (2017, March). Amplified fluorescent sensing of DNA using luminescent carbon dots and AuNPs/GO as a sensing platform: A novel coupling of FRET and DNA hybridization for homogeneous HIV-1 gene detection at femtomolar level. *Biosensors and Bioelectronics, 89*, 773–780. https://doi.org/10.1016/j.bios.2016.10.033

Qiu, B., Wang, Y., Zhang, D., & Chu, H. (2024, October). Microwave-assisted pyrolysis of biomass to high-value products: Factors assessment, mechanism analysis, and critical issues proposal. *Chemical Engineering Journal, 498*, 155362. https://doi.org/10.1016/j.cej.2024.155362

Ravi, S., & Vadukumpully, S. (2016). Sustainable carbon nanomaterials: Recent advances and its applications in energy and environmental remediation. *Journal of Environmental Chemical Engineering, 4*(1), 835–856. https://doi.org/10.1016/j.jece.2015.11.026

Ruoff, R. S., Tse, D. S., Malhotra, R., & Lorents, D. C. (1993). Solubility of fullerene (C60) in a variety of solvents. *The Journal of Physical Chemistry, 97*(13), 3379–3383. https://doi.org/10.1021/j100115a049

Serp, P. (2003). Carbon nanotubes and nanofibers in catalysis. *Applied Catalysis A: General, 253*(2), 337–358. https://doi.org/10.1016/S0926-860X(03)00549-0

Shi, Z., Lian, Y., Zhou, X., Gu, Z., Zhang, Y., Iijima, S., Zhou, L., Yue, K. T., & Zhang, S. (1999). Mass-production of single-wall carbon nanotubes by arc discharge method[1]. This work was supported by the National Natural Science Foundation of China, No. 29671030. *Carbon, 37*(9), 1449–1453. https://doi.org/10.1016/S0008-6223(99)00007-X

Shi, Z., Lian, Y., Liao, F. H., Zhou, X., Gu, Z., Zhang, Y., Iijima, S., Li, H., Yue, K. T., & Zhang, S.-L. (2000). Large scale synthesis of single-wall carbon nanotubes by arc-discharge method. *Journal of Physics and Chemistry of Solids, 61*(7), 1031–1036. https://doi.org/10.1016/S0022-3697(99)00358-3

Shi, J., Xu, J., Cui, H., Yan, N., Zou, J., Liu, Y., & You, S. (2023, October). Synthesis of highly porous N-doped hollow carbon nanospheres with a combined soft template-chemical activation method for CO_2 capture. *Energy, 280*, 128172. https://doi.org/10.1016/j.energy.2023.128172

Shin, H.-J., Choi, W. M., Choi, D., Han, G. H., Yoon, S.-M., Park, H.-K., Kim, S.-W., et al. (2010). Control of electronic structure of graphene by various dopants and their effects on a nanogenerator. *Journal of the American Chemical Society, 132*(44), 15603–15609. https://doi.org/10.1021/ja105140e

Singh, S., Tiwari, H., Verma, A., Gupta, P., Chattopadhaya, A., Singh, A., Singh, S., et al. (2024). Sustainable synthesis of novel green-based nanoparticles for therapeutic interventions and environmental remediation. *ACS Synthetic Biology, 13*(7), 1994–2007. https://doi.org/10.1021/acssynbio.4c00206

Slepičková Kasálková, N., Slepička, P., & Švorčík, V. (2021). Carbon nanostructures, nanolayers, and their composites. *Nanomaterials, 11*(9), 2368. https://doi.org/10.3390/nano11092368

Soni, S. K., Thomas, B., & Kar, V. R. (2020, December). A comprehensive review on CNTs and CNT-reinforced composites: Syntheses, characteristics and applications. *Materials Today Communications, 25*, 101546. https://doi.org/10.1016/j.mtcomm.2020.101546

Stetzer, M. R., Heiney, P. A., Fischer, J. E., & McGhie, A. R. (1997). Thermal stability of solid C60. *Physical Review B, 55*(1), 127–131. https://doi.org/10.1103/PhysRevB.55.127

Tao, Z.-G., Deng, S., Prezhdo, O. V., Xiang, H., Chu, W., & Gong, X.-G. (2024). Tunable ultrafast charge transfer across homojunction interface. *Journal of the American Chemical Society, 146*(34), 24016–24023. https://doi.org/10.1021/jacs.4c07454

Ugarte, D. (1992). Curling and closure of graphitic networks under electron-beam irradiation. *Nature, 359*(6397), 707–709. https://doi.org/10.1038/359707a0

Vaddiraju, S., Tomazos, I., Burgess, D. J., Jain, F. C., & Papadimitrakopoulos, F. (2010). Emerging synergy between nanotechnology and implantable biosensors: A review. *Biosensors and Bioelectronics, 25*(7), 1553–1565. https://doi.org/10.1016/j.bios.2009.12.001

Veeman, D., Varsha Shree, M., Sureshkumar, P., Jagadeesha, T., Natrayan, L., Ravichandran, M., & Paramasivam, P. (2021, September). Sustainable development of carbon nanocomposites: Synthesis and classification for environmental remediation. *Journal of Nanomaterials, 2021*, 1–21. https://doi.org/10.1155/2021/5840645

Wang, Y., & Hu, A. (2014). Carbon quantum dots: Synthesis, properties and applications. *Journal of Materials Chemistry C, 2*(34), 6921. https://doi.org/10.1039/C4TC00988F

Wang, M., Cai, L., Wang, Y., Zhou, F., Xu, K., Tao, X., & Chai, Y. (2017). Graphene-draped semiconductors for enhanced photocorrosion resistance and photocatalytic properties. *Journal of the American Chemical Society, 139*(11), 4144–4151. https://doi.org/10.1021/jacs.7b00341

Wang, F., Wang, K., Zheng, B., Dong, X., Mei, X., Lv, J., Duan, W., & Wang, W. (2018). Laser-induced graphene: Preparation, functionalization and applications. *Materials Technology, 33*(5), 340–356. https://doi.org/10.1080/10667857.2018.1447265

Wang, P., Cao, Q., Wang, H., Liu, S., Chen, Y., & Peng, Q. (2021, April). CNT-sandwiched copper composites as super thermal conductors for heat management. *Physica E: Low-dimensional Systems and Nanostructures, 128*, 114557. https://doi.org/10.1016/j.physe.2020.114557

Wazeer, A., Das, A., Sinha, A., & Karmakar, A. (2023). Nanomaterials synthesis via laser ablation in liquid: A review. *Journal of The Institution of Engineers (India): Series D, 104*(1), 413–426. https://doi.org/10.1007/s40033-022-00370-w

Wei, G., Quan, X., Fan, X., Chen, S., & Zhang, Y. (2017, July). Carbon-nanotube-based sandwich-like hollow fiber membranes for expanded microcystin-LR removal applications. *Chemical Engineering Journal, 319*, 212–218. https://doi.org/10.1016/j.cej.2017.02.125

Wu, M.-S., Lee, J.-T., Chiang, P.-C. J., & Lin, J.-C. (2007). Carbon-nanofiber composite electrodes for thin and flexible lithium-ion batteries. *Journal of Materials Science, 42*(1), 259–265. https://doi.org/10.1007/s10853-006-1062-z

Wu, Z., Zhen, Z., Jiang, J.-H., Shen, G.-L., & Yu, R.-Q. (2009). Terminal protection of small-molecule-linked DNA for sensitive electrochemical detection of protein binding via selective carbon nanotube assembly. *Journal of the American Chemical Society, 131*(34), 12325–12332. https://doi.org/10.1021/ja9038054

Yang, S., Hu, J., Chen, C., Shao, D., & Wang, X. (2019). Correction to mutual effects of Pb(II) and humic acid adsorption on multiwalled carbon nanotubes/polyacrylamide composites from aqueous solutions. *Environmental Science & Technology, 53*(22), 13546–13546. https://doi.org/10.1021/acs.est.9b06367

Yang, X., Liu, T., Li, R., Yang, X., Lyu, M., Fang, L., Zhang, L., et al. (2021). Host–guest molecular interaction enabled separation of large-diameter semiconducting single-walled carbon nanotubes. *Journal of the American Chemical Society, 143*(27), 10120–10130. https://doi.org/10.1021/jacs.1c02245

Yuge, R., Miyawaki, J., Ichihashi, T., Kuroshima, S., Yoshitake, T., Ohkawa, T., Aoki, Y., Iijima, S., & Yudasaka, M. (2010). Highly efficient field emission from carbon nanotube−nanohorn hybrids prepared by chemical vapor deposition. *ACS Nano, 4*(12), 7337–7343. https://doi.org/10.1021/nn102452q

Zhang, R., Wen, Q., Qian, W., Su, D. S., Zhang, Q., & Wei, F. (2011). Superstrong ultralong carbon nanotubes for mechanical energy storage. *Advanced Materials, 23*(30), 3387–3391. https://doi.org/10.1002/adma.201100344

Zhang, L., Zhang, Z.-Y., Liang, R.-P., Li, Y.-H., & Qiu, J.-D. (2014). Boron-doped graphene quantum dots for selective glucose sensing based on the 'abnormal' aggregation-induced photoluminescence enhancement. *Analytical Chemistry, 86*(9), 4423–4430. https://doi.org/10.1021/ac500289c

Zhang, Y., Rhee, K. Y., Hui, D., & Park, S.-J. (2018, June). A critical review of nanodiamond based nanocomposites: Synthesis, properties and applications. *Composites Part B: Engineering, 143*, 19–27. https://doi.org/10.1016/j.compositesb.2018.01.028

Zhang, Z., Wu, C., Hu, J., Li, C., Liu, Y., Lei, B., & Zheng, M. (2025, January). Recent advances of carbon dots: Synthesis, plants applications, prospects, and challenges. *ACS Applied Bio Materials, 8*, 935. https://doi.org/10.1021/acsabm.4c01785

Zhao, H., Liu, X., Cao, Z., Zhan, Y., Shi, X., Yang, Y., Zhou, J., & Xu, J. (2016, June). Adsorption behavior and mechanism of chloramphenicols, sulfonamides, and non-antibiotic pharmaceuticals on multi-walled carbon nanotubes. *Journal of Hazardous Materials, 310*, 235–245. https://doi.org/10.1016/j.jhazmat.2016.02.045

Zheng, M., & Semke, E. D. (2007). Enrichment of single chirality carbon nanotubes. *Journal of the American Chemical Society, 129*(19), 6084–6085. https://doi.org/10.1021/ja071577k

Zheng, M., Jagota, A., Semke, E. D., Diner, B. A., Mclean, R. S., Lustig, S. R., Richardson, R. E., & Tassi, N. G. (2003). DNA-assisted dispersion and separation of carbon nanotubes. *Nature Materials, 2*(5), 338–342. https://doi.org/10.1038/nmat877

Zieleniewska, A., Lodermeyer, F., Roth, A., & Guldi, D. M. (2018). Fullerenes—How 25 years of charge transfer chemistry have shaped our understanding of (interfacial) interactions. *Chemical Society Reviews, 47*(3), 702–714. https://doi.org/10.1039/C7CS00728K

Chapter 3
Bulk Carbon Materials for the Environment: Synthesis Routes and Properties

K. E. Merin Rose, Mary Soumya, Meha Mohanan, Hanna J. Maria, and Sabu Thomas

Abstract Bulk carbon materials are widely applicable in environment-related functions due to their unique, versatile properties and functionality profiles. This chapter covers different categories of bulk carbon materials, namely graphite, bio-char, activated carbon, carbon fibres, and several other evolving forms, discussing how they contributed to sustainable approaches. Synthetic methods involve hydro-thermal carbonization, pyrolysis, and physical and chemical activation, which are debated on the influences on the characteristics of the obtained material. These materials exhibit important qualities such as high surface area, abundant functional groups, strong adsorption capacity, good thermal stability, and regeneration potential, which are explained. These properties are essential for how effectively the material protects the environment. Its use in water purification, air filtration, and soil

K. E. Merin Rose · M. Mohanan · H. J. Maria
School of Energy Materials, Mahatma Gandhi University, Kottayam, Kerala, India

M. Soumya
International and Inter University Centre for Nanoscience and Nanotechnology, Mahatma Gandhi University, Kottayam, Kerala, India

James Cook University, Townsville, QLD, Australia

S. Thomas (✉)
School of Energy Materials, Mahatma Gandhi University, Kottayam, Kerala, India

International and Inter University Centre for Nanoscience and Nanotechnology, Mahatma Gandhi University, Kottayam, Kerala, India

School of Nanoscience and Nanotechnology, Mahatma Gandhi University, Kottayam, Kerala, India

Trivandrum Engineering, Science and Technology (TrEST) Research Park, Trivandrum, India

School of Polymer Science and Technology, Mahatma Gandhi University, Kottayam, Kerala, India

School of Chemical Sciences, Mahatma Gandhi University, Kottayam, Kerala, India

Department of Chemical Sciences, University of Johannesburg, Johannesburg, South Africa
e-mail: sabuthomas@mgu.ac.in

A. M. Parambil et al. (eds.), *Carbon*,
https://doi.org/10.1007/978-3-031-90613-8_3

remediation demonstrates how these materials can help address hazardous environmental problems. However, the large-scale production, regeneration, and ecological effects must be studied more.

Keywords Carbon materials · Carbon properties · Synthesis techniques · Environmental applications · Pollutant removal

Abbreviations

AC	Activated charcoal
BC	Biochar
CBMs	Carbon-based materials
CD	Carbon dots
CF	Carbon fibres
CNT	Carbon nanotubes
CO-HTC	Co-hydrothermal carbonization
DWCNT	Double-walled carbon nanotubes
GO	Graphene oxide
HTC	Hydrothermal carbonization
MWCNT	Multi-walled carbon nanotubes
PAN	Polyacrylonitrile
SPV	Specific pore volume
SSA	Specific surface area
SWCNT	Single-walled carbon nanotubes

3.1 Introduction to Bulk Carbon Materials

In our day-to-day life, we deal with different types of environmental issues. There are numerous technologies and solutions available to address these challenges. Finding economical, effective, and increasingly sustainable solutions is necessary to address and meet the problems posed by the world's growing population's energy demands and the environmental effects of a civilization that still primarily relies on fossil fuels. Although many materials are accessible for various environmental applications, the sustainability point of view is presently the main focus. In the context of sustainable materials development, carbon is the most abundant element in the biosphere after oxygen. Carbon-based materials are safe for advancing many practical technologies. The adaptable materials used in the current renewable energy sector are carbon-based constructions. (i.e., in both generation and storage) and environmental science (e.g., purification/remediation) (Titirici et al., 2015; Ravi & Vadukumpully, 2016; Maduraiveeran & Jin, 2021). The following are the newly developed carbon materials. Most carbon materials are utilized in environmental

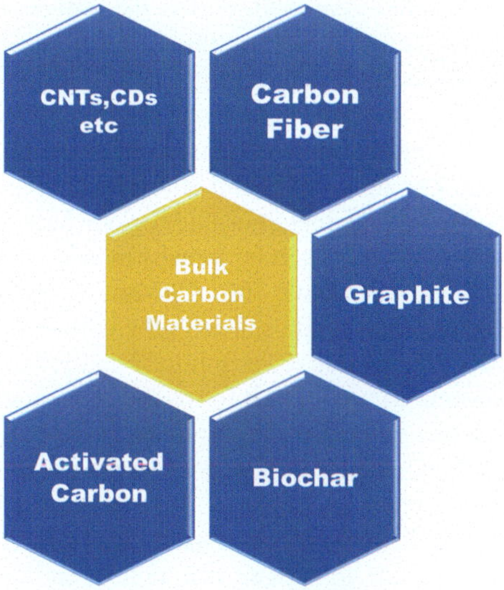

Fig. 3.1 List of bulk carbon materials

applications (Fig. 3.1), including activated carbon, biochar, carbon fibres, carbon dots, carbon nanotubes, carbon flowers, and amorphous carbon.

Various techniques are used to produce bulk carbon materials. Their structure, surface area, and porosity can be changed to suit specific environmental applications. These materials' characteristics can be finely tuned using several methods, from hydrothermal carbonization to pyrolysis, thermal activation, and chemical activation (Sosa et al., 2023). This carbon material synthesis process enables tuning several mechanical, surface, and property features for a wide range of applications.

This chapter reviews many of the synthesis methods employed to generate bulk carbon materials and describes how critical each of these processes is to the properties of the final product. We further discuss potential approaches to utilize the promising attributes of bulk carbon materials, i.e., their high surface area, regeneration potential, and thermal stability for environmental sustainability applications, such as air filtration, water purification, soil remediation, and sustainable energy generation.

3.1.1 Types of Bulk Carbon Materials in Environmental Applications

3.1.1.1 Graphite

Environmental protection and sustainable energy consumption are the pressing global society needs. Carbon materials are vital in the above issues due to their significant and multifaceted characteristics. Carbon materials are crucial in energy conservation and storage (Chen et al., 2020, 2024). Graphite is a crystalline form of carbon, like diamond, and is the natural allotrope of the elemental carbon. Each carbon atom bonds to the other three carbon atoms and forms a stable two-dimensional structure in graphite. The weak van der Waals forces hold these two layers so that one layer can slide over the other easily.

Natural graphite is mainly classified into three types. The vein is also known as lump graphite, flake, and amorphous graphite. The flake graphite could be seen in the high-grade metamorphic rocks where there is a chance for the organic carbon to be sediments. It can be transformed into graphite under pressure above 5 kbar and temperatures above 700 °C. Vein or lump graphite mostly occurs in the pipe- or fracture-filled bodies. There, the carbon fluids and the graphitic carbon drifted and then precipitated as graphite. Finally, anthracitic coal serves as a source for the amorphous graphite. It forms through variable degrees of graphitization during contact or regional metamorphism. A large amount of graphite is produced through synthetic routes. Amorphous carbon materials like calcinated petroleum coke are considered the prime source for the production of synthetic graphite (Keeling, 2017).

The different carbon-based materials, such as exfoliated graphite, graphene, graphene oxide, graphene nanoplates, carbon onions, and carbon nanotubes, can be reconditioned from graphite. Graphene takes up the prime role among the other derivatives of graphite. It excels in the industrial and academic fields due to its unique properties and wide range of applications (Kigozi et al., 2020). Several mechanical methods and oxidation can be used to produce graphene oxide (GO). The graphite flakes observed in the deposits are shown as the key source for the production of carbon-based materials and, in particular, graphene oxide. Thermally expanded graphene is widely used to manufacture different components for miscellaneous applications (Kigozi et al., 2020).

The low cost and excellent corrosion resistance make graphite a primary source. Its thermal and electrical conductivities enhance its usage. Graphite can also be used as a functional additive in different applications. Expanded graphite uses graphite as a raw material and intercalates appropriate chemical reagents between graphite layers using different methods to obtain expandable graphite (Zhang et al., 2023). The most explored and vitally used form is the thermally expanded graphite (TEG) (Murugan et al., 2021; Bai et al., 2024).

Graphite and its derivatives can be well employed in environmental applications. They can be used for the adsorption of organic pollutants, adsorption of metal ions, pollutant detection, desalination, and photocatalysis. Graphene, the most used derivative of graphite, is a suitable candidate for the detection of environmental toxin detection (Liu et al., 2021; Karthik et al., 2021).

3.1.1.2 Biochar

Biochar is a carbon-rich material widely used in various applications, such as wastewater treatment, soil amendment, and carbon sequestration due to its exceptional features like large surface area, high porosity, oxygen-containing functional groups, excellent adsorption capacity, and presence of mineral and trace elements. The properties of biochar mainly depend on the type of feedstocks and production conditions. Various feedstock such as agricultural waste, food waste, municipal solid waste, animal manure, industrial waste, and waste tires are used for the preparation. The conversion of waste biomass into making value-added products notably reduces waste production. Therefore, biochar production helps with waste management and improves global sanitation. In addition, it prevents contamination of groundwater, a notable contribution to environmental sustainability. Biochar became a suitable candidate in a circular economy because of the utilization of waste material for environmental remediation in different areas, such as soil, air, and water (Singh et al., 2022; Kumar & Bhattacharya, 2021).

Biochar is derived through the thermal decomposition of waste substances at elevated temperatures in the absence or limited amount of oxygen. Different types of thermal treatments are pyrolysis, gasification, torrefaction, combustion, carbonization, and slow and fast pyrolysis. Among them, slow pyrolysis is the simplest and most effective method for biochar production. Pyrolysis temperature plays a crucial role in determining the properties and yield of biochar. Biochar yield decreases with an increase in temperature. Moreover, carbon content increases, while oxygen and hydrogen content declines with temperature rise (Kumar & Bhattacharya, 2021).

Due to physiochemical surface characteristics, biochar is a green adsorbent with strong adsorption capacity for contaminants from the environment. Furthermore, biochar is enriched with carbon and other minerals, such as nitrogen and phosphorous, improving soil fertility and producing higher crop yields. Moreover, biochar can hold carbon for a long period. Hence, it helps to minimize the change in climatic conditions through carbon sequestration and mitigating greenhouse gas emissions (Hu et al., 2022; Bhattacharyya et al., 2024).

Biochar is a highly carbonaceous substance extracted from biomass feedstock, and they are highly known for its rich mineral content, stability, environmental friendliness, low cost, easy availability, and physicochemical features. Additionally, biochar can be further modified to enhance its properties, making them a suitable candidate for multifunctional applications (Wang et al., 2024a).

3.1.1.3 Activated Carbon

Activated carbon, often called activated charcoal, is a black porous carbonaceous material. It exhibits a large specific surface area (SSA), appropriate pore size, and high surface reactivity. AC has a turbostratic structure with amorphous and nongraphitic microcrystalline forms. Each microcrystal is between 1 and 3 nm in size and between 1 and 1.3 nm in thickness (Gao et al., 2020). Nearly 100,000 tonnes of activated carbon are produced each year worldwide. Wood, bituminous and anthracite charcoal, lignite, peat shells, and coconut are the most widely used commercial sources of activated carbon. Almond and olive shells are also utilized as substitute supplies. These materials have a density of 0.4–1.45 g/m^3 and a carbon content of 40–90% (wt.) (Heidarinejad et al., 2020). The direct carbonization product's structure is inadequate for many applications due to its low specific surface area and underdeveloped microporosity. It can be enhanced through activation. The procedure for activated carbon contains two phases: carbonization of the conversion of carbonaceous substances into carbon and activation for the SSA and specific pore volume (SPV) expansion of the converted carbon (Xia & Shi, 2016). The physical or chemical activation procedure is carried out to improve the properties mentioned above. In the physical activation process, gases like carbon dioxide, water vapour, air, or combinations of these gases are commonly used. Incorporating chemical agents such as KOH, K_2CO_3, H_3PO_4, and $ZnCl_2$ with AC is known as chemical activation (Serafin & Dziejarski, 2024). AC has tiny, low-volume pores that improve the surface area, which helps chemical reactions like adsorption so that it can be used in various applications, such as removing contaminants from air, water, and soil (Ganjoo et al., 2023). Additionally, it is used as a catalyst for biodiesel production (Teimouri et al., 2024). As a perfect adsorbent, AC has the notable ability to adsorb various polluting materials. This ability is promising and provides the foundation for accidental economic developments; it has a good future for tackling significant global challenges like air, water, and soil pollution. Still, AC has its drawbacks, such as expensive thermal/chemical regeneration methods and the loss of a significant amount of adsorbent during regeneration. Also, the adsorption mechanism of AC is subject to numerous aspects such as diffusive, electrostatic, and chemical interactions, as well as essential properties of the solute and adsorbent, and hence, the interaction between the pollutants and AC is complex to predict (Ahmed et al., 2022).

3.1.1.4 Carbon Fibres

Carbon fibres (CF) are carbonaceous objects with a 5–10 μm diameter and fibrous structure. According to the International Union of Pure and Applied Chemistry (IUPAC), CF is described as filaments, lines, fibres, or reals that, often in a nongraphitic form, contain at least 92% carbon. The nongraphitic state is a long-range two-dimensional arrangement of carbon atoms in planar hexagonal networks without any crystallographic order in the third direction other than roughly parallel stacking (Hassan et al., 2020). Substantial problems with energy security uncertainties and environmental concerns have prompted people worldwide to give clean, sustainable energy sources and solutions. Carbon fibres (CFs) are among the best options for energy and environmental applications because of their high mechanical strength, low density, flexibility, freestanding nature, good chemical stability, and superior thermal and electrical conductivity (Liu et al., 2022a). Annually, around 100,000 tonnes of carbon fibres are manufactured in large quantities. Over 90% of these fibres are made from the PAN, while the pitch-based CFs account for the rest. The production of CF consists of five steps: fibre spinning, carbonization, stabilization, surface treatment, and precursor synthesis (Choi et al., 2019). Techniques like surface modification and composite fabrication were used to modify the characteristics of carbon fibre, which can be used for different applications (Alshammari et al., 2021). Primary environmental uses include the removal of contaminants from soil, water, and air due to its large surface area and the presence of functional groups.

Furthermore, carbon fibres have a remarkable capacity for recycling; according to studies, efficient recycling techniques can reduce CO_2 emissions by 90–95% compared to manufactured fresh carbon fibre. In addition to reducing landfill waste, this recycling capability promotes resource sustainability in ecological fields (Ateeq, 2023).

3.1.1.5 Other Emerging Carbon Materials

Carbon-based materials (CBMs) have drawn interest due to their significant contribution to the advancement of materials research. This emerging group of CBMs, includes carbon nanotubes (CNTs) with single or multiple walls, carbon nanofibres (CNF), carbon dots, etc. (Egbedina et al., 2022). Table 3.1 explains different bulk carbon materials and their properties. Carbon nanotubes (CNTs) are a series of hollow cylindrical nanostructures made of sp^2-hybridized carbon; they are composed of carbon atoms with a diameter of 3–15 nm and arranged in hexagons and pentagons. Since the 1990s, CNTs have attracted attention from academics and businesses in various sectors due to their unique mechanical, physicochemical, and electrical characteristics (Kim et al., 2024; Rathinavel et al., 2021). Nanotubes are categorized based on the number of walls. CNTs generally fall into three categories: Single-walled CNTs (SWCNTs), double-walled CNTs (DWCNTs), and multi-walled CNTs (MWCNTs) (Yahyazadeh et al., 2024). A single-wall carbon nanotube (SWCNT) consists of a single cylindrical graphene tube (SWCNT), the most basic type of carbon nanotube. Double-wall CNTs (DWCNTs) and multiwall CNTs (MWCNTs) are made of two or more tubes wrapped in a scroll pattern or nested concentrically. SWCNTs typically have a diameter of 0.4–2 nm, whereas MWCNTs can have diameters of 10 s of nanometres, and both ends are usually capped by domes that resemble fullerene (Hughes et al., 2024). Carbon nanotubes (CNTs) help store hydrogen and solar energy. This will help alleviate the energy crisis issue to a certain degree. Because of their inert nature and strong electrical and thermal conductivity, CNTs were ideal for this use. CNTs are also used to eliminate heavy metals (such as mercury, lead, and cadmium) and organic chemicals (such as benzene, chlorinated solvents, and toluene) are substances from soils, sediment, and

Table 3.1 Different bulk carbon materials and their properties

Bulk carbon material	Pore volume (cm^3/g)	Surface area (m^2/g)	Reference
Biochar	0.016–0.083	8–132	Leng et al. (2021)
Porous carbon	–	3077	Avval et al. (2022)
Reduced graphene oxide	0.1–2.5	170.98	Attia et al. (2024)
Carbon nanotubes	–	209	Attia et al. (2024)
Graphitic carbon nitride (g-C_3N_4)	–	89	Attia et al. (2024)
Biochar from corn starch	0.0110	23.50	Choudhury and Lansing (2020)
Graphite-type activated carbon	1.09	1998	Keppetipola et al. (2021)

groundwater. Despite this, CNTs have been used to adsorb many trace pollutants. Three-dimensional CNT devices with extremely porous surfaces enable wastewater filtration to eliminate viruses and harmful microorganisms (Siti Norhafizan et al., 2013). As a novel form of carbon-based nanomaterial, carbon dots (CDs) have garnered attention from researchers for years (Liu et al., 2020). Carbon-based quantum dots, including carbon quantum dots (CQDs, C-dots, or CDs) and graphene quantum dots (QGDs), are a novel family of carbon nanomaterials that have diameters smaller than 10 nm (Gao et al., 2020). CDs were initially identified in 2004 during the separation of single-walled carbon nanotubes from the constituents of luminous nanoparticles. These carbon nanoparticles were referred to as "carbon quantum dots" until 2006 when Sun et al. suggested the technique for creating CDs using straightforward surface passivation and chemical modification to increase fluorescence emission. Since then, CDs have been considered because of their distinctive structure and intriguing qualities in numerous sectors (Mao et al., 2021). Simple one-pot synthesis methods regulate CDs' physical, chemical, and optical characteristics. Furthermore, various applications are made possible by their non-toxicity, biocompatibility, chemical and physical responsiveness, resistance to chemical and photobleaching, and affordability (Ozyurt et al., 2023). CDs have the potential for photocatalytic dye and emerging pollutant degradation (Hebbar et al., 2023). Carbon nanoflowers (Ma & Yuan, 2009), carbon nanofiber (Painuli et al., 2020), and carbon nanoscaffolds are promising materials in the environmental field, and research in this area is still emerging.

3.2 Synthesis Routes of Bulk Carbon Materials

3.2.1 Hydrothermal Carbonization

The European Union has adopted a circular economy-based society plan that mainly focuses on replacing conventional and non-renewable resources with renewable-based approaches. Hydrothermal carbonization (HTC) is a promising waste management technique among the different ways to create renewable carbon-based materials (González-Arias et al., 2022). This procedure involves introducing the feedstock into the water at a moderate temperature, often between 180 and 250 °C, and pressure (normally autogenous). As the reaction progresses, a combination of other reactions, including hydrolysis, dehydration, decarbonylation, decarboxylation, polymerization, re-condensation, etc., converts biomass into hydrochar (Fig. 3.2a).

Compared to other thermochemical processes, HTC has the significant advantage of not requiring expensive drying procedures beforehand; also, residual moisture frequently serves as both a solvent and a catalyst for HTC (Román et al., 2018). The conversion of the feedstock into carbon-rich hydrochar with specific surface area, porosity, and OFGs on the hydrochar's surface (Supee & Zaini, 2024) is

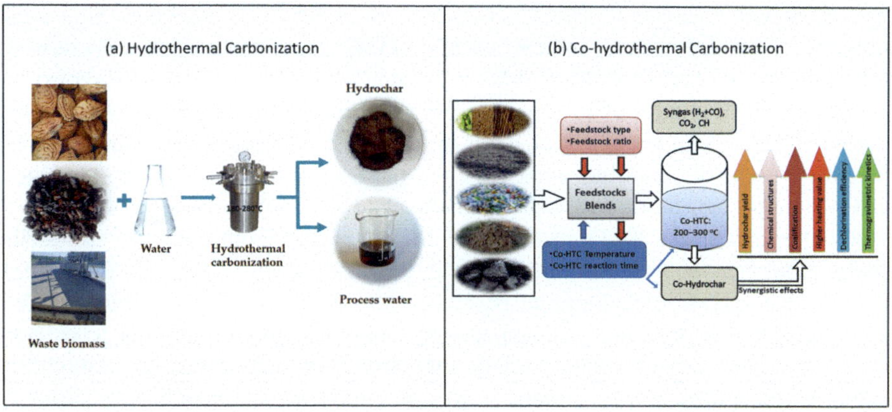

Fig. 3.2 Schematic representation of the HTC and Co-HTC process. 2024. (Reproduced with permission from Bardhan et al. Copyright 2021, Elsevier. Reproduced with permission from Petrović et al. Copyright 2024, MDBI)

significantly influenced by the operating parameters: the temperature applied during the HTC process, the residence time, and the feedwater conditions. As the temperature and residence time increase, the yield rate of hydrochar decreases, whereas the carbon content in hydrochar is high, and the higher heating value is high, which is conducive to reducing the hydrogen and oxygen content. The HTC process is typically fed with one feedstock type, but this enhances the uncertainty about the sustainability of industrial production processes.

Hydrothermal carbonization (HTC) procedures show that increasing the process temperature enhances biomass degradation, which leads to a decrease in hydrochar yield (Petrović et al., 2024). For example, Islam et al. observed that during the carbonization of banana stalks, extending the reaction time from 60 to 180 min at 200 °C reduced the hydrochar yield from 61.8% to 57.8%. However, the higher heating value (HHV) and fixed carbon content increased from 18.7 to 18.9 MJ/kg and 35.0% to 44.3%, respectively. Similarly, when the reaction temperature increased from 160 to 200 °C (with a reaction time of 180 min), the same study found a reduction in hydrochar yield from 72.8% to 57.8%, while the HHV and fixed carbon content increased from 18.4 to 18.9 MJ/kg and from 22.5% to 44.3%. This underscores the need to fine-tune residence time for given temperatures to produce hydrochar with the desired properties (Islam et al., 2019). HTC converts organic solid residues, such as biomass, sewage sludge, plastics, or livestock manure, into value-added products such as hydrochar. Nonetheless, the disadvantages of hydrochar produced by single-feedstock HTC can be overcome through co-hydrothermal carbonization (co-HTC) (Fig. 3.2b). There are several benefits for co-HTC compared to single-feedstock HTC: It is not limited to a feedstock type or a water requirement (wet wastes) and addresses the challenges posed by high inorganic contents in some wastes. For instance, co-pyrolyzing inorganic-rich wastes (e.g., sewage sludge) with inorganic-poor wastes (e.g., sawdust) produces a hydrochar with superior properties compared to single feedstock containing high inorganic content. Research indicates that co-HTC can enhance carbon sequestration, increase hydrochar yield, and be associated with higher HHVs. Moreover, it increases the efficiency of processes such as dechlorination and desulfurization relative to HTC with a single feedstock (Ho et al., 2024). The synergistic effect is the interaction of multiple feedstocks, which affects the efficiency of co-HTC. Synergy can be of two types: beneficial or harmful. A positive synergistic effect (Bardhan et al., 2021) occurs when the value from combining several feedstocks is greater than the sum of each value (Cavali et al., 2025). Hydrothermal carbonization (HTC) development is contingent on complete information on the tactics involved in forming carbon nano- and microspheres in addition to carbon dots. These mechanisms remain largely unexplored, and uncovering them ought to enable us to optimize HTC situations for our preferred merchandize. Additional studies on the carbon transformation happening at some stage in HTC may also lead to the improvement of complete hypotheses, thereby enhancing our information and optimizing the process (Ischia et al., 2024).

3.2.2 Pyrolysis

Pyrolysis is a simple and effective method of producing high-value carbonaceous materials such as carbon nanotubes, biochar, and graphene from carbon sources such as biomass, CH_4, CO_2, and plastics. It is a thermochemical decomposition process of carbon-based material at elevated temperatures without oxygen, converting organic–inorganic materials into solid, liquid, and gaseous products with multiple applications. The solid products are utilized as bioadsorbent, fertilizer, soil amendment, and solid fuel. Liquid products serve as biochemical sources and can be refined as liquid fuel. In contrast, gaseous products can either be recirculated to sustain the pyrolysis process or used as fuel for heat and power generation (Mong et al., 2022). Pyrolysis is a promising waste management technique and is considered the best way to convert waste to energy and other value-added products. This process involves heating various waste materials, including tires, plastics, medical waste, and lignocellulosic biomass, in a closed chamber like a furnace or kiln in an anaerobic environment, ranging from 3000 to 12,000 °C. This controlled setup results in only negligible emission of air pollutants and greenhouse emissions, while conventional methods utilize high energy to emit greenhouse gases. Therefore, pyrolysis is a versatile process that addresses current issues, such as energy consumption, greenhouse gas emissions, and waste management. This approach of producing value-added products from waste material is sustainable and has been considered a vital tool for the circular economy (Andooz et al., 2023; Chew et al., 2021).

The key properties of carbon material developed through the pyrolysis process are primarily based on pyrolysis temperature, heating rate, and reaction time.

According to this reaction condition, pyrolysis can be categorized into slow, fast, and flash. A comparative analysis of these pyrolysis methods considering quantitative factors such as temperature ranges and product yields provides insight into their efficiency and suitability for specific applications. In slow pyrolysis, the temperature range is from 300 to 500 °C, and residence time takes minutes to hours. It produces charcoal or biochar as the main product, with slow residence time and slow heating rate, while fast pyrolysis is carried out at moderate temperature (400–6000 °C), moderate heating rate, and short residence time; the major product is biofuels. Following this, flash pyrolysis occurs at high temperatures (>1000 °C), a high heating rate, and a minimal residence time, producing bio-oil. The yield of biocarbon varies by process; the flash process provides a high yield for fast pyrolysis, around 40%, whereas in slow pyrolysis, the yield is significantly lower, at approximately 25–30% (Adelawon et al., 2022).

The quantity and quality of the product are determined by the type of feedstock used and the pyrolysis process. Apart from this, with increasing temperature, yield decreases, whereas the carbon content and purity of the product improve due to the reduction in oxygen content. More than this, some crucial alterations of the other characteristics, including porosity, surface area, thermal stability, and electrical conductivity, made it an emergent potential material for many applications (Adelawon et al., 2022, Agweh et al., 2024). Depending on the heating source and additives, pyrolysis is classified into different types: vacuum plasma, microwave, catalytic, and co-pyrolysis. Among them, catalytic pyrolysis is one of the most common forms of pyrolysis. Catalysts are mainly used to improve the yield and quality of pyrolysis products and mitigate the energy consumption in the pyrolysis process. In a recent study, waste residues such as spent FCC catalyst, red mud, and fly ash obtained from a different production technique are utilized as catalysts to improve the properties of the products during the pyrolysis of waste tires. This method has received great attention due to its low cost and environmental benefits. Integration of the catalyst in the waste material has many advantages, including reducing pyrolysis temperature, increasing thermal degradation rate, lowering reaction complexity, and reducing energy consumption (Wang et al., 2024b).

Pyrolysis is an efficient way of producing biochar, bio-oil, and biogas from various waste materials. Figure 3.3 depicts the general outline of the pyrolysis treatment of different feedstock. The process involves three main steps: feedstock pretreatment, conversion, and up-grading and refining of the end products. Firstly, the pretreatment process differs for different feedstock; e.g., lignocellulosic biomass undergoes thermal treatment to remove moisture content, plastic and tires crumble into small particles, and medical waste is disinfected before use. The pretreatment causes variations in moisture content, volatile components, and other physiochemical properties, determining the end product's quantity and quality. Secondly, pyrolysis occurs inside the reactor, where feedstock is introduced. In this stage, the parameters such as temperature, heating rate, and residence time can be controlled depending on the desired properties of the end product. After this, char, oil, and gas have to be collected, which are then converted to useful products (Chew et al., 2021; Srivastava et al., 2021).

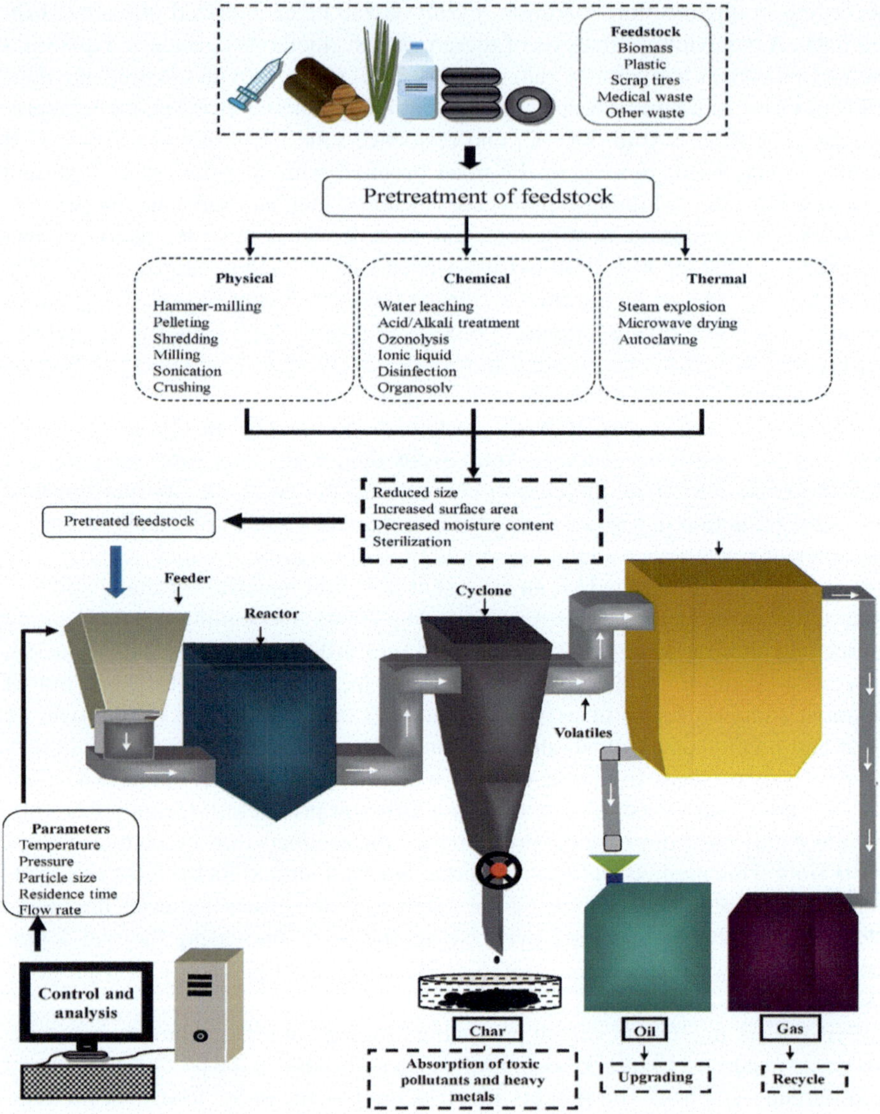

Fig. 3.3 General schematic diagram of the pyrolysis process. (Reproduced with permission from Chew et al. Copyright 2021, Elsevier)

3.2.3 *Physical Activation*

Activation is a common technique for modifying the properties of carbon materials. It can be classified mainly into physical and chemical activations. Physical activation involves two steps: first, the material is carbonized using gas, and then the activation is carried out (Din et al., 2017). Gases such as steam and carbon dioxide are employed as physical activators (Yang et al., 2021).

In the particular case of gasification with steam, the global reaction can be expressed as:

$$C + H_2O \rightarrow CO + H_2 \tag{3.1}$$

Here, there are two equivalent mechanisms:

1. Oxygen exchange model.

$$H_2O + C \leftrightarrow H_2 + C(O) \tag{3.2}$$

$$C(O) \rightarrow CO + C \tag{3.3}$$

2. Hydrogen inhibition model

$$H_2O + C \rightarrow H_2 + C(O) \tag{3.4}$$

$$C(O) \rightarrow CO + C \tag{3.5}$$

$$H_2 + C \leftrightarrow C(H_2) \tag{3.6}$$

where C is a carbon-free active site, C(O) is the oxygen surface complex, and $C(H_2)$ is the adsorbed hydrogen.

Both processes start with the adsorption and disassociation of water in the carbon surfaces' active sites, followed by the production of CO and H_2. It is more energetically challenging to remove the C(O) complexes from the surface (González-García, 2018). When the carbon in the chart reacts with the oxidizing gases, atoms are selectively removed from the material's surface, forming a network of pores. If carbon atoms were extracted from the inside of the still-developing pores, it would result in devolatilization during carbonization. This reaction would cause the opened micropores to widen and the closed micropores to open. This technique creates micropores, essential for adsorption applications, and increases the carbon's surface area. Physical activation's main benefit is creating microporous material with a large surface area, usually between 500 and 1500 m^2/g (Mamatha, 2023). Burning off from the outside of the particle does not create additional porosity, but it does help reduce the size of the particle. Physical activation is ecologically friendly since it uses gaseous activation chemicals and does not generate wastewater. However, there is a problem with this technique. Using physical activation techniques to produce microporous-activated carbon requires a lot of energy and time. This method's intrinsic disadvantage is that it eliminates a significant proportion of internal carbon mass to achieve a well-developed pore structure. Therefore, if one follows this path, one must be content with modest carbon yields (Viswanathan et al., 2009). Analysis of bulk carbon material fabrication processes shows each process's pros and cons.

Table 3.2 Analysis of the characteristics of bulk carbon materials

Synthesis method	Yield	Cost	Advantages	Disadvantages	References
Hydrothermal carbonization	30–60%	€157 per ton	Utilizing maximum feedstock, water reuse, heat recovery No harsh chemicals are used Low energy consumption	High cost	Ho et al. (2024), Lachos-Perez et al. (2022), Lucian and Fiori (2017)
Pyrolysis	20–40%	€164 per ton	CO_2 emission reduction Renewable Sustainable	High cost of production High energy consumption	Ning et al. (2013), Faragò et al. (2022)
Physical activation	Depends on gases	–	No use of chemicals High porous materials Easy process	Low yield High energy consumption	Bergna et al. (2018), Lucian and Fiori (2017)
Chemical activation	Depends on the chemicals used	–	Low activation temperature High carbon efficiency A large number of activating agent	Use of toxic chemicals	Heidarinejad et al. (2020)

Certain processes are more conducive to low-cost synthesis, while others can provide higher purity and greater yield, as outlined in Table 3.2.

3.2.4 Chemical Activation

Activation is the most exploited and effective strategy to produce carbonaceous materials. It improves the specific surface area and the pore volume of the particular material. There are three basic activation methods: physical, chemical, and physio-chemical. Chemical activation is considered the most effective method due to its advantages, such as lower processing temperature, higher yield, shorter duration, and excellent control over the surface area and pore volume. The main factors determining chemical activation's efficiency and outcome are the selection of activating agents, the heating approach, the mixing method, and the mass ratio of activating agents (Gao et al., 2020).

Various models have been suggested for the chemical activation. However, the current studies propose the microdomain model as the most fitting approach. The spherical carbon (SC) particles developed via a microdomain model were analysed. The physical properties, crystallinities, and morphologies were studied. The resulting SC chemically activated samples were found to excel in their porosity even in a lower activation temperature.

Wet oxidation, a different term for chemical activation, uses raw materials that contain cellulose. The raw materials saturate with highly oxidizing and degrading chemical agents in the early stages. The resulting suspension later dries and is heated to some particular temperatures as per the specifications of the desired product. Activation agents are the most decisive agents for the improvement of the properties of carbon-based materials. Sodium hydroxide, potassium hydroxide, phosphoric acid, and zinc chloride are the renowned chemical activation agents. In recent years, potassium hydroxide has been the activation agent most used. It can effectively participate in different reactions and selectively eliminate nongraphitic components. The resulting activated carbons are found to be equipped with a larger surface area and acceptable pore size under comparable conditions. It is also seen as less corrosive, less polluting, and far more cost-effective (Yang et al., 2021; Heidarinejad et al., 2020).

Even though KOH is an exceptional activating agent, corrosiveness and equipment deterioration limit its application. Its high boiling point (1327 °C) approximates that its miscellanies can be left behind even after the carbonization and activation processes. The K-salt-based activation (Fig. 3.4) can be a better option for chemical activation. It includes the direct one-step method and the conventional

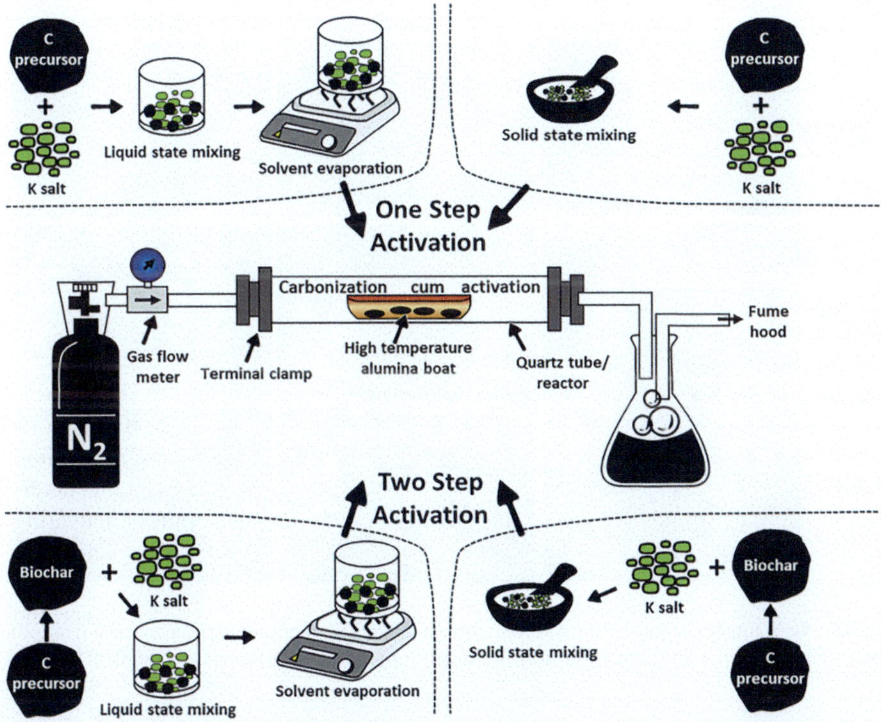

Fig. 3.4 The two types of mechanisms observed in K-salt activation. (Reproduced with permission from Singh et al. Copyright 2023, Elsevier)

two-step method. Both procedures can give different products in terms of porosity, and the condition is that they should be selected carefully for any of the desired carbon-containing precursors (Singh et al., 2023).

The activation of cellulose using $KHCO_3$ has become the focal point of much research. The higher surface area and the increased functional entities of the formed activated carbon increase its significance. While comparing the features of $KHCO_3$, KOH, and KCl, it is found that the pore size increases as the concentration of the activators increases. The $KHCO_3$ activation induces more uniformly distributed and well-formed pores. Additionally, the carbonization temperature also influences the specific surface area. As the temperature rises from 600 to 800 °C, an abrupt rise can be seen in the surface area. The carbonization temperature and surface area collaboratively motivate the nanopore development (Fig. 3.5) in the sample. Since the activators become gradually active at higher temperatures, it leads to the development of additional micropores and mesopores, further enhancing the specific surface area. The $KHCO_3$ and KOH activators highlight the enhanced pore-making ability of the activators at higher carbonization temperatures (Xu et al., 2022). All these factors effectively enhance the application, such as air filtration.

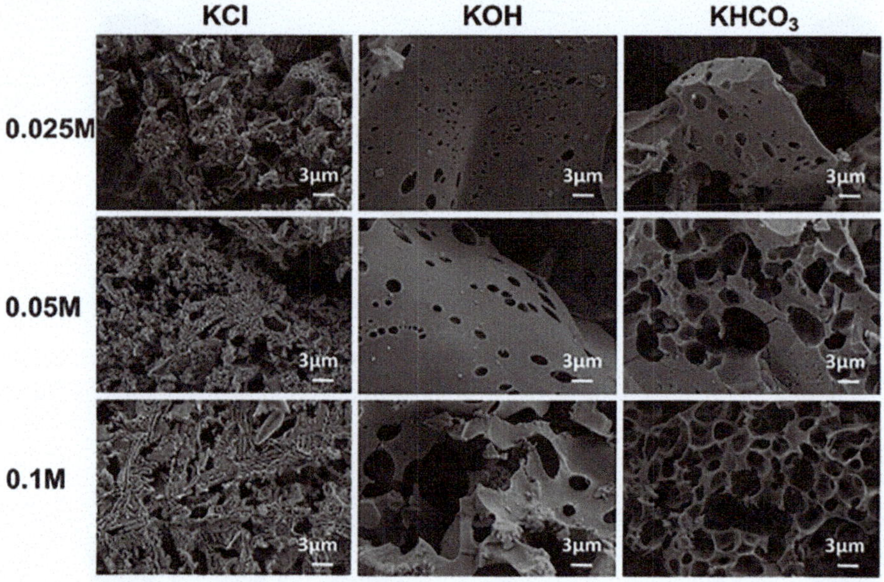

Fig. 3.5 SEM images of the different K-salts and KOH-activated porous carbon material from silk waste. (Reproduced with permission from Xu et al. Copyright 2022, American Chemical Society)

3.3 Key Properties of Bulk Carbon Materials

3.3.1 *Porosity and Surface Area*

Porosity and the surface area are the key factors that equip the carbon material to provide frequent adsorption sites and thus efficiently eliminate the pollutants (Zhang, 2024). The surface area and the porosity of carbon materials such as biochar are decisive in enhancing the chief applications like cation exchange ability, adsorption, and water-holding capacity. The macropores can be conductive to diffusion, the mesopores act as the mass transfer agents, and the micropores can contribute to the trapping space (Leng et al., 2021). The carbonaceous nanomaterials with a high surface area to volume ratio and controlled pore size can overcome the intrinsic constraint of the carbonaceous materials, such as huge dimensions of traditional sorbents, slow kinetics, and the nonequilibrium of sorption in heterogenous systems. It has also been noted that aggregation state and solvent chemistry can mark carbonaceous materials' surface area and pore volume (Mauter & Elimelech, 2008).

The synthesis methods can suggestively affect carbon materials'pore structure and properties. It has been observed that a series of reactions occur through high-temperature pyrolysis, such as dehydration, depolymerization, cross-linking, and isomerization. The decomposition of the carbon source and the gas flow creates many pores, and a highly ordered and stable carbon framework has been obtained (Zuo et al., 2023). Electrostatic spinning produces nanofibres with membranes that display accurate porosity, well-structured morphology, and a particular surface area. The chemical activation process can significantly alter the morphological features of the carbon materials. The K-salt-based activation can etch the carbon skeleton by the oxidation–reduction reactions and cause carbon lattice expansion. The porosity of carbon materials can increase further (Chen et al., 2020). The study of the difference in the activating agents showcases the influence of the surface properties of the resulting materials. $KHCO_3$ and KOH can produce an apparent pore structure, while KCl is a weaker activating agent (Xu et al., 2022).

The pyrolysis temperature and the type of carbon source significantly affect the resulting materials' specific surface area and pore volume. In the case of spherical biochar (SBC), the pyrolysis temperature rises above 500 °C, showing better surface area and pore volume than the conventional biochar systems. Regular morphology and the large pore volume could enable the material to hold green gases like CH_4 and remediate the effect of global warming (Qin et al., 2020).

3.3.2 *Functional Groups*

Materials made of pristine carbon, such as graphite, biochar, and activated carbon, have constraints when functionalizing their properties. A simple and effective method is required to enhance the material's quality. However, because of the well-ordered porous structures of these carbon materials, new functionalities like organic, inorganic, or biomaterials can be integrated into the porous channels or on the surface of the carbon walls. This dramatically improves their performance in a variety of applications (Benzigar et al., 2018). Surface functionalization denotes a procedure that alters the material's surface chemistry to give it particular qualities. Various carbon species are functionalized using different techniques (Benzigar et al., 2018). Oxidation, nitrogenation, and sulfuration are frequent modification processes that create functional groups by introducing oxygen, nitrogen, and sulphur heteroatoms onto the carbon surface. Figure 3.6 depicts the chemical agents used in each process and the matching functional groups that are generated (Yang et al., 2019).

For practical uses, graphene sheets must be functionalized and dispersed. Chemical functionalization preserves graphene's intrinsic qualities by allowing it to be treated with solvent assistance and preventing single-layer graphene from clumping together during reduction (Kuila et al., 2012). Graphene and similar substances have two central target locations appropriate for covalent functionalization. First,

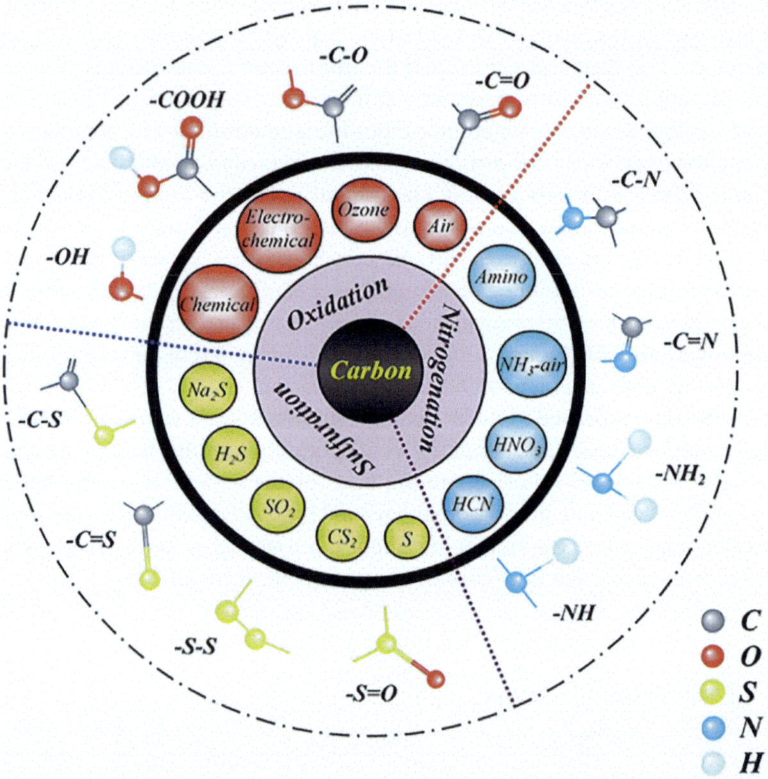

Fig. 3.6 Modification techniques to functionalized carbon materials with various functional groups. (Reproduced with permission from X. Yang et al. Copyright 2019, Elsevier)

new moieties can be added to the oxygenated groups (−OH, −COC on the basal plane, and −COOH at the margins) by esterification, condensation, or processes that target the oxygen atoms' lone pair electrons. The second location is the −C═C bonds on graphene's basal plane, which are susceptible to reactions with dienophiles or free radicals (Gong et al., 2016). A covalent organic polymer (AC-COP), amine groups (AC-NH_2 and AC-APTES), sulfonic groups (AC-MPTMS), and other functional groups were successfully added to paper mill sludge-based activated carbon (AC) by Jaria et al. (2020). Except for AC-MPTMS, where only the micropore surface area (Smicro) decreased, these functional groups reduced the specific surface area (SBET) and Smicro of all materials compared to the parent AC. The functionalized AC demonstrated good adsorption patterns when used to remove six pharmaceuticals (CBZ, LOR, SMX, PIR, PAR, and VEN) from ultrapure water, ultrapure water at pH 7.6, and effluent from a wastewater treatment plant (WWTP) (Jaria et al., 2020). Significant improvements in carbon nanotube (CNT)-based sensors and adsorbents for environmental protection have resulted from the functionalization of CNTs (Ogunsola et al., 2024). Functional oxygen groups can be added to CNT surfaces through oxidation or oxygen-plasma treatment. Further, CNTs can be functionalized by covering them with metal oxides like Al_2O_3, MnO_2, and Fe_3O_4 in addition to these surface changes, increasing effectiveness in removing contaminants (Aslam et al., 2021). Another study used the HNO_3–H_2SO_4 and NaOH–H_2O_2 systems to chemically oxidize old wheat straw biochar. The treatment enhanced oxygen-containing functional groups on the surface, especially carboxyl groups. For biochar treated with 40% HNO_3–H_2SO_4, the specific surface area increased by 226%, and for biochar treated with NaOH–H_2O_2, the mesopores were more developed. The highest sorption capacity of the biochar for cadmium increased by 21.2% due to this surface improvement (Fan et al., 2018).

3.3.3 Adsorption Capacity

Carbon-based materials exhibit the highest adsorption capacity for carbon dioxide and various organic (dyes) and inorganic (heavy metal ions, rare earth elements) pollutants from the environment. Adsorption is effective due to its cost-effectiveness, environmental friendliness, and reusability. It is a surface phenomenon, and adsorption occurs through chemisorption or physisorption. The adsorption capacity of a material primarily depends on the specific surface area, surface functional group, pore volume, and pore size distribution (Li et al., 2020; Sharma et al., 2022; Gunawardene et al., 2022).

Carbonaceous materials such as biochar, activated carbon, carbon nanotubes, and graphite show excellent adsorption capacity due to their high specific surface area, good mechanical properties, thermal and chemical stability, and plenty of surface functional groups. These properties can be tailored using various feedstock, changing reaction conditions, preparation techniques, and modification methods. Furthermore, carbon nanotube-based adsorbents are demonstrated as excellent adsorbents due to easy chemical activation, numerous internal and external adsorption sites, and functionalization (Mashkoor et al., 2020).

The presence of various oxygen-containing functional groups such as carboxyl, carbonyl, and ketone groups acts as primary active sites for the adsorption of pollutants. Moreover, enhancing such functional groups improves the properties of the materials and, most notably, enhances the interaction with pollutants. Therefore, in recent studies, surface modification has been carried out to introduce an active site, which further enhances the adsorption efficiency (Qiu et al., 2022; Togibasa et al., 2021).

Biosorption is an effective and sustainable approach to removing toxic pollutants from the environment. Biochar is a highly efficient bioadsorbent due to its superior properties and excellent adsorption performance. Biochar exhibits different adsorption mechanisms for pollutants due to the presence of numerous functional groups present on the surface of biochar. Adsorption mainly took place through physiochemical interactions, including electrostatic interaction, ion exchange, physical sorption, and complexation between metal ions and surface functional groups of biochar shown in Fig. 3.7 (Mergbi et al., 2023; Nguyen et al., 2023).

3.3.4 Thermal Stability

Many research endeavours are focused on understanding the thermal stability of carbonaceous materials, with attention on the individual thermokinetic spectra of classes of nanostructured carbon allotropes, such as graphene, fullerenes, nanotubes, and graphene-based materials. However, stability analysis is often complicated by numerous factors, including the temperature regime, heating rate, gas environment, and type and amount of the nanostructured carbon being investigated. This aligns with the specific requirements for using the material (Sádovská et al., 2023).

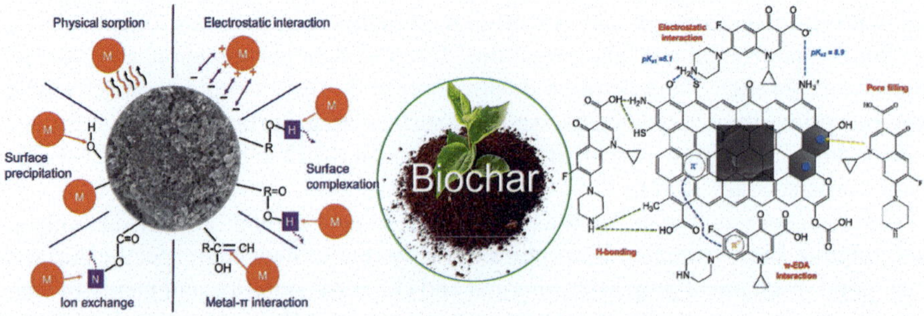

Fig. 3.7 Major adsorption mechanism in biochar for heavy metal ion. (Reproduced with permission from Nguyen et al. Copyright 2023, Elsevier)

Thermogravimetric analysis (TGA) is a familiar and easy approach for the characterization of the thermal properties of carbon materials (Farivar et al., 2021). Research on carbon nanomaterials has highlighted that stacking graphene layers regularly (i.e., having long-range order) and having parallel layers increase the starting oxidation temperature (Ton)—this is the most critical property for stability. The temperatures for graphite can reach up to 800 °C, and for graphene, they usually start around 530 °C. Because of its 3D structure, bent stress forms in the graphene layers, which causes defects and crumbling. This also creates many unsaturated carbon atoms at the edges, which can handle short-term high temperature heating better than disordered amorphous materials like fullerenes, zeolite templates, or other carbon based products. The oxidation of these compounds occurs between 485 and 530 °C, which is a rather narrow interval. Importantly, for materials with oxidized edges, a few oxygen-containing functional groups can decrease the stability by up to 150 °C and among them are hydroxyl groups (see Fig. 3.8) (Sádovská et al., 2023).

TGA was also employed in a recent study to evaluate the thermal stability of sawdust and A-CS-1000. Regarding thermal behaviour, the sawdust and A-CS-1000 show distinct loss of weight trends. As the temperature steadily rises to 150 °C, sawdust experiences a minor weight loss of roughly 5%. However, between 240 and 400 °C, a significant weight decrease of up to 72% indicates the thermal breakdown of the raw material's main constituents, cellulose and hemicellulose. It is well-known that cellulose, hemicellulose, and lignin commonly break down between 200 and 700 °C (Foorginezhad et al., 2024). Multi-walled carbon nanotubes (MWNTs) exhibit significant thermal stability, especially when cemented using techniques like spark plasma sintering. Because of their disordered structure, these materials retain their structural integrity at high temperatures, but their electrical conductivity exhibits a weak dependence on temperature changes (Farivar et al., 2021). In a study by Hu et al., Al-doped polysiloxane was used as the precursor to creating three-dimensional braided carbon fibre-reinforced SiAlOC (C/SiAlOC) composites. To

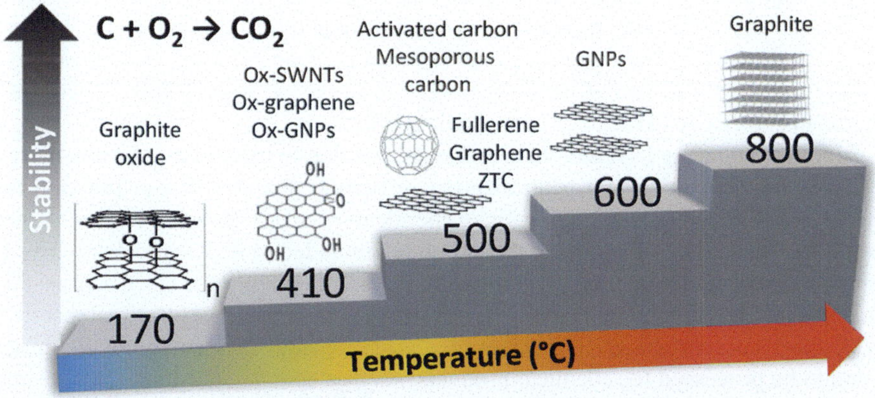

Fig. 3.8 Illustration of the relationship between temperature and the stability of each material. (Reproduced with permission from Sádovská et al. Copyright 2023, Elsevier)

examine the composites' thermal stability in an inert environment and vacuum, they were annealed at temperatures between 1200 and 1600 °C. According to the findings, the C/SiAlOC composites maintained their stability at temperatures up to 1400 °C in a vacuum and 1500 °C in an inert environment. The C/SiAlOC composites were more thermally stable, particularly in vacuum, due to the improved thermal stability of SiAlOC ceramics over SiOC ceramics (Hu et al., 2019). Thermal stability can be enhanced through material purification and nitrogen doping, making graphene suitable for applications where retarded combustion is desired (Martincic et al., 2024).

3.3.5 Chemical Reactivity

The capacity of carbonaceous materials to become anode is a related topic. Graphite is considered to be the most suitable material for this purpose. Carbonaceous materials are often conferred for efficient K^+ intercalation/deintercalation reversibility. The graphite can be well suited for the quicker electronic channel, whereas the amorphous carbon can host more K^+ ions (Wang et al., 2021c). The interactions between the carbon materials and the electrolytes substantially impact the reactivity while used as anodes. The different functional groups fitted in the graphene oxide (GO), like COOH, CO, COC, and OH, can improve the performance of the capacitors (Kigozi et al., 2020). Encouraging reactant adsorption and product diffusion is easier when we have a high surface area, chemical affinity for metal atoms, non-self-aggregation, and high porosity. Moreover, doping in metal clusters and defects formulates an unstable carbon support that can alter the selectivity and catalytic activity by shifting the localized electronic structure through charge transfer (Shi et al., 2020). Carbon nanomaterials can influence the efficiency of different materials' manufacturing procedures. Nanotubes can be used as catalysts, reagent scaffolds, or nanofluidics applicators for the process of fluids while they are integrated into membranes. CNT incorporated into the traditional iron-oxide catalyst enhances the production of styrene. The reaction itself can be changed from endothermic to exothermic with a lowering in the temperature range of around 200 °C with an increased selectivity (Mauter & Elimelech, 2008).

Biochar is alluring to researchers and scientists due to its activity and reactivity. It makes the utilization and neutralization of carbon possible. It can function to strengthen electron transfer and facilitate chemical reactions. It can be done with the help of the special components or the structures included in it. The biochar derived from the lignocellulosic biomass is found to be intermediating the electron transfer and enhancing the chemical reaction. It is a strong reference to the chemical reactivity of carbonaceous materials (Qin et al., 2022). A small percentage of lignin is found to be processed for chemical products, such as concrete slushing agents, surfactants, and binders. It can be noticeable to use it for further reactivity templates operating numerous functional groups (carbonyl, phenolic, sulfonic, methoxyl, and hydroxyl) into it (Wang et al., 2021a).

3.3.6 Regeneration Potential

Regeneration is a crucial strategy to improve waste management procedures and achieve a sustainable future. Thermal and chemical regeneration are the primary regeneration techniques most frequently employed (Larasati et al., 2021. The most common industrial regeneration method for spent carbon is thermal regeneration, which typically uses hot gases. This technique recovers heat-resistant fibres by pyrolyzing, calcining, or combining the two without using chemicals or mechanical damage. Pyrolysis and oxidation are the two primary mechanisms in thermal regeneration. The spent carbon is dried at 105 °C to eliminate moisture before regeneration. Thermal regeneration is the most feasible approach but has some significant drawbacks. Due to attrition, it reduces carbon capacity by 5–10%, and the process necessitates high temperatures (around 800 °C), which raises energy expenditures significantly (Larasati et al., 2021; Nistratov et al., 2022).

Chemical regeneration, however, desorbs the adsorbate from the carbon materials using solvents and chemical reagents. This technique works exceptionally well with organic adsorbents with low boiling temperatures and little heat stability (Alsawy et al., 2022). Several organic and inorganic solvents have been used to desorb organic or inorganic adsorbates. However, chemical regeneration may not be practical for large-scale industrial applications due to its cost, complexity, and environmental impact. Despite these challenges, it offers several advantages, such as recovering wet and end-of-life (EoL) wastes with higher mechanical properties compared to mechanical methods. It also allows for the recovery of carbon fibres with longer fibre lengths, which benefits applications requiring higher strength and performance (Heidarian et al., 2024). The thermal and chemical regeneration of MWCNTs was examined in one study. Since MWCNTs were significantly lost at higher temperatures, the thermal regeneration process was carried out for 2 h at 300 °C. After drying and thermal regeneration of the MWCNTs, these rinsed particles were fed to the influent of a wastewater treatment plant for a time period of 16 h for every contamination and regeneration cycle. The results proved that after five contamination and heat regeneration cycles, the sorption ability for three anticancer drugs—cyclophosphamide, ifosfamide, and 5-fluorouracil—was not changed. Furthermore, cyclophosphamide sorption decreased from approximately 50% (for untreated and thermally regenerated MWCNTs) to approximately 30% due to testing a chemical regeneration process employing 3 M HNO_3. This implies that the acid treatment changed the material's properties (Tośnski et al., 2021). However, these methods have several drawbacks: low efficiency, energy consumption, cost, little regeneration cycle, porosity decay, and material weight loss. As a result, many businesses are still using landfills or combustion to dispose of waste carbons. To move towards more sustainable development, it is important to find methods to recycle waste air conditioners with lower energy demand, cost, and higher (Wang et al., 2024c).

3.4 Environmental Applications of Bulk Carbon Materials

3.4.1 Air Filtration

The exemplary characteristics of carbon-based nanomaterials unlock novel technologies for solving environmental challenges, one of which is air filtration (Mauter & Elimelech, 2008). Since air pollution plays a hazardous role in environmental safety, different filtration tools have become an adequate necessity. Bulk carbon materials have proven themselves to be an effective means of solving the issue of environmental protection. Some of the air filters use activated carbon for this purpose. Fabricating the coronavirus filter with cellulose and activated carbon nanomaterial showed more than 95% exceptional particulate removal. The structure of the filter is designed by intertwining the carbon nanoparticles with fibres, producing highly porous air filter paper. It positively impacted the prevention of the hazardous coronavirus pandemic by removing more than 100 nm particulate matter. This air filter was capable of supplying incoming air flow from 10 to 200 L/min and pressure drop from 0 to 1200 Pa. The surface analysis showed the holes' size is less than 30 nm (Avval et al., 2022).

Additionally, a special interest has been bestowed upon graphene and the evolving carbon derivatives. Graphene aerogel provides exceptional modifiability and efficient adsorption. Due to these factors, it is considered to be a hopeful material for air filtration. Different types of particulate filtration membranes have been developed with the graphene incorporated polymers. Graphene aerogel has gained attention due to its high modifiability and exceptional adsorption capacity, making it a hopeful material for air filtration primarily occurring through adsorption. Graphene oxide incorporated with other polymers has created particulate matter filtration membranes. Polyvinylidene fluoride (PVDF), a high polymer, can form a composite with graphene and act as the nanofibrous air filtration membranes (Chen et al., 2021). Along with PVDF, polyacrylonitrile is used with graphene oxide to create air filtration membranes. Polyaniline/GO composites are also applied to remove particulate matter (PM) from the air. Both can eliminate the PM 2.5 specification. They could retain stable performance even after repeated usage cycles (Zhang, 2024). It is observed that the graphene/TiO_2 hybrid can operate a dual mechanism. It takes the role of an air filter and photodestructor. Added to TiO_2, ZnO and silver nanoparticles can be used for pollutant removal and antibacterial activity. Moreover, element-doped graphene nanosheets showcased enhanced adsorption ability. FeG (graphene doped with Fe) has served as an alternative for pollutant removal. The laser-induced graphene (LIG) exclusively used for air filters provides antiviral, antifouling, and antibacterial properties in addition (Zhang et al., 2020; Karanjikar et al., 2021; Gupta et al., 2021). Carbon nanotubes (CNTs) have a large surface area and are popular for air filtration. The high adsorption and catalytic efficiency of CNTs have rapidly removed gas and solid pollutants. Their reusability and enduring capacity make them promising candidates for air filtration. A recent study shows that activated carbon declines 5% after 10 cycles, and CNT loses only 2% efficacy after 15 usage cycles. The capacity of Graphene oxide loses 3% after 12 cycles of usage (Filtration, 2024). Bulk carbon materials are promising for environmental applications, specifically for air filtration.

3.4.2 Water Purification

Water pollution is a global issue that harms human health and the ecosystem. The presence of organic and inorganic pollutants in various water resources results in scarcity of clean drinking water and water-borne diseases. Carbon-based materials have emerged as a promising material for water purification due to their economic and environmental benefits. Different types of carbon material, including activated carbon, biochar, graphene, carbon nanotubes, and quantum dots, are best for applying in water purification due to their excellent properties. Some modifications have been made to enhance the properties of adsorbent material, depending on the type of pollutant targeted for elimination. The modification improves the adsorption efficiency of different pollutants (Beniwal et al., 2022; Tomczyk et al., 2023).

Biochar has been given huge attention as it is a good adsorbent for the removal of pollutants like organic pollutants, heavy metal ions, microplastics, and inorganic fertilizers from water sources due to its exceptional features. Due to many surface functional groups, biochar properties, and their effects, adsorption efficiency is further improved. Various metal oxides are used for biochar modification to produce biochar composite. In a recent study, incorporating biochar composite not only improves the adsorption capacity of organic dye but also makes them magnetic, making them easily recoverable and reusable (Majumder et al., 2023; Eniola et al., 2023). Blending of nanoparticles into biochar showed a major change in adsorption efficiency. Lui et al. synthesized a novel biochar hybrid composite to remove two heavy metal ions. First, the biochar is activated by KOH, and then the material is modified with MgO. As a result of KOH activation, pore diameter and surface area increased, which led to a higher adsorption performance. Further modification with MgO enlarged nanoporosity and adsorption for heavy metal ions, such as Pb^{2+} (1276.46 mg/g) and Cd^{2+} (861.82 mg/g) (Liu et al., 2023).

Graphene-based membranes have high adsorption performance. Later, it emerged as a promising material for wastewater treatment applications like desalination, adsorption of heavy metal ions and organic dyes, separating pollutants, and serving as a sensor (Kumari et al., 2021). A recent study found that incorporating silver (Ag) nanoparticles onto graphene oxide enhances the antimicrobial properties of the membrane. Further addition of catalytic functionalities enhanced adsorption performance. Also, it caused a reduction in the fouling of the membrane. As-synthesized graphene oxide membrane showed the highest rejection rate for organic molecules, such as bovine serum albumin (BSA) (98%), CR (95%), RhB (88%), and MB (85%) (Jose et al., 2023). Another study done by Wang and co-workers developed a series of graphene oxide nanofiltration membranes incorporated with polycyclic p-conjugated cations like toluidine blue O (TBO) for desalination. GO-TBO membranes showed high nanofiltration performance, rejecting salt and neutral solutes in various concentration ranges. Moreover, as-synthesized membranes have proven to be best material for desalination purpose due to their good chemical and mechanical strength and reliability in severe condition (Wang et al., 2021b).

3.4.3 Soil Remediation

Insufficient agricultural management has accelerated the decomposition of organic residual matter in the soil and raised greenhouse gas (GHG) emissions (Osman et al., 2021), contamination of soil by heavy metals (Cd, Pb, Cr, Cu, Zn, As, Co, Ni, etc.), and organic pollutants (pesticides, fertilizers, antibiotics, polycyclic aromatic hydrocarbons (PAHs), polychlorinated biphenyls (PCBs), etc.). In addition to affecting crop yields and soil quality, these pollutants—which are primarily introduced into the soil environment through sewage irrigation, atmospheric deposition, fertilizers, pesticides, industrial wastes, mineral resources, etc.—cause carcinogenesis, teratogenesis, mutagenesis, and genetic toxicity, which in turn endanger human health through the food chain. To prevent harmful impacts and successfully repair the ecosystem of contamination (Ji et al., 2022), adding carbon materials can increase the amount of organic carbon in the soil (Panchamoorthy et al., 2021). Because of their many uses and superior reclamation capabilities, carbon-based materials (CBMs) have drawn increasing attention to remediating contaminants from polluted soil (Fu et al., 2023). Promising substances for soil remediation include graphene (GN), carbon nanotubes (CNTs), biochar (BC), and activated carbon (AC). These carbon-based materials have a porous structure and a vast surface area. Their core structure is made up of aromatic rings composed of carbon atoms. Although they can be pretty successful at absorbing contaminants, several factors affect this ability, such as the characteristics of the soil, the presence of functional groups, and the surface chemistry and structure of the materials employed as sorbents (Bui et al., 2024; Liang et al., 2022).

Sabitov et al. (2024) reported that oil-contaminated soils could be successfully remedied using plant waste-activated carbons (ACs). Two-stage pyrolysis and chemical activation with potassium hydroxide produced ACs from rice husks, reed stalks, pine sawdust, and wheat straw, demonstrating exceptional hydrocarbon sorption capacity. Because of its extensive microporous and mesoporous features, rice husk AC, in particular, showed a significant sorption capacity. In field tests conducted at the "Zhanatalap" oil deposit, ACs achieved a 67% purification degree in 16 days by reducing the oil concentration in polluted soils from 79.2 to 2.6 g/kg. Biochar's (BC) surface area is between 100 and 400 m^2/g. Surface characteristics, such as morphology (porosity and surface area) and chemistry (surface functional groups, pH, chemical composition, and mineral content), are the main determinants of the adsorption efficiency of carbon materials. There are two types of adsorption mechanisms: chemical and physical. van der Waals forces acting on surface planes promote physical adsorption, but chemisorption occurs at edges when heterogeneous surface functional groups (SFG) are present, producing intermediate intermediates. These surface functional groups are the main force behind chemisorption. More effective adsorbents and modified carbon materials can be engineered to have particular surface functional groups for the remediation of particular contamination (Sumaraj & Padhye, 2017). Studies on using biochar derived from various materials, including wood, rice husks, and sunflower husks, for soil remediation have demonstrated that biochar dramatically improves the removal of metals from soil. Adding biochar significantly improved the metal removal efficiency (>77%) with higher removal percentages for Cu, Pb, and Zn (Burachevskaya et al., 2023). In another study, rather than extracting heavy metals from the soil, the biochar amendment immobilizes heavy metals and transforms contaminants into less soluble and bioavailable forms. This method is similar to chemical stabilization, where contaminated soils are treated with phosphate and carbonate chemicals to convert the heavy metal pollutants into precipitates (Guo et al., 2020). Microwave-assisted carbon nanotubes (CNTs) have also shown the ability to remove total petroleum hydrocarbons (TPH) in soil. Hundred grams of the CNTs under microwave irradiation for 60 s showed an 82% TPH removal (Baghel et al., 2022). Similarly, Zhang et al.'s study assessed the performance of carboxylate or hydroxylated multiwalled carbon nanotubes (MWCNT-COOH or MWCNT-OH) for the remediation of Cr(VI)-contaminated soil (Zhang et al., 2018). Bulk carbon products are thought to be viable options for soil recovery. Generally speaking, high carbon char does not immediately affect soil physical characteristics or crop yield. Since it takes time for the char particles to interact with the soil particles, it might eventually have a favourable impact on crop yield and soil characteristics.

3.5 Challenges and Future Perspective of Bulk Carbon Materials

Carbon-based functional materials (CBMs) are helpful and flexible in many areas (Ma et al., 2017). There is much interest in using CBMs for ecological applications because they have unique features like connected holes, a large surface area, and a particular shape (Wang et al., 2019). These qualities make CBMs very good for different conservational applications.

However, using CBMs more often means examining how they affect the environment and health. More use raises concerns about their safety and how they react to other things, which means we need better ways to eliminate and reuse them. A recent study showed that approximately 13.8 tons of CO_2 are produced during graphite manufacturing. Therefore, tracking the carbon footprint of all bulk carbon materials, optimizing their conditions, and implementing them in large-scale products are essential. Additionally, small amounts of CBMs in water, air, and soil show that our current treatment methods do not entirely remove these materials (Soffian et al., 2022).

Even though carbon-based nanomaterials have much potential, problems like high production costs and trouble making them in large amounts are significant obstacles to using them more widely. Currently, most work with these materials is done in labs, making it hard to produce them in enormous quantities and use them in real-world situations. To make these materials more useful, future research needs to focus on lowering costs and finding ways to produce them on a larger scale. Also, it is important to look at how these materials affect the environment since they will eventually end up in nature (Liu et al., 2022b). Better production methods and thorough checks on how they affect the environment will be key to using these materials in ways that help protect the environment.

Implementing AI assistant modelling and machine learning for carbon-based materials shows great promise for advanced research. It helps to minimize trial and error, thereby saving a considerable amount of energy. Furthermore, during the production of carbon-based materials, conducting a life cycle assessment is essential to comprehend the cost and energy invested and to quickly identify the carbon materials with lower carbon footprint values, thereby supporting a circular economy.

3.6 Conclusion

This chapter comprehensively compiles the synthesis of carbon materials and their properties for environmental remediation in condensed data. Many carbon materials have become indispensable multi-purpose tools in addressing environmental challenges. Because of its unique properties and adaptability, synthesis routes include hydrothermal carbonization and pyrolysis thermochemical activation. It also discusses various methods to tailor these materials for specific environmental applications and highlights their important properties. This determines the efficiency of many carbon materials in applications, such as water purification, air filtration, and soil remediation.

The application is vast, but challenges such as repeatability remain in optimizing production methods to balance cost, sustainability, and efficiency. The environmental impact of the synthesis process and the scalability of continuous innovation and interdisciplinary research are required. Moreover, as new carbon materials and hybrid composites emerge, the need to investigate their properties and uses is increasing, which is opening the way to future success. In short, carbon remains the foundation of state-of-the-art technology in environmental mitigation. With a solution to the problems of today and the technologies of tomorrow, these materials will ensure that environmental quality and sustainability are better than ever before in the near future, thanks to advancements in materials science.

References

Adelawon, B. O., Latinwo, G. K., Eboibi, B. E., Agbede, O. O., & Agarry, S. E. (2022). Comparison of the slow, fast, and flash pyrolysis of recycled maize-cob biomass waste, Box-Behnken process optimization and characterization studies for the thermal fast pyrolysis production of bioenergy. *Chemical Engineering Communications, 209*(9), 1246–1276. https://doi.org/10.1080/00986445.2021.1957851

Agweh, K., Snowdon, M. R., Kumar, R., Chen, G., Vivekanandhan, S., Mohanty, A. K., & Misra, M. (2024). Production and characterization of waste nutshells derived biocarbon through slow pyrolysis: An investigation on the effects of pyrolysis temperature. *Biomass Conversion and Biorefinery, 14*(22), 28619–28633. https://doi.org/10.1007/s13399-023-03851-4

Ahmed, M., Mavukkandy, M. O., Giwa, A., Elektorowicz, M., Katsou, E., Khelifi, O., Naddeo, V., & Hasan, S. W. (2022). Recent developments in hazardous pollutants removal from wastewater and water reuse within a circular economy. *npj Clean Water, 5*(1), 1–25. https://doi.org/10.1038/s41545-022-00154-5

Alsawy, T., Rashad, E., El-Qelish, M., & Mohammed, R. H. (2022). A comprehensive review on the chemical regeneration of biochar adsorbent for sustainable wastewater treatment. *npj Clean Water, 5*(1). https://doi.org/10.1038/s41545-022-00172-3

Alshammari, B. A., Alsuhybani, M. S., Almushaikeh, A. M., Alotaibi, B. M., Alenad, A. M., Alqahtani, N. B., & Alharbi, A. G. (2021). Comprehensive review of the properties and modifications of carbon fiber-reinforced thermoplastic composites. *Polymers, 13*(15), 1–32. https://doi.org/10.3390/polym13152474

Andooz, A., Eqbalpour, M., Kowsari, E., Ramakrishna, S., & Ansari Cheshmeh, Z. (2023, June 2022). A comprehensive review on pyrolysis from the circular economy point of view and its environmental and social effects. *Journal of Cleaner Production, 388*, 136021. https://doi.org/10.1016/j.jclepro.2023.136021

Aslam, M. M. A., Kuo, H. W., Den, W., Usman, M., Sultan, M., & Ashraf, H. (2021). Functionalized carbon nanotubes (CNTs) for water and wastewater treatment: Preparation to application. *Sustainability (Switzerland), 13*(10), 1–54. https://doi.org/10.3390/su13105717

Ateeq, M. (2023, October). A state of art review on recycling and remanufacturing of the carbon fiber from carbon fiber polymer composite. *Composites Part C: Open Access, 12*, 100412. https://doi.org/10.1016/j.jcomc.2023.100412

Attia, Y. A., Ezet, A. E., Saeed, S., & Galmed, A. H. (2024). Nano carbon-modified air purification filters for removal and detection of particulate matters from ambient air. *Scientific Reports, 14*(1), 1–12. https://doi.org/10.1038/s41598-023-50902-x

Avval, Y. F., Pour, G. B., & Aram, M. M. (2022). Fabrication of high efficiency coronavirus filter using activated carbon nanoparticles. *International Nano Letters, 12*(4), 421–426. https://doi.org/10.1007/s40089-022-00379-9

Baghel, P., Sakhiya, A. K., & Kaushal, P. (2022). Ultrafast growth of carbon nanotubes using microwave irradiation: Characterization and its potential applications. *Heliyon, 8*(10), e10943. https://doi.org/10.1016/j.heliyon.2022.e10943

Bai, M., Chen, R., Liu, X., Li, H., Li, J., Huang, H., Song, M., Zhang, Q., Su, Y., Wang, H., Xu, M., & Xie, G. (2024). An effective strategy for synthesizing high-performance photocatalyst by recycling the graphite target wastes. *Journal of Environmental Chemical Engineering, 12*(5), 113872. https://doi.org/10.1016/j.jece.2024.113872

Bardhan, M., Novera, T. M., Tabassum, M., Islam, M. A., Islam, M. A., & Hameed, B. H. (2021). Co-hydrothermal carbonization of different feedstocks to hydrochar as potential energy for the future world: A review. *Journal of Cleaner Production, 298*, 126734. https://doi.org/10.1016/j.jclepro.2021.126734

Beniwal, K., Kaur, H., Saini, A. K., & Siwal, S. S. (2022). Synthesis and applications of carbon porous nano-materials for environmental remediation. *Nanofabrication, 7*, 174–194. https://doi.org/10.37819/nanofab.007.249

Benzigar, M. R., Talapaneni, S. N., Joseph, S., Ramadass, K., Singh, G., Scaranto, J., Ravon, U., Al-Bahily, K., & Vinu, A. (2018). Recent advances in functionalized micro and mesoporous carbon materials: Synthesis and applications. *Chemical Society Reviews, 47*(8), 2680–2721. https://doi.org/10.1039/c7cs00787f

Bergna, D., Varila, T., Romar, H., & Lassi, U. (2018). Comparison of the properties of activated carbons produced in one-stage and two-stage processes. *C, 4*(3), 41. https://doi.org/10.3390/c4030041

Bhattacharyya, P. N., Protim, S., Bhaskar, S., & Abhay, S. (2024). Biochar as soil amendment in climate—Smart agriculture: Opportunities, future prospects, and challenges. *Journal of Soil Science and Plant Nutrition, 24*(1), 135–158. https://doi.org/10.1007/s42729-024-01629-9

Bui, T. H., Zuverza-Mena, N., Dimkpa, C. O., Nason, S. L., Thomas, S., & White, J. C. (2024, January). PFAS remediation in soil: An evaluation of carbon-based materials for contaminant sequestration. *Environmental Pollution, 344*, 123335. https://doi.org/10.1016/j.envpol.2024.123335

Burachevskaya, M., Minkina, T., Bauer, T., Lobzenko, I., Fedorenko, A., Mazarji, M., Sushkova, S., Mandzhieva, S., Nazarenko, A., Butova, V., Wong, M. H., & Rajput, V. D. (2023). Fabrication of biochar derived from different types of feedstocks as an efficient adsorbent for soil heavy metal removal. *Scientific Reports, 13*(1), 1–14. https://doi.org/10.1038/s41598-023-27638-9

Cavali, M., Hennig, T. B., Junior, N. L., Kim, B., Garnier, V., Benbelkacem, H., Bayard, R., Woiciechowski, A. L., & Matias, W. G. (2025). *Co-hydrothermal carbonization of sawdust and sewage sludge: Assessing the potential of the hydrochar as an adsorbent and the ecotoxicity of the process water*. 1–21.

Chen, Q., Tan, X., Liu, Y., Liu, S., Li, M., Gu, Y., Zhang, P., Ye, S., Yang, Z., & Yang, Y. (2020). Biomass-derived porous graphitic carbon materials for energy and environmental applications. *Journal of Materials Chemistry A, 8*(12), 5773–5811. https://doi.org/10.1039/c9ta11618d

Chen, M., Jiang, J., Feng, S., Low, Z. X., Zhong, Z., & Xing, W. (2021, May). Graphene oxide functionalized polyvinylidene fluoride nanofibrous membranes for efficient particulate matter removal. *Journal of Membrane Science, 635*, 119463. https://doi.org/10.1016/j.memsci.2021.119463

Chen, Z., Jiang, X., Boyjoo, Y., Zhang, L., Li, W., Zhao, L., Liu, Y., Zhang, Y., Liu, J., & Li, X. (2024). Nanoporous carbon materials derived from biomass precursors: Sustainable materials for energy conversion and storage. In *Electrochemical energy reviews* (Vol. 7, Issue 1). Springer. https://doi.org/10.1007/s41918-024-00223-y

Chew, K. W., Chia, S. R., Chia, W. Y., Cheah, W. Y., Munawaroh, H. S. H., & Ong, W. J. (2021). Abatement of hazardous materials and biomass waste via pyrolysis and co-pyrolysis for environmental sustainability and circular economy. *Environmental Pollution, 278*, 116836. https://doi.org/10.1016/j.envpol.2021.116836

Choi, D., Kil, H. S., & Lee, S. (2019). Fabrication of low-cost carbon fibers using economical precursors and advanced processing technologies. *Carbon, 142*, 610–649. https://doi.org/10.1016/j.carbon.2018.10.028

Choudhury, A., & Lansing, S. (2020, March). Biochar addition with Fe impregnation to reduce H_2S production from anaerobic digestion. *Bioresource Technology, 306*, 123121. https://doi.org/10.1016/j.biortech.2020.123121

Din, M. I., Ashraf, S., & Intisar, A. (2017). Comparative study of different activation treatments for the preparation of activated carbon: A mini-review. *Science Progress, 100*(3), 299–312. https://doi.org/10.3184/003685017X14967570531606

Egbedina, A. O., Bolade, O. P., Ewuzie, U., & Lima, E. C. (2022). Emerging trends in the application of carbon-based materials: A review. *Journal of Environmental Chemical Engineering, 10*(2), 107260. https://doi.org/10.1016/j.jece.2022.107260

Eniola, J. O., Sizirici, B., Khaleel, A., & Yildiz, I. (2023, July). Fabrication of engineered biochar-iron oxide from date palm frond for the effective removal of cationic dye from wastewater. *Journal of Water Process Engineering, 54*, 104046. https://doi.org/10.1016/j.jwpe.2023.104046

Fan, Q., Sun, J., Chu, L., Cui, L., Quan, G., Yan, J., Hussain, Q., & Iqbal, M. (2018). Effects of chemical oxidation on surface oxygen-containing functional groups and adsorption behavior of biochar. *Chemosphere, 207*, 33–40. https://doi.org/10.1016/j.chemosphere.2018.05.044

Faragò, M., Damgaard, A., Logar, I., & Rygaard, M. (2022). Life cycle assessment and cost-benefit analysis of technologies in water resource recovery facilities: The case of sludge pyrolysis. *Environmental Science and Technology, 56*(24), 17988–17997. https://doi.org/10.1021/acs.est.2c06083

Farivar, F., Lay Yap, P., Karunagaran, R. U., & Losic, D. (2021). Thermogravimetric analysis (TGA) of graphene materials: Effect of particle size of graphene, graphene oxide and graphite on thermal parameters. *C, 7*(2), 41. https://doi.org/10.3390/c7020041

Filtration, A. (2024). *Introduction, 02*(06), 79–86.

Foorginezhad, S., Zerafat, M. M., Asadnia, M., & Rezvannasab, G. (2024, November 2023). Activated porous carbon derived from sawdust for CO_2 capture. *Materials Chemistry and Physics, 317*, 129177. https://doi.org/10.1016/j.matchemphys.2024.129177

Fu, T., Zhang, B., Gao, X., Cui, S., Guan, C. Y., Zhang, Y., Zhang, B., & Peng, Y. (2023, September 2022). Recent progresses, challenges, and opportunities of carbon-based materials applied in heavy metal polluted soil remediation. *Science of the Total Environment, 856*, 158810. https://doi.org/10.1016/j.scitotenv.2022.158810

Ganjoo, R., Sharma, S., Kumar, A., & Daouda, M. M. A. (2023). Activated carbon: Fundamentals, classification, and properties. *Activated Carbon*, 1–22. https://doi.org/10.1039/bk9781839169861-00001

Gao, Y., Yue, Q., Gao, B., & Li, A. (2020). Insight into activated carbon from different kinds of chemical activating agents: A review. *Science of the Total Environment, 746*, 141094. https://doi.org/10.1016/j.scitotenv.2020.141094

Gong, X., Liu, G., Li, Y., Yu, D. Y. W., & Teoh, W. Y. (2016). Functionalized-graphene composites: Fabrication and applications in sustainable energy and environment. *Chemistry of Materials, 28*(22), 8082–8118. https://doi.org/10.1021/acs.chemmater.6b01447

González-Arias, J., Sánchez, M. E., Cara-Jiménez, J., Baena-Moreno, F. M., & Zhang, Z. (2022). Hydrothermal carbonization of biomass and waste: A review. *Environmental Chemistry Letters, 20*(1), 211–221. https://doi.org/10.1007/s10311-021-01311-x

González-García, P. (2018, August 2017). Activated carbon from lignocellulosics precursors: A review of the synthesis methods, characterization techniques and applications. *Renewable and Sustainable Energy Reviews, 82*, 1393–1414. https://doi.org/10.1016/j.rser.2017.04.117

Gunawardene, O. H. P., Gunathilake, C. A., Vikrant, K., & Amaraweera, S. M. (2022). Carbon dioxide capture through physical and chemical adsorption using porous carbon materials: A review. In *Atmosphere* (Vol. 13, Issue 3). https://doi.org/10.3390/atmos13030397

Guo, M., Song, W., & Tian, J. (2020, October). Biochar-facilitated soil remediation: Mechanisms and efficacy variations. *Frontiers in Environmental Science, 8*. https://doi.org/10.3389/fenvs.2020.521512

Gupta, A., Sharma, C. P., Thamaraiselvan, C., Pisharody, L., Powell, C. D., & Arnusch, C. J. (2021). Low-voltage bacterial and viral killing using laser-induced graphene-coated non-woven air filters. *ACS Applied Materials and Interfaces, 13*(49), 59373–59380. https://doi.org/10.1021/acsami.1c20198

Hassan, M. F., Sabri, M. A., Fazal, H., Hafeez, A., Shezad, N., & Hussain, M. (2020, May 2019). Recent trends in activated carbon fibers production from various precursors and applications—A comparative review. *Journal of Analytical and Applied Pyrolysis, 145*, 104715. https://doi.org/10.1016/j.jaap.2019.104715

Hebbar, A., Selvaraj, R., Vinayagam, R., Varadavenkatesan, T., Kumar, P. S., Duc, P. A., & Rangasamy, G. (2023, October 2022). A critical review on the environmental applications of carbon dots. *Chemosphere, 313*, 137308. https://doi.org/10.1016/j.chemosphere.2022.137308

Heidarian, P., Mokhtari, F., Naebe, M., Henderson, L. C., & Varley, R. J. (2024, January). Reclamation and reformatting of waste carbon fibers: A paradigm shift towards sustainable waste management. *Resources, Conservation and Recycling, 203*, 107465. https://doi.org/10.1016/j.resconrec.2024.107465

Heidarinejad, Z., Dehghani, M. H., Heidari, M., Javedan, G., Ali, I., & Sillanpää, M. (2020). Methods for preparation and activation of activated carbon: A review. *Environmental Chemistry Letters, 18*(2), 393–415. https://doi.org/10.1007/s10311-019-00955-0

Ho, T. T. T., Nadeem, A., & Choe, K. (2024). A review of upscaling hydrothermal carbonization. *Energies, 17*(8). https://doi.org/10.3390/en17081918

Hu, Z., Ma, Q., & Xu, T. (2019). High-temperature stability of carbon fiber reinforced polymer-derived SiAlOC composites under different environment. *Ceramics International, 45*(1), 1434–1438. https://doi.org/10.1016/j.ceramint.2018.09.218

Hu, J., Zhao, L., Luo, J., Gong, H., & Zhu, N. (2022). A sustainable reuse strategy of converting waste activated sludge into biochar for contaminants removal from water: modifications, applications and perspectives. *Journal of Hazardous Materials*, 438, 129437.

Hughes, K. J., Iyer, K. A., Bird, R. E., Ivanov, J., Banerjee, S., Georges, G., & Zhou, Q. A. (2024). Review of carbon nanotube research and development: Materials and emerging applications. *ACS Applied Nano Materials, 7*(16), 18695–18713. https://doi.org/10.1021/acsanm.4c02721

Ischia, G., Berge, N. D., Bae, S., Marzban, N., Román, S., Farru, G., Wilk, M., Kulli, B., & Fiori, L. (2024). Advances in research and technology of hydrothermal carbonization: Achievements and future directions. *Agronomy, 14*(5). https://doi.org/10.3390/agronomy14050955

Islam, M. A., Akber, M. A., Limon, S. H., Akbor, M. A., & Islam, M. A. (2019). Characterization of solid biofuel produced from banana stalk via hydrothermal carbonization. *Biomass Conversion and Biorefinery, 9*(4), 651–658. https://doi.org/10.1007/s13399-019-00405-5

Jaria, G., Lourenço, M. A. O., Silva, C. P., Ferreira, P., Otero, M., Calisto, V., & Esteves, V. I. (2020). Effect of the surface functionalization of a waste-derived activated carbon on pharmaceuticals' adsorption from water. *Journal of Molecular Liquids, 299*, 112098. https://doi.org/10.1016/j.molliq.2019.112098

Ji, M., Wang, X., Usman, M., Liu, F., Dan, Y., Zhou, L., Campanaro, S., Luo, G., & Sang, W. (2022, October 2021). Effects of different feedstocks-based biochar on soil remediation: A review. *Environmental Pollution, 294*, 118655. https://doi.org/10.1016/j.envpol.2021.118655

Jose, J. K., Mishra, B., Kootery, K. P., Cherian, C. T., Tripathi, B. P., Sarojini, S., & Balachandran, M. (2023, August). Fabrication of silver nanoparticle decorated graphene oxide membranes for water purification, antifouling and antibacterial applications. *Materials Science and Engineering: B, 297*, 116789. https://doi.org/10.1016/j.mseb.2023.116789

Karanjikar, S. R., Sena, A. S., Manekar, P., Mudagi, S., & Juneja, A. S. (2021, December 2022). Utilization of graphene and its derivatives for air & water filtration: A review. *Materials Today: Proceedings, 50*, 2007–2017. https://doi.org/10.1016/j.matpr.2021.09.346

Karthik, V., Selvakumar, P., Senthil Kumar, P., Vo, D. V. N., Gokulakrishnan, M., Keerthana, P., Tamil Elakkiya, V., & Rajeswari, R. (2021). Graphene-based materials for environmental applications: A review. *Environmental Chemistry Letters, 19*(5), 3631–3644. https://doi.org/10.1007/s10311-021-01262-3

Keeling, J. (2017, September). Graphite: Properties, uses and South Australian resources exploration background and properties. *28 MESA Journal, 84*, 2017–2020.

Keppetipola, N. M., Dissanayake, M., Dissanayake, P., Karunarathne, B., Dourges, M. A., Talaga, D., Servant, L., Olivier, C., Toupance, T., Uchida, S., Tennakone, K., Kumara, G. R. A., & Cojocaru, L. (2021). Graphite-type activated carbon from coconut shell: A natural source for eco-friendly non-volatile storage devices. *RSC Advances, 11*(5), 2854–2865. https://doi.org/10.1039/d0ra09182k

Kigozi, M., Koech, R. K., Kingsley, O., Ojeaga, I., Tebandeke, E., Kasozi, G. N., & Onwualu, A. P. (2020, June). Synthesis and characterization of graphene oxide from locally mined graphite flakes and its supercapacitor applications. *Results in Materials, 7*, 100113. https://doi.org/10.1016/j.rinma.2020.100113

Kim, M., Goerzen, D., Jena, P. V., Zeng, E., Pasquali, M., Meidl, R. A., & Heller, D. A. (2024). Human and environmental safety of carbon nanotubes across their life cycle. *Nature Reviews Materials, 9*(1), 63–81. https://doi.org/10.1038/s41578-023-00611-8

Kuila, T., Bose, S., Mishra, A. K., Khanra, P., Kim, N. H., & Lee, J. H. (2012). Chemical functionalization of graphene and its applications. *Progress in Materials Science, 57*(7), 1061–1105. https://doi.org/10.1016/j.pmatsci.2012.03.002

Kumar, A., & Bhattacharya, T. (2021). Biochar: A sustainable solution. *Environment, Development and Sustainability, 23*(5), 6642–6680. https://doi.org/10.1007/s10668-020-00970-0

Kumari, P., Tripathi, K. M., Jangir, L. K., Gupta, R., & Awasthi, K. (2021). Recent advances in application of the graphene-based membrane for water purification. *Materials Today Chemistry, 22*, 100597. https://doi.org/10.1016/j.mtchem.2021.100597

Lachos-Perez, D., César Torres-Mayanga, P., Abaide, E. R., Zabot, G. L., & De Castilhos, F. (2022, September 2021). Hydrothermal carbonization and liquefaction: Differences, progress, challenges, and opportunities. *Bioresource Technology, 343*, 126084. https://doi.org/10.1016/j.biortech.2021.126084

Larasati, A., Fowler, G. D., & Graham, N. J. D. (2021). Insights into chemical regeneration of activated carbon for water treatment. *Journal of Environmental Chemical Engineering, 9*(4), 105555. https://doi.org/10.1016/j.jece.2021.105555

Leng, L., Xiong, Q., Yang, L., Li, H., Zhou, Y., Zhang, W., Jiang, S., Li, H., & Huang, H. (2021). An overview on engineering the surface area and porosity of biochar. *Science of the Total Environment, 763*, 144204. https://doi.org/10.1016/j.scitotenv.2020.144204

Li, X., Zhang, L., Yang, Z., Wang, P., Yan, Y., & Ran, J. (2020, July 2019). Adsorption materials for volatile organic compounds (VOCs) and the key factors for VOCs adsorption process: A review. *Separation and Purification Technology, 235*, 116213. https://doi.org/10.1016/j.seppur.2019.116213

Liang, W., Wang, G., Peng, C., Tan, J., Wan, J., Sun, P., Li, Q., Ji, X., Zhang, Q., Wu, Y., & Zhang, W. (2022, September 2021). Recent advances of carbon-based nano zero valent iron for heavy metals remediation in soil and water: A critical review. *Journal of Hazardous Materials, 426*, 127993. https://doi.org/10.1016/j.jhazmat.2021.127993

Liu, J., Li, R., & Yang, B. (2020). Carbon dots: A new type of carbon-based nanomaterial with wide applications. *ACS Central Science, 6*(12), 2179–2195. https://doi.org/10.1021/acscentsci.0c01306

Liu, G., Xiong, Z., Yang, L., Shi, H., Fang, D., Wang, M., Shao, P., & Luo, X. (2021). Electrochemical approach toward reduced graphene oxide-based electrodes for environmental applications: A review. *Science of the Total Environment, 778*, 146301. https://doi.org/10.1016/j.scitotenv.2021.146301

Liu, C., Li, Q., Kang, W., Lei, W., Wang, X., Lu, C., & Naebe, M. (2022a). Structural design and mechanism analysis of hierarchical porous carbon fibers for advanced energy and environmental applications. *Journal of Materials Chemistry A, 10*(1), 10–49. https://doi.org/10.1039/d1ta08646d

Liu, Z., Ling, Q., Cai, Y., Xu, L., Su, J., Yu, K., Wu, X., Xu, J., Hu, B., & Wang, X. (2022b). Synthesis of carbon-based nanomaterials and their application in pollution management. *Nanoscale Advances, 4*(5), 1246–1262. https://doi.org/10.1039/d1na00843a

Liu, B., Zhang, L., Ning, K., & Yang, W. (2023). Biochar with nanoparticle incorporation and pore engineering enables enhanced heavy metals removal. *Journal of Environmental Chemical Engineering, 11*(5), 111056. https://doi.org/10.1016/j.jece.2023.111056

Lucian, M., & Fiori, L. (2017). Hydrothermal carbonization of waste biomass: Process design, modeling, energy efficiency and cost analysis. *Energies, 10*(2). https://doi.org/10.3390/en10020211

Ma, X., & Yuan, B. (2009). Fabrication of carbon nanoflowers by plasma-enhanced chemical vapor deposition. *Applied Surface Science, 255*(18), 7846–7850. https://doi.org/10.1016/j.apsusc.2009.03.061

Ma, Q., Yu, Y., Sindoro, M., Fane, A. G., Wang, R., & Zhang, H. (2017). Carbon-based functional materials derived from waste for water remediation and energy storage. *Advanced Materials, 29*(13). https://doi.org/10.1002/adma.201605361

Maduraiveeran, G., & Jin, W. (2021, June). Carbon nanomaterials: Synthesis, properties and applications in electrochemical sensors and energy conversion systems. *Materials Science and Engineering: B, 272*, 115341. https://doi.org/10.1016/j.mseb.2021.115341

Majumder, S., Sharma, P., Singh, S. P., Nadda, A. K., Sahoo, P. K., Xia, C., Sharma, S., Ganguly, R., Lam, S. S., & Kim, K. H. (2023). Engineered biochar for the effective sorption and remediation of emerging pollutants in the environment. *Journal of Environmental Chemical Engineering, 11*(2), 109590. https://doi.org/10.1016/j.jece.2023.109590

Mamatha, M. (2023). A review on activated carbon: Preparation, characterization and applications. *International Journal for Research in Applied Science and Engineering Technology, 11*(8), 372–378. https://doi.org/10.22214/ijraset.2023.55201

Mao, H., Xue, M., & Han, G. (2021). Synthesis, properties and applications of fluorescent carbon dots. *Gongneng Cailiao/Journal of Functional Materials, 52*(1), 01053–01063. https://doi.org/10.3969/j.issn.1001-9731.2021.01.009

Martincic, M., Sandoval, S., Oró-Solé, J., & Tobías-Rossell, G. (2024). *Thermal stability and purity of graphene and carbon nanotubes: Key parameters for their thermogravimetric analysis (TGA)*. www.preprints.org

Mashkoor, F., Nasar, A., & Inamuddin. (2020). Carbon nanotube-based adsorbents for the removal of dyes from waters: A review. *Environmental Chemistry Letters, 18*(3), 605–629. https://doi.org/10.1007/s10311-020-00970-6

Mauter, M. S., & Elimelech, M. (2008). Environmental applications of carbon-based nanomaterials: Critical review. *American Chemical Society, 42*(16), 5843–5859.

Mergbi, M., Greta, M., Dominic, G., Ehiaghe, A., Su, P., Ikram, B. M., Nabgan, W., Bedia, J., Ben, H., Sandra, A., Francisco, C., & Djellabi, R. (2023). Valorization of lignocellulosic biomass into sustainable materials for adsorption and photocatalytic applications in water and air remediation. *Environmental Science and Pollution Research, 30*(30), 74544–74574. https://doi.org/10.1007/s11356-023-27484-2

Mong, G. R., Chong, C. T., Chong, W. W. F., Ng, J. H., Ong, H. C., Ashokkumar, V., Tran, M. V., Karmakar, S., Goh, B. H. H., & Mohd Yasin, M. F. (2022, May). Progress and challenges in sustainable pyrolysis technology: Reactors, feedstocks and products. *Fuel, 324*, 124777. https://doi.org/10.1016/dj.fuel.2022.124777

Murugan, P., Nagarajan, R. D., Shetty, B. H., Govindasamy, M., & Sundramoorthy, A. K. (2021). Recent trends in the applications of thermally expanded graphite for energy storage and sensors—A review. *Nanoscale Advances, 3*(22), 6294–6309. https://doi.org/10.1039/d1na00109d

Nguyen, T. B., Sherpa, K., Bui, X. T., Nguyen, V. T., Vo, T. D. H., Ho, H. T. T., Chen, C. W., & Dong, C. D. (2023, September). Biochar for soil remediation: A comprehensive review of current research on pollutant removal. *Environmental Pollution, 337*, 122571. https://doi.org/10.1016/j.envpol.2023.122571

Ning, S. K., Hung, M. C., Chang, Y. H., Wan, H. P., Lee, H. T., & Shih, R. F. (2013). Benefit assessment of cost, energy, and environment for biomass pyrolysis oil. *Journal of Cleaner Production, 59*, 141–149. https://doi.org/10.1016/j.jclepro.2013.06.042

Nistratov, A. V., Klimenko, N. N., Pustynnikov, I. V., & Vu, L. K. (2022). Thermal regeneration and reuse of carbon and glass fibers from waste composites. *Emerging Science Journal, 6*(5), 967–984. https://doi.org/10.28991/ESJ-2022-06-05-04

Ogunsola, S. S., Oladipo, M. E., Oladoye, P. O., & Kadhom, M. (2024, January). Carbon nanotubes for sustainable environmental remediation: A critical and comprehensive review. *Nano-Structures and Nano-Objects, 37*, 101099. https://doi.org/10.1016/j.nanoso.2024.101099

Osman, A. I., Hefny, M., Abdel Maksoud, M. I. A., Elgarahy, A. M., & Rooney, D. W. (2021). Recent advances in carbon capture storage and utilisation technologies: a review. Environmental Chemistry Letters, 19(2), 797–849.

Ozyurt, D., Al Kobaisi, M., Hocking, R. K., & Fox, B. (2023, June). Properties, synthesis, and applications of carbon dots: A review. *Carbon Trends, 12*, 100276. https://doi.org/10.1016/j.cartre.2023.100276

Panchamoorthy, K., Dai, G., Vo, V. N., & Gnana, D. (2021). Environmental applications of carbon-based materials: A review. *Environmental Chemistry Letters, 19*(1), 557–582. https://doi.org/10.1007/s10311-020-01084-9

Petrović, J., Ercegović, M., Simić, M., Koprivica, M., Dimitrijević, J., Jovanović, A., & Janković Pantić, J. (2024). Hydrothermal carbonization of waste biomass: A review of hydrochar preparation and environmental application. *Processes, 12*(1). https://doi.org/10.3390/pr12010207

Qin, C., Wang, H., Yuan, X., Xiong, T., Zhang, J., & Zhang, J. (2020, September 2019). Understanding structure-performance correlation of biochar materials in environmental remediation and electrochemical devices. *Chemical Engineering Journal, 382*, 122977. https://doi.org/10.1016/j.cej.2019.122977

Qin, F., Zhang, C., Zeng, G., Huang, D., Tan, X., & Duan, A. (2022, September 2021). Lignocellulosic biomass carbonization for biochar production and characterization of biochar reactivity. *Renewable and Sustainable Energy Reviews, 157*, 112056. https://doi.org/10.1016/j.rser.2021.112056

Qiu, B., Shao, Q., Shi, J., Yang, C., & Chu, H. (2022, July). Application of biochar for the adsorption of organic pollutants from wastewater: Modification strategies, mechanisms and challenges. *Separation and Purification Technology, 300*, 121925. https://doi.org/10.1016/j.seppur.2022.121925

Rathinavel, S., Priyadharshini, K., & Panda, D. (2021, December 2020). A review on carbon nanotube: An overview of synthesis, properties, functionalization, characterization, and the application. *Materials Science and Engineering: B, 268*, 115095. https://doi.org/10.1016/j.mseb.2021.115095

Ravi, S., & Vadukumpully, S. (2016). Sustainable carbon nanomaterials: Recent advances and its applications in energy and environmental remediation. *Journal of Environmental Chemical Engineering, 4*(1), 835–856. https://doi.org/10.1016/j.jece.2015.11.026

Román, S., Libra, J., Berge, N., Sabio, E., Ro, K., Li, L., Ledesma, B., Alvarez, A., & Bae, S. (2018). Hydrothermal carbonization: Modeling, final properties design and applications: A review. *Energies, 11*(1), 1–28. https://doi.org/10.3390/en11010216

Sabitov, A., Atamanov, M., Doszhanov, O., Saurykova, K., Tazhu, K., Kerimkulova, A., Orazbayev, A., & Doszhanov, Y. (2024). Surface characteristics of activated carbon sorbents obtained from biomass for cleaning oil-contaminated soils. *Molecules, 29*(16). https://doi.org/10.3390/molecules29163786

Sádovská, G., Honcová, P., Morávková, J., Jirka, I., Vorokhta, M., Pilař, R., Rathouský, J., Kaucký, D., Mikysková, E., & Sazama, P. (2023, February). The thermal stability of carbon materials in the air: Quantitative structural investigation of thermal stability of carbon materials in air. *Carbon, 206*, 211–225. https://doi.org/10.1016/j.carbon.2023.02.042

Serafin, J., & Dziejarski, B. (2024). Activated carbons—Preparation, characterization and their application in CO_2 capture: A review. In *Environmental science and pollution research* (Vol. 31, Issue 28). Springer. https://doi.org/10.1007/s11356-023-28023-9

Sharma, A., Kumar, N., Sillanpää, M., Makgwane, P. R., Kumar, S., & Kumari, K. (2022, May). Carbon nano-structures and functionalized associates: Adsorptive detoxification of organic and inorganic water pollutants. *Inorganic Chemistry Communications, 141*, 109579. https://doi.org/10.1016/j.inoche.2022.109579

Shi, Z., Yang, W., Gu, Y., Liao, T., & Sun, Z. (2020). Metal-nitrogen-doped carbon materials as highly efficient catalysts: Progress and rational design. *Advanced Science, 7*(15). https://doi.org/10.1002/advs.202001069

Singh, E., Mishra, R., Kumar, A., Shukla, S. K., Lo, S. L., & Kumar, S. (2022, March). Circular economy-based environmental management using biochar: Driving towards sustainability. *Process Safety and Environmental Protection, 163*, 585–600. https://doi.org/10.1016/j.psep.2022.05.056

Singh, G., Maria Ruban, A., Geng, X., & Vinu, A. (2023). Recognizing the potential of K-salts, apart from KOH, for generating porous carbons using chemical activation. *Chemical Engineering Journal, 451*(P4), 139045. https://doi.org/10.1016/j.cej.2022.139045

Siti Norhafizan, H., Nurul Fadly, H., Nursyazwani, M. F., Auni, F., Nadia, C. D., & Farah Izzaida, M. Z. (2013). *Journal of Environmental Science, Computer Science and Engineering & Technology, 2*(1), 129–138. www.jecet.org

Soffian, M. S., Abdul Halim, F. Z., Aziz, F., Rahman, M. A., Mohamed Amin, M. A., & Awang Chee, D. N. (2022, June). Carbon-based material derived from biomass waste for wastewater treatment. *Environmental Advances, 9*, 100259. https://doi.org/10.1016/j.envadv.2022.100259

Sosa, J. A., Laines, J. R., García, D. S., Hernández, R., Zappi, M., & Espinosa de los Monteros, A. E. (2023). Activated carbon: A review of residual precursors, synthesis processes, characterization techniques, and applications in the improvement of biogas. *Environmental Engineering Research, 28*(3), 220100. https://doi.org/10.4491/eer.2022.100

Srivastava, M., Srivastava, N., & Singh, R. (2021). Bioenergy research: Integrative solution for existing roadblock. In *Pretreatment of biomass for efficient pyrolysis*.

Sumaraj, & Padhye, L. P. (2017). Influence of surface chemistry of carbon materials on their interactions with inorganic nitrogen contaminants in soil and water. *Chemosphere, 184*, 532–547. https://doi.org/10.1016/j.chemosphere.2017.06.021

Supee, A. H., & Zaini, M. A. A. (2024). Hydrothermal carbonization of biomass: A commentary. *Fullerenes, Nanotubes, and Carbon Nanostructures, 32*(2), 119–127. https://doi.org/10.1080/1536383X.2023.2270089

Teimouri, Z., Nanda, S., Abatzoglou, N., & Dalai, A. K. (2024). Application of activated carbon in renewable energy conversion and storage systems: A review. *Environmental Chemistry Letters, 22*(3), 1073–1092. https://doi.org/10.1007/s10311-023-01690-3

Titirici, M. M., White, R. J., Brun, N., Budarin, V. L., Su, D. S., Del Monte, F., Clark, J. H., & MacLachlan, M. J. (2015). Sustainable carbon materials. *Chemical Society Reviews, 44*(1), 250–290. https://doi.org/10.1039/c4cs00232f

Togibasa, O., Ansanay, Y. O., Dahlan, K., & Erari, M. (2021). Identification of surface functional group on activated carbon from waste sago. *Journal of Physics: Theories and Applications, 5*(1), 1. https://doi.org/10.20961/jphystheor-appl.v5i1.49885

Tomczyk, A., Kondracki, B., & Szewczuk-Karpisz, K. (2023). Chemical modification of biochars as a method to improve its surface properties and efficiency in removing xenobiotics from aqueous media. *Chemosphere, 312*(P1), 137238. https://doi.org/10.1016/j.chemosphere.2022.137238

Toński, M., Paszkiewicz, M., Dołżonek, J., Flejszar, M., Bielicka-Giełdoń, A., Stepnowski, P., & Białk-Bielińska, A. (2021, March). Regeneration and reuse of the carbon nanotubes for the adsorption of selected anticancer drugs from water matrices. *Colloids and Surfaces A: Physicochemical and Engineering Aspects, 618*, 126355. https://doi.org/10.1016/j.colsurfa.2021.126355

Viswanathan, B., Neel, P., & Varadarajan, T. (2009). *Methods of activation and specific applications of carbon materials. ... for Catalysis Research, Indian Institute of ..., February 2009*. http://scholar.google.com/scholar?hl=en&btnG=Search&q=intitle:Methods+of+Activation+and+Specific+Applications+of+Carbon+Materials#0

Wang, L., Wang, Y., Ma, F., Tankpa, V., Bai, S., Guo, X., & Wang, X. (2019). Mechanisms and reutilization of modified biochar used for removal of heavy metals from wastewater: A review. *Science of the Total Environment, 668*, 1298–1309. https://doi.org/10.1016/j.scitotenv.2019.03.011

Wang, T., Hu, S., Wu, D., Zhao, W., Yu, W., Wang, M., Xu, J., & Zhang, J. (2021a). Boosting the capacity of biomass-based supercapacitors using carbon materials of wood derivatives and redox molecules from plants. *Journal of Materials Chemistry A, 9*(19), 11839–11852. https://doi.org/10.1039/d1ta01542g

Wang, Z., Ma, C., Xu, C., Sinquefield, S. A., Shofner, M. L., & Nair, S. (2021b). Graphene oxide nanofiltration membranes for desalination under realistic conditions. *Nature Sustainability, 4*(5), 402–408. https://doi.org/10.1038/s41893-020-00674-3

Wang, B., Yuan, F., Yu, Q., Li, W., Sun, H., Zhang, L., Zhang, D., Wang, Q., Lai, F., & Wang, W. (2021c, December 2020). Amorphous carbon/graphite coupled polyhedral microframe with fast electronic channel and enhanced ion storage for potassium ion batteries. *Energy Storage Materials, 38*, 329–337. https://doi.org/10.1016/j.ensm.2021.03.021

Wang, C., Lin, X., Zhang, X., & Show, P. L. (2024a, November 2023). Research advances on production and application of algal biochar in environmental remediation. *Environmental Pollution, 348*, 123860. https://doi.org/10.1016/j.envpol.2024.123860

Wang, K., Shan, T., Li, B., Zheng, Y., Xu, H., Wang, C., & Tian, X. (2024b, August 2023). Study on pyrolysis characteristics, kinetics and thermodynamics of waste tires catalytic pyrolysis with low-cost catalysts. *Fuel, 356*, 129644. https://doi.org/10.1016/j.fuel.2023.129644

Wang, Y., Wu, G., Zhang, Y., Su, Y., & Zhang, H. (2024c, September). The deactivation mechanisms, regeneration methods and devices of activated carbon in applications. *Journal of Cleaner Production, 476*, 143751. https://doi.org/10.1016/j.jclepro.2024.143751

Xia, C., & Shi, S. Q. (2016). Self-activation for activated carbon from biomass: Theory and parameters. *Green Chemistry, 18*(7), 2063–2071. https://doi.org/10.1039/c5gc02152a

Xu, Q., Si, Y., He, R., Qi, X., Su, X., & Fu, Y. (2022). Silk-waste-derived porous carbon for fast electric heating under safe voltage. *ACS Applied Materials and Interfaces, 14*(4), 6005–6015. https://doi.org/10.1021/acsami.1c21874

Painuli, R., Yadav, P. K., Raghav, S., & Kumar, D. (2020). Synthesis of carbon nanofibers and its application in environmental remediation. *In Environmental Remediation Through Carbon Based Nano Composites* (pp. 325–342). Singapore: Springer Nature Singapore.

Yahyazadeh, A., Nanda, S., & Dalai, A. K. (2024). Carbon nanotubes: A review of synthesis methods and applications. *Reactions, 5*(3), 429–451. https://doi.org/10.3390/reactions5030022

Yang, X., Wan, Y., Zheng, Y., He, F., Yu, Z., Huang, J., Wang, H., Ok, Y. S., Jiang, Y., & Gao, B. (2019, January). Surface functional groups of carbon-based adsorbents and their roles in the removal of heavy metals from aqueous solutions: A critical review. *Chemical Engineering Journal, 366*, 608–621. https://doi.org/10.1016/j.cej.2019.02.119

Yang, I., Jung, M., Kim, M. S., Choi, D., & Jung, J. C. (2021). Physical and chemical activation mechanisms of carbon materials based on the microdomain model. *Journal of Materials Chemistry A, 9*(15), 9815–9825. https://doi.org/10.1039/d1ta00765c

Zhang, Y. (2024). *Study of Graphene Oxide/Polymer Composite Membranes in Air, 6*(9), 49–53.

Zhang, Y., Yang, J., Zhong, L., & Liu, L. (2018). Effect of multi-wall carbon nanotubes on Cr(VI) reduction by citric acid: Implications for their use in soil remediation. *Environmental Science and Pollution Research, 25*(24), 23791–23798. https://doi.org/10.1007/s11356-018-2438-8

Zhang, G. H., Zhu, Q. H., Zhang, L., Yong, F., Zhang, Z., Wang, S. L., Wang, Y., He, L., & Tao, G. H. (2020). High-performance particulate matter including nanoscale particle removal by a self-powered air filter. *Nature Communications, 11*(1). https://doi.org/10.1038/s41467-020-15502-7

Zhang, D., Zhang, W., Zhang, S., Ji, X., & Li, L. (2023, January). Synthesis of expanded graphite-based materials for application in lithium-based batteries. *Journal of Energy Storage, 60*, 106678. https://doi.org/10.1016/j.est.2023.106678

Zuo, Y., Feng, J., Soyol-Erdene, T. O., Wei, Z., Hu, T., Zhang, Y., & Tang, W. (2023). Recent advances in wood-derived monolithic carbon materials: Synthesis approaches, modification methods and environmental applications. *Chemical Engineering Journal, 463*, 142332. https://doi.org/10.1016/j.cej.2023.142332

Chapter 4
Colorimetric Detection Techniques for Environmental Contaminants Using Carbon Nanomaterials

Sanjay Yadav, Nishu Choudhary, and Alok Ranjan Paital

Abstract Carbon nanomaterials such as fullerenes, carbon nanotubes (CNTs), carbon dots, graphene, and reduced graphene oxide have emerged as up-and-coming tools in colorimetric detection systems of environmental contaminants. These nanomaterials possess unique optical and electronic properties for the detectability of heavy metals, pesticides, and pathogen pollutants in water, air, and soil. Carbon nanomaterial-mediated colorimetric detection mechanisms and sensitivity, selectivity, and real-time monitoring possibilities are discussed in this chapter. Aggregation, enzymatic biomimetic, and surface functionalization of carbon nanomaterials with particular ligands or nanoparticles may lead to color changes in the absence and presence of target contaminants, thus paving an easy, fast, and inexpensive route toward environmental diagnostics. Besides, this chapter presents the engineering of multifunctional colorimetric systems, portable sensing devices, and their applications in environmental monitoring. Such nanomaterials are considered promising as soon as they are incorporated into smart and scalable technologies aimed at pollution management on a large scale and the protection of public health, as well as their role in next-generation environmental monitoring tools.

Keywords Carbon nanomaterials · Quantum dots · Graphene oxide · Carbon nanotubes · Environmental pollutants · Sensor mechanisms

---

S. Yadav (✉) · N. Choudhary · A. R. Paital (✉)
Salt and Marine Chemicals Division, CSIR—Central Salt and Marine Chemicals Research Institute, Bhavnagar, Gujarat, India

Academy of Scientific and Innovative Research, Ghaziabad, India
e-mail: arpaital@csmcri.res.in

A. M. Parambil et al. (eds.), *Carbon*,
https://doi.org/10.1007/978-3-031-90613-8_4

Abbreviations

AuNPs	Gold nanoparticles
C_{60}	Buckminsterfullerene
CDs	Carbon dots
CNMs	Carbon nanomaterials
CNT	Carbon nanotubes
DLS	Dynamic light scattering
GO	Graphene dots
MWCNT	Multiwalled carbon nanotubes
N-CDs	Nitrogen-doped carbon dots
NIR	Near-infrared
NPs	Nanoparticles
PEI	Polyethylenimine
QDs	Quantum dots
rGO	Reduced graphene oxide
ROS	Reactive oxygen species
SPR	Surface plasmon resonance
SWCNT	Single-walled carbon nanotubes
TMB	Trimethylbenzene
UV-Vis	Ultraviolet-visible

4.1 Introduction

Colorimetric detection is a versatile and very sensitive technique utilized to determine the existence and concentration of various analytes, such as environmental contaminants, biological molecules, and metal ions. Fundamental principle toward colorimetric detection depends on the reaction toward a target substance, or analyte, and a reagent or a material that causes some kind of color change observable to the naked eye or can even be measured quantitatively with simple optical devices, such as spectrophotometers. The most important advantage of colorimetric detection is its inherent simplicity. The detection methods are as simple as adding a reagent to a sample and observing a color change in most colorimetric assays. Colorimetric assays require low technical expertise and very little instrumentation compared with more complex techniques, such as chromatography, mass spectrometry, or fluorescence-based detection. This makes it accessible for field-based applications and low-resource environments where such sophisticated equipment is not available (Sukma et al., 2024; Liu et al., 2020; Yang & Ge, 2023; Chen et al., 2023; Quang & Kim, 2010). In environmental monitoring, colorimetric sensors detect different kinds of organic and inorganic contaminants present in environment (Ajith et al., 2022; Priyadarshini et al., 2023). Such a system is valuable for site testing because the results can be obtained quickly without sending the samples to the laboratory.

The third merit is cost-effectiveness. Many of the analytical techniques used are expensive, and equipment and reagents may be too costly to possess or utilize on a large or continuous scale for monitoring purposes (Kaur et al., 2018; Aragay et al., 2011; Priyadarshini & Pradhan, 2017; Oliveira et al., 2015; Annadhasan et al., 2014). On the contrary, materials to be used in colorimetric detection include reagents, simple substrates like paper, and basic equipment, which can be bought inexpensively in large quantities. For example, most colorimetric tests can be envisioned to function on strips of paper, which are inexpensive, disposable, and can be mass-produced as needed, which can be applied to environmental monitoring (Oliveira et al., 2015; Annadhasan et al., 2014; Soda & Bakker, 2019; Sriram et al., 2017).

In this regard, such tests can affordably detect nitrates, phosphates, and heavy metals at a minute concentration in water. For example, colorimetric sensors can be part of low-cost, laboratory-based equipment to form test kits that can be utilized frequently in remote or underserved areas. This, particularly in the developing world where the availability of money and access to laboratories are a hindrance, can benefit everyone. The simplicity of the method is going to reduce the demand for expensive equipment for analytical purposes and skilled labor; as such, it is going to promote great use in environmental and clinical diagnostics. This universality of chemistry allows for rapid design and deployment of new colorimetric sensors tailored to specific analytes, such as ions, biomolecules, pollutants, and toxins. Colorimetric detection also boasts very fast response times, with results typically returned in minutes (Geng et al., 2023; Carneiro et al., 2023). This is particularly important when the information must be timely, such as identifying the presence of pollutants in water supplies or determining heavy metal contamination within soils. The visual basis of color change provides the opportunity for real-time monitoring without the often-encountered delays arising from sample processing and data analysis typical of laboratory-based techniques. The most commonly used principles of colorimetric detection include redox reaction, complexation, and aggregation of nanoparticles, which results in changes in the optical properties. It is the reason it is particularly helpful for field-based applications and point-of-care decision-making because it can be quickly and visually determined to have been contaminated (Kaur et al., 2018; Aragay et al., 2011; Priyadarshini & Pradhan, 2017). The adaptability of colorimetric detection has made it widespread in several sectors. Colorimetric tests are highly applied in medical diagnostics to detect biologically important carbohydrates, such as glucose, cholesterol, and proteins, enabling rapid diagnosis and monitoring of diseases. The most common application of colorimetric assays is in environmental monitoring, where it is mainly used to determine the existence of environmental pollutants that ensure the safety of natural resources. The application of colorimetric sensors ensures the quality control and process monitoring of industrial applications, thereby ensuring safety and efficiency in manufacturing processes. All things being equal, colorimetric detection systems have the merits of simplicity, cost-effectiveness, and speed, making them enormously useful for environmental monitoring, point-of-care diagnostics, and industrial applications. The direct visual output and the flexibility introduced by functionalized nanomaterials ensure that colorimetric detection will be an indispensable tool in many areas of analytical science. For colorimetric detection, the conventional materials range from organic dyes to metal complexes, nanoparticles, and polymers (Carneiro et al., 2023; Kangas et al., 2017; Yang & Ge, 2023).

Of the numerous organic dyes, the most available are versatile yet inexpensive compounds, such as phenolphthalein or bromothymol blue, with corresponding disadvantages like poor sensitivity and photobleaching. Metal complexes, particularly those involving a transition metal such as iron or copper, are extremely useful for sensing ions and small molecules through changes in color arising from electronic structure modifications within the metal. Many such systems have been identified as having poor response times and reduced stability. Nanoparticles, particularly gold and silver, are also popular because these systems can show color changes strictly tied to their aggregation states or size. These systems are sensitive but sometimes require very precise synthesis and control of stability. Polymeric materials, like polydiacetylene, can be applied in colorimetric sensors as they can easily change color in response to stimuli of pH or temperature but are generally slower in response (Liu et al., 2020; Kim et al., 2020; Zeebaree et al., 2023; Li et al., 2019a, 2019b; Chatterjee et al., 2022; Ali et al., 2023; Yapor et al., 2017; Dey et al., 2019).

In recent times, carbon dots (CDs), graphene oxide (GO), carbon nanotubes (CNTs), and fullerene carbon nanomaterials have appeared, which are superior alternatives to such conventional colorimetric sensing materials. The carbon dots are sensitive to environmental changes due to the effects of quantum confinement and surface chemistry tunability. The functionalization of carbon dots with specific molecules provides better low concentrations of analyte detection ability than that of many conventional dyes or metal complexes (Yadav et al., 2023a, 2023b, 2024a). Another benefit of carbon-based materials is their eco-friendliness and biocompatibility. Although some metal-based systems may introduce toxicity, carbon dots are commonly derived from benign sources with low cytotoxicity; thus, they can be applied for medical purposes like biosensing and bio-imaging. Another strength of carbon-based materials is that their optical property can be tuned. Carbon dots, for example, have easily controllable fluorescence properties by adjusting their size, shape, or surface chemistry, thus allowing the development of tailored sensors for specific analytes. Carbon materials are also extremely stable. These can tolerate extreme environmental conditions, such as extreme pH, temperature, or humidity, much better than organic dyes or metal-based systems, which often degrade or lose activity. Moreover, carbon materials such as graphene oxide and carbon dots can be easily functionalized, making them very versatile for detecting targets toward heavy metals and biological molecules. Carbon dots can be synthesized to respond to multiple analytes with different color changes, a relatively less common phenomenon in traditional colorimetric systems. The distinguishing advantages related to enhanced sensitivity, tunability, robustness, eco-friendliness, and multiplexing capabilities make carbon-based materials suitable for many practical applications (Pandey & Chusuei, 2021; Li et al., 2019b; He et al., 2022; Ghosal & Ghosh, 2019).

Modern unmet critical challenges involve using advanced functional materials to remediate toxic organic and inorganic analytes and gaseous forms (Choudhary et al., 2024a, 2024b; Tsiotsias et al., 2025; Yadav et al., 2021, 2022, 2024b, 2024c). Carbon-based colorimetric sensors, including graphene oxide-based sensing materials, have already been widely used to detect environmental pollutants such as heavy metals, pesticides, and organic compounds in water and air. Carbon-based colorimetric sensors detect biomarkers, such as glucose, cholesterol, and proteins, in medical diagnostics against diseases like cancer. Graphene oxide-based sensors also detect food contamination by detecting pesticides and pathogenic bacteria. Carbon-based sensors are also applied in industrial process monitoring, which yields real-time analysis and ensures quality control. In short, carbon-based colorimetric sensing materials have established outstanding benefits compared with traditional systems. Enhanced sensitivity, stability, tunability, and eco-friendly character make them exceptional alternatives for an entire scope of applications that cover the scope of environmental and industrial monitoring through to medical diagnostics. These are materials ready to significantly enhance real-time detection in many fields while ensuring greater safety (Khan et al., 2023; Shahdeo et al., 2020; Jin et al., 2024; Gosai et al., 2021). Carbon-based colorimetric sensing materials have displayed an impressive superiority over traditional systems in many aspects. Their increased sensitivity, stability, tunability, and eco-friendliness make them outstanding alternatives for various applications, from environmental and industrial monitoring to medical diagnosis. This material will be the future generation of colorimetric sensing technologies with possibilities that advance real-time detection through superior safety ramifications in many industries.

Therefore, this chapter emphasizes the role of carbon-based nanomaterials in developing advanced colorimetric sensors to detect various environmental pollutants.

4.2 Graphene and Graphene Oxide

Graphene, composed of a single atomic layer of carbon in a two-dimensional lattice, became an all-around material in a number of sensing applications due to its extraordinary physical and chemical properties (Zhu et al., 2010; Eigler & Hirsch, 2014). Colorimetric sensing is an important application of graphene since it also allows the possibility to detect environmental contaminants more rapidly, sensitively, and cheaply by using visible color changes. These tend to happen as a result of the interaction of graphene with target analytes, thus altering the optical properties of the material and, hence, making it easily detectable on-site. Graphene's exceptional nature makes it an extremely good candidate for colorimetric sensing applications. Its remarkably high surface area, reaching up to 2630 m^2/g, provides immense active sites for interaction with target contaminants, significantly increasing sensitivity since even trace contaminants could give significant optical changes in graphene-based systems. Other optical properties of graphene, including transparency and the ability to absorb light at a wide range of wavelengths, also make graphene well-suited to integration with chromogenic reagents. An interaction with chromogenic reagents often results in visible color changes in response to the presence of contaminants, thus facilitating relatively easy detection of the latter. Besides its optical properties, graphene possesses outstanding electrical conductivity, contributing to catalyst-specific colorimetric reactions. This can, for instance, be a catalyst or an electron mediator that modulates redox reactions responsible for causing color changes in the sensing media, thereby increasing the detection abilities. Functionalized graphene improves the selectivity of colorimetric sensors to detect specific contaminants, heavy metals, pesticides, or toxins, even when highly diluted in environmental samples (Yadav et al., 2023b). These combined properties make graphene an excellent material with massive potential in pollutant detection through colorimetric sensing. Graphene oxide (GO) is an oxidized form of graphene, which is made up of a single layer of sp^2-hybridized carbon atoms arranged in a hexagonal lattice decorated with oxygen functional groups. The presence of these groups disrupts the π-conjugation in graphene, and GO is endowed with hydrophilic properties, which makes it dispersible in water solutions and highly reactive to contaminants (Kaur et al., 2017; Song et al., 2011; Li et al., 2023; Zhou et al., 2016; Liu et al., 2014).

Key functional properties include:

(a) High surface area: Providing extensive contact for contaminant binding.
(b) Oxygen-containing groups: Promoting hydrogen bonding, van der Waals forces, and electrostatic interactions.
(c) Electronic properties: Making GO a desirable candidate for sensing applications.

GO and its composites are used for the optical detection of heavy metal ions along with colorimetric detections. The interaction is often based on the agglomeration of GO upon interaction with the heavy metal ions or due to the catalytic activity expressed by GO and its composites that can be enhanced or inhibited concerning some heavy metal ions and can lead to colorimetric detection. Further, the surface functionalization of GO with hematoporphyrin or gold nanoparticles may increase the sensitivity and overall selectivity of these detection methods. Graphene oxide (GO) aggregation in the interaction with heavy metal ions is caused by the related alterations in its surface chemistry and the change in charge distribution. GO bears a negatively charged surface with rich oxygen-containing functionalities like carboxyl, hydroxyl, and epoxy groups. These groups maintain GO in stabilized colloidal dispersion under aqueous conditions using electrostatic repulsion prohibiting aggregation (Chen et al., 2012; Brisebois & Siaj, 2020; Dreyer et al., 2010). In the case of heavy metal ions, such as copper (Cu^{2+}), lead (Pb^{2+}), or mercury (Hg^{2+}), these interact with functional groups through electrostatic attraction, coordination bonding, or ion-exchange mechanisms. As the metal ions associate with the negatively charged sites on GO, they neutralize the surface charge, resulting in less electrostatic repulsion between GO sheets. Increased noncovalent interactions (van der Waals interactions) between the GO sheets enhance the aggregation. Moreover, the potential of metal ions to act as cross-linkers, in the form of coordinating with oxygen-containing groups from multiple sheets, contributes to the crosslinking effect, which effectively accelerates the aggregation process. The extent of aggregation is predicted by the type and concentration of analytes present, as well as the pH of the solution, which affects the charge density on GO surfaces. Visually, aggregation of GO in the presence of analytes (heavy analytes) could be monitored by detecting variations in the turbidity of the solution or even its color changes, depending on the specific sensing system applied. Quantification of the extent of aggregation by shifts in absorbance or distribution in particle size may be achieved with the use of instrumental techniques such as UV-Vis and dynamic light scattering (DLS).

One such study demonstrates that composite formation or amalgamation with GO induces several features for calorimetric detection. For instance, composite amalgamation-based optical and colorimetric sensing of mercury (II) ions with silver@graphene oxide (Ag@GO) composite material is a highly sensitive detection method, as depicted in Fig. 4.1a (Zangeneh Kamali et al., 2016). Graphene oxide is a stable dispersing platform for silver nanoparticles and enhances their interaction with mercury ions. When Hg^{2+} ions interact with the silver nanoparticles, their high affinity leads to amalgamation, forming an alloy at the nanoscale. This process alters the structure and size of the silver nanoparticles, changing their optical properties. Initially, the silver nanoparticles exhibit strong surface plasmon resonance (SPR), which gives the solution its characteristic color. Interaction with Hg^{2+} causes a shift in the SPR peak, and the color change is visible, usually from yellowish to colorless or another color, depending on the mercury concentration. As the amalgam forms, the silver nanoparticles aggregate, resulting in a loss of colloidal stability. This aggregation increases the intensity of the optical signal change, thus increasing the assay's sensitivity and enabling even trace amounts of Hg^{2+} to cause a visible color shift. Graphene oxide serves as a stable dispersion matrix for silver nanoparticles, increases their interaction with mercury ions, and helps preserve nanoparticle dispersion from premature aggregation until the presence of Hg^{2+} ions. Another similar work demonstrates a highly sensitive, label-free detection method

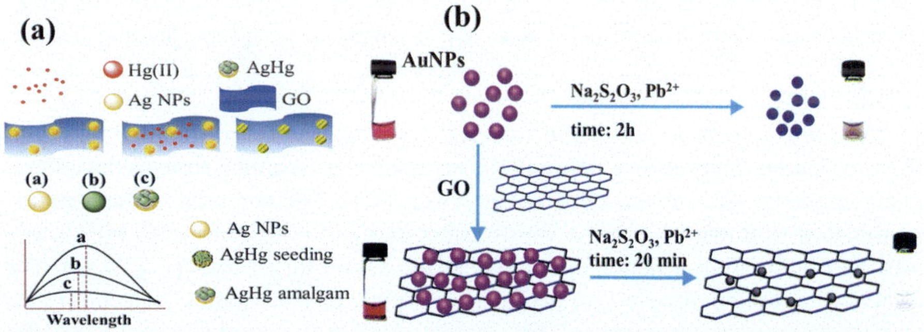

Fig. 4.1 (**a**) The representation of the optical activity of the material (Ag@GO) toward Hg^{2+} ions. (Reproduced with permission from Zangeneh Kamali et al. Copyright 2016, Elsevier). (**b**) The Au@GO composite material for the optical detection of Pb^{2+}. (Reproduced with permission from Xinhao Shi et al. Copyright 2015, Elsevier)

for lead ions (Pb^{2+}) by employing a GO-based sensor coupled with AuNPs, as shown in Fig. 4.1b (Shi et al., 2015). It also saves one step of processing as detection would only be carried out with interaction between Pb^{2+} and gold-graphene composite. A stepwise process can be highlighted with key elements regarding detection mechanisms: initially, there are colloidal dispersions of gold nanoparticles on graphene oxide to create colloidal stability.

Upon exposure to Pb^{2+}, the lead ions interact with the gold nanoparticles, accelerating their leaching. This leaching reduces the nanoparticle size, altering their surface plasmon resonance (SPR) properties, which results in a visible color change in the solution.

As lead ions destabilize the colloidal system, they promote nanoparticle aggregation. The decreased distance between nanoparticles due to aggregation further shifts the SPR peak, intensifying the optical signal. The aggregation mechanism increases the extent of colorimetric response, which amplifies the sensing ability of the sensor to even diluted concentrations of Pb^{2+}. Graphene oxide is used to stabilize the dispersibility of nanoparticles. At the same time, the aggregation toward the nanoprobe is induced solely by the challenge of Pb^{2+} that can be quantitatively revealed through color transformations from red-purple or colorless. The spectrophotometer can accurately measure the degree of aggregation and optical modification.

GO and its composites are highly catalytically active for the colorimetric detection of heavy metal ions. GO can mimic natural enzymes, thus significantly enhancing the oxidation of chromogenic substrates such as 3,3′,5,5′-tetramethylbenzidine in the presence of hydrogen peroxide (H_2O_2). Such enzymatic-like activity can be sensitive to detection as interactions with heavy metal ions, such as lead (Pb^{2+}) or mercury (Hg^{2+}), can modulate the catalytic efficiency and thus be detectable through color changes. Functional groups on GO allow for effective interaction and bonding with heavy metals, which causes structural changes that affect catalytic properties. Moreover, when mixed with metallic nanoparticles, the composites utilize synergistic effects to enhance catalytic performance and amplify optical signals through surface plasmon resonance (SPR). These mechanisms often involve electron transfer processes where heavy metal ions act as electron acceptors, leading to non-radiative decay and significant colorimetric changes (Lv et al., 2023; Zarlaida & Adlim, 2017).

An innovation is a composite material for visual detection of mercury ions Hg^{2+}, based on integrating CuS and NiS nanoparticles within porous-reduced graphene oxide (p-rGO) sheets (Borthakur et al., 2021). The potential of the composite material is drawn from the porous characteristic of rGO, enhancing the surface area and facilitating a better interactive feature with target ions (Fig. 4.2a). The CuS and NiS nanoparticles are artificial enzymes, or nanozymes, that mimic the activity of natural peroxidases. This characteristic is essential because it enables the composite to catalyze the oxidation of various substrates. In the presence of hydrogen peroxide (H_2O_2), these nanozymes oxidize the chromogenic substrates like 3,3′,5,5′-tetramethylbenzidine (TMB) and result in a color change from colorless to blue. In this respect, the color is measurable by the absorbance at particular wavelengths. If the Hg^{2+} ions are introduced into the solution, the addition of this

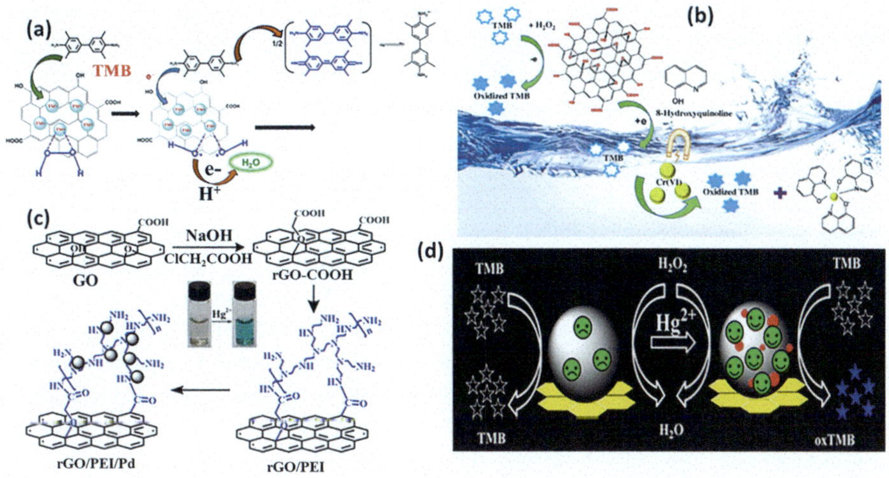

Fig. 4.2 (**a**) The catalytic activity mechanism of the probe materials in detecting Hg^{2+} ions. (Reproduced with permission from Borthakur et al. Copyright 2021, American Chemical Society). (**b**) The colorimetric detection of Cr^{6+} via mimetic catalyst. (Reproduced with permission from Nghia et al. Copyright 2019, Elsevier). (**c** and **d**) The synthetic route and catalytic activity of rGO/PEI/Pd nanohybrids toward Hg^{2+}. (Reproduced with permission from Zhang et al. Copyright 2017, Elsevier)

will result in interaction between the CuS and NiS nanoparticles, which affects their catalytic functions. The mechanism of electron transfer during the oxidation of TMB may be affected, thus altering the color intensity. The degree of change of color in this instance is directly proportional to the concentration of Hg^{2+} ions present in the sample being studied. Therefore, the higher mercury concentrations indicate a more profound color change brought about by higher substrate oxidation. With such detecting shocking limits, it is unsurprising that the analytical study reported such sensitivity at as low as 13.05 nM for CuS and 48.73 nM for NiS. Such sensitivity is vital for environmental analysis, mainly because traces of mercury are usually present. Additionally, the immobilization of nanozymes in paper strips makes the detection assay facile to perform on an analytical scale, making it possible to conduct at-site analyses. With this design, tests can be run without demanding complex laboratory apparatus; hence, the field tests can be done promptly.

Similarly, in another study, Nghia et al. presented a colorimetric method for Cr^{6+} detection based on the peroxidase-like activity of graphene oxide (GO) nanoparticles (Nghia et al., 2019). In this method, GO catalyzes the oxidation of colorless TMB into blue oxTMB in the presence of H_2O_2. However, the addition of 8-hydroxyquinoline inhibits this reaction, likely due to strong π–π interactions and hydrogen bonding between 8-hydroxyquinoline and GO, which block the active sites on the GO surface (Fig. 4.2b). When Cr^{6+} is introduced into the system, it oxidizes the 8-hydroxyquinoline, restoring the catalytic activity of GO. This "On–Off–On" mechanism allows for the efficient colorimetric detection of Cr^{6+}, with a detection limit of 5.8 nM and a linear range of 70–430 nM. The method successfully detected Cr^{6+} in lake water from industrial areas.

Consequently, Lu et al. developed 2D rGO/PEI/Pd nanohybrids using polyethyleneimine (PEI) as a coupling agent to crosslink reduced graphene oxide (rGO) nanosheets with Pd nanoparticles (PdNPs). PdNPs were anchored onto GO nanosheets via Pd–N coordination bonds, enhancing their catalytic properties for TMB oxidation in H_2O_2 (Fig. 4.2c) (Zhang et al., 2017). The improved catalytic activity was due to reduced PdNP agglomeration, the large surface area of GO, and abundant anchoring sites from PEI. Upon adding Hg^{2+}, citrate reduced the Hg^{2+}, forming a thin Pd–Hg amalgam layer on the PdNPs, further enhancing the peroxidase-like activity (Fig. 4.2d). This system attained ultra-sensitive colorimetric detection of Hg^{2+} with detection limits of 0.39 nM in the aqueous solution and 1.0 nM in wastewater and human serum samples. Surface functionalization of graphene oxide (GO) for calorimetric detection of heavy metal ions involves several key steps and considerations. The functionalization process enhances the interaction between GO and the target metal ions, improving sensitivity and selectivity. GO can be functionalized via several chemical approaches by introducing a particular functional group that can complex with heavy metal ions (Dong et al., 2014; Farooq et al., 2022; Song et al., 2020; Georgakilas et al., 2012, 2016).

Covalent Functionalization

(a) *Esterification:* Reacting carboxylic acid groups on GO with alcohols to form esters can improve heavy metals' binding capacity.
(b) *Amine Functionalization:* Amines can be grafted onto GO through a reaction with carboxyl groups, enhancing the affinity for metal ions.

Noncovalent Functionalization

(a) *Self-Assembly:* The adsorption of organic molecules or polymers on the GO surface through π–π interactions or hydrogen bonding. This can create a selective environment for metal ion binding.
(b) *Ionic Interactions:* Adding polyelectrolytes or surfactants that can create charged sites on the GO surface to attract cationic metal ions.

4.2.1 Colorimetric Detection of Pathogens

Graphene oxide (GO) and its composites have emerged as powerful tools for the colorimetric detection of pathogens. Their unique properties, including high surface area, tunability, and functionalization capabilities, allow for significant interactions with various biological entities, leading to easily observable color changes. Mechanisms of calorimetric detections with GO and composites are illustrated below.

4.2.1.1 Adsorption and Aggregation for Interaction-Induced Color Change

GO has a high surface area and varied functionalities, enhancing the strong adsorption of bacterial cells. Once adsorbed, GO sheets can be aggregated, which is further affected by the ionic strength and pH of the solution. This aggregation changes the turbidity of the surrounding medium, leading to color shifts. For instance, the initial light yellow color of a GO suspension may change to a darker color upon the introduction of *E. coli* due to enhanced scattering of light, and the release of adenine and proteins from bacterial cells upon interaction with GO results in changes in the absorbance characteristics of the solution, observable as a color shift (Nanda et al., 2016). GO's high surface area and rich functionalities enhance bacterial cells' strong adsorption ability. After the adsorption process, GO sheet aggregation may occur, which is influenced by the ionic strength and pH of the solution. This aggregation changes the surrounding medium's turbidity, leading to distinct color shifts. For example, when *Escherichia coli* is introduced to a GO suspension, the initial light yellow color can shift to a darker hue as the bacteria aggregate on the surface of the GO sheets, induced by enhanced scattering of light (Fig. 4.3). The unique sharp edges of GO nanosheets can pierce bacterial membranes, causing intracellular components to leak into the medium. This leakage alters the solution's refractive index, which can lead to visible color changes. For instance, the release of adenine

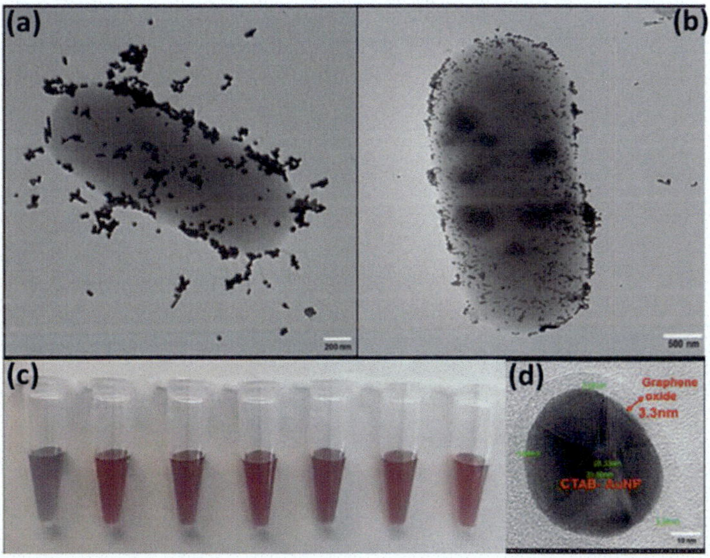

Fig. 4.3 (**a–c**) The aptamer conjugates composite material; aptamer citrate composite gold nanoparticles in the presence of *E. coli* cells; UV-Vis spectrum analysis of the aptamer–GO–AuNPs nanoprobe in the presence of both positive and negative bacterial strains, along with a colorimetric depiction of the nanoparticle aggregation; HR-TEM images of the GO coated CTAB gold nanoparticles. (**d**) HRTEM image of GO coated CTAB AuNPs. (Reproduced with permission from Gupta et al. Copyright 2021, Elsevier)

and proteins from bacterial cells upon interaction with GO induces changes in the absorbance characteristics of the solution, observable as a color shift (Gupta et al., 2021).

4.2.1.2 GO–Enzyme Complex for Bacterial Detection

A notable example involves using ionic positively charged GO–enzyme complexes for bacterial detection, which can generate colored products through enzymatic reactions (Fig. 4.4a). The detection system leverages the enzyme's catalytic ability to produce products that change color when bacteria are present. This method has been shown to allow for rapid detection of bacteria with high sensitivity and selectivity, demonstrating a practical application of GO in clinical diagnostics. As pathogens interact with the functional groups on GO, such as hydroxyl and carboxyl groups, it causes agglomeration and color shifts due to alterations in scattering and

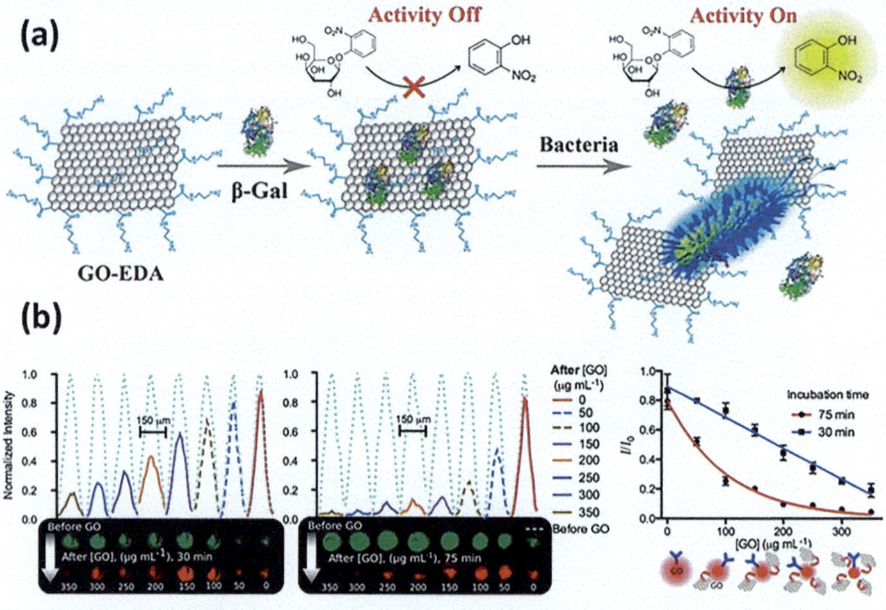

Fig. 4.4 (**a**) The enzymatic catalytic activity representation toward the bacteria and the representation of drug screening between the positively charged GO and b-Gal complex and the plot of bacteria (*E. coli*) versus drug dose. (Reproduced with permission from Li et al. Copyright 2013, Royal Society of Chemistry). (**b**) Concentration and time-dependent quenching of Ab-QD microspots by GO; spot profiles illustrating the quenching of Ab-QD microspots at varying GO concentrations (0–350 μg mL^{-1}) after 30 and 75 min incubation. The original profile of the analyzed spots is displayed alongside the final profiles of the quenched spots. Calibration curves for the quenching experiment are plotted against the same GO concentrations and incubation times. Error bars were derived from a parallel assay of ten microspots. I/I0 represents the ratio of final intensity to original intensity. (Reproduced with permission from Morales-Narváez et al. Copyright 2013, Wiley)

absorbance of light. The binding of the pathogens can thus be visually indicated by a noticeable change in the solution's color, allowing for rapid contamination assessment (Li et al., 2013). In this method, GO acts as a platform that enhances enzymatic reactions necessary for detecting bacterial presence. Its positively charged nature promotes efficient binding and immobilization with negatively charged bacterial cells. The complex contains an enzyme, typically horseradish peroxidase, which catalyzes the oxidation of substrates, thus causing a detectable colorimetric change. Specifically, the detection mechanism relies on the oxidation of chromogenic substrates, such as TMB, by the enzyme in the presence of hydrogen peroxide (H_2O_2).

The presence of bacteria modulates the enzyme's activity, resulting in a change in color intensity that can be quantitatively measured through spectrophotometry, allowing for sensitive and specific detection of bacterial cells. This method is tailored for high-throughput applications, making it ideal for screening multiple samples simultaneously—particularly beneficial for drug screening, where rapid evaluation of antibacterial activity is essential. Its straightforward colorimetric approach facilitates quick visual inspections or automated absorbance measurements. Furthermore, the method's versatility enables the detection of various bacterial strains by adjusting the substrate and enzyme used, catering to diverse applications in food safety, clinical diagnostics, and environmental monitoring. It is also applicable for high-throughput screening of antibacterial compounds, making it a valuable tool in the quest for new antibiotics or antimicrobial agents.

4.2.1.3 GO with Nanoparticles for Enhanced Detection

Graphene oxide composites with gold nanoparticles (AuNPs) reveal advanced colorimetric detection techniques. Here, the aggregation of AuNPs on GO surfaces due to the presence of bacteria results in a dramatic visible color change observable by naked eye detection. Such a technique will combine the AuNP-based surface plasmon resonance effects with the enhanced surface area associated with GO; this will improve the sensitivity of biosensors against pathogens, such as *Salmonella* and *E. coli*, significantly (Kaushal et al., 2021; Rose et al., 2021).

This approach is designed to enhance bacterial detection by focusing on a rapid and sensitive bacterial detection method further enhanced using near-infrared (NIR) light to boost antibacterial activity by coating gold nanoparticles with antibody-assisted graphene oxide (GO). The other method includes using gold nanoparticles (AuNPs) conjugated with antibodies targeting bacterial cells, such as *E coli*. The antibodies linked to the AuNPs ensure high specificity, with the help of which the biosensor can selectively target bacteria in a sample. This specificity reduces false positives, thus making the detection more reliable. Once the antibodies bind to the bacteria, this induces a change in the gold nanoparticle's aggregation state. This agglomeration alters the SPR of the AuNPs and is associated with the effect of a color change that could be detected by the naked eye or easily measured by simple optics. NIR illumination enhanced the antibacterial activity of graphene oxide-coated AuNPs as well. The nanomaterials were known to produce heat from NIR irradiation, which might further help in bacterial killing. So, it becomes an additive photothermal component of the detection protocol. While the detection of microscopic entities is followed by activation of NIR, it becomes possible to assess quite rapidly whether there is bacterial presence and viability, making it very suitable for food safety and clinical diagnostics applications.

4.2.1.4 Antibody-Conjugate GO

Specific antibodies to target pathogens improve the selectivity of GO-based biosensors. For instance, antibody-conjugated GO can be designed to identify *E. coli* O157:H7. Such biosensors provide clear colorimetric signals because of the strong affinity between antibodies and bacterial antigens. A recent report on a very sensitive biosensor that can detect *E. coli* in different environments with excellent selectivity and sensitvity. The Ab-QD probes exploit the unique properties of graphene oxide, featuring a two-dimensional structure that quenches fluorescence (Morales-Narváez et al., 2013). In such a biosensing system, pathogens do not induce a response but instead enhance the system's sensitivity once it has sensed an abnormality. The graphene oxide plays an integral role in the resonance energy transfer processes, which are fundamental to the mechanism of detection and act as a platform for immobilizing the Ab-QD probes (Fig. 4.4b). The presence of *E. coli* will reduce the degree of quenching and, thus, increase the fluorescence intensity that can be measured. In contrast, the specific binding of antibodies to the bacteria is enhanced. This allows for a "digital-like" response where the signal associated with detection gets highly amplified in the presence of pathogens. The method promises to bring fast and sensitive detection, finding applications in clinical diagnostics and environmental monitoring.

4.3 Carbon Dots (CDs)

Carbon dots, abbreviated as CDs, are a class of zero-dimensional carbon nanomaterials that has lately attracted abundant attention due to their distinctive optical properties, biocompatibility, and varied applications in diverse fields. It is generally spherical nanoparticles with diameters less than 10 nm. The constituent element mainly in carbon dots is carbon atoms that are either graphitic or amorphous. In the form of heteroatoms, they could include oxygen and nitrogen and many other heteroatoms that impart them with surface functionalities and their optical properties. Carbon dots typically have the core as sp^2-hybridized carbon atoms and hydroxyl (–OH), carboxyl (–COOH), and amine ($–NH_2$) functionalities at the surface, which play a central role in the solubility, stability, and reactivity of carbon dots. CDs have emerged as versatile nanomaterials of different applications, especially in UV-Vis colorimetric sensing. The special photophysical properties, including strong absorption, luminescence, and tunability through surface functionalization, make carbon dots especially suited for UV-Vis-based color change applications (Parambil & Rajamani, 2024). CDs can be both absorptive and emissive materials in these systems, exhibiting visible changes in their optical behavior to environmental or chemical stimuli (Rajput et al., 2021). This makes them efficient for detecting many analytes, such as metal ions, biomolecules, and pollutants. Carbon dots' optical properties mostly depend on the size, surface functional groups, and chemical environment. CDs can absorb light throughout the UV-Vis spectrum and generally exhibit strong absorption peaks at 260–350 nm, depending on the composition. Their size and functionalization enable modulation of their absorbance and fluorescence properties, making them highly tunable for specific detection tasks. CDs typically respond to an analyte by altering light absorption or fluorescence emission. These changes can be quantified based on shifts in wavelength or intensity, providing a straightforward mechanism for detection (Yadav et al., 2024d, 2025). The intense light absorption of CDs and their potential for surface modifications make them sensitive to small concentrations of analytes, often exhibiting distinct and observable color changes. The color change mechanism in carbon dot-based UV-Vis sensing primarily relies on their interaction with analytes, which induces shifts in their optical properties (Arcudi et al., 2019; Cadranel et al., 2019; Wu et al., 2017; Echeverry-Gonzalez & Kouznetsov, 2020).

4.3.1 Several Mechanisms Drive These Interactions, Leading to Color Changes

4.3.1.1 Aggregation-Induced Absorption Shifts

In some cases, CDs aggregate near metal ions or certain organic compounds. Such aggregation changes the electronic structure; consequently, its light absorption properties change, giving a color output. For instance, in the presence of heavy metal ions such as lead (Pb^{2+}) or mercury (Hg^{2+}), CDs aggregate and shift the absorbance from UV to the visible range. The color change occurring upon this aggregation is detectable, even without some specialist instrument.

4.3.1.2 Charge Transfer and Complex Formation

In the presence of analytes, these may undergo charge transfer (CT) or form complexes with target molecules. The interaction will alter the electronic configuration of CDs, thus shifting their absorption and/or emission spectra. For instance, carbon dots functionalized with a specific ligand can form a complex with metal ions or organic molecules, thereby invoking a bathochromic (red) shift in the UV-Vis absorption spectrum as an indicator of the presence of the analyte.

4.3.1.3 pH-Sensitive Color Change

CDs can be designed to change color at changes in pH, making them very useful for applications in which pH monitoring is critical, such as environmental monitoring or biochemical processes. Functionalization of CDs with pH-sensitive groups makes the CDs' fluorescence or absorbance shift pH-dependent, thus visibly changing the solution's color. These pH-dependent color changes can thus be detected by UV-Vis spectroscopy as changes in peaks associated with absorbance or emission with different pH values.

4.3.2 Advantages of Carbon Dots in UV-Vis Colorimetric Sensing

4.3.2.1 High Sensitivity

Carbon dots are highly sensitive toward a broad spectrum of analytes due to their extraordinary surface-to-volume ratio and tunable optical properties. This allows them to detect trace substances, even at nanomolar concentrations.

4.3.2.2 Real-Time and Rapid

Interaction times between CDs and analytes are usually short; thus, real-time monitoring of the events that require immediate feedback is also done.

4.3.2.3 Eco-Friendliness

CDs are commonly synthesized from benign and environment-friendly precursors, making them sustainable alternatives for other nanoparticle-based sensing systems, like those using toxic metal nanoparticles.

4.3.2.4 Versatility and Tunability

CDs can be easily functionalized with a variety of functional ligands, probes, or molecules, thus allowing customization of the sensors to detect specific target molecules. Their optical properties can be readily modified by changing their size or surface chemistry or even during synthesis.

4.3.2.5 Cost-Effectiveness

Unlike other sensing materials, gold nanoparticles, or organic dyes, carbon dots can be synthesized by a low-cost method; this renders them excellent candidates for large-scale applications.

One such study is the green synthesis of carbon dots functionalized silver nanoparticles for the colorimetric detection of phoxim, which attempts to reveal a sophisticated and eco-friendly approach toward pesticide phoxim detection (Fig. 4.5a, b). Here, the method deals with environmental concerns using green synthesis techniques and exploits the optical features of CDs and AgNPs for sensitive and low-cost rapid detection. Here, Lycii fructus is a natural precursor for synthesizing carbon dots. The idea of using such plant material also envisages green chemistry through the avoidance of the utilization of heavy amounts of dangerous chemicals. The CDs were synthesized using a straightforward method,

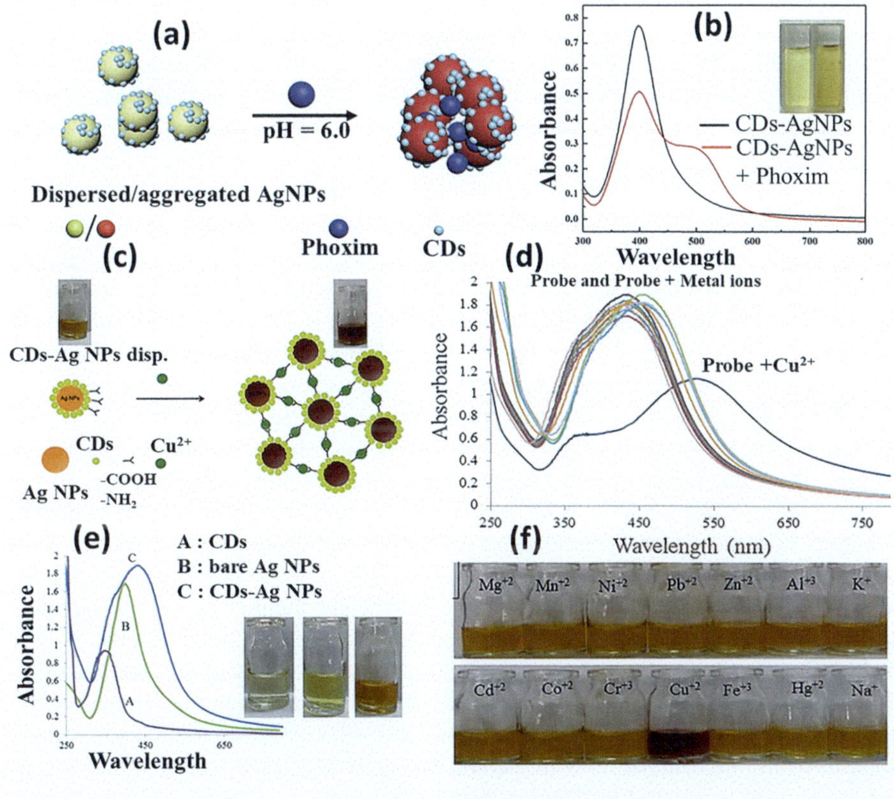

Fig. 4.5 (**a**, **b**) The synthetic scheme and the colorimetric change for the detection of phoxim using CD–AgNPs. (Reproduced with permission from Mingda et al. Copyright 2018, Elsevier). (**c–f**) The UV-Vis selectivity profile toward Cu^{2+} and of the bare CDs, Ag, and final composite material selectivity and the colorimetric change pictures toward Cu^{2+}. (Reproduced with permission from Beiraghi & Najibi-Gehraz Copyright 2017, Elsevier)

where $AgNO_3$ was reduced in the presence of CDs to form AgNPs. The carbon dots play a very significant role in stabilizing the silver nanoparticles. CDs are rich in surface functional groups (such as hydroxyl and carboxyl groups), which serve as binding sites for silver ions, which reduce Ag^+ to Ag^0 and inhibit agglomeration. This stable CD–AgNP complex is essential in using these particles for sensing applications. The detection mechanism relies on the good interaction between the CDs–AgNPs and phoxim, an organophosphate pesticide. For the plasmon resonance peak characteristic, the solution of CD–AgNPs under normal conditions is yellow. The UV-Vis spectrum from this particle size indicates a peak at about 400 nm. Introducing phoxim to the solution leads to aggregation of the CD–AgNPs (Fig. 4.5b). This may result from an interaction between the organophosphate groups in phoxim with surface functional groups of carbon dots and silver nanoparticles; particle clustering occurs. This aggregation results in a red shift of the surface plasmon resonance from 400 to 525 nm and thus leads to a visible color change from yellow to red. Therefore, this optical change is highly important for the practical application of this system as a colorimetric sensor. The visible shift enables the qualitative determination of phoxim, but the quantitative analysis is performed with the help of the absorbance ratio (A_{525}/A_{400}). This sensor can sense the phoxim concentration in the range of 0.1–100 μM linearly and presents its relation to the concentration of phoxim. The sensitivity of this sensor is determined to be approximately 0.04 μM. It is, therefore, susceptible, rendering it capable of detecting phoxim concentrations below the maximum residue limit tolerated in many regions. The electrostatic and covalent interactions between phoxim and the CDs–AgNPs enable the detection of phoxim. Reactive sites of phoxim include the phosphorus-oxygen bond that allows it to associate with silver surfaces as well as carbon dots. This aggregation of nanoparticles results in the color change observed. One effect of particular interest concerning colorimetric sensing with nanoparticles is aggregation. Aggregated AgNPs have a characteristic yellow-colored solution when dispersed; their plasmon resonance peak is located within the visible region. Upon aggregation, the plasmon resonance shifts toward longer wavelengths; thus, this phenomenon may be followed as a visible color change. This is because the electromagnetic field between the aggregated particles has changed, changing how the solution absorbs and scatters light. Several factors, like pH, temperature, and the concentration of CDs and AgNPs, are controlled to maximize the sensor's performance. Since these factors could affect stability as well as the interaction of CD–AgNP complexes with phoxim, controlling them is crucial. For instance, optimal pH allows for an interaction between the available surface functional groups of CDs and phoxim, whereas temperature affects the kinetics of the aggregation. Furthermore, it demonstrated the selectivity of the sensor toward phoxim from other interferents that could probably affect the sensor. When a sensor is exposed to several molecules in the environment or biological conditions, selective recognition within the sensing platform is required. In this regard, the CDs' surface functionalities proved to be quite important, as they possess specific binding sites for phoxim; the adsorption mechanism should be highly selective so that the interference of other chemicals obtains no spurious results. The sensor was tested in real samples, such as environmental water and fruit extracts, to validate its practicability. The recovery of phoxim in these samples was found to be between 87% and 110%. The sensor functioned suitably in complex matrices, showing no significant interference; this result promises good use in environmental monitoring and food safety where on-site, rapid pesticide residue determination is critical. Using green-synthesized CDs and AgNPs further boosts environmental sustainability for this sensor. Traditional nanoparticle synthesis methods typically involve harmful chemicals and high-energy processes. Still, in this regard, the green synthesis method takes its place using natural, renewable materials and benign reaction conditions. Therefore, the sensor is effective and environmentally friendly (Zheng et al., 2018).

Another study similarly uses carbon dots@modified silver nanoparticles (CDs–AgNPs) as colorimetric sensors, which detect the cupric ions (Cu^{2+}), an innovative approach by each of the materials' distinctive optical properties and surface, wherein carbon dots and silver nanoparticles in the system are integrated for the sensitization and selectivity of the sensor toward Cu^{2+} ions (Fig. 4.5c–f). In constructing this colorimetric sensor, carbon dots reduced silver nitrate ($AgNO_3$) to form CDs–AgNPs. The reduction leads to a stable colloidal solution of silver nanoparticles decorated with carbon dots. The SPR absorption peak of their material is around 430 nm, which typically points to the presence of silver nanoparticles (Fig. 4.5d). The carbon dots are crucial in stabilizing the functional groups to agglutinate the nanoparticles and interact with Cu^{2+} ions. The detection mechanism here involves the interaction of Cu^{2+} ions with the functional groups present on the surface of the carbon dots in the CDs–AgNPs complex. When the Cu^{2+} ions are exposed to the sensor, they bond with these functional groups, initiating a change in the local environment surrounding the silver nanoparticles. Aggregation of such nanomaterials leads to a greater shift in the optical property of the solution. Specifically, it results in a color change of the CDs–AgNPs dispersion to red-brown from orange following Cu^{2+} binding to the carbon dots (Fig. 4.5e, f). The optical color change is due to the aggregation of silver nanoparticles changing its SPR. The SPR changed the position of the absorption peak from 430 nm toward the higher wavelength corresponding to the red-brown color.

4.3.3 Factors Influence the Performance of the Sensor

4.3.3.1 pH

The pH of the solution is extremely important while fine-tuning its interaction with the functional group on carbon dots in terms of optimization of the interaction between Cu^{2+} ions. The performance of sensors is well optimized by varying pH, and ideal conditions are established for detecting Cu^{2+} ions.

4.3.3.2 CDs Concentration

The concentration of carbon dots in the silver nanoparticle matrix greatly influences the sensor's sensitivity. Higher concentrations of CDs provide more binding sites for Cu^{2+} ions, hence a higher sensor sensitivity.

4.3.3.3 Ionic Strength

It also affects the sensor's working since interfering ions are always in the solution. Controlling the ionic strength of the medium ensures selective detection of Cu^{2+}.

4.3.3.4 Interaction Time

The interaction time granted between the Cu^{2+} ions and the CDs–AgNPs determines the colorimetric response of this colorimetric assay. Ideally, there should be a reasonable enough interaction period so that the change of color has become notable and perceivable.

Under optimized conditions, it was possible to sense the presence of these molecules with a high sensitivity toward Cu^{2+} ions within a detection range of 0.3–8 μM, and the lowest detection limit was 0.037 μM. The concentration of Cu^{2+} can be quantitatively determined using the ratio between the absorbance at 525 nm, the peak after the aggregation process, and the absorbance at 430 nm, the original SPR peak. The sensor shows high selectivity toward Cu^{2+} over other metal ions and can be optimum for water quality testing and environmental monitoring. The practical applicability of the CDs–AgNPs sensor has been demonstrated by successfully detecting Cu^{2+} in real tap and river water samples. As it detects as little as low concentrations of Cu^{2+} in a simple, cost-effective, and fast colorimetric method, it is an effective tool for on-site monitoring of water sources contaminated by copper. Overall, the functionalized silver nanoparticles with carbon dots can serve as a good colorimetric probe for the selective detection of Cu^{2+} ions based on the higher synergy of the optical property of AgNPs in combination with the functionalization capability of carbon dots. The visually detectable color change resulting from the induced aggregation of the silver nanoparticles during the detection of Cu^{2+} ions through the solution can also be quantitatively monitored using UV-Vis spectroscopy. It has a promising approach for environmental monitoring due to its excellent sensitivity, selectivity, and simplicity of its operation (Beiraghi & Najibi-Gehraz, 2017).

4.4 Carbon Nanotubes (CNTs)

Carbon nanotubes (CNTs) are nanostructures of carbon atoms rolled into a hollow, tubular hexagonal lattice. They come in two major forms: single-walled carbon nanotubes (SWCNTs), a cylindrical nanostructure wherein carbon atoms are rolled into a tube from a single layer, and multiwalled carbon nanotubes (MWCNTs), which consist of multiple layers of graphene rolled concentrically. CNTs are characterized by the large aspect ratio, with lengths that may reach micrometer lengths and diameters ranging from 1 to 100 nm. Their unique cylindrical structure confers on its high mechanical strength, elasticity, and electrical conductivity. CNTs possess very strong absorption in the UV-visible spectrum and exhibit particularly efficient Raman scattering. These properties are useful in colorimetric detection since they allow one to visualize the color shift related to the response of the target analyte. CNTs can be functionalized with different chemical functional groups, enhancing their properties and interactions with specific analytes.

Functionalization allows the creation of composites of CNTs with other materials, enhances solubility, and introduces active sites for applications in sensing. Calorimetric sensing involving CNTs relies on the interaction of target analytes with the CNT surfaces, resulting in heat changes that can be measured. When molecules with high binding affinity adsorb onto CNTs, calorimetric methods can detect the heat changes associated with these interactions, facilitating the sensing process. CNTs'ability to act as scaffolds allows for the immobilization of various functional groups, thereby enhancing the specificity and sensitivity of these calorimetric sensors for multiple targets, including metal ions and organic pollutant. For example, a report, "The Colorimetric Nanozyme Sensor Array Based on Metal Nanoparticle-Decorated CNTS for Quantification of Pesticides in Real Water and Soil Sample," has been reported that delves into the development of an innovative sensor array using carbon nanotubes (CNTs) decorated with metal nanoparticles (Fig. 4.6). These nanoparticles exhibit nanozyme activity, mimicking natural enzymes to catalyze reactions that generate colorimetric changes. When pesticides

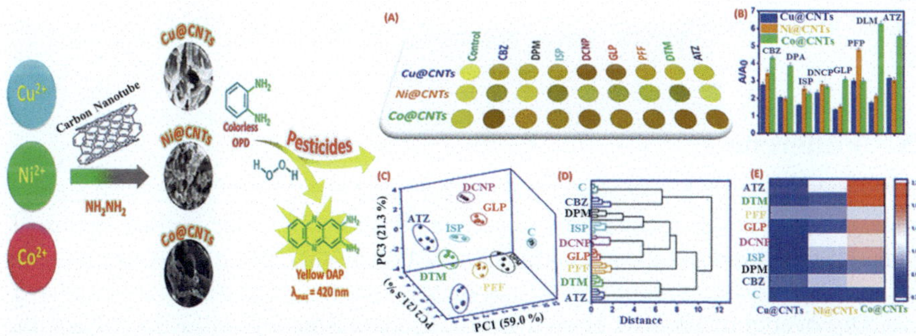

Fig. 4.6 The schematic diagram demonstrates the design of a nanozyme sensor array using metal-decorated carbon nanotubes (MDC) for pesticide identification. The MDC sensor array enables pesticide differentiation through the following: (**a**) Color variation patterns resulting from OPD oxidation catalyzed by the MDC sensor array when exposed to different pesticides. (**b**) Absorbance signal patterns (A/A_o) at 420 nm in the presence of various pesticides. (**c**) A canonical score plot for the first three factors of OPD absorbance change patterns showcasing the discrimination of eight pesticides at a 10 μM concentration using the MDC sensor array (n = 5). (**d**) Hierarchical cluster analysis (HCA) plot for segregation of the eight pesticides. (**e**) A heat map was generated from the absorbance responses of the eight pesticides. (Reproduced with permission from Kumar et al. Copyright 2024, American Chemical Society)

interact with the nanozyme, a distinct color shift occurs, which is measurable using UV-Vis spectroscopy. The sensor array demonstrates high sensitivity and selectivity for different pesticides in complex real-world matrices, including water and soil. This sensor capitalizes on the catalytic abilities of metal nanoparticles to oxidize substrates in the presence of hydrogen peroxide (H_2O_2).

The reaction produces reactive oxygen species (ROS) or hydroxyl radicals, which then interact with chromogenic substrates such as 3,3′,5,5′-tetramethylbenzidine (TMB) or similar dyes. These dyes are frequently used in colorimetric sensors since they significantly change color when oxidized. Without pesticides, the reaction goes at a baseline rate and is lightly colored. When a pesticide interacts with the nanoparticles, the catalytic rate is much faster, significantly enhancing the color change. Carbon nanotubes play an important role because they act as a support matrix for the metal nanoparticles, providing better dispersion of the nanoparticles and making catalytic sites more accessible. The higher surface area of CNTs increases the chance of interaction between pesticides and catalytic sites, thereby increasing the sensitivity. The electrical conductivity and adsorption of the overall system are also increased by CNTs, promoting pesticide interaction. The surface chemistry changes upon binding of pesticides to the metal nanoparticles, leading to changes in electron density and aggregation effects. These changes enhance the catalytic activity, thereby accelerating the oxidation reaction. The more pesticide molecules that bind, the faster the nanoparticles catalyze the reaction, leading to a more intense color shift. This nanozyme-based sensor has significant potential in both the environmental and agricultural sectors. It provides a strong monitoring device for pesticide residues in water bodies, which would be vital in ensuring the safety of the drinking water supply and preventing ecological damage. Likewise, in agricultural applications, the sensor array may be used in assessing soil contamination, which may help farmers and regulatory bodies ensure safe pesticide application and avoid over-application that causes harm (Kumar et al., 2024).

4.5 Fullerene

Fullerenes represent a distinct class of carbon allotropes, forming hollow, spherical, or tubular structures composed entirely of carbon atoms. The most well-known is C_{60}, a soccer ball-like structure of 60 carbon atoms arranged in pentagons and hexagons. Discovered in 1985 by Harold Kroto, Richard Smalley, and Robert Curl, the identification of fullerenes earned them the Nobel Prize in Chemistry in 1996. A large variety of sizes and forms were found for fullerenes, ranging from C_{60}, C_{70}, and C_{84} to larger Bucky balls and carbon nanotubes, among others. The fullerenes contain a carbon atom framework that consists of hexagons and pentagons, and they contain something close to an enormous cage, contributing significantly to their stability. In addition, they can trap other atoms or molecules within their cage and be subjected to many chemical reactions, such as oxidation and reduction. Moreover, fullerenes are freely soluble in organic solvents to functionalize nanotechnology and medicine applications. During these applications, fullerenes, particularly C_{60} and its derivatives, present promising nanomaterials that may be exploited to develop colorimetric sensors to detect toxic pollutants (Shetti et al., 2021).

Due to their electronic properties and the property to form a stable complex with analytes, fullerene derivatives can sense mercury (Hg^{2+}), lead (Pb^{2+}), cadmium (Cd^{2+}), and other metals by inducing changes in optical properties. The colorimetric changes typically result from the charge transfer processes by which the heavy metal ions form a complex with fullerene derivatives. The formation alters the electronic structure of the fullerene derivatives, which it expresses as a color change. Also, due to the surface plasmon resonance, the hybridization of fullerenes with metal nanoparticles results in color change, for example, gold or silver, which will enhance the signal by the optical signal and can, therefore, highly enhance the response toward the binding of metal ions. Functionalized fullerenes are very sensitive to the typical values of nanomolar concentrations and have very low interference with other substances. Stability and selectivity favor them for real-time environmental monitoring, particularly for detecting toxic metals (Wu et al., 2023).

For example, a novel colorimetric method for detecting hexavalent chromium (Cr (VI)) ions in water using N-doped carbon dots derived from C_{60} fullerene has been demonstrated. This method is significant due to the toxicity and carcinogenic properties of Cr (VI), making its detection crucial for health and environmental safety. The one-pot oxidative method has been employed to open the fullerene cage of C_{60}, producing nitrogen-doped carbon dots (N-CDs). In this synthesis, C_{60} was dissolved in a mixture of water and tetrahydrofuran (THF), followed by the addition of hydrogen peroxide and ammonium hydroxide (Fig. 4.6a). The reaction was conducted at elevated temperatures to facilitate the formation of stable suspensions of N-CDs.

After synthesis, the N-CDs were characterized using techniques such as dynamic light scattering (DLS) to assess their size distribution and UV-Vis spectrophotometry to determine their optical properties. The colorimetric sensing mechanism relies on the interaction between the N-CDs and Cr (VI) ions. When Cr (VI) ions are introduced to the N-CD solution, a distinct absorption band appears around 550 nm, indicating a change in the photophysical properties of the N-CDs (Fig. 4.7b, c). This band is significant as it lies within the visible region, making it amenable for colorimetric application. The increase in absorbance was found to be linearly correlated with the concentration of Cr (VI) ions in the range of 1–100 μM, thus showing the sensitivity of the method (Fig. 4.7d). The detection process is highly efficient, occurring within just 1 min and without the need for additional reagents or pH adjustments. These conditions of fast response and easy usage are highly critical for field applications, particularly during field testing. The prepared sensing system demonstrated excellent stability for 9 months, allowing long-term storage and use. The researchers investigated the selectivity of N-CDs toward Cr (VI) and other metal ions, such as Fe (III), Cu (II), and Pb (II). The N-CDs showed excellent selectivity toward Cr (VI) with minimal interference from these typical heavy metals, as shown in Fig. 4.7c. Such selectivity is essential for achieving reliable measurements in the presence of interfering species in complex environmental matrices. This

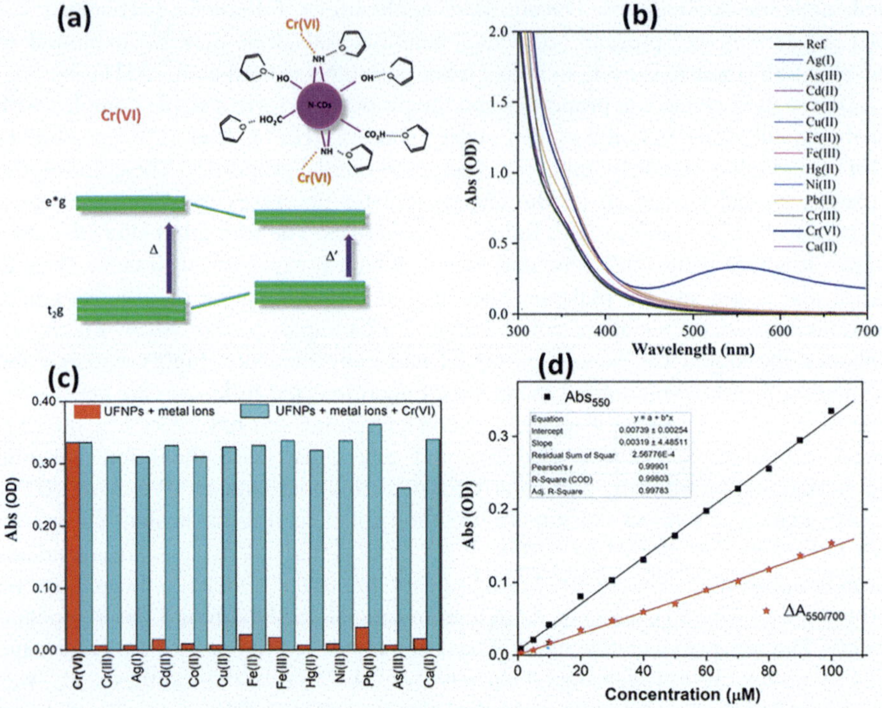

Fig. 4.7 (**a–d**) The Cr^{6+} and the energy gap correspond to the transition energy, selective interaction of Cr^{6+} toward N-CDs hydrogen bond with THF, and the reduction of the energy gap, the UV-Vis selectivity profile, and the linearity behavior at a particular wavelength toward Cr^{6+} ions. (Reproduced with permission from Babazadeh et al. Copyright 2021, MDBI)

approach has several merits: economical, green, and material-friendly. Long-term storage is possible for the N-CDs up to 9 months, and the fast response time makes this method more useful for field-based applications. The simplicity of the preparation and the operational ease make this colorimetric sensor a promising tool for detecting toxic metals in water, potentially facilitating regular monitoring in various settings.

Overall, unfolded fullerene-derived carbon nanoparticles represent an innovative approach to environmental sensing, particularly for hazardous substances like Cr (VI). The study highlights the potential application of carbon-based nanomaterials in developing advanced, sensitive, and selective detection methods for environmental pollutants (Babazadeh et al., 2021). Their outstanding features are expected to improve sensitivity, selectivity, and efficiency in sensing applications in all sectors (Table 4.1).

Table 4.1 The performance comparison of different carbon nanomaterials with pollutants

Material	Selectivity	Detection range	LOD	Mechanism
Silver@graphene oxide nanocomposite	Hg^{2+}	10–200 μM	338 nM	Surface plasmon resonance
Graphene oxide	Pb^{2+}	0.1–20 μM	0.05 μM	Surface plasmon resonance/ etching
CuS and NiS nanoparticle-modified porous-reduced graphene oxide sheets	Hg^{2+}	0–120 nM	50 nM	Peroxidase mimic activity
Graphene oxide nanoparticles	Cr^{6+}	0.07–0.43 μM	5.8 nM	Peroxidase mimetic catalyst
rGO/PEI/Pd nanohybrids	Hg^{2+}	0.1–25 nM	0.39 nM	Peroxidase mimetic catalyst
Graphene–enzyme complex	*E. coli, S. aureus*	1 × 102 to 1 × 106 cfu mL^{-1}	100 cfu mL^{-1}, 200 cfu mL^{-1}	Enzymatic/ electrostatic
Carbon dots functionalized silver nanoparticles	Phoxim	0.1–100 μM	0.04 μM	Aggregation
Carbon dots-modified silver nanoparticles	Cu^{2+}	0.3–8 μM	0.037 μM	Aggregation
Metal nanoparticle-decorated CNTs	Specific pesticides (carbendazim (CBZ), isoproturon (ISP), glyphosate (ATZ), deltamethrin (DTM))	1–8 μM	10.8 nM for CBZ, 28.8 nM for DTM, and 16.8 nM	Peroxidase mimetic catalyst
Fullerene (carbon nanoparticles)	Cr^{6+}	1–100 μM	300 nM	Donor atoms coordination

4.6 Carbon Nanomaterials and Miniaturized Sensing Devices Interfacing

These materials' distinct electronic, optical, and mechanical properties contribute to enhanced sensor functionality. Recent developments have targeted the use of extremely lightweight, flexible devices with capabilities to function well under a variety of environmental conditions. Fabrication has been reported to include wearable sensors that could be used to monitor health metrics or identify environmental pollutants (Cano et al., 2016; Chen, 2024). Progress has been made in integrating CNMs with microfluidic systems, where the high surface area and excellent conductivity support reliable sensing performances. The use of CNMs alongside the conventional sensing approaches can also enable the possibility of faster response and enhanced sensitivity as well, which make them suitable for applications ranging from personal health monitoring to environmental analysis (Shaffaf et al., 2021; Mangaiyarkkarasi & Shanthalakshmi Revathy, 2024).

4.7 Advances in Multifunctional Colorimetric Systems

Carbon nanomaterials have been reported to have achieved considerable milestones in colorimetric sensing, particularly regarding the development of multi-analyte detection systems capable of simultaneously detecting more than one analyte. Generally, these systems rely on the photophysical properties of the nanomaterials—the surface plasmon resonance, which changes color upon reacting with the target analytes. Another improvement in such a system is the enzyme-like activity of carbon nanomaterials that can be exploited to sense a wide variety of biochemicals and pathogens with very high sensitivity. Other recent advancements include paper-based colorimetric sensors that utilize carbon dot technology for fast testing. These systems are primarily appropriate for low cost, ease of use in field applications, and timely and effective monitoring of the environment or pollutants, pathogens, or even food safety indicators. With further development, we could soon find ourselves at the cusp of transitioning to more automated and user-friendly devices that offer real-time data across several fronts.

4.8 Challenges and Future Directions

Despite the promising prospect, some challenges still lie in embracing CNMs with most sensing technologies. Scalability and cost-effectiveness, along with safety aspects, must be overcome to make nanomaterials more acceptable for use in consumer and industrial applications. Key to reliability and safety will be the development of standard protocols for synthesizing and evaluating CNMs. The improvement in selectivity and sensitivity of CNM-based sensors will also be critical in establishing these technologies as robust solutions that can function properly in such complex environmental matrices. Functionalization techniques and hybrid nanocomposites continue to be developed, and research in these areas will have pivotal implications in overcoming many current limitations. Interdisciplinary research will further drive innovation in smart sensing technologies to lead toward solutions that can address the demands of a sustainable future. In such directions, carbon nanomaterials are well-positioned to revolutionize the realms of sensing and environmental monitoring through their ability to provide efficient solutions for pollution management and health safety at a global level. Given that scientific technology advances, utilizing these materials is likely to introduce new avenues of innovation and increase the sensitivity in detecting environmental contaminants.

Acknowledgments S.Y, N.C, and A.R.P acknowledge the Council of Scientific and Industrial Research (CSIR), Govt. of India for the financial support. AcSIR, CSIR, and CSIR-CSMCRI are acknowledged for fellowship and Ph.D. degrees.

References

Ajith, M. P., Pardhiya, S., Prabhakar, A. K., & Rajamani, P. (2022, September). Ag@CDs nanohybrid: Fabrication, design of a multi-mode chemosensory probe for selective Fe^{3+} detection and logic gate operation. *Chemosphere, 303*, 135090. https://doi.org/10.1016/j.chemosphere.2022.135090

Ali, S., Chen, X., Shi, W., Huang, G., Yuan, L.-m., Meng, L., Chen, S., Zhonghao, X., & Chen, X. (2023). Recent advances in silver and gold nanoparticles-based colorimetric sensors for heavy metal ions detection: A review. *Critical Reviews in Analytical Chemistry, 53*(3), 718–750. https://doi.org/10.1080/10408347.2021.1973886

Annadhasan, M., Muthukumarasamyvel, T., Babu, V. R. S., & Rajendiran, N. (2014). Green synthesized silver and gold nanoparticles for colorimetric detection of Hg^{2+}, Pb^{2+} , and Mn^{2+} in aqueous medium. *ACS Sustainable Chemistry & Engineering, 2*(4), 887–896. https://doi.org/10.1021/sc400500z

Aragay, G., Pons, J., & Merkoçi, A. (2011). Recent trends in macro-, micro-, and nanomaterial-based tools and strategies for heavy-metal detection. *Chemical Reviews, 111*(5), 3433–3458. https://doi.org/10.1021/cr100383r

Arcudi, F., Đorđević, L., & Prato, M. (2019). Design, synthesis, and functionalization strategies of tailored carbon nanodots. *Accounts of Chemical Research, 52*(8), 2070–2079. https://doi.org/10.1021/acs.accounts.9b00249

Babazadeh, S., Bisauriya, R., Carbone, M., Roselli, L., Cecchetti, D., Bauer, E. M., Sennato, S., Prosposito, P., & Pizzoferrato, R. (2021). Colorimetric detection of chromium(VI) ions in water using unfolded-fullerene carbon nanoparticles. *Sensors, 21*(19), 6353. https://doi.org/10.3390/s21196353

Beiraghi, A., & Najibi-Gehraz, S. A. (2017, December). Carbon dots-modified silver nanoparticles as a new colorimetric sensor for selective determination of cupric ions. *Sensors and Actuators B: Chemical, 253*, 342–351. https://doi.org/10.1016/j.snb.2017.06.049

Borthakur, P., Boruah, P. K., & Das, M. R. (2021). CuS and NiS nanoparticle-decorated porous-reduced graphene oxide sheets as efficient peroxidase nanozymes for easy colorimetric detection of Hg(II) ions in a water medium and using a paper strip. *ACS Sustainable Chemistry & Engineering, 9*(39), 13245–13255. https://doi.org/10.1021/acssuschemeng.1c04203

Brisebois, P. P., & Siaj, M. (2020). Harvesting graphene oxide—Years 1859 to 2019: A review of its structure, synthesis, properties and exfoliation. *Journal of Materials Chemistry C, 8*(5), 1517–1547. https://doi.org/10.1039/C9TC03251G

Cadranel, A., Margraf, J. T., Strauss, V., Clark, T., & Guldi, D. M. (2019). Carbon nanodots for charge-transfer processes. *Accounts of Chemical Research, 52*(4), 955–963. https://doi.org/10.1021/acs.accounts.8b00673

Cano, A. M., Kohl, K., Deleon, S., Payton, P., Irin, F., Saed, M., Shah, S. A., Green, M. J., & Cañas-Carrell, J. E. (2016, June). Determination of uptake, accumulation, and stress effects in corn (Zea mays L.) grown in single-wall carbon nanotube contaminated soil. *Chemosphere, 152*, 117–122. https://doi.org/10.1016/j.chemosphere.2016.02.093

Carneiro, S. V., Oliveira, J. J. P., Rodrigues, V. S. F., Lima, J. P. O., do Nascimento, J. H. O., Santos-Oliveira, R., Fechine, L. M. U. D., Freire, R. M., & Fechine, P. B. A. (2023, June). Recent advances in nanostructured materials: A look at the applications in optical chemical sensing. *Materials Today Nano, 22*, 100345. https://doi.org/10.1016/j.mtnano.2023.100345

Chatterjee, S., Lou, X.-Y., Liang, F., & Yang, Y.-W. (2022, May). Surface-functionalized gold and silver nanoparticles for colorimetric and fluorescent sensing of metal ions and biomolecules. *Coordination Chemistry Reviews, 459*, 214461. https://doi.org/10.1016/j.ccr.2022.214461

Chen, S. (2024). Carbon nanomaterials in flexible wearable sensors. *Applied and Computational Engineering, 85*(1), 120–126. https://doi.org/10.54254/2755-2721/85/20240827

Chen, D., Feng, H., & Li, J. (2012). Graphene oxide: Preparation, functionalization, and electrochemical applications. *Chemical Reviews, 112*(11), 6027–6053. https://doi.org/10.1021/cr300115g

Chen, Z., Zhang, Z., Qi, J., You, J., Ma, J., & Chen, L. (2023, January). Colorimetric detection of heavy metal ions with various chromogenic materials: Strategies and applications. *Journal of Hazardous Materials, 441*, 129889. https://doi.org/10.1016/j.jhazmat.2022.129889

Choudhary, N., Yadav, S., Bhai, S., Sonpal, V., Ganguly, B., & Paital, A. R. (2024a). Ionic liquid-based chemodosimetric probe: The selective detection and removal of bisulfite from a pure aqueous system and potential uses in biosensing. *Materials Advances, 5*(12), 5298–5306. https://doi.org/10.1039/D4MA00198B

Choudhary, N., Yadav, S., Patel, T. R., Wakchaure, P. D., Sonpal, V., Ganguly, B., & Paital, A. R. (2024b). A regenerative dual-functional platform combining dendritic silica and anthraquinone amide: Advancing seawater lithium detection and recovery with biosensing capabilities. *Journal of Materials Chemistry A, 12*(40), 27340–27354. https://doi.org/10.1039/D4TA05025H

Dey, N., Biswakarma, D., & Bhattacharya, S. (2019). Metal complex as an optical sensing platform for rapid multimodal recognition of a pathogenic biomarker in real-life samples. *ACS Sustainable Chemistry & Engineering, 7*(1), 569–577. https://doi.org/10.1021/acssuschemeng.8b04107

Dong, Z., Wang, D., Liu, X., Pei, X., Chen, L., & Jin, J. (2014). Bio-inspired surface-functionalization of graphene oxide for the adsorption of organic dyes and heavy metal ions with a superhigh capacity. *Journal of Materials Chemistry A, 2*(14), 5034–5040. https://doi.org/10.1039/C3TA14751G

Dreyer, D. R., Park, S., Bielawski, C. W., & Ruoff, R. S. (2010). The chemistry of graphene oxide. *Chemical Society Reviews, 39*(1), 228–240. https://doi.org/10.1039/B917103G

Echeverry-Gonzalez, C. A., & Kouznetsov, V. V. (2020). Carbon dots: An insight into their application in heavy metal sensing. *Recent Progress in Materials, 03*(02), 1. https://doi.org/10.21926/rpm.2102015

Eigler, S., Andreas H. 2014. "Chemistry with Graphene and Graphene Oxide—Challenges for Synthetic Chemists." Angewandte Chemie International Edition 53 (30): 7720–38. https://doi.org/10.1002/anie.201402780.

Farooq, S., Aziz, H., Ali, S., Murtaza, G., Rizwan, M., Saleem, M. H., Mahboob, S., Al-Ghanim, K. A., Riaz, M. N., & Murtaza, B. (2022). Synthesis of functionalized carboxylated graphene oxide for the remediation of Pb and Cr contaminated water. *International Journal of Environmental Research and Public Health, 19*(17), 10610. https://doi.org/10.3390/ijerph191710610

Geng, Z., Miao, Y., Zhang, G., & Liang, X. (2023, January). Colorimetric biosensor based on smartphone: State-of-art. *Sensors and Actuators A: Physical, 349*, 114056. https://doi.org/10.1016/j.sna.2022.114056

Georgakilas, V., Otyepka, M., Bourlinos, A. B., Chandra, V., Kim, N., Kemp, K. C., Hobza, P., Zboril, R., & Kim, K. S. (2012). Functionalization of graphene: Covalent and noncovalent approaches, derivatives and applications. *Chemical Reviews, 112*(11), 6156–6214. https://doi.org/10.1021/cr3000412

Georgakilas, V., Tiwari, J. N., Kemp, K. C., Perman, J. A., Bourlinos, A. B., Kim, K. S., & Zboril, R. (2016). Noncovalent functionalization of graphene and graphene oxide for energy materials, biosensing, catalytic, and biomedical applications. *Chemical Reviews, 116*(9), 5464–5519. https://doi.org/10.1021/acs.chemrev.5b00620

Ghosal, K., & Ghosh, A. (2019, March). Carbon dots: The next generation platform for biomedical applications. *Materials Science and Engineering: C, 96*, 887–903. https://doi.org/10.1016/j.msec.2018.11.060

Gosai, A., Khondakar, K., Ma, X., & Ali, M. (2021). Application of functionalized graphene oxide based biosensors for health monitoring: Simple graphene derivatives to 3D printed platforms. *Biosensors, 11*(10), 384. https://doi.org/10.3390/bios11100384

Gupta, R., Kumar, A., Kumar, S., Pinnaka, A. K., & Singhal, N. K. (2021, February). Naked eye colorimetric detection of Escherichia coli using aptamer conjugated graphene oxide enclosed gold nanoparticles. *Sensors and Actuators B: Chemical, 329*, 129100. https://doi.org/10.1016/j.snb.2020.129100

He, Y., Hu, C., Li, Z., Wu, C., Zeng, Y., & Peng, C. (2022, March). Multifunctional carbon nanomaterials for diagnostic applications in infectious diseases and tumors. *Materials Today Bio, 14*, 100231. https://doi.org/10.1016/j.mtbio.2022.100231

Jin, J., Guo, J., Guo, J., & Li, D. (2024). Carbon-based biosensor in point of care setting. *Advanced Sensor Research, 3*(10). https://doi.org/10.1002/adsr.202400037

Kangas, M. J., Burks, R. M., Atwater, J., Lukowicz, R. M., Williams, P., & Holmes, A. E. (2017). Colorimetric sensor arrays for the detection and identification of chemical weapons and explosives. *Critical Reviews in Analytical Chemistry, 47*(2), 138–153. https://doi.org/10.1080/10408347.2016.1233805

Kaur, M., Mehta, S. K., & Kansal, S. K. (2017, June). A fluorescent probe based on nitrogen doped graphene quantum dots for turn off sensing of explosive and detrimental water pollutant, TNP in aqueous medium. *Spectrochimica Acta Part A: Molecular and Biomolecular Spectroscopy, 180*, 37–43. https://doi.org/10.1016/j.saa.2017.02.035

Kaur, B., Kaur, N., & Kumar, S. (2018, March). Colorimetric metal ion sensors—A comprehensive review of the years 2011–2016. *Coordination Chemistry Reviews, 358*, 13–69. https://doi.org/10.1016/j.ccr.2017.12.002

Kaushal, S., Pinnaka, A. K., Soni, S., & Singhal, N. K. (2021, February). Antibody assisted graphene oxide coated gold nanoparticles for rapid bacterial detection and near infrared light enhanced antibacterial activity. *Sensors and Actuators B: Chemical, 329*, 129141. https://doi.org/10.1016/j.snb.2020.129141

Khan, A., DeVoe, E., & Andreescu, S. (2023). Carbon-based electrochemical biosensors as diagnostic platforms for connected decentralized healthcare. *Sensors & Diagnostics, 2*(3), 529–558. https://doi.org/10.1039/D2SD00226D

Kim, D.-H., Cha, J.-H., Lim, J. Y., Bae, J., Lee, W., Yoon, K. R., Kim, C., Jang, J.-S., Hwang, W., & Kim, I.-D. (2020). Colorimetric dye-loaded nanofiber yarn: Eye-readable and weavable gas sensing platform. *ACS Nano, 14*(12), 16907–16918. https://doi.org/10.1021/acsnano.0c05916

Kumar, M., Kaur, N., & Singh, N. (2024). Colorimetric nanozyme sensor array based on metal nanoparticle-decorated CNTs for quantification of pesticides in real water and soil samples. *ACS Sustainable Chemistry & Engineering, 12*(2), 728–736. https://doi.org/10.1021/acssuschemeng.3c04153

Li, J., Wu, L.-J., Guo, S.-S., Fu, H.-E., Chen, G.-N., & Yang, H.-H. (2013). Simple colorimetric bacterial detection and high-throughput drug screening based on a graphene–enzyme complex. *Nanoscale, 5*(2), 619–623. https://doi.org/10.1039/C2NR32704J

Li, Z., Askim, J. R., & Suslick, K. S. (2019a). The optoelectronic nose: Colorimetric and fluorometric sensor arrays. *Chemical Reviews, 119*(1), 231–292. https://doi.org/10.1021/acs.chemrev.8b00226

Li, M., Chen, T., Gooding, J. J., & Liu, J. (2019b). Review of carbon and graphene quantum dots for sensing. *ACS Sensors, 4*(7), 1732–1748. https://doi.org/10.1021/acssensors.9b00514

Li, G., Liu, Z., Gao, W., & Tang, B. (2023, March). Recent advancement in graphene quantum dots based fluorescent sensor: Design, construction and bio-medical applications. *Coordination Chemistry Reviews, 478*, 214966. https://doi.org/10.1016/j.ccr.2022.214966

Liu, Z., Liu, B., Ding, J., & Liu, J. (2014). Fluorescent sensors using DNA-functionalized graphene oxide. *Analytical and Bioanalytical Chemistry, 406*(27), 6885–6902. https://doi.org/10.1007/s00216-014-7888-3

Liu, B., Zhuang, J., & Wei, G. (2020). Recent advances in the design of colorimetric sensors for environmental monitoring. *Environmental Science: Nano, 7*(8), 2195–2213. https://doi.org/10.1039/D0EN00449A

Lv, Q., Xia, N., Gao, L., & Han, B. (2023). Detection of mercury ions using graphene oxide sensors assisted by Ag@SiO_2. *Applied Nanoscience, 13*(6), 3787–3796. https://doi.org/10.1007/s13204-022-02555-x

Mangaiyarkkarasi, J., & Shanthalakshmi Revathy, J. (2024). Harnessing carbon nanomaterials for next-generation environmental sensing. In *Environmental applications of carbon-based materials* (pp. 144–166). https://doi.org/10.4018/979-8-3693-3625-0.ch006

Morales-Narváez, E., Hassan, A.-R., & Merkoçi, A. (2013). Graphene oxide as a pathogen-revealing agent: Sensing with a digital-like response. *Angewandte Chemie International Edition, 52*(51), 13779–13783. https://doi.org/10.1002/anie.201307740

Nanda, S. S., Yi, D. K., & Kim, K. (2016). Study of antibacterial mechanism of graphene oxide using Raman spectroscopy. *Scientific Reports, 6*(1), 28443. https://doi.org/10.1038/srep28443

Nghia, N. N., Huy, B. T., & Lee, Y.-I. (2019). Colorimetric detection of chromium(VI) using graphene oxide nanoparticles acting as a peroxidase mimetic catalyst and 8-hydroxyquinoline as an inhibitor. *Microchimica Acta, 186*(1), 36. https://doi.org/10.1007/s00604-018-3169-8

Oliveira, E., Núñez, C., Santos, H. M., Fernández-Lodeiro, J., Fernández-Lodeiro, A., Capelo, J. L., & Lodeiro, C. (2015, June). Revisiting the use of gold and silver functionalised nanoparticles as colorimetric and fluorometric chemosensors for metal ions. *Sensors and Actuators B: Chemical, 212*, 297–328. https://doi.org/10.1016/j.snb.2015.02.026

Pandey, R. R., & Chusuei, C. C. (2021). Carbon nanotubes, graphene, and carbon dots as electrochemical biosensing composites. *Molecules, 26*(21), 6674. https://doi.org/10.3390/molecules26216674

Parambil, A. M., & Rajamani, P. (2024). Carbon dots: A promising path towards environmental sustainability. *Environmental Science: Advances, 3*(11), 1513–1523. https://doi.org/10.1039/D4VA00273C

Priyadarshini, E., & Pradhan, N. (2017, January). Gold nanoparticles as efficient sensors in colorimetric detection of toxic metal ions: A review. *Sensors and Actuators B: Chemical, 238*, 888–902. https://doi.org/10.1016/j.snb.2016.06.081

Priyadarshini, E., Parambil, A. M., Rajamani, P., Ponnusamy, V. K., & Chen, Y.-H. (2023, May). Exposure, toxicological mechanism of endocrine disrupting compounds and future direction of identification using nano-architectonics. *Environmental Research, 225*, 115577. https://doi.org/10.1016/j.envres.2023.115577

Quang, D. T., & Kim, J. S. (2010). Fluoro- and chromogenic chemodosimeters for heavy metal ion detection in solution and biospecimens. *Chemical Reviews, 110*(10), 6280–6301. https://doi.org/10.1021/cr100154p

Rajput, A., Raj, S. K., Lebedeva, O. V., Chesnokova, A. N., Raskulova, T. V., & Kulshrestha, V. (2021, January). Functionalized carbon dots composite cation exchange membranes: Improved electrochemical performance and salt removal efficiency. *Colloids and Surfaces A: Physicochemical and Engineering Aspects, 609*, 125677. https://doi.org/10.1016/j.colsurfa.2020.125677

Rose, L., Anitha Mary, X., Johnson, I., Srinivasan, G., Priya, L., & Bhagavathsingh, J. (2021). Polyaza functionalized graphene oxide nanomaterial based sensor for Escherichia coli detection in water matrices. *Scientific Reports, 11*(1), 16872. https://doi.org/10.1038/s41598-021-96539-6

Shaffaf, T., Forouhi, S., & Ghafar-Zadeh, E. (2021). Towards fully integrated portable sensing devices for COVID-19 and future global hazards: Recent advances, challenges, and prospects. *Micromachines, 12*(8), 915. https://doi.org/10.3390/mi12080915

Shahdeo, D., Roberts, A., Abbineni, N., & Gandhi, S. (2020). Graphene based sensors. In *Comprehensive analytical chemistry* (pp. 175–199). https://doi.org/10.1016/bs.coac.2020.08.007

Shetti, N. P., Mishra, A., Basu, S., & Aminabhavi, T. M. (2021, June). Versatile fullerenes as sensor materials. *Materials Today Chemistry, 20*, 100454. https://doi.org/10.1016/j.mtchem.2021.100454

Shi, X., Gu, W., Zhang, C., Zhao, L., Peng, W., & Xian, Y. (2015). A label-free colorimetric sensor for Pb^{2+} detection based on the acceleration of gold leaching by graphene oxide. *Dalton Transactions, 44*(10), 4623–4629. https://doi.org/10.1039/C4DT03883E

Soda, Y., & Bakker, E. (2019). Quantification of colorimetric data for paper-based analytical devices. *ACS Sensors, 4*(12), 3093–3101. https://doi.org/10.1021/acssensors.9b01802

Song, Y., Wei, W., & Qu, X. (2011). Colorimetric biosensing using smart materials. *Advanced Materials, 23*(37), 4215–4236. https://doi.org/10.1002/adma.201101853

Song, M. G., Choi, J., Jeong, H. E., Song, K. H., Jeon, S., Cha, J. H., Baeck, S.-H., Shim, S. E., & Qian, Y. (2020, November). A comprehensive study of various amine-functionalized graphene oxides for room temperature formaldehyde gas detection: Experimental and theoretical approaches. *Applied Surface Science, 529*, 147189. https://doi.org/10.1016/j.apsusc.2020.147189

Sriram, G., Bhat, M. P., Patil, P., Uthappa, U. T., Jung, H.-Y., Altalhi, T., Kumeria, T., et al. (2017, August). Paper-based microfluidic analytical devices for colorimetric detection of toxic ions: A review. *TrAC Trends in Analytical Chemistry, 93*, 212–227. https://doi.org/10.1016/j.trac.2017.06.005

Sukma, W., Boonyakong, C., Chaiphet, T., & Chamni, S. (2024). A simple and rapid colorimetric detection of methanol, ethanol, and isopropanol and its application toward alcohol-based hand sanitizer quality control. *Green Chemistry Letters and Reviews, 17*(1). https://doi.org/10.1080/17518253.2023.2298982

Tsiotsias, A. I., Harkou, E., Charisiou, N. D., Sebastian, V., Naikwadi, D. R., van der Linden, B., Bansode, A., et al. (2025, March). Very low Ru loadings boosting performance of Ni-based dual-function materials during the integrated CO_2 capture and methanation process. *Journal of Energy Chemistry, 102*, 309–328. https://doi.org/10.1016/j.jechem.2024.11.001

Wu, Y., Jing F., Guang H., En Z., Huan-Huan Y. 2023. "Colorimetric Sensors for Chemical and Biological Sensing Applications." Sensors 23 (5): 2749. https://doi.org/10.3390/s23052749.

Wu, Z. L., Liu, Z. X., & Yuan, Y. H. (2017). Carbon dots: Materials, synthesis, properties and approaches to long-wavelength and multicolor emission. *Journal of Materials Chemistry B, 5*(21), 3794–3809. https://doi.org/10.1039/C7TB00363C

Yadav, S., Choudhary, N., Bhai, S., Bhojani, G., Chatterjee, S., Ganguly, B., & Paital, A. R. (2021). Recyclable functionalized material for sensitive detection and exceptional sorption of hexavalent chromium and permanganate ions with biosensing applications. *ACS Applied Bio Materials, 4*(8), 6430–6440. https://doi.org/10.1021/acsabm.1c00609

Yadav, S., Choudhary, N., Dash, M. R., & Paital, A. R. (2022, December). High surface area dendritic silica pairing with anthraquinone derivative: A promising single platform for dual applications of detection and remediation of nitroaromatics and copper ion. *Chemical Engineering Journal, 450*, 138042. https://doi.org/10.1016/j.cej.2022.138042

Yadav, S., Choudhary, N., Sonpal, V., & Paital, A. R. (2023a, September). Engineering excitation-independent turn-on fluorescent probe for mercury: Functionalized dendritic silica doped with red-emissive carbon dots towards simultaneous detection and remediation with biosensing application. *Chemical Engineering Journal, 471*, 144715. https://doi.org/10.1016/j.cej.2023.144715

Yadav, S., Choudhary, N., & Paital, A. R. (2023b, March). Tailored graphene/silica functional composite as signaling and adsorbent material: A sequential excimer probe with a fluoro-switching response with zinc and adenosine triphosphate. *Carbon, 205*, 527–539. https://doi.org/10.1016/j.carbon.2023.01.063

Yadav, S., Choudhary, N., & Paital, A. R. (2024a). Carbon within carbon: Growth of excitation-independent CDs within functional mesoporous carbon towards detection and adsorption of a specific nitrofuran class of antibiotics. *Journal of Materials Chemistry A, 12*(1), 202–213. https://doi.org/10.1039/D3TA05273G

Yadav, S., Choudhary, N., Vasave, A. T., Sonpal, V., & Paital, A. R. (2024b). Perylene diimide functionalized nano-silica: Green emissive material for selective probing and remediation of 4-nitrocatechol, Ru^{3+} , and Cu^{2+} with biosensing applications. *Materials Advances, 5*(22), 8937–8952. https://doi.org/10.1039/D4MA00862F

Yadav, S., Kachgunde, H. G., Choudhary, N., Wanole, G. H., Ravi, K., Biradar, A. V., & Paital, A. R. (2024c). Metal-free bi-functional cooperative catalysis: Amine and quaternary amine-functionalized dendritic fibrous nanosilica as heterogeneous catalysts for the henry reaction and CO_2 conversion. *Journal of Materials Chemistry A, 12*(37), 25022–25034. https://doi.org/10.1039/D4TA03980G

Yadav, S., Choudhary, N., Sonpal, V., & Paital, A. R. (2024d). Carbon dots-embedded silica tubes: An excitation-independent yellow-emitting turn-on probe for simultaneous detection and removal of inorganic arsenic with in vivo tracking in living organisms. *Small, 20*(11). https://doi.org/10.1002/smll.202307491

Yadav, S., Choudhary, N., Sonpal, V., Vyas, B. G., & Paital, A. R. (2025). Sunlight-mediated photocatalytic uranium extraction from seawater: A bioresistant heterojunction for highly selective uranium sensing and extraction with in vivo application. *Journal of Materials Chemistry A*. https://doi.org/10.1039/D4TA08292C

Yang, F.-Q., & Ge, L. (2023). Colorimetric sensors: Methods and applications. *Sensors, 23*(24), 9887. https://doi.org/10.3390/s23249887

Yapor, J. P., Alharby, A., Gentry-Weeks, C., Reynolds, M. M., Mashud Alam, A. K. M., & Li, Y. V. (2017). Polydiacetylene nanofiber composites as a colorimetric sensor responding to Escherichia coli and PH. *ACS Omega, 2*(10), 7334–7342. https://doi.org/10.1021/acsomega.7b01136

Zangeneh Kamali, K., Pandikumar, A., Jayabal, S., Ramaraj, R., Lim, H. N., Ong, B. H., Bien, C. S. D., Kee, Y. Y., & Huang, N. M. (2016). Amalgamation based optical and colorimetric sensing of mercury(II) ions with silver@graphene oxide nanocomposite materials. *Microchimica Acta, 183*(1), 369–377. https://doi.org/10.1007/s00604-015-1658-6

Zarlaida, F., & Adlim, M. (2017). Gold and silver nanoparticles and indicator dyes as active agents in colorimetric spot and strip tests for mercury(II) ions: A review. *Microchimica Acta, 184*(1), 45–58. https://doi.org/10.1007/s00604-016-1967-4

Zeebaree, A. Y. S., Zebari, O. I. H., Zeebaree, S. Y. S., Thanoon, R. D., & Yaseen, Y. N. (2023). Current contributions of novel nanoparticles-based colorimetric sensors for detection of SCN ions in different aqueous models. A review. *Arabian Journal of Chemistry, 16*(12), 105297. https://doi.org/10.1016/j.arabjc.2023.105297

Zhang, S., Zhang, D., Zhang, X., Shang, D., Xue, Z., Shan, D., & Lu, X. (2017). Ultratrace naked-eye colorimetric detection of Hg^{2+} in wastewater and serum utilizing mercury-stimulated peroxidase mimetic activity of reduced graphene oxide-PEI-Pd nanohybrids. *Analytical Chemistry, 89*(6), 3538–3544. https://doi.org/10.1021/acs.analchem.6b04805

Zheng, M., Chenge W., Yingying W., Wei W., Shuang M., Xiaohan S., Jiang H. 2018. "Green Synthesis of Carbon Dots Functionalized Silver Nanoparticles for the Colorimetric Detection of Phoxim." Talanta 185 (August):309–15. https://doi.org/10.1016/j.talanta.2018.03.066.

Zhou, S., Xu, H., Gan, W., & Yuan, Q. (2016). Graphene quantum dots: Recent progress in preparation and fluorescence sensing applications. *RSC Advances, 6*(112), 110775–110788. https://doi.org/10.1039/C6RA24349E

Zhu, Y., Murali, S., Cai, W., Li, X., Suk, J. W., Potts, J. R., & Ruoff, R. S. (2010). Graphene and graphene oxide: Synthesis, properties, and applications. *Advanced Materials, 22*(35), 3906–3924. https://doi.org/10.1002/adma.201001068

Chapter 5
Carbon-Nano-Architectonic for Fluorescence-Based Detection of Environmental Contaminants and Logic Gate Operations

Prayasee Baruah, Jouharsha Afthab, Ramjayakumar Venkatesh, and Mohamed Nabeel Mattath

Abstract Carbon nano-architectures, which comprise a myriad variety of graphene, graphene oxide, and carbon nanotubes, are emerging as powerful resources in the development of advanced sensing technologies owing to their distinctive properties, such as high surface area, tunable electronic and optical properties, and environmental compatibility. Because of their exceptional characteristics, they are excellent candidates for fluorescence-based detection of environmental contaminants and the design of computational operations toward environmental monitoring. The precise sensing abilities of carbon nano-architectures can be attributed to different fluorescence-based sensing mechanisms, such as intramolecular charge transfer (ICT), excited-state intramolecular proton transfer (ESIPT), photoinduced electron transfer (PET), and Förster resonance energy transfer (FRET), which are pivotal in the detection of common environmental toxicants like heavy metals, organic pollutants, and per- and polyfluoroalkyl substances (PFAS). Carbon nanomaterials also enable the application of molecular computing principles to create elementary logic gates to advanced logic devices or smart technologies that can accomplish complex environmental detection tasks in real-time. Furthermore, these nano-architectures also serve as catalytic scaffolds for promulgating different chemical and biological reactions vital for environmental preservation and healthcare. However, in spite of their potential, the practical application of carbon nano-architectures remains hindered by issues of stability, scalability, and matrix effects. To overcome these hur-

P. Baruah
Department of Chemistry, IIT Hyderabad, Kandi, Telangana, India

J. Afthab · M. N. Mattath (✉)
School of Chemical Science and Engineering, Tongji University, Shanghai, P. R. China
e-mail: mattathnabeel@tongji.edu.cn

R. Venkatesh
Bioorganic Chemistry Laboratory, New Chemistry Unit, Jawaharlal Nehru Centre for Advanced Scientific Research (JNCASR), Bengaluru, Karnataka, India

A. M. Parambil et al. (eds.), *Carbon*,
https://doi.org/10.1007/978-3-031-90613-8_5

dles, future efforts must focus on improving functionalization strategies and integrating these systems with smart technologies to enhance their sensitivity and reliability, improve their synthesis and fabrication process, and augment overall performance in environmental sensing applications. Their integration into economic and portable sensor systems holds the potential for real-time monitoring of ecological and physiological parameters, resulting in sustainable and more targeted advancements.

Keywords Carbon nanomaterials · Fluorescence sensing · Environmental monitoring · Logic gate operations · Pollutant detection · Energy transfer mechanisms

Abbreviations

Ag@CDs	Silver carbon dot nanohybrid
CDs	Carbon dots
CNO	Carbon nano-onions
CNPs	Carbon nanoparticles
CNTs	Carbon nanotubes
D-A	Donor–acceptor
DET	Dexter energy transfer
dex-fl	Dextran fluorescein
DMA	Dimethylacridine
DOTA	Tetraxetan
DPA	Diphenylacridine
EDTA	Ethylenediaminetetraacetic acid
ESIPT	Excited-state intramolecular proton transfer
Fe_3O_4@NH_2&F_{13}	Surface-functionalized magnetic nanoparticles
FRET	Förster resonance energy transfer
FW-CDs	Flowers of wintersweet carbon dots
GO	Graphene oxide
GQDs	Graphene quantum dots
GRH	Graphene-rhodamine
HOMO	Highest occupied molecular orbital
ICT	Intramolecular charge transfer
IFE	Inner filter effect
IoT	Internet of things
LOD	Limit of detection
LUMO	Lowest unoccupied molecular orbital
MWCNTs	Multi-walled carbon nanotubes
PEG	Polyethylene glycol
PEI	Polyethyleneimine
PET	Photoinduced electron transfer
PFA	Perfluoroalkyl substances
PFAS	Per- and polyfluoroalkyl substances

PF-DBT-Im	Conjugated polymer
PFOS	Perfluorooctanesulfonate
PIFE	Primary inner filter effect
PyDMA	Pyridine dimethylacridine
PyDPA	Pyridine diphenylacridine
rGO	reduced graphene oxide
RPT	Reversible photon transfer
SET	Surface electron transfer
SIFE	Secondary inner filter effect
SWCNT	Single-walled carbon nanotubes
TTHA	Triethylenetetramine-*N*, *N*, *N*′, *N*″, *N*‴, *N*‴-hexaacetic acid

5.1 Introduction

5.1.1 Overview of Carbon-Based Nanomaterials

Carbon, a cornerstone of Earth's civilization and one of the most prevalent substances in the human body and environment, remains distinguished due to its incomparable ability to form chemical bonds. The development of nanomaterials based on carbon has revealed their potential across multidisciplinary domains, including energy storage and production, water purification, and biomedical innovation. Carbon nanomaterials encompass a myriad variety of versatile materials that are characterized by their nanoscale dimensions and fascinating physical, chemical, and mechanical properties. In the past decade, these materials have gained prominence due to their multipronged applicability in the fields of nanotechnology, materials science, electronics, biomedicine, and environmental redress (Zhang et al., 2016; Rao et al., 2021; Priyadarshini et al., 2023). Graphene, a single-atom-thick layer of carbon atoms arranged in a two-dimensional honeycomb lattice, displays exceptional strength and is a precursor for advanced carbon derivatives. However, graphene oxide (GO), which is produced when graphene is oxidized, is further reduced to synthesize graphene quantum dots (GQDs) and reduced graphene oxide (rGO) (Li et al., 2021). Each of these nanomaterials is meticulously tailored to suit specific applications. The large surface area, strong adsorption capacity, and high conductivity of graphene nanomaterials make them useful candidates for biosensing, in vitro and in vivo imaging, detection of environmental contaminants, drug and gene delivery, photothermal therapy, and antibacterial and antimicrobial applications (Gu et al., 2022). However, despite the implementation of functional groups on GO/rGO to enhance water solubility, targeting, and biocompatibility, apprehensions regarding their in vivo and in vitro toxicity effects and their protein adsorption and hemolysis capacities remain a challenge. Functional modifications, enhanced solubility, and synthesis of novel derivatives hold promise for mitigating these limitations (Izquierdo-García et al., 2024).

Another revolutionary carbon nanomaterial is the carbon nanotubes (CNTs), whose remarkable chemical and physical characteristics are closely linked to those of fullerenes and graphene, despite differences in their synthesis routes (Takakura et al., 2019). With the rising demand for post-silicon electronics, the fabrication and purification of single-walled carbon nanotubes (SWCNTs) by chirality, which dictates their atomic and electronic properties, is gaining critical importance. In recent work, alkyl cellulose has been used as a dispersant in organic solvents to select high-quality semiconducting SWCNTs (Yagi et al., 2024). Fullerenes, the earliest discovered carbon nanostructures, have paved the way for numerous derivatives and composites. However, the successful synthesis of monolayer fullerene networks represents a potential advancement in photocatalytic materials due to their large surface area with abundant active sites, compatibility with 2D materials for heterojunction formation, and the hydrogen storage potential of the C_{60} cages (Hou et al., 2022). Carbon nanoparticles (CNPs), with their tunable surface properties and high surface area, have enabled applications in imaging, drug delivery, and catalysis (AlMalki et al., 2022). CNTs and graphene remain at the forefront of research in nanotechnology, holding immense promise across a diverse range of applications. Activated carbon, yet another versatile form, finds significant applications in adsorption, separation, and catalytic processes (Li et al., 2024). Together, these carbon-based materials exemplify the extraordinary adaptability, transformative potential, and widespread utility of carbon chemistry.

5.1.2 Properties of Carbon Nanomaterials

Carbon dots (CDs), GQDs, fluorographenes, and CNTs are excellent candidates for environmental and biomedical applications, especially sensing. Their high surface area, which paves the way for significant interactions with target molecules, electric and optical properties that can be significantly modulated to tailor to specific applications, and their strength and durability contribute to their emergence as key resources in the sensing and diagnostic milieu (Wang et al., 2023). The high surface area CNTs, which contain a tubular structure, provide an expansive interface for gas adsorption and pollutant binding, thereby augmenting their sensitivity to environmental sensing (Norizan et al., 2020). GQDs and CNPs, however, demonstrate substantial surface-to-volume ratios, making them efficient scaffolds for chemical and biological functionalization, detecting contaminants and toxicants even at trace levels (Speranza, 2021).

Carbon nanomaterials, such as CDs and GQDs, exhibit low toxicity, biocompatibility, and water solubility, making them ideal for environmentally friendly applications like water quality monitoring (Đorđević et al., 2022; Praseetha et al., 2024). Through surface modifications and controlled synthesis, they can be tailored to meet high environmental safety standards, while fluorographenes provide a stable, degradation-resistant platform for prolonged sensing applications. CDs and GQDs exhibit strong photoluminescence, which can be tuned through variations in their polarity, nearby environment, quantum confinement effects, and surface passivation. A measurable response to their interaction with analytes, such as quenching or enhancement in their fluorescence intensity or a change in their fluorescence lifetime upon contact with specific analytes like heavy metals, organic pollutants, or pesticides (Zong et al., 2014; LeCroy et al., 2014; Parambil et al., 2024) makes them well-suited for sensing applications. Fluorographenes, which have a modified electronic structure due to the doping of graphene with fluorine, display altered properties that make them more appropriate for detecting oil-based pollutants and volatile organic compounds (Nair et al., 2010). These optical features contribute to developing efficient, low-cost, rapid detection systems for environmental monitoring. These properties position carbon nanomaterials as pivotal elements in developing advanced environmental sensing technologies.

5.1.3 Summary of the Growing Significance of Environmental Sensing

Carbon nanomaterials by virtue of their nanoscale dimensions and functional versatility are revolutionizing the field of environmental sensing, addressing the increasing demand for accurate and real-time monitoring of contaminants in air, water, and soil (Parambil & Rajamani, 2024). The formation of these nanomaterials occurs predominantly through multiple intramolecular and intermolecular bonds that carbon is capable of forming as well as through various allotropes of carbon. They can also be synthesized through various structural and functional modifications in the form of nanospheres, nanocapsules, nanotubes, and nanogels apart from CDs and GQDs (Fig. 5.1).

CDs and GQDs were among the first nanomaterials to gain prominence because of their excellent photoluminescence, water solubility, and biocompatibility properties. Their ability to generate rapid fluorescence responses due to their tunable optical properties and quantum confinement effect makes them excellent sensors for detecting specific analytes. This property has been used particularly in detecting heavy metals, like lead and mercury, and organic pollutants in waterbodies (Kumar et al., 2021; Xu et al., 2022b). Similarly, fluorographenes display modulated electronic properties and hydrophobicity, making them well-suited for sensing oil-based and volatile organic contaminants. This modified chemistry of their surface areas boosts selectivity and detection efficiency, especially in detecting industrial emissions and oil spills (Elhenawy et al., 2022). The tubular architecture with a high surface area of CNTs enables enhanced signal transduction, which results in remarkable electrical conductivity and mechanical stability, making them indispensable tools in electrochemical sensing. CNT-based sensors are highly efficient for the selective detection of air borne pollutants like ammonia, nitrogen dioxide, and carbon monoxide, thereby providing rapid, low-cost, and real-time solutions for monitoring air quality (Srinivasreddy & Kadeppagari, 2021). In parallel, CNPs, with

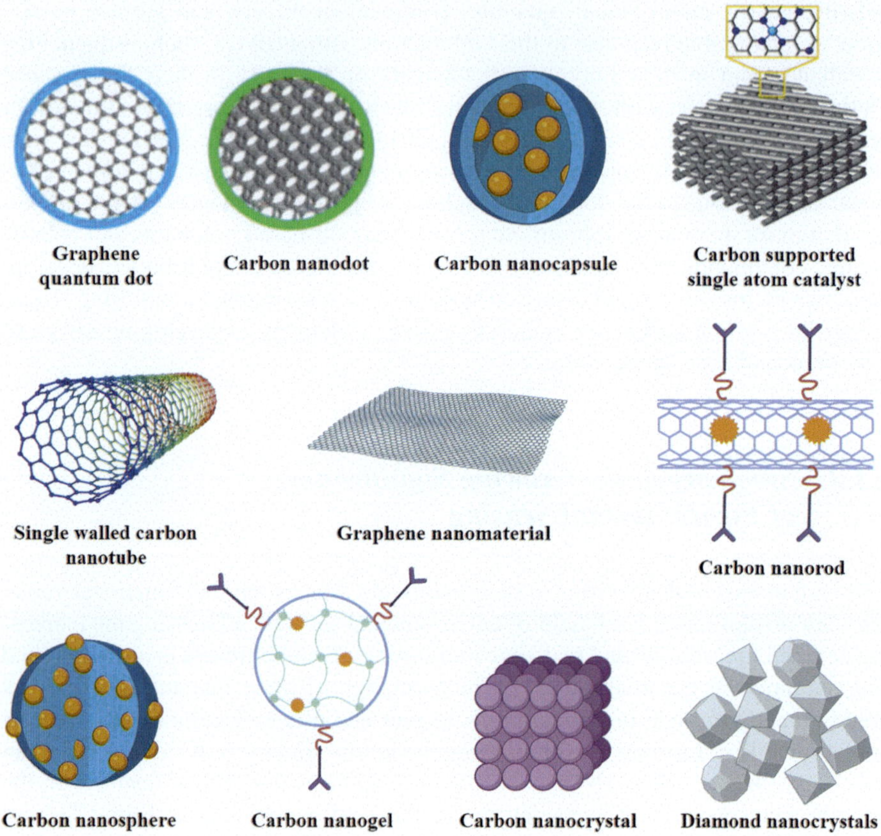

Fig. 5.1 Types of commonly synthesized and utilized carbon nanomaterials

their large surface-to-volume ratio and functionalization flexibility, bind selectively to specific analytes. These have been judiciously leveraged in detecting organic contaminants and pesticides in soil, thereby aiding in soil health assessment and contributing to agricultural redress.

The growing applicability and implications of carbon nanostructures outspread their sensitivity and selectivity. Their reasonably low cost, toxicity nature, scalability, biocompatibility, and ease of amalgamation into portable devices make them idyllic contenders for sensing technologies. Fluorescence-based systems using CDs and GQDs are increasingly employed for on-site water quality testing, and CNT-based gas sensors have enabled the development of compact air monitoring devices (Li et al., 2023). Additionally, fluorographenes and CNTs are highly efficient in precisely detecting oil spill contamination in marine ecosystems (Nair et al., 2010). The quest for environmental sustainability, being the need of the hour and the centerpiece of momentous research for many decades, emphasizes carbon nanomaterials' potential to address environmental challenges. As progress is made in material engineering and surface functionalization, the abilities of carbon nano-architectures are expected to rise further, aiding multi-analyte detection and augmenting real-time detection systems. Integrating these nanomaterials into environmental sensing platforms implies a technological marvel which would result in efficient, low-cost, and scalable solutions for pollution control and environmental sustainability. Their multifunctional nature and adaptability will ensure that they remain the centerpiece of integrated and innovative concepts to safeguard ecosystems and healthcare.

5.2 Fundamentals of Fluorescence-Based Detection

5.2.1 Introduction to Fluorescence in Detection

Fluorescence-based probes detect target compounds through weak molecular interactions or chemical reactions with the analyte. Fluorescence-based detection strategies are widespread analytical techniques that apply the principles of photoluminescence to quantify target molecules with high sensitivity and remarkable specificity. In principle, photon absorption using an appropriate light source induces a femtosecond transition of an electron from the ground state to an excited singlet state. The excited molecule relaxes and returns to the energetically favorable ground state either through non-radiative decay in the form of vibrational relaxation or by radiative decay, emitting a photon. The efficacy of fluorescence emission can be measured by two critical parameters: quantum yield and excited-state fluorescence lifetime (Valeur & Berberan-Santos, 2013; Lakowicz, 2006). Quantum yield is a quantitative measure of a molecule's fluorescence efficiency and is denoted by the ratio of emitted photons to absorbed photons.

Conversely, fluorescence lifetime indicates the time duration an excited molecule remains in the excited state before its return to the ground state through radiative decay. Both properties depend on the local molecular environment and are sensitive to changes in pH, solvent polarity and viscosity, ion concentration, or the presence of specific molecules that can act as quenchers (Lakowicz, 2006; Rohman et al., 2019a, 2019b; Rajendran et al., 2024). This sensitivity of probes to their local environment positions fluorescence as a powerful tool for probing chemical and biological interactions with remarkable precision.

Fluorescent techniques such as Förster resonance energy transfer (FRET), fluorescence quenching, and wavelength shifts form the backbone of fluorescence-based applications. When a quencher molecule binds to the probe, forming a complex or colliding with it in a dynamic fashion, the fluorescence of the parent probe decreases, a phenomenon called fluorescence quenching. Conversely, fluorescence enhancement arises when pollutants induce structural or electronic changes in the fluorophore, increasing emission intensity (Chen et al., 2023b; Ariga et al., 2017; Baruah et al., 2022). Both fluorescence quenching and enhancement play important roles in detecting environmental contaminants, offering rapid, efficient, sensitive, and selective approaches for environmental monitoring (Li et al., 2022). These processes provide a quantifiable signal where the degree of quenching or enhancement corroborates directly with the concentration of the pollutant, even in trace amounts. Furthermore, the selectivity of fluorescence-based sensors can be tuned by functionalizing fluorophores with specific moieties to target specific pollutants, enabling rapid, cost-effective, and in situ monitoring of environmental hazards.

Along with sensing environmental perils, these mechanisms also allow the detection of trace concentrations of analytes in complex biological or chemical systems. The scope of fluorescence-based detection techniques has significantly increased in recent years, attributed to advancements in fluorophore design, including organic dyes, quantum dots, and genetically encoded fluorescent proteins. The structure of fluorescent probes can be rationally modified to exhibit highly emissive spectrum with photostability, tunable optical properties, thereby augmenting their utility in medical diagnostics, environmental sensing and high-throughput screening. Consequently, fluorescence-based detection has become an indispensable tool in modern analytical, environmental, and biological sciences because it can efficiently harness fluorescence to report molecular events.

5.2.2 Fluorescence Detection Mechanisms

By utilizing the fundamental molecular interactions that define the behavior of the fluorophore and its response to external triggering stimuli, many fluorescence-based mechanisms have been designed for detection or sensing applications. One pivotal mechanism is intramolecular charge transfer (ICT), which occurs when a fluorophore experiences an uneven redistribution of electron density between electron-donating and electron-accepting moieties in the same molecule (Fig. 5.2a). When excited, ICT causes changes in the electronic environment around the fluorophore, resulting in a shift in the fluorescence emission wavelength or intensity (Zhang et al., 2016). An electron in the excited-state in a fluorophore can undergo ICT, traversing across from an electron-donating group to an electron-accepting group within the same molecule, thereby affecting its fluorescence properties (Felouat et al., 2015). ICT-based systems are particularly useful for detecting changes in polarity, pH, or ion concentrations, which make them appropriate for sensing applications. ICT can substantially boost the sensitivity of carbon nanomaterials. The electron transfer within a molecule alters the electronic structure of the nanomaterial like CDs and GQDs, often resulting in changes in emission properties like emission wavelength shift or enhanced/quenched fluorescence intensity.

In the context of contaminant detection, when nanomaterials are exposed to pollutants like heavy metals or organic compounds, their electronic properties suffer a change due to interactions with the contaminant, triggering a measurable fluorescence response that is susceptible to disruption by metal ions, such as copper or mercury. Donor–acceptor (D-A) type organic luminescent materials, known for

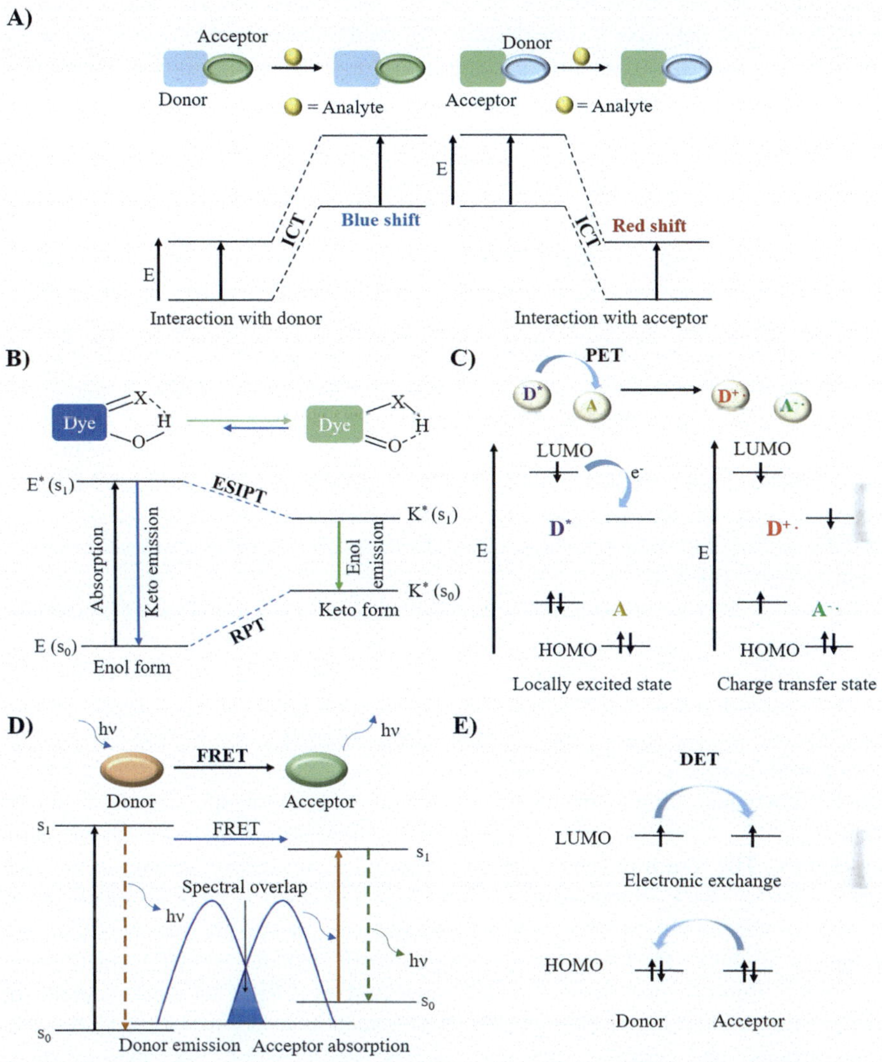

Fig. 5.2 Electron transfer mechanisms involved in various fluorescence processes of (**a**) ICT (intramolecular charge transfer), (**b**) ESIPT (excited-state intramolecular proton transfer), (**c**) PET (photoinduced electron transfer), (**d**) FRET (Förster resonance energy transfer), and (**e**) DET (Dexter energy transfer). *RPT* reversible photon transfer, *LUMO* lowest unoccupied molecular orbital, *HOMO* highest occupied molecular orbital

their ICT properties, are commonly used to create ratiometric fluorescence probes. The interaction of analytes with these probes modulates ICT, leading to changes in emission properties due to changes in charge separation. Recently, two D-A fluorescent probes, pyridine diphenylacridine (PyDPA) and pyridine dimthylacridine (PyDMA), were synthesized using diphenylacridine (DPA) and dimethylacridine

(DMA) as donor moieties (Miao et al., 2024). The design of PyDPA and PyDMA facilitated highly selective visible detection of methamphetamine, a nervous system stimulant, making them promising candidates for on-site, real-time methamphetamine monitoring.

Surface modifications of carbon nanomaterials allow for the construction of sensitive and selective sensors that can detect trace levels of different groups of contaminants at the same time. Moreover, ICT is a powerful tool in developing high-efficiency, sensitive sensing platforms for environmental analysis, which can demonstrate high specificity, rapid response, and low effects from complex sample matrices. Doping of CDs with heteroatoms such as N, P, S, or B can improve the electronic properties of their surface functionalities by making intramolecular charge transfer more efficient and promoting charge displacement to neighboring carbon atoms. Researchers employed phosphorus-doped carbon nano-onions (CNOs) to construct highly sensitive NH_3 sensing devices (Raghu et al., 2019). Chemisorbed oxygen on the P-doped CNOs played a key role in the sensing mechanism by facilitating charge transfer from carbon to oxygen. This model, which was proposed since the discovery of CDs and is widely acknowledged, remains supported in theory, but definitive experimental validation that substantiates this theory remains sparse till date.

Excited-state intramolecular proton transfer (ESIPT) is a significant photophysical mechanism in fluorescence-based sensing, particularly for the selective detection of environmental contaminants (Fig. 5.2b). ESIPT refers to the transfer of a proton from a donor group, like a hydroxyl or amino group, to an acceptor group, such as a carbonyl or imine group, within a molecule (Padalkar & Seki, 2016). The fluorophore undergoes intramolecular hydrogen bonding-mediated ultrafast proton transfer when absorbing a photon, giving rise to a tautomeric species with different photophysical properties, such as changes in fluorescence intensity or emission wavelength. The return of the proton to the initial donor occurs through reversible proton transfer (RPT). An advantage of the ESIPT process is its high sensitivity to the local environment, such as pH, solvent polarity, or specific analytes that can disrupt or modulate the hydrogen bonding network critical for ESIPT, resulting in measurable fluorescence changes and thereby making ESIPT an ideal mechanism for detecting contaminants (Yang et al., 2016), such as heavy metal ions, aromatic hydrocarbons, and reactive chemical species. ESIPT-based fluorophores often exhibit dual-emission characteristics corresponding to the normal and tautomeric forms, enhancing their detection capabilities by providing ratiometric signals for more accurate quantification. The robustness of ESIPT systems stems from their high sensitivity, large Stokes shift, environmental responsiveness, minimal interference from background fluorescence, and ability to detect even in complex media like water, soil, or industrial effluents. By modulating hydrogen bonding sites in ESIPT-active probes, the selective and sensitive detection of specific environmental contaminants can be achieved, making ESIPT invaluable for environmental sustainability and pollution control. A recent study observed the ESIPT process in resorcinol-based CDs and 2,3-dichloro-5,6-dicyano-*p*-benzoquinone (Bai et al., 2022b). Using ESIPT-induced fluorescence sensitivity, a photoswitchable anti-counterfeiting platform responsive to dimethyl sulfoxide and ammonia vapor was developed with a highly selective and sensitive picric acid sensing system.

Photoinduced electron transfer (PET) is another fluorescence-based mechanism that had widespread use in diagnostic mediums like cancer diagnosis, wherein electron transfer occurs between a fluorophore and analyte or within different moieties within the same molecule upon photoexcitation (Fig. 5.2c) (Escudero, 2016). PET is widely employed to design "turn-on" or "turn-off" fluorescence sensors to detect environmental pollutants. When heavy metal ions, pesticides, or organic toxicants interact with a fluorophore, they either enhance or inhibit PET, resulting in quantifiable fluorescence intensity changes. The presence of metal ions can inhibit PET by coordinating with electron-donating groups (fluorescence "turn-on"), thereby acting as sensors for these toxicants (Zhang, 2018). Carbon nanomaterials, being eco-friendly, cost-effective, and exhibiting remarkable photostability, tunable emission profiles, and large surface areas to accommodate analyte interactions, are particularly well-suited for detecting pollutants with high sensitivity and selectivity, contributing efficiently toward environmental monitoring and remediation. Researchers achieved control over biodistribution and tissue-selective targeting using inorganic SWNTs noncovalently coated with polyethylene glycol (PEG) and functionalized with cyclic Arg-Gly-Asp peptides, DOTA (tetraxetan) for Cu chelation and PET imaging (Welch et al., 2009). In this study, PET results showed effective targeting of $\alpha v\beta 3$ integrin-positive tumors in mice, with significant liver uptake.

Forster resonance energy transfer (FRET), involving non-radiative energy transfer between fluorophores, finds application in the detection of various contaminants, such as heavy metals and organic pollutants (Fig. 5.2d) (Ha et al., 2024). FRET-based sensors operate on the principle of the donor molecule absorbing light and subsequently transferring energy to an adjacent acceptor molecule within an allowed distance of 1–10 nm. The distance between the donor and acceptor and the relative orientation of their dipoles significantly influence this energy transfer, making it ideal for detecting molecular interactions caused by pollutants or environmental changes (Algar et al., 2019). FRET can monitor interactions of contaminants with receptors or sensors. Heavy metal binding can induce conformational changes, moving the position of the donor and acceptor to make them more favorable toward FRET and producing a fluorescence signal. Similarly, organic pollutants can affect the fluorescence properties of FRET systems. This enables the design of multiplexed sensors that can simultaneously detect and distinguish different contaminants based on their unique effects on energy transfer. These attributes of FRET-based sensing modalities highly suitable for real-time, in situ environmental monitoring of complex pollutant mixtures. In a recent study, researchers developed a user-friendly, solution-based aptasensor for rapid, sensitive detection of low-level lactoferrin. This biosensor utilizes a FRET quenching mechanism, in which CDs are used as the FRET donor, while GO nanosheets, with excellent quenching properties, are employed as the fluorescence quencher (Zhang & Zhang, 2022).

Another mechanism of energy transfer exhibited by molecular systems is Dexter energy transfer (DET), also known as spin-forbidden energy transfer (Fig. 5.2e). DET involves the exchange of excited-state energy between two molecules through direct electron exchange, needing significant orbital overlap (Kancherla et al., 2022; Bai et al., 2020). This short-range mechanism decays exponentially with distance and typically occurs over a short range of less than 10 Å. Although this mechanism is often confused with and sometimes considered a subset of FRET, unlike FRET, DET does not rely on dipole–dipole interactions and instead requires close molecular proximity. DET is mostly seen in organic photovoltaics and triplet–triplet energy transfer processes. Surface electron transfer (SET) is another mechanism that occurs when energy is transferred between a donor molecule and a nearby metallic surface, such as gold or silver nanoparticles, via long-range dipole–surface interactions (Yun et al., 2005). SET is a non-radiative and typically follows a distance dependence, making it highly sensitive to the separation between the donor and the metallic surface.

During a fluorescence measurement, the observed fluorescence signal is sometimes disturbed due to an artifact called the inner filter effect (IFE) in fluorescence spectroscopy. IFE is an optical phenomenon that affects fluorescence measurements in samples. It is a consequence of light absorption (either excitation or emission wavelength) by the fluorophore or other absorbing species in the solution (Xu et al., 2018; Baruah et al., 2019). IFE presents a significant challenge to the accuracy of obtained spectra when the absorption and emission spectra of the fluorophore overlay, especially in solutions with high concentrations of fluorophores or other absorbing entities. The effect can be categorized into two types: the primary inner filter effect (PIFE), where the excitation light is absorbed before reaching all the fluorophores, and the secondary inner filter effect (SIFE), where the emitted fluorescence is reabsorbed by the fluorophore itself or other absorbing moieties present in the solution (Lakowicz, 2006; Valeur & Berberan-Santos, 2013). Both PIFE and SIFE result in a reduction in fluorescence signal, perturbing and leading to incorrect conclusions about the quantitative relationships between fluorescence intensity and concentration. Different strategies, such as sample dilution, front-face geometry, or mathematical corrections based on absorbance, mitigate IFE and ensure more accurate and precise fluorescence measurements. A well-planned experimental design is imperative to minimize this artifact in fluorescence spectroscopy.

5.2.3 Fluorescence-Based Detection of Contaminants

Heavy metals such as cadmium (Cd^{2+}), mercury (Hg^{2+}), nickel (Ni^{2+}), lead (Pb^{2+}), chromium (Cr) (particularly Cr^{6+} and Cr^{3+}), arsenic (As) (although technically a metalloid, it is often included), copper (Cu^{2+}), zinc (Zn^{2+}), cobalt (Co^{2+}), manganese (Mn^{2+}), and iron (Fe^{3+}) are dangerous for their detrimental effects on environmental and human health, necessitating the development of spontaneous, feasible, and efficient techniques for their detection. Heavy metal ions like Cd^{2+}, Hg^{2+}, Ni^{2+}, and Pb^{2+} are present in water resources, food grains, and the environment, posing serious threats to human health. Cd^{2+} exposure is particularly alarming due to its accumulation in water and food sources, leading to kidney dysfunction, bone damage, and increased cancer risk (Kumar & Sharma, 2019). Hg^{2+} exists in organic and inorganic forms, with methylmercury being the environment's most toxic and commonly encountered form (Rana, 2008). Ni^{2+}, used extensively in battery manufacturing and catalytic processes, contaminates waterbodies and, upon exposure, can cause dermatitis, respiratory failure, and cardiovascular diseases. Pb^{2+} pollution from industrial emissions and old water pipes affects the nervous system, kidneys, liver, and heart, leading to cognitive impairments and cardiovascular issues (Wani et al., 2015).

Multi-mode sensing has recently gained significant attention. A study shows that a sustainable and green-synthesized CD probe derived from waste biomass, specifically the flowers of wintersweet (FW-CDs), was synthesized and employed for the parallel and semi-quantitative visual detection of Cr(VI) and Fe^{3+} (Xia et al., 2022) (Fig. 5.3a). The system exhibited wide linearity ranges and a limit of detection (LOD) of 0.07 μM for Cr(VI) and 0.15 μM for Fe^{3+}. The quenching mechanism for Cr(VI) was identified as an IFE, while for Fe^{3+}, it was due to ground-state complex formation. The quenching of FW-CD-based hydrogel fluorescence by Cr(VI) and Fe^{3+} was reversed by the subsequent addition of Ba^{2+} and F^-, respectively. The FW-CD nanoprobe was highly effective and reliable for quantitatively determining Cr(VI) in industrial effluent and Fe^{3+} in oral iron supplements.

A study described the development of a novel fluorescent "turn-on" method for detecting organic pollutants in wastewater by using a composite of reduced graphene oxide (rGO) and polystyrene sulfonate (Mitra & Saha, 2017). Dextran fluorescein (dex-FL) acts as a fluorescent probe, which is initially quenched due to energy transfer while being attached to the rGO surface via weak interactions (Fig. 5.3b). When organic pollutants are present, π–π interactions of the pollutants with rGO compete for the binding site, displacing dex-FL. This prevents energy transfer, thus causing the deactivated dex-FL fluorescence to activate back, and the intensity is proportional to the concentration of pollutants. In turn, the electronic properties of the pollutant affect the strength of this interaction and, hence, the fluorescence recovery. Therefore, electron-deficient molecules such as picric acid show strong interaction with rGO, while electron-rich molecules like 1-naphthol interact weakly, allowing for selective detection. This approach presents an innovative and integrated approach for sensitive and potentially selective monitoring of organic contaminants in wastewater.

Per- and polyfluoroalkyl substances (PFAS), a class of fluorinated surfactants, are commonly used in industrial applications, such as pesticide formulations, textile waterproofing, oil extraction, and in the manufacture of cosmetics, firefighting foams, and food packaging (Evich et al., 2022). Their widespread use can be attributed to the strength of the carbon-fluorine bond, their chemical and thermal stability, and their oil- and water-repellent properties. Nevertheless, emerging evidence of probable adverse effects of PFAS on human health and the environment has brought forth the need for advanced research on these chemicals and stricter regulatory oversight. To address these challenges, researchers developed a novel triple-channel optical assay for perfluorooctanesulfonate (PFOS) using CDs (Chen et al., 2019c). This strategy exploits the unique optical properties of CDs, which demonstrate measurable changes in fluorescence signals, UV-Vis absorption, and resonance light

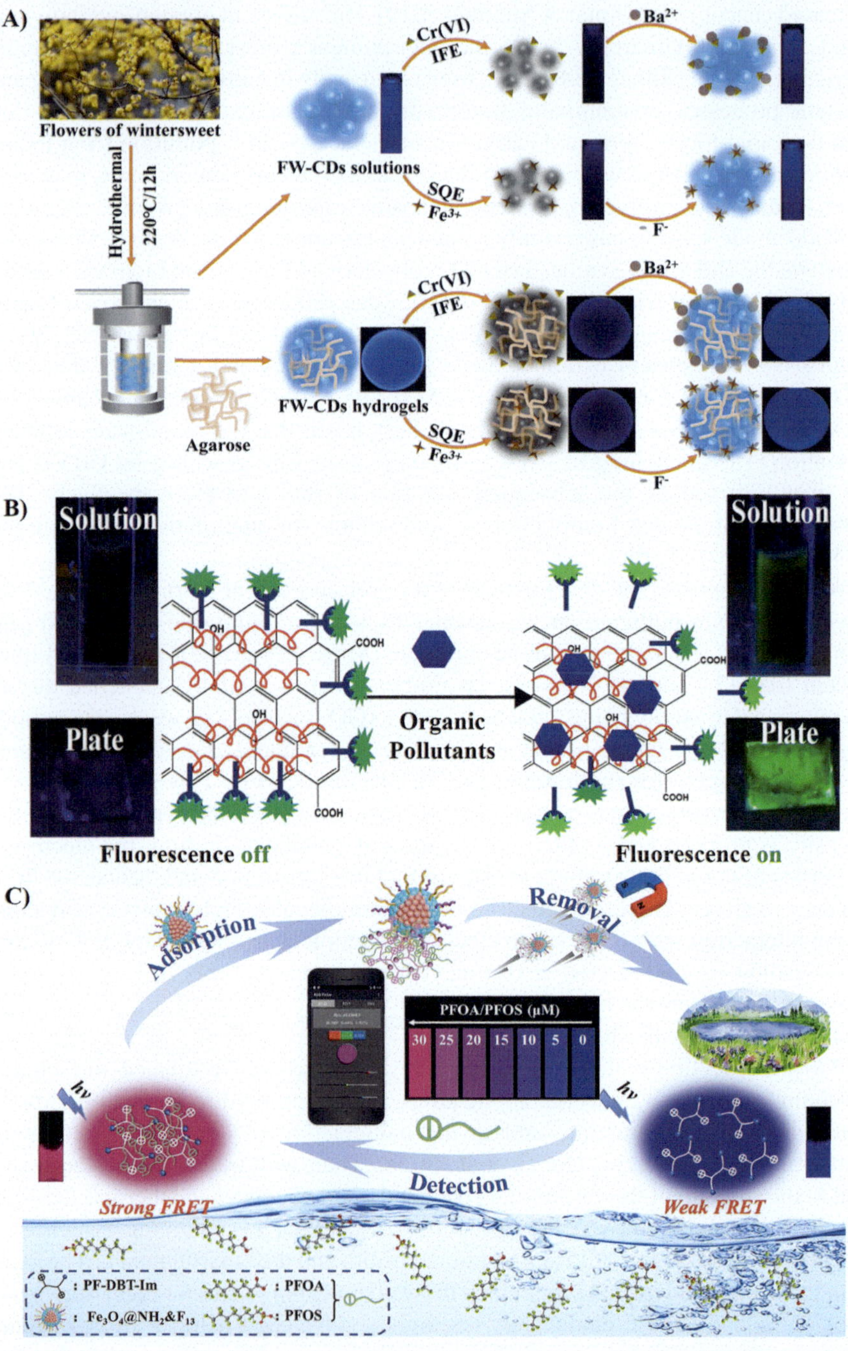

Fig. 5.3 (**a**) Green-synthesized CD probe derived from waste biomass, the FW-CDs for the parallel and semi-quantitative visual detection of Cr(VI) and Fe^{3+}. (Reproduced with permission from

Xia et al. Copyright 2022, MDPI). (**b**) Graphene-based composite with dextran fluorescein as a fluorescent "turn-on" probe for detecting organic pollutants in wastewater. (Reproduced with permission from Mitra and Saha Copyright 2017, American Chemical Society). (**c**) A two-in-one platform utilizing a conjugated polymer (PF-DBT-Im) and surface-functionalized magnetic nanoparticles (Fe_3O_4@NH_2&F_{13}) for the detection and removal of PFOA and PFOS from environmental water. (Reproduced with permission from X. Chen et al. Copyright 2023a, Elsevier)

scattering upon interaction with PFOS. This assay exhibits high sensitivity with aa detection limit of 18.27 nmol/L and displays selectivity for PFOS over other perfluorinated compounds. The interaction between PFOS and CD involves the formation of a ground-state complex between them, resulting in static quenching of fluorescence, which was further confirmed by no significant change in fluorescence lifetime measurements. A combination of electrostatic interactions, hydrogen bonding, and the strong affinity between PFOS and the nitrogen-donor atoms in the CDs drives this complexation. The concentration of PFOS can be accurately determined by quantifying the change in fluorescence intensity. The CD-based assay offers a simpler, faster, and more cost-effective approach than traditional analytical techniques. Its successful application to real-world water samples highlights its potential for effective environmental monitoring of PFOS contamination.

A recent study described a two-in-one platform using a conjugated polymer (PF-DBT-Im) and surface-functionalized magnetic nanoparticles (Fe_3O_4@NH_2&F_{13}) for the detection and removal of perfluorooctanoic acid (PFOA) and PFOS from water (Chen et al., 2023a). PF-DBT-Im, a conjugated polymer, was used as the probe, exhibiting a blue-to-magenta color change upon binding with PFOA/PFOS, which enables visual and amplified detection as a consequence of polymer chain aggregation, which enhances inter-chain FRET. This highly sensitive method has detection limits of 6.12 nM for PFOA and 14.3 nM for PFOS. To add more robustness to the system and improve on-site detection, its integration with a smartphone portable device was done. The magnetic adsorbent, Fe_3O_4@NH_2&F_{13}, is further used to remove PFOA/PFOS from water by magnetic separation with removal efficiencies of 95.68% for PFOA and 96.99% for PFOS. The selectivity of the fluorescent probe is also confirmed, and the system is successfully applied to real water samples. The integration of computational mechanisms into biosensor technologies promises to address various key challenges by using principles of decision-making processes toward the development of highly selective and sensitive detection systems (Tregubov et al., 2018; Mattath et al., 2022b). These smart systems function by processing inputs, such as the presence or absence of specific analytes, and generating a corresponding output, similar to the decision-making processes in advanced electronic circuits.

5.3 Logic Gate Operations for Environmental Monitoring

5.3.1 Overview of Logic Gates and Operational Mechanism

Nanomaterial computing represents an unconventional approach to nano-architecture engineering, incorporating Boolean logic principles. The concept originated with George Boole's mapping of "FALSE" and "TRUE" to the binary digits 0 and 1 for the purpose of performing logical operations (de Silva & Uchiyama, 2007; de Silva et al., 1993; Andréasson & Pischel, 2010, 2015). Computing devices convert input information into digital signals through mathematical processing. This signal serves as a code representing physical cues, assuming discrete values. For instance, in binary code, the fundamental unit of information consists of a sequence of "0" and "1" digits, representing the two states of a logic circuit. A defined threshold distinguishes these states based on outputs such as visual, fluorescent, chemiluminescent, or electro-chemiluminescent signals; values below or above this threshold correspond to "0" and "1," respectively. Logic devices can be broadly classified into several categories: elementary logic gates, advanced logic devices tailored for specific arithmetic functions or non-arithmetic information processing, other generalized cascaded/concatenated logic circuits, and other miscellaneous categories. Elementary logic gates, such as AND, OR, and NOT, form the foundation of digital circuits by performing basic binary operations. Sequential logic gates connect one after another, making their outputs dependent on both current and past inputs. Combinational logic gates combine multiple gates to create and generate outputs based solely on input values. Integrative or cascaded logic gates connect multiple gates in series, enabling the construction of complex circuits and advanced functions. These logic gates are essential for designing and building digital systems, from simple devices to advanced computing architectures (Mattath et al., 2025).

The basic logic gates serve as the keystone of molecular computing and serve as essential constituents in the development of advanced logic devices. Among the commonly used logic gates are the PASS 0, YES, AND, OR, INHIBIT, XOR, and MAJORITY gates, as well as their counterparts with opposite functions such as PASS 1, NOT, NAND, NOR, IMPLICATION, XNOR, MINORITY, and others (Li et al., 2015; Fan et al., 2022; Parambil et al., 2023). Table 5.1 provides a detailed illustration of their logical operating patterns. The two simplest logic gates in the input-independent group are PASS 0 and PASS 1, which produce the same outputs (0 and 1, respectively), irrespective of the input value. The YES gate exhibits an identity function, wherein the output is equivalent to the input; however, the NOT gate performs logical negation, producing the inverse of the input. In an AND gate, the output will be 1 only if all inputs are present (1,1), otherwise, the output is 0. An OR gate produces an output of 1 whenever one or more inputs are 1. Only when all inputs are 0 will the output be 0. An INHIBIT gate has one input that inhibits; it will only produce output (1) when that inhibitory input is absent, and all other inputs are present. The XOR gate's input–output pattern is unique. If any input is present (0,1

Table 5.1 Conventional representations of Boolean logic gates: Symbols and truth tables

	Output	PASS 0	PASS 1	YES1	YES2	NOT1	NOT2	OR	AND	INHIBIT1
		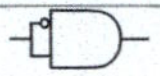	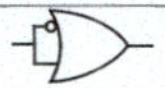	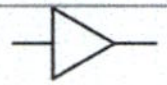	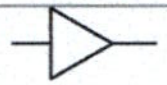	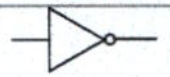	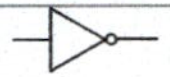	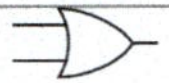	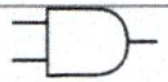	
Input										
0	0	0	1	0	0	1	1	0	0	0
0	1	0	1	0	1	1	0	1	0	1
1	0	0	1	1	0	0	1	1	0	0
1	1	0	1	1	1	0	0	1	1	0

	Output	INHIBIT2	IMPLICATION1	IMPLICATION2	XOR	NOR	XNOR	NAND
		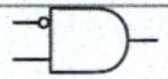	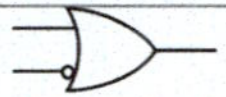	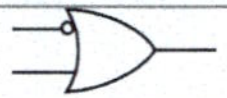	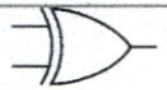	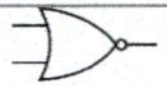	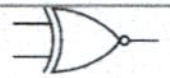	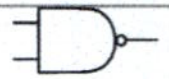
Input								
0	0	0	1	1	0	1	1	1
0	1	0	0	1	1	0	0	0
1	0	1	1	0	1	0	0	0
1	1	0	1	1	0	0	1	0

or 1,0), the output will be 1, but if all inputs are either absent or present (0,0 or 1,1), the output will be 0. In fault-tolerant computing, the MAJORITY gate is designed so that its output is 1 only when most of its inputs are 1. If the number of 1 input is equal to or less than half the total number of inputs, the output is 0. These examples show how positive logic gates work. The input–output relationship of each of the NAND, NOR, IMP, XNOR, and MINORITY gates is the opposite of its corresponding gates the AND, OR, INH, XOR, and MAJORITY, respectively. These various elementary logic gates serve as the fundamental building blocks for advanced logic devices, both non-arithmetic and arithmetic, each with its unique functions. More complex computing tasks can be executed through concatenated circuits by integrating or cascading them appropriately. We have provided a universal classification of advanced logic devices based on their distinct construction principles (Table 5.2). This classification is predicated on a systematic elaboration within the field.

Building on the foundation of basic logic gates and combining them can lead to the development of more complex logic devices. When contemplating "computation," we frequently gravitate toward "arithmetic," wherein adders and subtractors emerge as prominent tools for arithmetic operations. This classification encompasses half-adders/subtractors and full-adders/subtractors, illustrating the evolution from basic to more intricate computing systems (Fan et al., 2020, 2022). When various logic devices are functionally integrated to form concatenated logic circuits, they unlock the ability to execute complex functions that a single device could not manage on its own. Importantly, in these concatenated setups, the output from one device in the sequence serves as an input for the next device, allowing for a chain of operations that enhance the overall functionality.

Table 5.2 Classification of non-arithmetic and arithmetic logic devices

Advanced logic devices	Logical functions or constituents	
Non-arithmetic devices	Multiplexer/demultiplexer	Data compression/decompression
	Encoder/decoder	Binary data/code conversion
	Parity generator/checker (pG/pC)	Error detection through data transmission
	Keypad lock	Sequential logic
	Voter	Majority gate
	Ternary logic gate	Three states (0/1/2)
	Others	–
Arithmetic devices	Half-adder	AND + XOR
	Half-subtractor	INH + XOR
	Full-adder	XOR + AND + OR
	Full-subtractor	XOR + INH + OR
	Others	–
Concatenated circuits	Reasonable integration of different logic devices	
Others	–	

These devices are categorized by their respective functionalities of multiplexers and demultiplexers, encoders and decoders, as well as parity generators and checkers used for error detection (Arugula et al., 2010; Wu et al., 2015; Andréasson et al., 2011; Bälter et al., 2013; Fan et al., 2015). Multiplexers and demultiplexers act as control switches generally used to compress and decompress data in different signal and electronic processing applications. A 2-to-1 multiplexer leads two input signals to one output channel, while a 1-to-2 demultiplexer does the opposite, distributing a single input signal across two separate output channels (Arugula et al., 2010; Wu et al., 2015). Analogous to multiplexers and demultiplexers, scientists have also observed extensive use of encoders and decoders in molecular computing, which are essential for converting binary data into code and enabling various functions. Encoders and decoders have inverse functions where encoders scan compressed data for storage or transmission by converting binary information into a coded format, while decoders carry out the reverse operation, reconstructing the original binary data from the code (Fan et al., 2016). Along with the data conversion logic devices discussed earlier, parity generators and checkers perform important functional role in detecting errors occurring during data transmission. Given the inevitability of bit errors during the transmission of binary data, these devices are employed to identify such errors, which could otherwise compromise computation. A parity generator is placed at the transmitting end and a parity checker at the receiving end (Bälter et al., 2013; Mattath et al., 2022a, 2023).

Beyond the functional devices previously mentioned in the context of information conversion, compression, and transmission, there is a range of other devices with specialized and often intriguing functions. Examples include voter circuits, keypad locks, and ternary logic gates. For voters, the function is similar to using a Majority logic gate from Boolean logic (Fan et al., 2015). This gate says "yes" (or makes it TRUE) if over half the inputs are on board, leading to a "PASS" on whatever is being proposed. If that is not the case, it swings the other way, hitting a "no" (or FALSE), which means the proposal gets a thumbs down, or "Denied." The keypad lock acts as a sequential logic device designed to safeguard information against unauthorized access (Rout et al., 2013; Jiang & Ng, 2014; Parambil et al., 2023). This lock will only disengage if the correct "password" is entered in the precise sequence. Unlike other logic devices where the order of inputs does not impact functionality, the outcome of the keypad lock depends on both the specific inputs and their sequence. Only the exact "password," keyed in the right order, will unlock it; if not, it remains securely locked. Unlike logic devices that work on the principle of Boolean logic with just two states (0 or 1), ternary logic gates introduce a third state, making the inputs and outputs more varied with 0, 1, and 2 (Fan et al., 2017). This, in turn, enables ternary gates to handle even imprecise molecular information and substantially enhance computing power. Because of their exemplary characteristics, these advanced logic devices have broadened the dimensions of non-arithmetic devices and paved the way for the discovery of new devices with enhanced potential applications.

5.3.2 *Application of Logic Gates in Environmental Sensing*

Researchers developed a graphene-rhodamine (GRH) array for dual colorimetric and fluorometric sensing of Hg^{2+} ions (Mohan & Kizhakayil, 2016). The study relies on the principle of dual functionality of graphene, which interacts with rhodamine via π–π and dispersive forces and selectively adsorbs Hg^{2+}. The absorbance of rhodamine undergoes a reduction upon interaction with graphene, but this decrease is reversed upon exposure to Hg^{2+} ions. The efficiency of this mechanism was established using ethylenediaminetetraacetic acid (EDTA) as a chelating ligand, wherein all observed optical changes occurred within the visible spectrum. The colorimetric GRH array responds linearly to Hg^{2+} concentration from 0 to 1000 nM, with the absorbance maxima being at 554 nm and a limit of detection (LOD) of 2 ppb. A reversible INH logic gate was developed, displaying both inhibitory and activating functionalities, thereby expanding the potential for automated chemical systems (Fig. 5.4a). The study also explored fluorometric detection using FRET of graphene, achieving an even lower LOD of 380 ppt, which signifies that the system can be used even for trace detection of Hg^{2+} ions. This study validates the GRH array as a simple, economically viable, sensitive, and selective probe for Hg^{2+} detection, without the additional hassle of chemical modifications, offering colorimetric and fluorometric sensing options. This demonstrates a novel approach to logic gate design and acts as a centerpiece in the trace detection of mercury ions.

A recent study designed a multi-mode sensor for selective Fe^{3+} detection by using a hydrothermally synthesized silver–carbon dot nanohybrid (Ag@CDs) (Ajith et al., 2022). The synthesized CDs surface contained multiple oxygen-containing (–OH and $–COO^-$) and nitrogen-containing ($–NH_2$) functional groups, facilitating interaction with Fe^{3+}. Based on Fourier transform infrared spectroscopy and observed selectivity for Fe^{3+} over Fe^{2+} and Cr^{3+}, it was hypothesized that $–NH_2$ groups bind to the silver nanoparticle surfaces, while –OH and $–COO^-$ groups stabilize the Ag@CDs in aqueous solution via electrostatic repulsion. The selective interaction with Fe^{3+} is likely due to the electronegativity difference between Fe^{3+} and the –OH groups at neutral pH. The –OH groups react with Fe^{3+} to form iron hydroxide ($Fe(OH)_3$ or FeO(OH)), leading to Ag@CDs aggregation. This aggregation results in a bathochromic shift of the surface plasmon resonance absorption band from 415 to 555 nm, a color change from dark brown to yellowish-brown, and fluorescence quenching. The Ag@CDs exhibited an LOD of 0.76 ppm for Fe^{3+}, even in natural river water sample interferents. Mechanistic studies confirmed that Fe^{3+} interacts with the Ag@CDs' surface functional groups, causing aggregation and quenching. Furthermore, the Ag@CDs were incorporated into an IMP-OR concatenated logic gate with ascorbic acid and urea as additional inputs (Fig. 5.4b). The synthesized Ag@CDs serve as a multi-mode sensor and logic system because of their simple synthesis, high selectivity, and rapid reaction time, showing promise for heavy metal detection. The statistical analysis (Table 5.3) shows the C-dots-based platforms used for environmental sensing and development of logic systems reported to date.

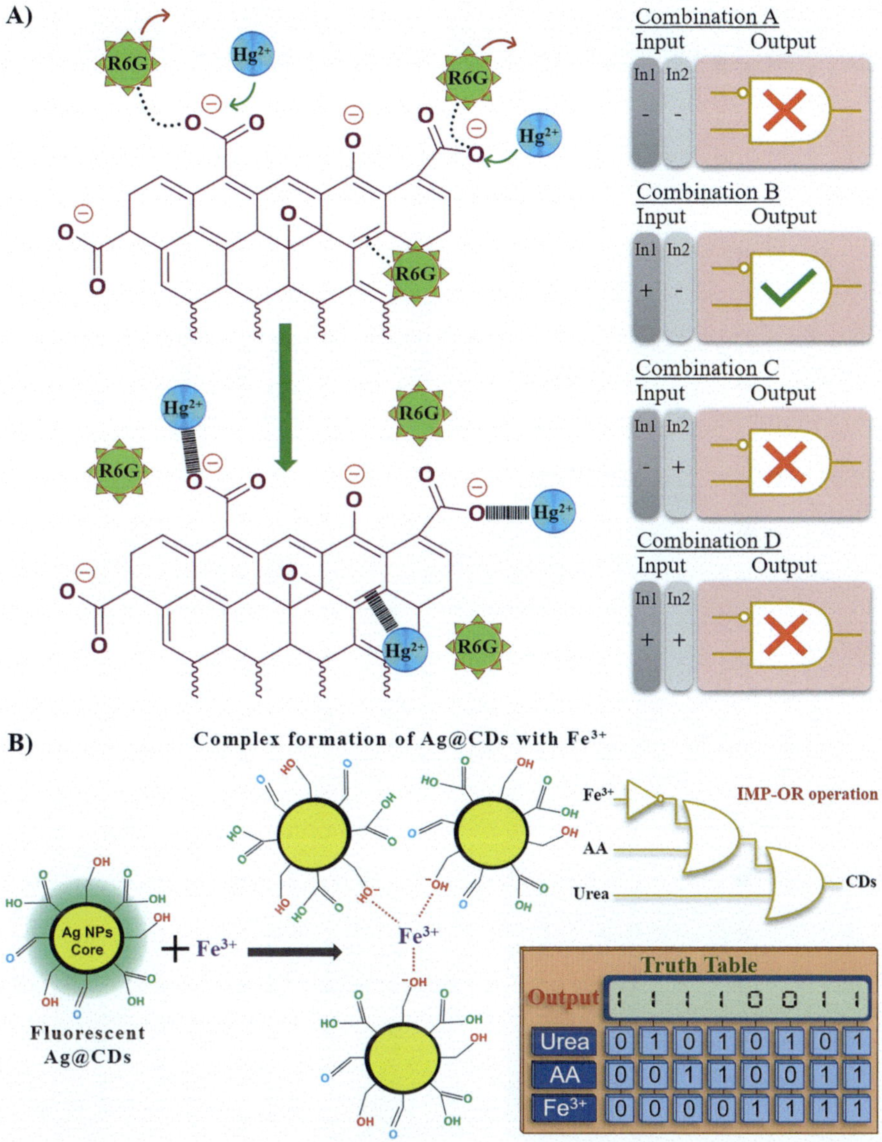

Fig. 5.4 (**a**) The interaction of Hg^{2+} with GN, mediated by both soft–soft and cationic–π interactions, results in the displacement of the R6G molecule. An INHIBIT logic gate system predicated on GRH has been implemented, utilizing Hg^{2+} (In1) and EDTA (In2) as input variables. (**b**) Schematic representation elucidating the mechanism of Fe^{3+} detection by the Ag@CDs nanostructure. The logical symbol and associated truth table for a ternary IMPLICATION-OR gate implemented using Ag@CDs are presented. (Reproduced with permission from Ajith et al. Copyright 2022, Elsevier)

Table 5.3 Fluorescent C-dots-based platforms for environmental sensing and logic systems

C-dots platform	Synthesis route and precursor	Logic operations	Applications	Reference
Acid and amine-functionalized C-dots	Biogenic (rice)	Elementary (YES, OR, XOR, NOT, IMP)	Detection of cations and anions	Dhenadhayalan and Lin (2015)
Carbon nano-dots (CNDs)	Chemogenic (chitosan)	Elementary (AND)	Arginine and Cu^{2+} detection	Lu et al. (2017)
C-dots	Biogenic (citrus limetta)	Elementary (NOT, PASS1, NOR, IMP)	BSA detection	Swaminathan and Balasubramanian (2019)
C-dots	Chemogenic (*p*-phenylenediamine)	Elementary (AND)	Cr^{6+} and Cys detection	Gao et al. (2018)
C-dots	Chemogenic (citric acid)	Elementary (AND)	Hg^{2+} andglutathione detection	Bai et al. (2018)
C-dots	Chemogenic (salicylic acid)	Elementary (AND)	Fe^{3+} and F^- detection	Li et al. (2020)
C-dots	Chemogenic (citric acid and dopamine)	Combinational (XOR-AND, INH-XOR)	Intracellular pH sensing	Lan et al. (2017)
C-dots	Biogenic (pineapple peel)	Elementary (NOT, IMP), keypad lock, memory device	Cellular imaging and detection of Hg^{2+} ion	Vandarkuzhali et al. (2018)
C-dots	Chemogenic (sodium citrate)	Elementary (AND)	D-Penicillamine detection	Yuan et al. (2016)
C-dots	Chemogenic (citric acid, Tris)	Elementary (AND)	Glyphosate detection	Yuan et al. (2017)
C-dots	Chemogenic (TTHA)	Sequential (NOR-AND), keypad lock	Auramine *O* detection	Zhu et al. (2015)
C-dots	Chemogenic (citric acid)	Elementary and sequential (NOT, OR, AND, NOR, NAND, NOT-NAND)	Picric acid, Fe^{3+}, Fe^{2+}, and H_2O_2 detection	Sk et al. (2015)
C-dots	Chemogenic (1,3,5-benzenetricarboxylic acid)	Elementary (YES)	LED and chemical sensing	Tong et al. (2022)
C-dots	Chemogenic (diammonium phosphate, sodium citrate)	Elementary (NOT, PASS1, NOR, NAND, IMP)	Information storage and extraction	Xiao et al. (2021)

Silver nanoparticles/N-doped C-dots nanocomposites	Biogenic (*Siraitia grosvenorii*)	Elementary (AND)	SERS	Su et al. (2016)
Nitrogen, chlorine co-doped C-dots	Chemogenic (urea, choline chloride, alanine)	Elementary (IMP, INH)	Glutathione detection	Bai et al. (2022a)
C-dots-gold nanoparticles	Chemogenic (citric acid)	Combination (INH-OR)	Detection of pesticides	Korram et al. (2019)
Nitrogen co-doped C-dots (NCDs)	Chemogenic (glucose and L-asparagine)	Elementary (AND)	Cu^{2+} and GSH detection	Gao et al. (2019)
Nitrogen-co-doped C-dots	Biogenic (kiwi fruit juice)	Elementary (NOT, IMP)	Fe^{3+} detection, fluorescent inks	Xu et al. (2022a)
Nitrogen-doped C-dots	Chemogenic (citric acid)	Elementary (IMP)	Hg^{2+} and biothiols detection	Liao et al. (2019)
Nitrogen-doped C-dots	Chemogenic (citric acid and 2,6-diaminopyridine)	Elementary (IMP)	Hg^{2+} and glutathione detection, cell imaging	Singh et al. (2019)
Nitrogen-doped C-dots	Chemogenic (PEI)	Elementary (AND)	Cu^{2+} and L-Cys detection	Liu et al. (2020)
Nitrogen-doped C-dots	Chemogenic (ascorbic acid)	Integrative (IMP-AND-OR)	Creatinine detection	Kainth et al. (2020)
Nitrogen-doped C-dots	Chemogenic (catechol)	Elementary (AND)	Fe^{3+} and AA detection	Du et al. (2018)
Nitrogen-doped C-dots	Chemogenic (citric acid)	Elementary (AND)	Chlorpromazine hydrochloride detection	Zhang et al. (2019)
Amino acid-derivatized C-dots	Chemogenic (citric acid)	Elementary (AND, INH)	Detection of Pb^{2+}, Hg^{2+}, Fe^{3+}, Zr^{2+}, Cr^{3+}, Cu^{2+}	Cheng et al. (2017)
Polyethyleneimine-capped C-dots (PEI-CDs)	Chemogenic (citric acid and PEI)	Elementary (INH)	Cu^{2+} and H_2S detection	Chen et al. (2019b)
Magnesium and nitrogen co-doped C-dots	Chemogenic (citric acid)	Elementary (IMP)	Hg^{2+} and Cys detection	Liu et al. (2016)
Ag-CD nanohybrid	Chemogenic (glucose)	Elementary (IMP)	AA detection	Tian et al. (2017)

(continued)

Table 5.3 (continued)

C-dots platform	Synthesis route and precursor	Logic operations	Applications	Reference
N, S-codoped C-dots	Chemogenic (cigarette butts)	Elementary (NOT, IMP)	Fe^{2+}, Fe^{3+}, and AA detection, film, security ink, and bioimaging	Bandi et al. (2018)
Nitrogen, sulfur co-doped C-dots	Chemogenic (ammonium citrate, cysteamine hydrochloride)	Elementary (AND)	Cr^{6+} and AA detection	Yang et al. (2018)
Nitrogen, sulfur co-doped C-dots	Biogenic (*Rosa indica*)	Elementary (AND)	Au^{3+} and S^{2-} detection	Sharma et al. (2018)
Surface quaternized cationic C-dots	Chemogenic (betaine hydrochloride and tris(hydroxymethyl)aminomethane)	Elementary and combinational (YES, INH-INH)	Phosphate detection	Xie et al. (2016)
Spermine-functionalized C-dots	Biogenic (D-glucose, spermine)	Elementary and sequential (AND, NAND, AND-INH, NAND-INH)	B to Z DNA transition	Feng et al. (2013)
Nucleic acid functionalized C-dots	Chemogenic (citric acid)	Elementary and sequential (OR, INH, OR-INH)	Room-temperature phosphorescence (RTP) logic gates	Gui et al. (2015)
Polyamine coated C-dots	Chemogenic (citric acid, PEI)	Combinational and integrative (NOR-INH, IMP-NOR-AND)	Zn^{2+}, Cu^{2+}, S^{2-}, and H^+ detection	Zou et al. (2018)
Holmium (III)-doped C-dots	Chemogenic (citric acid, PEI)	Integrative (NOR–INH, XOR-INH-OR)	FL/MR imaging	Fang et al. (2020a)
Benzalkonium chloride functionalized C-dots	Biogenic (kappa carrageenan, lemon juice)	Elementary (AND)	Antibacterial activity and Cr^{6+} sensing	Das et al. (2019)
C-dots-MnO_2 adduct	Chemogenic (ascorbic acid)	Elementary (YES, NOT, AND, IMP)	NAC detection	Jana et al. (2017)
N, S, P co-doped C-dots	Biogenic (*Saccharomyces cerevisiae*)	Elementary (AND)	Mn(VII) and L-AA detection	Gong et al. (2017)

Gadolinium (III)-doped C-dots	Chemogenic (citric acid, urea)	Elementary (IMP, NOR)	Intracellular imaging	Fang et al. (2020b)
r-C-dots	Chemogenic (1,3-dihydroxynaphthalene)	Combinational (AND-INH-IMP), keypad lock	Chemical/molecular keypad lock	Li et al. (2018)
Tb^{3+}-doped carbon dots	Chemogenic (ammonium citrate)	Elementary (INH)	Oxidase-based biosensor	Chen et al. (2019a)
DNA–CuNC/C-dots composite	Chemogenic (chitosan and polyethylene amine)	Elementary (INH)	Arginine and acetaminophen detection	Bu et al. (2020)
Y-C-dots	Chemogenic (*O*-phenylenediamine and ethylene glycol)	Elementary (IMP)	Au^{3+} and biothiol detection	Xu et al. (2021)
C-dots/gold nanoparticles	Chemogenic (3-aminophenylboronic acid)	Integrative (AND)	Fluoride anion detection	Feng et al. (2020)
C-dots/CdTe QDs	Chemogenic (citric acid, urea)	Combinational (OR-NOT-OR)	Spermine and spermidine detection	Abbasi-Moayed et al. (2022)
Europium-based aminoclay-C dots	Chemogenic (citric acid, urea)	Elementary (NOT, YES, PASS0)	Tetracycline detection	Jia et al. (2022)
Ag@C-dots nanohybrid	Biogenic (*F. religiosa*)	Sequential (IMP-OR)	Fe^{3+} detection	Ajith et al. (2022)
C-dots/TC-dots/AC-dots	Biogenic (*F. religiosa*)	Elementary, sequential, parity checker, keypad lock	Construction of logic library	Parambil et al. (2023)

5.4 Challenges and Future Perspective of Carbon Nano-Architectures

Carbon nano-architectures are attracting considerable attention for environmental sensing due to their superior sensitivity, selectivity, low operating temperatures, and recyclability compared to traditional metal oxide sensors. However, despite their promising attributes, their large-scale commercialization faces several challenges stemming from their intrinsic properties and the complexity involved in their synthesis and fabrication (Dariyal et al., 2021; Liu et al., 2012). Carbon nanomaterials exhibit remarkable performance in controlled laboratory environments, but they often suffer from a loss of efficiency in real-world applications. This is partly due to the difficulty of reproducibility in developing uniform material properties at a larger scale since the control over nanoscale features reduces during bulk material downsizing.

The stability of functional groups and the occurrence of defects during synthesis can also compromise the strength and performance of the synthesized nanomaterials. In practical applications, factors such as interference from the sample matrix and exposure to harsh environmental conditions, like temperature fluctuations, UV radiation, humidity, and pollutants, also make long-term usage of these materials challenging. For example, GO is vulnerable to degradation in humid environments since moisture can block active sites in hydrophilic materials, affecting their reusability and effectiveness. Advanced strategies to improve the strength, structural integrity, recyclability, and resistance to prolonged exposure to pollutants of the materials must be developed for them to act as reliable and commercially viable sensing platforms (Wu et al., 2020; Dimiev et al., 2013; Charitidis et al., 2014).

Fluorescence-based detection probes made up of carbon nanomaterials have certain limitations when the analyte in consideration is heavy metals, organic pollutants, and emerging contaminants like PFAs. The interference caused by heavy metal ions present in the environment as well as pH and salinity results in a decrease in both selectivity and accuracy by reducing binding efficiency. Organic pollutants, such as pesticides and endocrine disruptors, often need improvements and surfaces to enable the formation of strong interactions in aqueous systems. Conversely, these modifications can be the main reason for the loss of fluorophore stability and efficiency. Moreover, a low concentration of pollutants brings amplification techniques into play, leading to an increased complexity of the detection process. PFAs pose additional challenges because of their chemical inertness, hydrophobicity, and trace presence in samples, making it challenging to accomplish and maintain enough of them to achieve sufficient binding or sensitivity. Strong fluorescence quenching in real-world systems and the absence of specific functional groups for effective interaction further hinder the applicability of these methods. Advanced synthesis and functionalization strategies, hybrid systems, enhanced matrix compatibility, and resistance to degradation to improve reliability and sensitivity are needed to enable real-time applications of carbon-based nanomaterials. Recent developments in hybrid nanocomposites, like carbon-metal oxide composites and polymer-inorganic hybrid structures, have shown better and improved properties, such as better mechanical and physical stability (Monteiro & Simões, 2024). In situ CVD and additive manufacturing techniques result in better integration and uniform distribution of nanomaterials within functional matrices, increasing stability and resistance to environmental wear and tear. Surface functionalization, heteroatom doping, and protective coatings of hydrophobic polymers, titanium dioxide, etc. can ameliorate degradation and improve the durability and selectivity of carbon nanomaterials in harsh environments (Speranza, 2021).

Looking towards the future, carbon nanomaterials combined with smart technologies look promising. Miniaturized sensors made from rGO or multi-walled carbon nanotubes (MWCNTs) are increasingly being embedded in wearable devices and Internet of things (IoT) platforms, allowing for real-time environmental monitoring (Mallakpour & Soltanian, 2016; Santhiran et al., 2021; Du et al., 2023). These sensors can detect environmental toxicants such as ammonia, hydrogen sulfide, or ozone and provide valuable data for monitoring air quality and environmental sustainability. Combining carbon nanomaterials with smart sensor networks and IoT could revolutionize monitoring and managing environmental pollution over large geographic ranges.

Expanding environmental sensing and carbon nano-architectures are promising candidates for molecular computing and information processing. The development of molecular-level sensors and logic gates powered by carbon nanomaterials can greatly augment the efficiency and sophistication of computational platforms. Future research focusing on integrating carbon nano-architectures into logic computing systems could enable real-time, autonomous pollutant detection and response. Enhancing these sensors' stability, scalability, and computational capabilities would allow for smarter and more efficient environmental monitoring, resulting in an approach that could revolutionize pollution detection, offering more accurate, dynamic, and timely solutions for environmental redress and sustainability. Continual efforts toward this milieu would drive new paradigms in computing, diagnostics, and environmental monitoring by using carbon nanomaterials, resulting in breakthroughs on the near horizon.

References

Abbasi-Moayed, S., Bigdeli, A., & Hormozi-Nezhad, M. R. (2022). Determination of spermine and spermidine in meat with a ratiometric fluorescence nanoprobe and a combinational logic gate. *Food Chemistry, 384*, 132459. https://doi.org/10.1016/j.foodchem.2022.132459

Ajith, M. P., Pardhiya, S., Prabhakar, A. K., & Rajamani, P. (2022). Ag@CDs nanohybrid: Fabrication, design of a multi-mode chemosensory probe for selective Fe^{3+} detection and logic gate operation. *Chemosphere, 303*, 135090. https://doi.org/10.1016/j.chemosphere.2022.135090

Algar, W. R., Hildebrandt, N., Vogel, S. S., & Medintz, I. L. (2019). FRET as a biomolecular research tool—Understanding its potential while avoiding pitfalls. *Nature Methods, 16*(9), 815–829. https://doi.org/10.1038/s41592-019-0530-8

AlMalki, F. A., Khashan, K. S., Jabir, M. S., Hadi, A. A., Sulaiman, G. M., Abdulameer, F. A., Albukhaty, S., Al-Karagoly, H., & Albaqami, J. (2022). Eco-friendly synthesis of carbon nanoparticles by laser ablation in water and evaluation of their antibacterial activity. *Journal of Nanomaterials, 2022*(1), 7927447. https://doi.org/10.1155/2022/7927447

Andréasson, J., & Pischel, U. (2010). Smart molecules at work—Mimicking advanced logic operations. *Chemical Society Reviews, 39*(1), 174–188. https://doi.org/10.1039/b820280j

Andréasson, J., & Pischel, U. (2015). Molecules with a sense of logic: A progress report. *Chemical Society Reviews, 44*(5), 1053–1069. https://doi.org/10.1039/c4cs00342j

Andréasson, J., Pischel, U., Straight, S. D., Moore, T. A., Moore, A. L., & Gust, D. (2011). All-photonic multifunctional molecular logic device. *Journal of the American Chemical Society, 133*(30), 11641–11648. https://doi.org/10.1021/ja203456h

Ariga, G. G., Naik, P. N., Nandibewoor, S. T., & Chimatadar, S. A. (2017). Quenching of fluorescence by meclizine, a probe study for structural and conformational changes in human serum albumin. *Journal of Biomolecular Structure and Dynamics, 35*(14), 3161–3175. https://doi.org/10.1080/07391102.2016.1245159

Arugula, M. A., Bocharova, V., Halámek, J., Pita, M., & Katz, E. (2010). Enzyme-based multiplexer and demultiplexer. *The Journal of Physical Chemistry B, 114*(15), 5222–5226. https://doi.org/10.1039/c5cc05565b

Bai, Z., Yan, F., Xu, J., Zhang, J., Wei, J., Luo, Y., & Chen, L. (2018). Dual-channel fluorescence detection of mercuric (II) and glutathione by down-and up-conversion fluorescence carbon dots. *Spectrochimica Acta Part A: Molecular and Biomolecular Spectroscopy, 205*, 29–39. https://doi.org/10.1016/j.saa.2018.07.012

Bai, S., Zhang, P., & Beratan, D. N. (2020). Predicting dexter energy transfer interactions from molecular orbital overlaps. *The Journal of Physical Chemistry C, 124*(35), 18956–18960. https://doi.org/10.1021/acs.jpcc.0c06694

Bai, F., Wang, H., Lin, L., & Zhao, L. (2022a). A ratiometric fluorescence platform composed of MnO_2 nanosheets and nitrogen, chlorine co-doped carbon dots and its logic gate performance for glutathione determination. *New Journal of Chemistry, 46*, 1972. https://doi.org/10.1039/d1nj05210a

Bai, J., Zhu, W., He, F., Cheng, Y., Meng, X., Xu, H., Xu, Y., Zhang, W., Fang, X., & Li, H.-B. (2022b). Excited-state intramolecular proton-transfer-induced dual fluorescence emission in 2,3-dichloro-5,6-dicyano-1,4-benzoquinone and resorcinol-based carbon dots. *Optical Materials, 123*, 111845. https://doi.org/10.1016/j.optmat.2021.111845

Bälter, M., Li, S., Nilsson, J. R., Andréasson, J., & Pischel, U. (2013). An all-photonic molecule-based parity generator/checker for error detection in data transmission. *Journal of the American Chemical Society, 135*(28), 10230–10233. https://doi.org/10.1021/ja403828z

Bandi, R., Devulapalli, N. P., Dadigala, R., Gangapuram, B. R., & Guttena, V. (2018). Facile conversion of toxic cigarette butts to N, S-codoped carbon dots and their application in fluorescent film, security ink, bioimaging, sensing and logic gate operation. *ACS Omega, 3*(10), 13454–13466. https://doi.org/10.1021/acsomega.8b01743

Baruah, P., Basumatary, G., Yesylevskyy, S. O., Aguan, K., Bez, G., & Mitra, S. (2019). Novel coumarin derivatives as potent acetylcholinesterase inhibitors: Insight into efficacy, mode and site of inhibition. *Journal of Biomolecular Structure and Dynamics, 37*(7), 1750–1765. https://doi.org/10.1080/07391102.2018.1465853

Baruah, P., Paul, D., Doshi, J., & Mitra, S. (2022). Elevated fibrinogen level reduces therapeutic efficiency of AD drugs: Biophysical insights into the interaction of FDA-approved cholinesterase inhibitors with human fibrinogen. *The Journal of Physical Chemistry B, 126*(1), 30–43. https://doi.org/10.1021/acs.jpcb.1c07495

Bu, X., Fu, Y., Jiang, X., Jin, H., & Gui, R. (2020). Self-assembly of DNA-templated copper nanoclusters and carbon dots for ratiometric fluorometric and visual determination of arginine and acetaminophen with a logic-gate operation. *Microchimica Acta, 187*(3), 1–10. https://doi.org/10.1007/s00604-020-4146-6

Charitidis, C. A., Georgiou, P., Koklioti, M. A., Trompeta, A.-F., & Markakis, V. (2014). Manufacturing nanomaterials: From research to industry. *Manufacturing Review, 1*,11. https://doi.org/10.1051/mfreview/2014009

Chen, Z.-H., Han, X.-Y., Deng, L.-X., Lin, Z.-Y., Mu, F.-Y., Zhang, S., Shi, G., & Zhang, M. (2019a). A self-calibrating logic system and oxidase-based biosensor using Tb^{3+}-doped carbon dots/DNA conjugates. *Talanta, 191*, 235–240. https://doi.org/10.1016/j.talanta.2018.08.035

Chen, Z.-H., Han, X.-Y., Lin, Z.-Y., Fan, Y.-L., Shi, G., Zhang, S., & Zhang, M. (2019b). Facile reflux synthesis of polyethyleneimine-capped fluorescent carbon dots for sequential bioassays toward Cu^{2+}/H_2S and its application for a logic system. *Biotechnology and Applied Biochemistry, 66*(3), 426–433. https://doi.org/10.1002/bab.1739

Chen, Q., Zhu, P., Xiong, J., Gao, L., & Tan, K. (2019c). A sensitive and selective triple-channel optical assay based on red-emissive carbon dots for the determination of PFOS. *Microchemical Journal, 145*, 388–396. https://doi.org/10.1016/j.microc.2018.11.003

Chen, X., Hussain, S., Tang, Y., Chen, X., Zhang, S., Wang, Y., Zhang, P., Gao, R., Wang, S., & Hao, Y. (2023a). Two-in-one platform based on conjugated polymer for ultrasensitive ratiometric detection and efficient removal of perfluoroalkyl substances from environmental water. *Science of the Total Environment, 860*, 160467. https://doi.org/10.1016/j.scitotenv.2022.160467

Chen, Z., Lu, Y.-L., Wang, L., Xu, J., Zhang, J., Xu, X., Cheng, P., Yang, S., & Shi, W. (2023b). Efficient recognition and removal of persistent organic pollutants by a bifunctional molecular material. *Journal of the American Chemical Society, 145*(1), 260–267. https://doi.org/10.1021/jacs.2c09866

Cheng, H.-J., Kao, C.-L., Chen, Y.-F., Huang, P.-C., Hsu, C.-Y., & Kuei, C.-H. (2017). Amino acid derivatized carbon dots with tunable selectivity as logic gates for fluorescent sensing of metal cations. *Microchimica Acta, 184*(9), 3179–3187. https://doi.org/10.1007/s00604-017-2336-7

Dariyal, P., Sharma, S., Chauhan, G. S., Singh, B. P., & Dhakate, S. R. (2021). Recent trends in gas sensing via carbon nanomaterials: Outlook and challenges. *Nanoscale Advances, 3*(23), 6514–6544. https://doi.org/10.1039/d1na00707f

Das, P., Ganguly, S., Bose, M., Ray, D., Ghosh, S., Mondal, S., Aswal, V. K., Das, A. K., Banerjee, S., & Das, N. C. (2019). Surface quaternized nanosensor as a one-arrow-two-hawks approach for fluorescence turn "on–off–on" bifunctional sensing and antibacterial activity. *New Journal of Chemistry, 43*(16), 6205–6219. https://doi.org/10.1039/c8nj06308g

de Silva, A. P., & Uchiyama, S. (2007). Molecular logic and computing. *Nature Nanotechnology, 2*(7), 399–410. https://doi.org/10.1038/nnano.2007.188

de Silva, P. A., Gunaratne, N. H. Q., & McCoy, C. P. (1993). A molecular photoionic and gate based on fluorescent signalling. *Nature, 364*, 42–44. https://doi.org/10.1038/364042a0

Dhenadhayalan, N., & Lin, K.-C. (2015). Chemically induced fluorescence switching of carbon-dots and its multiple logic gate implementation. *Scientific Reports, 5*(1), 1–10. https://doi.org/10.1038/srep10012

Dimiev, A. M., Alemany, L. B., & Tour, J. M. (2013). Graphene oxide. Origin of acidity, its instability in water, and a new dynamic structural model. *ACS Nano, 7*(1), 576–588. https://doi.org/10.1021/nn3047378

Đorđević, L., Arcudi, F., Cacioppo, M., & Prato, M. (2022). A multifunctional chemical toolbox to engineer carbon dots for biomedical and energy applications. *Nature Nanotechnology, 17*(2), 112–130. https://doi.org/10.1038/s41565-021-01051-7

Du, F., Gong, X., Lu, W., Liu, Y., Gao, Y., Shuang, S., Xian, M., & Dong, C. (2018). Bright-green-emissive nitrogen-doped carbon dots as a nanoprobe for bifunctional sensing, its logic gate operation and cellular imaging. *Talanta, 179*, 554–562. https://doi.org/10.1016/j.talanta.2017.11.030

Du, Y., Wang, M., Ye, X., Liu, B., Han, L., Jafri, S. H., Liu, W., Zheng, X., Ning, Y., & Li, H. (2023). Advances in the field of graphene-based composites for energy–storage applications. *Crystals, 13*(6). https://doi.org/10.3390/cryst13060912

Elhenawy, S., Khraisheh, M., AlMomani, F., Hassan, M. K., Al-Ghouti, M. A., & Selvaraj, R. (2022). Recent developments and advancements in graphene-based technologies for oil spill cleanup and oil–water separation processes. *Nanomaterials, 12*(1). https://doi.org/10.3390/nano12010087

Escudero, D. (2016). Revising intramolecular photoinduced electron transfer (PET) from first-principles. *Accounts of Chemical Research, 49*(9), 1816–1824. https://doi.org/10.1021/acs.accounts.6b00299

Evich, M. G., Davis, M. J. B., McCord, J. P., Acrey, B., Awkerman, J. A., Knappe, D. R. U., Lindstrom, A. B., Speth, T. F., Tebes-Stevens, C., & Strynar, M. J. (2022). Per-and polyfluoroalkyl substances in the environment. *Science, 375*(6580), eabg9065. https://doi.org/10.1126/science.abg9065

Fan, D., Wang, K., Zhu, J., Xia, Y., Han, Y., Liu, Y., & Wang, E. (2015). DNA-based visual majority logic gate with one-vote veto function. *Chemical Science, 6*(3), 1973–1978. https://doi.org/10.1039/c4sc03495c

Fan, D., Zhu, J., Liu, Y., Wang, E., & Dong, S. (2016). Label-free and enzyme-free platform for the construction of advanced DNA logic devices based on the assembly of graphene oxide and DNA-templated AgNCs. *Nanoscale, 8*(6), 3834–3840. https://doi.org/10.1039/c6nr00032k

Fan, D., Wang, E., & Dong, S. (2017). Simple, fast, label-free, and nanoquencher-free system for operating multivalued DNA logic gates using polythymine templated CuNPs as signal reporters. *Nano Research, 10*, 2560–2569. https://doi.org/10.1007/s12274-017-1458-x

Fan, D., Wang, J., Wang, E., & Dong, S. (2020). Propelling DNA computing with materials' power: Recent advancements in innovative DNA logic computing systems and smart bio-applications. *Advanced Science, 7*(24), 2001766. https://doi.org/10.1002/advs.202001766

Fan, D., Wang, J., Han, J., Wang, E., & Dong, S. (2022). Engineering DNA logic systems with non-canonical DNA-nanostructures: Basic principles, recent developments and bio-applications. *SCIENCE CHINA Chemistry, 65*(2), 284–297. https://doi.org/10.1007/s11426-021-1131-1

Fang, Y., Zhou, L., Yang, J., Zhao, J., Zhang, Y., & Yi, C. (2020a). Multilevel, dual-readout logic operations based on pH-responsive holmium (III)-doped carbon nanodots. *ACS Applied Bio Materials, 3*(6), 3761–3769. https://doi.org/10.1021/acsabm.0c00356

Fang, Y., Zhou, L., Zhao, J., Zhang, Y., Yang, M., & Yi, C. (2020b). Facile synthesis of pH-responsive gadolinium (III)-doped carbon nanodots with red fluorescence and magnetic resonance properties for dual-readout logic gate operations. *Carbon, 166*, 265–272. https://doi.org/10.1016/j.carbon.2020.05.060

Felouat, A., D'Aléo, A., Charaf-Eddin, A., Jacquemin, D., Le Guennic, B., Kim, E., Lee, K. J., Woo, J. H., Ribierre, J.-C., Wu, J. W., & Fages, F. (2015). Tuning the direction of intramolecular charge transfer and the nature of the fluorescent state in a T-shaped molecular dyad. *The Journal of Physical Chemistry A, 119*(24), 6283–6295. https://doi.org/10.1021/acs.jpca.5b03699

Feng, L., Zhao, A., Ren, J., & Qu, X. (2013). Lighting up left-handed Z-DNA: Photoluminescent carbon dots induce DNA B to Z transition and perform DNA logic operations. *Nucleic Acids Research, 41*(16), 7987–7996. https://doi.org/10.1093/nar/gkt575

Feng, D.-Q., Liu, G., Chen, Z., Lu, H., Gao, Y., & Fang, X. (2020). A logic gate for fluoride anion detection based on carbon dots/gold nanoparticles. *Microchemical Journal, 157*, 104977. https://doi.org/10.1016/j.microc.2020.104977

Gao, Y., Jiao, Y., Lu, W., Liu, Y., Han, H., Gong, X., Xian, M., Shuang, S., & Dong, C. (2018). Carbon dots with red emission as a fluorescent and colorimetric dual-readout probe for the detection of chromium (VI) and cysteine and its logic gate operation. *Journal of Materials Chemistry B, 6*(38), 6099–6107. https://doi.org/10.1039/c8tb01580e

Gao, Y., Zhang, H., Shuang, S., Han, H., & Dong, C. (2019). An "on–off–on" fluorescent nanoprobe for recognition of Cu^{2+} and GSH based on nitrogen co-doped carbon quantum dots, and its logic gate operation. *Analytical Methods, 11*(20), 2650–2657. https://doi.org/10.1039/c9ay00424f

Gong, X., Li, Z., Hu, Q., Zhou, R., Shuang, S., & Dong, C. (2017). N, S, P co-doped carbon nanodot fabricated from waste microorganism and its application for label-free recognition of manganese (VII) and L-ascorbic acid and AND logic gate operation. *ACS Applied Materials & Interfaces, 9*(44), 38761–38772. https://doi.org/10.1021/acsami.7b11170

Gu, Y., Qiu, Z., & Müllen, K. (2022). Nanographenes and graphene nanoribbons as multitalents of present and future materials science. *Journal of the American Chemical Society, 144*(26), 11499–11524. https://doi.org/10.1021/jacs.2c02491

Gui, R., Jin, H., Wang, Z., Zhang, F., Xia, J., Yang, M., Bi, S., & Xia, Y. (2015). Room-temperature phosphorescence logic gates developed from nucleic acid functionalized carbon dots and graphene oxide. *Nanoscale, 7*(18), 8289–8293. https://doi.org/10.1039/c4nr07620f

Ha, T., Fei, J., Schmid, S., Lee, N. K., Gonzalez, R. L., Paul, S., & Yeou, S. (2024). Fluorescence resonance energy transfer at the single-molecule level. *Nature Reviews Methods Primers, 4*(1), 21. https://doi.org/10.1038/s43586-024-00306-6

Hou, L., Cui, X., Guan, B., Wang, S., Li, R., Liu, Y., Zhu, D., & Zheng, J. (2022). Synthesis of a monolayer fullerene network. *Nature, 606*(7914), 507–510. https://doi.org/10.1038/s41586-022-04771-5

Izquierdo-García, P., Fernández-García, J. M., & Martín, N. (2024). Twenty years of graphene: From pristine to chemically engineered nano-sized flakes. *Journal of the American Chemical Society, 146*(47), 32222–32234. https://doi.org/10.1021/jacs.4c12819

Jana, J., Aditya, T., Ganguly, M., & Pal, T. (2017). Carbon dot-MnO_2 FRET system for fabrication of molecular logic gates. *Sensors and Actuators B: Chemical, 246*, 716–725. https://doi.org/10.1016/j.snb.2017.02.129

Jia, L., Chen, X., Xu, J., Zhang, L., Dong, X., Zhao, D., Bi, N., Zhu, T., & Zhao, T. (2022). Europium-based aminoclay containing carbon dots: A new visual fluorescence platform for visual point-of-care testing of tetracycline in various real samples. *Journal of Luminescence, 241*, 118497. https://doi.org/10.1016/j.jlumin.2021.118497

Jiang, X.-J., & Ng, D. K. P. (2014). Sequential logic operations with a molecular keypad lock with four inputs and dual fluorescence outputs. *Angewandte Chemie International Edition, 53*(39), 10481–10484. https://doi.org/10.1002/ange.201406002

Kainth, S., Maity, B., & Basu, S. (2020). Label-free detection of creatinine using nitrogen-passivated fluorescent carbon dots. *RSC Advances, 10*(60), 36253–36264. https://doi.org/10.1039/d0ra06512a

Kancherla, R., Muralirajan, K., Maity, B., Karuthedath, S., Kumar, G. S., Laquai, F., Cavallo, L., & Rueping, M. (2022). Mechanistic insights into photochemical nickel-catalyzed cross-couplings enabled by energy transfer. *Nature Communications, 13*(1), 2737. https://doi.org/10.1038/s41467-022-30278-8

Korram, J., Dewangan, L., Nagwanshi, R., Karbhal, I., Ghosh, K. K., & Satnami, M. L. (2019). A carbon quantum dot–gold nanoparticle system as a probe for the inhibition and reactivation of acetylcholinesterase: Detection of pesticides. *New Journal of Chemistry, 43*(18), 6874–6882. https://doi.org/10.1039/c9nj00555b

Kumar, S., & Sharma, A. (2019). Cadmium toxicity: Effects on human reproduction and fertility. *Reviews on Environmental Health, 34*(4), 327–338. https://doi.org/10.1515/reveh-2019-0016

Kumar, V., Meenakshi, Shukla, S. K., & Thakur, N. (2021). Carbon-based materials approach for environmental sensing. In C. M. Hussain, S. K. Shukla, & G. M. Joshi (Eds.), *Functionalized nanomaterials based devices for environmental applications* (pp. 77–106). Elsevier. https://doi.org/10.1016/B978-0-12-822245-4.00015-5

Lakowicz, J. R. (2006). *Principles of fluorescence spectroscopy* (p. 132). University of Maryland School of Medicine Baltimore. https://doi.org/10.1007/978-0-387-46312-4

Lan, C., Zhang, L., Shi, B., Chen, D., Zhao, L., & Zhao, S. (2017). Fluorescent carbon dots with tunable emission by dopamine for sensing of intracellular pH, elementary arithmetic operations and a living cell imaging based INHIBIT logic gate. *Journal of Materials Chemistry B, 5*(26), 5265–5271. https://doi.org/10.1039/c7tb01072a

LeCroy, G. E., Sonkar, S. K., Yang, F., Veca, L. M., Wang, P., Tackett, K. N., II, Yu, J.-J., Vasile, E., Qian, H., Liu, Y., Luo, P., & Sun, Y.-P. (2014). Toward structurally defined carbon dots as ultra-compact fluorescent probes. *ACS Nano, 8*(5), 4522–4529. https://doi.org/10.1021/nn406628s

Li, H., Guo, S., Liu, Q., Qin, L., Dong, S., Liu, Y., & Wang, E. (2015). Implementation of arithmetic functions on a simple and universal molecular Beacon platform. *Advanced Science, 2*(5), 1500054. https://doi.org/10.1002/advs.201500054

Li, M., Wang, Z., Liang, J., Yao, H., Shen, L., Liu, H., & Fan, L. (2018). A chemical/molecular 4-input/2-output keypad lock with easy resettability based on red-emission carbon dots–Prussian blue composite film electrodes. *Nanoscale, 10*(16), 7484–7493. https://doi.org/10.1039/c8nr01258j

Li, L., Shi, L., Jia, J., Jiao, Y., Gao, Y., Liu, Y., Dong, C., & Shuang, S. (2020). "On-off-on" detection of Fe^{3+} and F^{-}, biological imaging, and its logic gate operation based on excitation-independent blue-fluorescent carbon dots. *Spectrochimica Acta Part A: Molecular and Biomolecular Spectroscopy, 227*, 117716. https://doi.org/10.1016/j.saa.2019.117716

Li, J., Zeng, H., Zeng, Z., Zeng, Y., & Xie, T. (2021). Promising graphene-based nanomaterials and their biomedical applications and potential risks: A comprehensive review. *ACS Biomaterials Science & Engineering, 7*(12), 5363–5396. https://doi.org/10.1021/acsbiomaterials.1c00875

Li, X., David, A. H. G., Zhang, L., Song, B., Jiao, Y., Sluysmans, D., Qiu, Y., Wu, Y., Zhao, X., Feng, Y., Mosca, L., & Fraser Stoddart, J. (2022). Fluorescence quenching by redox molecular pumping. *Journal of the American Chemical Society, 144*(8), 3572–3579. https://doi.org/10.1021/jacs.1c12480

Li, G., Liu, Z., Gao, W., & Tang, B. (2023). Recent advancement in graphene quantum dots based fluorescent sensor: Design, construction and bio-medical applications. *Coordination Chemistry Reviews, 478*, 214966. https://doi.org/10.1016/j.ccr.2022.214966

Li, Y., Yan, X., Cui, Z., Yuan, J., Xu, B., & Yang, G. (2024). Design and preparation of activated carbon with high specific surface area and porosity through an organic activator coupled with CO_2 activation. *Advanced Materials Interfaces, 12*(3), 2400450. https://doi.org/10.1002/admi.202400450

Liao, S., Li, X., Yang, H., & Chen, X. (2019). Nitrogen-doped carbon dots rapid and selective detection of mercury ion and biothiol and construction of an IMPLICATION logic gate. *Talanta, 194*, 554–562. https://doi.org/10.1016/j.talanta.2018.09.114

Liu, X., Cheng, S., Liu, H., Hu, S., Zhang, D., & Ning, H. (2012). A survey on gas sensing technology. *Sensors, 12*(7), 9635–9665. https://doi.org/10.3390/s120709635

Liu, T., Li, N., Dong, J. X., Luo, H. Q., & Li, N. B. (2016). Fluorescence detection of mercury ions and cysteine based on magnesium and nitrogen co-doped carbon quantum dots and IMPLICATION logic gate operation. *Sensors and Actuators B: Chemical, 231*, 147–153. https://doi.org/10.1016/j.snb.2016.02.141

Liu, X., Zhang, S., Xu, H., Wang, R., Dong, L., Gao, S., Tang, B., Fang, W., Hou, F., & Zhong, L. (2020). Nitrogen-doped carbon quantum dots from poly (ethyleneimine) for optical dual-mode determination of Cu^{2+} and L-cysteine and their logic gate operation. *ACS Applied Materials & Interfaces, 12*(42), 47245–47255. https://doi.org/10.1021/acsami.0c12750

Lu, W., Gao, Y., Jiao, Y., Shuang, S., Li, C., & Dong, C. (2017). Carbon nano-dots as a fluorescent and colorimetric dual-readout probe for the detection of arginine and Cu^{2+} and its logic gate operation. *Nanoscale, 9*(32), 11545–11552. https://doi.org/10.1039/c7nr02336g

Mallakpour, S., & Soltanian, S. (2016). Surface functionalization of carbon nanotubes: Fabrication and applications. *RSC Advances, 6*(111), 109916–109935. https://doi.org/10.1039/c6ra24522f

Mattath, M. N., Ghosh, D., Dong, C., Govindaraju, T., & Shi, S. (2022a). Mercury mediated DNA–Au/Ag nanocluster ensembles to generate a gray code encoder for biocomputing. *Materials Horizons, 9*(8), 2109–2114. https://doi.org/10.1039/d2mh00598k

Mattath, M. N., Ghosh, D., Pratihar, S., Shi, S., & Govindaraju, T. (2022b). Nucleic acid architectonics for pH-responsive DNA systems and devices. *ACS Omega, 7*(4), 3167–3176. https://doi.org/10.1021/acsomega.1c06464

Mattath, M. N., Zhang, H., Ghosh, D., Govindaraju, T., & Shi, S. (2023). Nanoclusters with specific DNA overhangs: Modifying configurability, engineering contrary logic pairs and the parity generator/checker for error detection. *Nanoscale, 15*(43), 17386–17397. https://doi.org/10.1039/d3nr04167k

Mattath M.N., Pratihar S., Govindaraju T., & Shi S. (2025). Advanced logic computing and hybrid crypto-steganography for molecular information coding using gold/silver nanoclusters. ACS Applied Materials & Interfaces, 17, 18731–18740. https://doi.org/10.1021/acsami.4c21488

Miao, X., Wu, C., Xia, Y., Yu, S., Li, F., & Zhang, M. (2024). Ratiometric fluorescence probes for visible detection and accurate identification of MPEA vapor. *Nature Communications, 15*(1), 10641. https://doi.org/10.1038/s41467-024-55011-5

Mitra, R., & Saha, A. (2017). Reduced graphene oxide based "turn-on" fluorescence sensor for highly reproducible and sensitive detection of small organic pollutants. *ACS Sustainable Chemistry & Engineering, 5*(1), 604–615. https://doi.org/10.1021/acssuschemeng.6b01971

Mohan, A., & Kizhakayil, R. N. (2016). Graphene-rhodamine nanoprobe for colorimetric and fluorimetric Hg^{2+} ion assay. *ACS Applied Materials & Interfaces, 8*(22), 14125–14132. https://doi.org/10.1021/acsami.6b03904

Monteiro, B., & Simões, S. (2024). Recent advances in hybrid nanocomposites for aerospace applications. *Metals, 14*(11). https://doi.org/10.3390/met14111283

Nair, R. R., Ren, W., Jalil, R., Riaz, I., Kravets, V. G., Britnell, L., Blake, P., Schedin, F., Mayorov, A. S., Yuan, S., Katsnelson, M. I., Cheng, H.-M., Strupinski, W., Bulusheva, L. G., Okotrub, A. V., Grigorieva, I. V., Grigorenko, A. N., Novoselov, K. S., & Geim, A. K. (2010). Fluorographene: A two-dimensional counterpart of Teflon. *Small, 6*(24), 2877–2884. https://doi.org/10.1002/smll.201001555

Norizan, M. N., Moklis, M. H., Demon, S. Z. N., Halim, N. A., Samsuri, A., Mohamad, I. S., Knight, V. F., & Abdullah, N. (2020). Carbon nanotubes: Functionalisation and their application in chemical sensors. *RSC Advances, 10*(71), 43704–43732. https://doi.org/10.1039/d0ra09438b

Padalkar, V. S., & Seki, S. (2016). Excited state intramolecular proton transfer (ESIPT)-inspired solid state emitters. *Chemical Society Reviews, 45*(1), 169–202. https://doi.org/10.1039/c5cs00543d

Parambil, A. M., & Rajamani, P. (2024). Carbon dots: A promising path towards environmental sustainability. *Environmental Science: Advances, 3*(11), 1513–1523. https://doi.org/10.1039/d4va00273c

Parambil, A. M., Mattath, M. N., Rajamani, P., Pham, P. V., Kumar, G., & Ponnusamy, V. K. (2023). Biogenic fluorescent carbon dots modulated fabrication of concatenate logic library and pattern-mediated molecular keypad lock for chemical sensing application. *Chemical Engineering Journal, 463*, 142354. https://doi.org/10.1016/j.cej.2023.142354

Parambil, A. M., Prasad, A., Tomar, A. K., Ghosh, I., & Rajamani, P. (2024). Biogenic carbon dots: A novel mechanistic approach to combat multidrug-resistant critical pathogens on the global priority list. *Journal of Materials Chemistry B, 12*(1), 202–221. https://doi.org/10.1039/d3tb02374e

Praseetha, P. K., Jari Litany, R. I., Alharbi, H. M., Khojah, A. A., Akash, S., Bourhia, M., Mengistie, A. A., & Shazly, G. A. (2024). Green synthesis of highly fluorescent carbon quantum dots from almond resin for advanced theranostics in biomedical applications. *Scientific Reports, 14*(1), 24435. https://doi.org/10.1038/s41598-024-75333-0

Priyadarshini, E., Parambil, A. M., Rajamani, P., Ponnusamy, V. K., & Chen, Y.-H. (2023). Exposure, toxicological mechanism of endocrine disrupting compounds and future direction of identification using nano-architectonics. *Environmental Research, 225*, 115577. https://doi.org/10.1016/j.envres.2023.115577

Raghu, A. V., Karuppanan, K. K., & Pullithadathil, B. (2019). Highly surface active phosphorus-doped onion-like carbon nanostructures: Ultrasensitive, fully reversible, and portable NH_3 gas sensors. *ACS Applied Electronic Materials, 1*(11), 2208–2219. https://doi.org/10.1021/acsaelm.9b00412

Rajendran, K., Nabeel, M. M., Pandian, K., Manikandan, V., Ganesan, S., Lo, H.-M., Rajendiran, N., Liu, X., & Oluwafemi, O. S. (2024). Fabrication of silicon quantum dots-methyl viologen nanohybrids: Turn-on-off-on fluorescence nanoprobe for the detection of *D*-penicillamine. *Microchemical Journal, 197*, 109803. https://doi.org/10.1016/j.microc.2023.109803

Rana, S. V. S. (2008). Metals and apoptosis: Recent developments. *Journal of Trace Elements in Medicine and Biology, 22*(4), 262–284. https://doi.org/10.1016/j.jtemb.2008.08.002

Rao, N., Singh, R., & Bashambu, L. (2021). Carbon-based nanomaterials: Synthesis and prospective applications. *Materials Today: Proceedings, 44*, 608–614. https://doi.org/10.1016/j.matpr.2020.10.593

Rohman, M. A., Baruah, P., Bhatta, A., & Mitra, S. (2019a). Deciphering the interactions of substituted chromones possessing acetylcholinesterase inhibition activity in human serum albumin matrix. *Journal of Molecular Liquids, 290*, 111210. https://doi.org/10.1016/j.molliq.2019.111210

Rohman, M. A., Baruah, P., Bhattacharjee, D., Myrboh, B., & Mitra, S. (2019b). Fluorescence solvatochromism and modulated anticholinergic activity of novel coumarin compounds sequestered in human serum albumin nanocavities. *New Journal of Chemistry, 43*(47), 18713–18725. https://doi.org/10.1039/c9nj03293b

Rout, B., Milko, P., Iron, M. A., Motiei, L., & Margulies, D. (2013). Authorizing multiple chemical passwords by a combinatorial molecular keypad lock. *Journal of the American Chemical Society, 135*(41), 15330–15333. https://doi.org/10.1021/ja4081748

Santhiran, A., Iyngaran, P., Abiman, P., & Kuganathan, N. (2021). Graphene synthesis and its recent advances in applications—A review. *C, 7*(4). https://doi.org/10.3390/c7040076

Sharma, V., Kaur, N., Tiwari, P., Saini, A. K., & Mobin, S. M. (2018). Multifunctional fluorescent "off-on-off" nanosensor for Au^{3+} and S^{2-} employing N, S co-doped carbon–dots. *Carbon, 139*, 393–403. https://doi.org/10.1016/j.carbon.2018.07.004

Singh, V. K., Singh, V., Yadav, P. K., Chandra, S., Bano, D., Koch, B., Talat, M., & Hasan, S. H. (2019). Nitrogen doped fluorescent carbon quantum dots for on-off-on detection of Hg^{2+} and glutathione in aqueous medium: Live cell imaging and IMPLICATION logic gate operation. *Journal of Photochemistry and Photobiology A: Chemistry, 384*, 112042. https://doi.org/10.1016/j.jphotochem.2019.112042

Sk, M. P., Sailapu, S. K., & Chattopadhyay, A. (2015). Luminescent carbon dots for logic operations in two phases. *ChemPhysChem, 16*(4), 723–727. https://doi.org/10.1002/cphc.201500136

Speranza, G. (2021). Carbon nanomaterials: Synthesis, functionalization and sensing applications. *Nanomaterials, 11*(4). https://doi.org/10.3390/nano11040967

Srinivasreddy, R. K., & Kadeppagari, R.-K. (2021). Functionalized carbon nanotubes for ammonia sensors. *Environmental applications of carbon nanomaterials-based devices* (pp. 251–263). https://doi.org/10.1002/9783527830978.ch10

Su, Y., Shi, B., Liao, S., Zhao, J., Chen, L., & Zhao, S. (2016). Silver nanoparticles/N-doped carbon-dots nanocomposites derived from Siraitia grosvenorii and its logic gate and surface-enhanced Raman scattering characteristics. *ACS Sustainable Chemistry & Engineering, 4*(3), 1728–1735. https://doi.org/10.1021/acssuschemeng.5b01698

Swaminathan, H., & Balasubramanian, K. (2019). Fluorescence on-off switching, Boolean logic gates like behavior of carbon quantum dots and highly sensitive bovine serum albumin sensing. *Journal of Applied Physics, 126*(8), 084503. https://doi.org/10.1063/1.5099464

Takakura, A., Beppu, K., Nishihara, T., Fukui, A., Kozeki, T., Namazu, T., Miyauchi, Y., & Itami, K. (2019). Strength of carbon nanotubes depends on their chemical structures. *Nature Communications, 10*(1), 3040. https://doi.org/10.1038/s41467-019-10959-7

Tian, T., Zhong, Y., Deng, C., Wang, H., He, Y., Ge, Y., & Song, G. (2017). Brightly near-infrared to blue emission tunable silver-carbon dot nanohybrid for sensing of ascorbic acid and construction of logic gate. *Talanta, 162*, 135–142. https://doi.org/10.1016/j.talanta.2016.10.021

Tong, Y.-J., Yu, L.-D., Huang, Y., Li, Y., Li, N., Fu, Q., Ye, Y.-X., Zhu, F., Pawliszyn, J., & Xu, J. (2022). High-quality full-color carbon quantum dots synthesized under an unprecedentedly mild condition. *iScience, 25*, 104421. https://doi.org/10.1016/j.isci.2022.104421

Tregubov, A. A., Nikitin, P. I., & Nikitin, M. P. (2018). Advanced smart nanomaterials with integrated logic-gating and biocomputing: Dawn of theranostic nanorobots. *Chemical Reviews, 118*(20), 10294–10348. https://doi.org/10.1021/acs.chemrev.8b00198

Valeur, B., & Berberan-Santos, M. N. (2013). *Molecular fluorescence: Principles and applications*. 2nd Edition, Wiley-VCH, Weinheim.

Vandarkuzhali, S. A. A., Natarajan, S., Jeyabalan, S., Sivaraman, G., Singaravadivel, S., Muthusubramanian, S., & Viswanathan, B. (2018). Pineapple peel-derived carbon dots: Applications as sensor, molecular keypad lock, and memory device. *ACS Omega, 3*(10), 12584–12592. https://doi.org/10.1021/acsomega.8b01146

Wang, M.-W., Fan, W., Li, X., Liu, Y., Li, Z., Jiang, W., Wu, J., & Wang, Z. (2023). Molecular carbons: How far can we go? *ACS Nano, 17*(21), 20734–20752. https://doi.org/10.1021/acsnano.3c07970

Wani, A. L., Ara, A., & Usmani, J. A. (2015). Lead toxicity: A review. *Interdisciplinary Toxicology, 8*(2), 55–64. https://doi.org/10.1515/intox-2015-0009

Welch, M. J., Hawker, C. J., & Wooley, K. L. (2009). The advantages of nanoparticles for PET. *Journal of Nuclear Medicine, 50*(11), 1743. https://doi.org/10.2967/jnumed.109.061846

Wu, C., Wang, K., Fan, D., Zhou, C., Liu, Y., & Wang, E. (2015). Enzyme-free and DNA-based multiplexer and demultiplexer. *Chemical Communications, 51*(88), 15940–15943. https://doi.org/10.1039/c5cc05565b

Wu, J., Wu, Z., Ding, H., Wei, Y., Huang, W., Yang, X., Li, Z., Qiu, L., & Wang, X. (2020). Three-dimensional graphene hydrogel decorated with SnO_2 for high-performance NO_2 sensing with enhanced immunity to humidity. *ACS Applied Materials & Interfaces, 12*(2), 2634–2643. https://doi.org/10.1021/acsami.9b18098

Xia, L., Li, X., Zhang, Y., Zhou, K., Yuan, L., Shi, R., Zhang, K., & Fu, Q. (2022). Sustainable and green synthesis of waste-biomass-derived carbon dots for parallel and semi-quantitative visual detection of Cr(VI) and Fe^{3+}. *Molecules, 27*(4). https://doi.org/10.3390/molecules27041258

Xiao, M., Zhou, Q., Zhang, H., Zhou, L., Ma, J., & Yi, C. (2021). Logic gate design using multi-color fluorescent carbon nanodots for smartphone-based information extraction. *ACS Applied Nano Materials, 4*(8), 8184–8191. https://doi.org/10.1021/acsanm.1c01413

Xie, Y.-J., Wu, W.-Y., Chen, H., Li, X., Zhang, H.-L., Liu, L.-L., Shao, X.-X., Shan, C.-F., Liu, W.-S., & Tang, Y. (2016). An elaborate supramolecular assembly for a smart nanodevice for ratiometric molecular recognition and logic gates. *Chemistry—A European Journal, 22*(24), 8339–8345. https://doi.org/10.1002/chem.201505082

Xu, J. X., Vithanage, B. C. N., Athukorale, S. A., & Zhang, D. (2018). Scattering and absorption differ drastically in their inner filter effects on fluorescence, resonance synchronous, and polarized resonance synchronous spectroscopic measurements. *Analyst, 143*(14), 3382–3389. https://doi.org/10.1039/c8an00790j

Xu, H., Liu, X., Wang, R., Gao, S., Hou, F., Liang, K., & Luo, S. (2021). A yellow-emitting carbon quantum dot-based fluorescent logic gate for the continuous detection of Au^{3+} and biothiols. *Chemical Communications, 57*(87), 11549–11552. https://doi.org/10.1039/d1cc04935f

Xu, J., Cui, K., Gong, T., Zhang, J., Zhai, Z., Hou, L., uz Zaman, F., & Yuan, C. (2022a). Ultrasonic-assisted synthesis of N-doped, multicolor carbon dots toward fluorescent inks, fluorescence sensors, and logic gate operations. *Nanomaterials, 12*(3), 312. https://doi.org/10.3390/nano12030312

Xu, X., Yang, S., Wang, Y., & Qian, K. (2022b). Nanomaterial-based sensors and strategies for heavy metal ion detection. *Green Analytical Chemistry, 2*, 100020. https://doi.org/10.1016/j.greeac.2022.100020

Yagi, T., Yoshida, K., Sakurai, S., Kawai, T., & Nonoguchi, Y. (2024). Semiconducting carbon nanotube extraction enabled by alkylated cellulose wrapping. *Journal of the American Chemical Society, 146*(30), 20913–20918. https://doi.org/10.1021/jacs.4c05468

Yang, Y., Zhao, J., & Li, Y. (2016). Theoretical study of the ESIPT process for a new natural product quercetin. *Scientific Reports, 6*(1), 32152. https://doi.org/10.1038/srep32152

Yang, H., He, L., Long, Y., Li, H., Pan, S., Liu, H., & Hu, X. (2018). Fluorescent carbon dots synthesized by microwave-assisted pyrolysis for chromium (VI) and ascorbic acid sensing and logic gate operation. *Spectrochimica Acta Part A: Molecular and Biomolecular Spectroscopy, 205*, 12–20. https://doi.org/10.1016/j.saa.2018.07.015

Yuan, Y., Zhao, X., Liu, S., Li, Y., Shi, Y., Yan, J., & Hu, X. (2016). A fluorescence switch sensor used for D-penicillamine sensing and logic gate based on the fluorescence recovery of carbon dots. *Sensors and Actuators B: Chemical, 236*, 565–573. https://doi.org/10.1016/j.snb.2016.06.007

Yuan, Y., Jiang, J., Liu, S., Yang, J., Yan, J., & Hu, X. (2017). Fluorescent carbon dots for glyphosate determination based on fluorescence resonance energy transfer and logic gate operation. *Sensors and Actuators B: Chemical, 242*, 545–553. https://doi.org/10.1016/j.snb.2016.11.050

Yun, C. S., Javier, A., Jennings, T., Fisher, M., Hira, S., Peterson, S., Hopkins, B., Reich, N. O., & Strouse, G. F. (2005). Nanometal surface energy transfer in optical rulers, breaking the FRET barrier. *Journal of the American Chemical Society, 127*(9), 3115–3119. https://doi.org/10.1021/ja043940i

Zhang, X.-F. (2018). BODIPY photosensitizers based on PET and heavy atom effect: A comparative study on the efficient formation of excited triplet state and singlet oxygen in BODIPY dimers and monomers. *Journal of Photochemistry and Photobiology A: Chemistry, 355*, 431–443. https://doi.org/10.1016/j.jphotochem.2017.07.019

Zhang, Y., & Zhang, J. (2022). Fluorescence resonance energy transfer-based aptasensor made of carbon-based nanomaterials for detecting lactoferrin at low concentrations. *ACS Omega, 7*(42), 37964–37970. https://doi.org/10.1021/acsomega.2c05129

Zhang, X., Hou, L., & Samorì, P. (2016). Coupling carbon nanomaterials with photochromic molecules for the generation of optically responsive materials. *Nature Communications, 7*(1), 11118. https://doi.org/10.1038/ncomms11118

Zhang, L., Qin, J., Yang, Q., Wei, S., & Yang, R. (2019). Cost-effective and facile fluorescent probes for label-free recognition of chlorpromazine hydrochloride and logic gate operation. *Journal of Photochemistry and Photobiology A: Chemistry, 382*, 111918. https://doi.org/10.1016/j.jphotochem.2019.111918

Zhu, J.-H., Li, M.-M., Liu, S.-P., Liu, Z.-F., Li, Y.-F., & Hu, X.-L. (2015). Fluorescent carbon dots for auramine *O* determination and logic gate operation. *Sensors and Actuators B: Chemical, 219*, 261–267. https://doi.org/10.1016/j.snb.2015.05.032

Zong, J., Yang, X., Trinchi, A., Hardin, S., Cole, I., Zhu, Y., Li, C., Muster, T., & Wei, G. (2014). Carbon dots as fluorescent probes for "off–on" detection of Cu^{2+} and L-cysteine in aqueous solution. *Biosensors and Bioelectronics, 51*, 330–335. https://doi.org/10.1016/j.bios.2013.07.042

Zou, W.-S., Zhao, Q.-C., Kong, W.-L., Wang, X.-F., Chen, X.-M., Zhang, J., & Wang, Y.-Q. (2018). Multi-level fluorescent logic gate based on polyamine coated carbon dots capable of responding to four stimuli. *Chemical Engineering Journal, 337*, 471–479. https://doi.org/10.1016/j.cej.2017.12.123

Chapter 6
Carbon-Based Nanomaterials in Enhancing the Performance of Electrochemical Sensors for Environmental Monitoring

Megha A. Deshmukh, Hanuman N. Thorat, and Namrata S. Gajmal

Abstract This chapter explores the critical role of carbon-based nanomaterials in advancing electrochemical sensors for environmental monitoring, emphasizing their applications in detecting a wide range of contaminants, including heavy metals, organic pollutants, endocrine disruptors, and perfluoroalkyl and polyfluoroalkyl substances (PFAS). It begins by underscoring the importance of environmental contaminant detection and the advantages of electrochemical methods, such as high sensitivity, selectivity, and rapid response times. This chapter then provides an overview of various carbon-based nanomaterials—carbon dots, graphene, carbon nanotubes, activated carbon, and carbon-supported single-atom catalysts (CSACs)—highlighting their unique properties, such as high surface area, conductivity, and catalytic activity, which enhance electrode performance in electrochemical sensors. Focusing on performance enhancement mechanisms, the discussion covers how these nanomaterials facilitate increased contaminant detection through improved electrochemical activity and stability. Additionally, specific sensor applications are reviewed for detecting established and emerging contaminants, while key challenges in sensor design, such as stability, reproducibility, and miniaturization, are also discussed. This chapter concludes with an outlook on future advancements, discussing the potential of carbon-based electrochemical sensors for real-time environmental monitoring and effective pollution control.

Keywords Electrochemical sensing · Heavy metal ions · PFAS · Endocrine disruptors · Carbon nanomaterials · Nanocomposites

M. A. Deshmukh (✉) · H. N. Thorat
Nanotechnology Centre, Centre of Energy and Environmental Technologies, VŠB–Technical University of Ostrava, Ostrava-Poruba, Czech Republic

N. S. Gajmal
Government Institute of Science, Aurangabad, Maharashtra, India

A. M. Parambil et al. (eds.), *Carbon*,
https://doi.org/10.1007/978-3-031-90613-8_6

Abbreviations

ACC	Activated carbon cloth
ASV	Anodic stripping voltammetry
BPA	Bisphenol A
CAC	Colloidal activated carbon
CBZ	Carbendazim
CDs	Carbon dots
CFP	Carbon fiber paper
CN	Carbon nitride
CNTs	Carbon nanotubes
CQD-AT	Carbon quantum dots-aminothiazole
CSACs	Carbon-supported single-atom catalysts
Cu-CN	Copper(I)-carbon nitride
DPASV	Differential pulse anodic stripping voltammetry
DPV	Differential pulse voltammetry
EDCs	Endocrine-disrupting chemicals
EDS	Energy-dispersive spectroscopy
EDTA	Ethylenediaminetetraacetic acid
EE2	17α-ethinylestradiol
Fe_3O_4@C	Magnetic carbon
GO	Graphene oxide
HEF	Heterogeneous electro-Fenton
HRP	Horseradish peroxidase
MAC	Macadamia nutshell activated carbon
MOF	Metal-organic framework
MSPE	Magnetic solid-phase extraction
MWCNTs	Multiwalled carbon nanotubes
MWNTs	Multiwalled carbon nanotubes
N-GQDs	Nitrogen-graphene quantum dots
NSAIDs	Non-steroidal anti-inflammatory drugs
PAHs	Polycyclic aromatic hydrocarbons
PANI	Polyaniline
PCBs	Polychlorinated biphenyls
PEI-CNTs	Polyethyleneimine-functionalized carbon nanotubes
PFAS	Perfluoroalkyl and polyfluoroalkyl substances
PFOA	Perfluorooctanoic acid
PFOS	Perfluorooctane sulfonate
RSD	Relative standard deviations
RSSCTs	Rapid small-scale column tests
SEM	Scanning electron microscope
SWASV	Square wave anodic stripping voltammetry
SWNTs	Single-walled carbon nanotubes
TEM	Transmission electron microscope
TOF	Turnover frequency

6.1 Introduction

Detecting environmental contaminants has become increasingly critical as the global population grows, industrial activities expand, and urbanization strengthens (Awewomom et al., 2024). Contaminants, ranging from heavy metals and pesticides to organic pollutants, can severely impact ecosystems and human health (Ali et al., 2019a). Effective monitoring is essential to prevent or mitigate the adverse consequences of pollution, ensuring safe drinking water, air quality, and agricultural products (Ali et al., 2019a). Environmental contaminants can originate from industrial processes, agricultural practices, waste disposal, and other human activities (Wang et al., 2024). For example, heavy metals like mercury, lead, and cadmium have serious health risks, including neurological damage and cancer, when they accumulate in the food chain or water supplies (Deshmukh et al., 2018a). Although essential for crop protection, pesticides often leach into soil and water, affecting aquatic ecosystems and non-target species (Orou-Seko et al., 2024). Similarly, organic pollutants such as polychlorinated biphenyls (PCBs) and polycyclic aromatic hydrocarbons (PAHs) have long-lasting effects on ecosystems and biodiversity due to their persistence and bioaccumulation (Cui et al., 2022; Wang et al., 2024). Endocrine disruptors interfere with hormone systems, leading to developmental, reproductive, and neurological issues, while perfluoroalkyl and polyfluoroalkyl substances (PFAS), often referred to as "forever chemicals," are highly persistent and linked to various health problems, including cancer and immune system disorders (Hu et al., 2023). The growing concern over these pollutants is compounded by their often-undetectable nature, especially at low concentrations (Hu et al., 2023; Cui et al., 2022; Wang et al., 2024). Traditional monitoring methods often require time-consuming and expensive procedures, which can delay timely interventions (Deshmukh et al., 2018c). Therefore, real-time, accurate, and sensitive detection methods are necessary to provide actionable data to help environmental agencies and industries with immediate corrective measures (Olawade et al., 2024; Wang et al., 2024; Ali et al., 2019a; Awewomom et al., 2024).

Electrochemical detection methods provide a promising solution to this challenge and are popular in environmental monitoring due to their compactness, ease of miniaturization, and potential for real-time monitoring (Deshmukh et al., 2018b). When combined with modern electrode materials, such as carbon-based nanomaterials, metal nanoparticles, and conducting polymers, electrochemical sensors exhibit excellent performance in detecting various contaminants, even at trace levels (Zhu et al., 2015; Sahragard et al., 2023; Kumar et al., 2024). One of the most important advantages of electrochemical detection methods is their high sensitivity (Arafa et al., 2024; Hassan et al., 2022). The high sensitivity of electrochemical sensors arises from their direct interaction with analytes on the electrode surface, which leads to an amplified response in the presence of even small quantities of contaminants (Arafa et al., 2024; Hassan et al., 2022). Modifying the electrode surface with selective coatings or nanomaterials can further enhance the signal, improving the detection limits of the sensors (Deshmukh et al., 2018c, 2020). Another critical advantage of electrochemical methods is their high selectivity. Environmental samples often contain a complex mixture of substances, including target contaminants and non-hazardous species (Manasa & Rout, 2024; Kamel & Abd-Rabboh, 2024; Hassan et al., 2022; Shanbhag et al., 2023; Barhoum et al., 2023). Electrochemical sensors can be tailored to distinguish specific analytes from a mixture by functionalizing the electrode surface with selective recognition elements, such as enzymes, antibodies, or molecularly imprinted polymers (Manasa & Rout, 2024; Kamel & Abd-Rabboh, 2024; Hassan et al., 2022; Shanbhag et al., 2023; Barhoum et al., 2023).

Traditional laboratory-based methods, such as chromatography or spectrophotometry, usually require lengthy sample preparation, transportation, and analysis, which delays the availability of results (Deshmukh et al., 2018a). In contrast, electrochemical detection methods offer a rapid response, often producing results within seconds to minutes of sampling (Hassan et al., 2022; Barhoum et al., 2023). This real-time capability is essential for early warning systems in water treatment plants, industrial discharge monitoring, or agricultural pesticide runoff assessment (Singh et al., 2023). Electrochemical sensors can also be integrated into portable, on-site devices, allowing immediate field testing. This portability reduces the reliance on central laboratories, enables continuous monitoring, and ensures that environmental data can be collected in real time (Huang et al., 2023; Maduraiveeran, 2022; Windmiller & Wang, 2013; Sampath et al., 2024; Aryal et al., 2024; Srinivasan & Tung, 2015). The fast response times of these sensors make them suitable for dynamic environments where pollutant levels may fluctuate rapidly. Beyond sensitivity, selectivity, and rapid response, electrochemical detection methods offer additional advantages such as cost-effectiveness and scalability (Huang et al., 2023; Windmiller & Wang, 2013). Electrochemical sensors are generally inexpensive to produce, especially when using common electrode materials like carbon or metals (Bhimaraya et al., 2023). They require minimal reagent use and have low energy consumption, making them highly practical for widespread use in environmental monitoring programs (Shanbhag et al., 2023).

Carbon nanomaterials such as carbon dots, graphene, carbon nanotubes (CNTs), activated carbon, and CSACs are crucial in improving the performance of electrochemical sensors. These materials enhance sensor performance by increasing sensitivity, enabling the detection of low concentrations of analytes (Porto et al., 2020; Pandey & Chusuei, 2021). They also improve selectivity through functionalization, allowing the sensor to target specific contaminants (Porto et al., 2020; Pandey & Chusuei, 2021). Their excellent electrical conductivity also accelerates electron transfer, leading to faster response times. Overall, these nanomaterials greatly boost the efficiency and effectiveness of electrochemical sensors for environmental monitoring and other applications (Gabrielle Sutanto et al., 2024; Asadian et al., 2019; Yang et al., 2015).

6.2 Types of Carbon Nanomaterials and Their Applications in Electrochemical Sensing

Carbon nanomaterials, including carbon dots (CDs), graphene, CNTs, activated carbon, and CSACs, have emerged as highly versatile materials in electrochemical sensing (Asadian et al., 2019; Gabrielle Sutanto et al., 2024; Yang et al., 2015). Their unique properties, such as large surface area, excellent electrical conductivity, chemical stability, and ease of functionalization, make them ideal for detecting a wide range of environmental contaminants, biomolecules, and industrial pollutants (Fig. 6.1). Each type of carbon nanomaterial offers specific advantages, allowing for

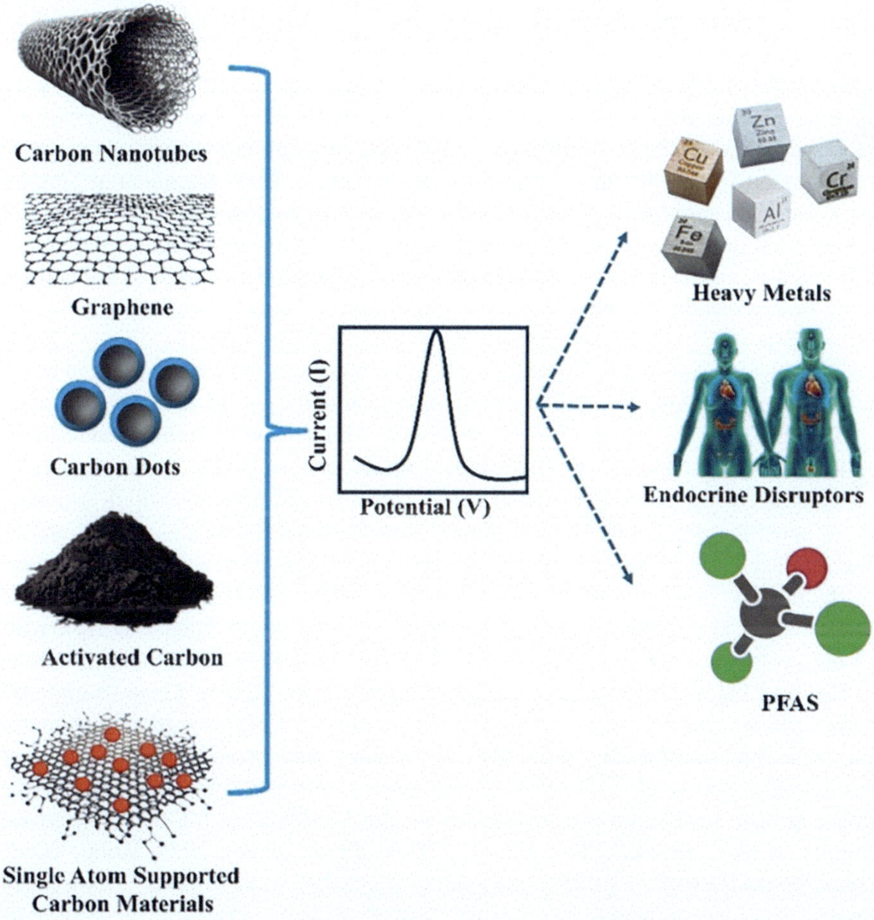

Fig. 6.1 Carbon nanomaterials and their applications as electrochemical sensors for the detection of industrial and biological effluents

tailored applications in different electrochemical sensors, thereby enhancing their sensitivity, selectivity, and response time across various fields.

6.2.1 Carbon Nanotubes

CNTs are highly valued materials in electrochemical sensing due to their unique structure, high conductivity, and adaptable surface chemistry (Yáñez-Sedeño et al., 2010; Zhang et al., 2022; Park et al., 2020). Composed of rolled graphene sheets, CNTs offer an exceptionally high surface area that supports abundant active sites and effective interactions with target molecules, making them highly sensitive to environmental changes (Yáñez-Sedeño et al., 2010). Their outstanding electrical conductivity enhances electron transfer efficiency, leading to rapid and strong electrochemical signals, which is a crucial advantage in detecting low analyte concentrations (Sánchez-Romate et al., 2022; Lekawa-Raus et al., 2014). Furthermore, CNTs are chemically stable and easily functionalized with molecules, enzymes, or nanoparticles, allowing for selective analyte detection by tailoring the surface to interact specifically with target species (Schroeder et al., 2019; Saguin et al., 2024). This versatility, combined with their robustness, fast response times, and signal amplification properties, makes CNTs ideal for developing advanced, reliable electrochemical sensors across various applications, including detecting biological molecules, heavy metals, and environmental pollutants (Schroeder et al., 2019; Saguin et al., 2024).

CNTs play a critical role in detecting heavy metal ions in electrochemical sensing, thanks to their high surface area, excellent conductivity, and ability to be modified for selective detection (Wu et al., 2019; Venkateswara Raju et al., 2023; Deshmukh et al., 2018a, 2018c). The large surface area of CNTs allows for a high density of active sites that can adsorb heavy metal ions effectively, which is essential for sensitive detection. Their superior conductivity enhances electron transfer between the electrode and the analyte, leading to a strong electrochemical response that facilitates the accurate measurement of heavy metal ion concentrations (Deshmukh et al., 2018b, 2018c). Additionally, CNTs can be functionalized with specific groups or chelating agents that selectively bind heavy metal ions, such as Pb, Hg, and Cd (Patil et al., 2018a, 2018b; Park et al., 2020). These functional groups enhance selectivity by increasing the affinity of CNTs for particular metal ions, reducing interference from other substances, and improving measurement accuracy. The ability of CNTs to form complexes with metal ions and to amplify signals at low concentrations makes them highly effective for detecting trace amounts of toxic metals (Morton et al., 2009; Yuwen et al., 2023; Zhou, 2024; Venkateswara Raju et al., 2023; Schroeder et al., 2019). This property is especially valuable for environmental monitoring, where accurate, real-time detection of heavy metals in water or soil is essential for assessing pollution levels and protecting public health.

A sensitive voltammetric method for detecting trace heavy metal ions using cysteine-modified CNT electrodes was presented by Morton et al. (2009). Cysteine, an amino acid with strong affinities for heavy metals, was covalently attached to CNTs and cast onto electrode surfaces. Heavy metal ions were accumulated on the cysteine-modified CNT surface before undergoing differential pulse anodic stripping voltammetry (DPASV). The resulting peak currents showed a linear relationship with metal ion concentrations, achieving detection limits of 1 ppb for Pb^{2+} and 15 ppb for Cu^{2+} (Morton et al., 2009). This technique was successfully applied to detect Pb^{2+} and Cu^{2+} in spiked lake water, with average recoveries of 96.2% and 94.5% and relative standard deviations (RSD) of 8.43% and 7.53%, respectively. Additionally, the modified CNTs demonstrated potential for the simultaneous detection of multiple heavy metal ions (Morton et al., 2009). A sensitive sensor for the detection of Pb(II) ions in water was developed using self-assembled polyethyleneimine-functionalized carbon nanotubes (PEI-CNTs) and graphene oxide (GO) films deposited onto gold interdigitated electrodes (Jimenez et al., 2021). This graphene-based nanostructure displayed resistive properties and enabled Pb(II) detection through impedance spectroscopy, amperometry, and potentiometric stripping analysis. The sensor achieved a detection limit of 36 pmol/L and a sensitivity of 4.3 μA L/μmol, with high repeatability. The detection mechanism relies on the strong affinity of Pb(II) ions for the functional groups on PEI-CNTs and GO, offering a promising approach for monitoring Pb contamination in drinking water. The electrochemical detection of Cu(II), Pb(II), and Hg(II) ions using a nanocomposite sensor made of polyaniline (PANI), single-walled carbon nanotubes (SWNTs), and ethylenediaminetetraacetic acid (EDTA) as a chelating ligand was reported. Differential pulse voltammetry (DPV) was employed to determine Cu(II), Pb(II), and Hg(II) concentrations, achieving detection limits of 0.08, 1.65, and 0.68 μM, respectively, with good sensitivity and selectivity for each ion (Deshmukh et al., 2018a). In another example, the study reports the electrochemical synthesis of PANI and its composite with SWNTs. The incorporation of SWNTs was analyzed for its impact on the electrochemical properties of PANI using techniques like electrochemical impedance spectroscopy (EIS) and cyclic voltammetry (CV) (Patil et al., 2018a). The PANI/SWNTs composite displayed a good electroactive performance and therefore was opted as a sensing platform for the detection of Ni ions (Patil et al., 2018a). Further research efforts were explored to study the impact of 100 MeV oxygen ion irradiation on the electrochemically synthesized PANI/SWNTs composite electrode (Patil et al., 2018b). The irradiated composite exhibited enhanced electroactive properties and was used as an electrochemical sensor to detect Co(II) ions. The sensor demonstrated high sensitivity, detecting concentrations as low as 0.355 μM or 0.1 mg/L, meeting the US EPA's maximum contaminant level (MCL) for Co(II) ions (Patil et al., 2018b).

Detecting endocrine disruptors is crucial because they can interfere with hormone systems, potentially causing reproductive, developmental, and immune issues in humans and wildlife (Marlatt et al., 2022). These chemicals, often present in trace amounts in water and food, can accumulate over time and disrupt biological processes even at low concentrations (Marlatt et al., 2022). Rapid and sensitive detection methods are essential for monitoring and mitigating their effects on health and the environment (Marlatt et al., 2022). Traditional methods like enzyme-linked immunosorbent assay (ELISA) and high-performance liquid chromatography (HPLC) are commonly employed for detecting endocrine-disrupting chemicals (EDCs) (Jaffrezic-Renault et al., 2020). While these techniques are known for their accuracy and sensitivity, they tend to be expensive, intricate, and unsuitable for field applications. In contrast, electrochemical methods offer a promising alternative due to their affordability, simplicity, and potential for integration into portable devices (Jaffrezic-Renault et al., 2020). A simple and cost-effective biosensor based on a screen-printed electrode was developed for the sensitive detection of alkylphenols, using horseradish peroxidase (HRP) entrapped in a nanocomposite gel made of SWCNTs and the ionic liquid [BMIM][PF6] (Gurban et al., 2011). The HRP-based biosensor demonstrated high sensitivity and stability, detecting 4-t-octylphenol and 4-n-nonylphenol in a linear range of 5.5–97 and 5.5–140 μM, respectively, with a rapid response time of 5 s. Detection limits were 1.1 μM for 4-t-octylphenol and 0.4 μM for 4-n-nonylphenol (S/N = 3) (Gurban et al., 2011). Y. C. Wang et al. fabricated a highly sensitive and selective electrochemical sensor for the determination of BPA in aqueous solution using reduced graphene oxide (rGO), CNT, and gold nanoparticle (AuNP)-modified screen-printed electrode (SPE) (Wang et al., 2015). The reported electrochemical sensor produced linear current responses for BPA concentrations of 1.45–20 and 20–1490 nM, with an ultralow detection limit of 800 pM. The sensor response was unpretentious by the presence of interferents, such as phenol, *p*-nitrophenol, pyrocatechol, 2,4-dinitrophenol, and hydroquinone (Wang et al., 2015). Recently, an electrochemical sensor based on a metal-organic framework (Cu, Zr-BTC) combined with CNT (Cu, Zr-BTC/CNT) was developed for rapid on-site detection of BPA (Tien et al., 2023). The employment of highly conductive CNT improved the recorded electrochemical signals from 2 to 50 μM, with a detection limit down to 0.5 μM (Tien et al., 2023). A composite of graphene oxide and β-cyclodextrin-functionalized multiwalled CNTs was used as a sensor electrode, leveraging the large surface area of graphene oxide and CNTs along with the excellent host–guest interaction properties of β-cyclodextrin (Alam & Deen, 2020). The electrochemical detection of BPA was enabled by a diffusion-controlled oxidation reaction involving equal protons and electrons. The sensor exhibited a two-step linear detection range of 0.05–5 μM and 5–30 μM, with a detection limit of 6 nM (Alam & Deen, 2020). The removal of (PFOA) and (PFOS) from water has earned substantial consideration due to their persistence in the environment, bioaccumulative nature, and potential toxicity.

While several destructive methods have been developed, achieving effective removal of PFOX (X = A or S) under mild conditions remains a priority (Li et al., 2011). In these regards, X. Li et al. applied multiwalled carbon nanotubes (MWNTs) to remove PFOX via electrochemically assistance adsorption. Electrosorption kinetics and isotherms were explored relative to open circuit (OC) adsorption and adsorption on powder MWNTs. Compared with powder MWNT adsorption, the initial adsorption rate ($\upsilon 0$) of 100 μg/L PFOX at 0.6 V increased 60-fold (PFOA) and 41-fold (PFOS) according to the pseudosecond-order kinetics model and the maximum electrosorption capacity (qm) of PFOX (50 μg/L to 10 mg/L) was enlarged 150-fold (PFOA) and 94-fold (PFOS), as simulated by the Langmuir model (Li et al., 2011). These significant improvements were assumed to benefit from enhanced electrostatic attraction under electrochemical assistance (Li et al., 2011).

6.2.2 Carbon Dots

CDs have acquired interest for their adaptable electronic and optical properties, which depend on the choice of precursor and synthesis method. Recently, a growing emphasis has been on utilizing natural materials for CD production. CDs are zero-dimensional materials with a diameter of less than 10 nm that hold multipurpose physical, optical, and chemical properties. Recently, Ismail et al. reported CDs synthesized from curry leaves using a microwave-assisted approach (Ismail et al., 2024). The spectroscopic and structural analysis of the material confirmed the existence of carbon graphitic elements with carboxyl and hydroxyl functional groups on the surfaces of CDs. AuNPs/CDs/GS electrodes in ferricyanide revealed that as-synthesized CDs formed, using a lower heating power of 700 W, demonstrate noticeable electrocatalytic activity with sluggish electron transfer kinetics.

Conversely, as-synthesized CDs created with a higher heating power of 800 W demonstrate reduced electrocatalytic activity but rapid electron transfer kinetics for electrochemical detection of Pb(II) (Ismail et al., 2024). The separate peaks also appeared in mixture solutions, with a minor drop in peak current density advocating the selectivity of the AuNPs/CDs/GS electrodes toward Pb(II) and Hg(II) ion detection (Ismail et al., 2024). Nitrogen-doped carbon dots (NCDs) are used as efficient fluorescent probes for the detection of dual heavy metals, including Fe(III) and Hg(II), in real samples. Further, the metal ion sensing is also validated by electrochemical CV measurements (Yadav et al., 2024). Real water sample analysis in pond, sea, and well water without treatment confirmed 90–100% recovery (Yadav et al., 2024).

An electrochemical sensor for simultaneous detection of Pb(II) and Hg(II) ions has been developed by Keerthana et al. using anodic stripping voltammetry (ASV) with a carbon quantum dots-aminothiazole (CQD-AT) composite coated on a carbon fiber paper electrode (CQD-AT/CFP) (Keerthana et al., 2024). The reported sensor exhibited enhanced sensitivity through its high electroactive surface area and abundant functional groups, enabling efficient adsorption of metal ions. This sensor demonstrates excellent performance, with a wide linear range (0.6×10^{-11}–160×10^{-6} M) and low detection limits (3.0 pM for Pb(II) and 6.2 pM for Hg(II)), making it suitable for detecting these metals in water samples (Keerthana et al., 2024). An integrated 1D/0D/1D hybrid nanomaterial was prepared by Bhuvaneswari et al., from MWNT-supported CQDs@MnO_2 nanomaterial for sensitive and selective electrochemical detection of Cd and Cr (Bhuvaneswari et al., 2023). The observed DPV results revealed that prepared MWNT (0.05 wt%) and CQD (0.1 wt%) supported MnO_2 nanoparticles offered effective detection response for Cr (IV) metal ion. The combination of 0D CQD, 1D MWCNT, and MnO_2 hybrid nanostructures created a synergistic effect, enhancing the prepared materials' electrochemical performance detecting target metal ions (Bhuvaneswari et al., 2023). A sulfur co-doped nitrogen-graphene quantum dot (S,N-GQD) was used for the simultaneous detection of Cd(II), Pb(II), and Hg(II), showing distinct and well-separated peaks (Saisree et al., 2023). The control N-GQD (without sulfur doping) only responded to Cd(II). The current response for Cd(II) on S,N-GQD was approximately seven times higher than on N-GQD. The S,N-GQD demonstrated the lowest detection limits reported so far for Cd(II), Pb(II), and Hg(II), with values of 1, 10, and 1 pM, respectively, for simultaneous metal ion sensing (Saisree et al., 2023).

Conversely, aloe vera gel (*Aloe barbadensis* Miller) derived self-assembled flower-shaped CQDs interface for ultra-sensitive electrochemical detection of the endocrine disruptor BPA displayed significant electrochemical behavior in the presence of tyrosinase enzyme (Srivastava et al., 2024). Two analytical methods were elaborated using electrochemical and spectrometric techniques to simultaneously detect the endocrine disruptors triclosan and methylparaben in personal care products (Santana & Spinelli, 2020). A sensitive electrode based on graphene QDs supported by chitosan was used. Under optimized conditions, with working potentials of +0.60 V for triclosan and +0.81 V (vs. Ag/AgCl) for methylparaben, linear calibration curves were attained through DPV in the range of 0.10–10.0 μmol/L. The detection limits were 0.03 μmol/L for triclosan and 0.04 μmol/L for methylparaben (Santana & Spinelli, 2020).

6.2.3 Activated Carbon

Activated carbon is highly effective for detecting and removing heavy metals, organic pollutants, and emerging contaminants like endocrine disruptors and PFAS (Rodríguez-Hernández et al., 2022; Li et al., 2024). Its high surface area and porous structure enable it to adsorb a wide range of pollutants, including toxic metals and organic chemicals (Satyam & Patra, 2024). For heavy metals, activated carbon captures ions through electrostatic interactions or complexation, while for organic pollutants like BPA, its functional groups facilitate adsorption (Murtaza et al., 2024). Modified versions, enhanced with specific chemical groups or nanoparticles, improve selectivity for PFAS and other contaminants. This makes activated carbon a versatile, cost-effective material for environmental monitoring and pollution control (Cantoni et al., 2021). Guan et al. proposed a kelp-derived activated porous carbon material (KPC) with an exclusive highly disordered graphite layer structure as an outstanding sensor via facile KOH activation method (Guan et al., 2020). The unique structure of KPC not only provided massive active sites but also displayed high sensitivity in the detection of heavy metal by square wave anodic stripping voltammetry (SWASV), with Pb^{2+} at 53.4 μA μM − 1 and Cd^{2+} at 26.5 μA μM − 1 in simultaneous detection (Guan et al., 2020). Pd nanoparticles supported on porous activated carbon (Pd@PAC), prepared using KOH activation and thermal reduction, were utilized for detecting trace levels of Cd^{2+}, Pb^{2+}, and Cu^{2+} ions through SWASV (Zhang et al., 2019). The sensor demonstrated high sensitivity and selectivity for simultaneous and individual detection of heavy metal ions, along with excellent anti-interference properties, reproducibility, repeatability, and stability under consistent experimental conditions. The porous structure of activated carbon, combined with Pd nanoparticles, enhanced electrochemical activity, while the interconnected micropores and mesopores improved mass diffusion and facilitated the deposition-stripping process of the metal ions (Zhang et al., 2019). The study involved the fabrication of an electrochemical sensor for the simultaneous detection of Pb^{2+} and Cd^{2+} ions using modified activated carbon paste electrodes (mCPE) (Aravind et al., 2024). AC-graphite, AC-reduced graphene oxide (RGO), and AC-RGO-chitosan were the types of mCPEs that were utilized. Though all the three mCPEs detected Cd^{2+} and Pb^{2+} in the simulated wastewater, the CPE with RGO and AC could detect Cd^{2+} as low as 10.91 μg L^{-1} and Pb^{2+} as low as 14.01 μg L^{-1}. The work explored the possibility of using AC as a potential sustainable substitute for graphite in CPE (Aravind et al., 2024). A carbon paste electrode (CPE) incorporating biomass-derived orange peel activated carbon (ACOP) and MWNTs was developed for Cd^{2+} detection using SWV and CV (Beas-Bernuy et al., 2023). The ACOP/MWCNT-CPE sensor showed high sensitivity, with an R^2 of 0.9907, a detection limit of 0.91 ± 0.79 μmol/L, and a quantification limit of 3.00 ± 2.60 μmol/L. It selectively detected Cd^{2+} even in the presence of other metal ions (Zn^{2+}, Pb^{2+}, Ni^{2+}, Co^{2+}, Cu^{2+}, and Fe^{2+}) and demonstrated excellent stability and reproducibility. Successfully applied to river water samples, it achieved recovery rates of 97.33–115.6%, proving its potential for cadmium analysis in environmental matrices (Beas-Bernuy et al., 2023).

Hayat et al. developed a nanocomposite of titanium oxide (TiO_2) and activated carbon, which was used to modify the surface of a screen-printed carbon electrode for constructing an electrochemical sensor (Hayat et al., 2016). This modified sensor was employed to detect four phenolic endocrine disruptors: *p*-nitrophenol, hydroquinone, catechol, and 1-naphthol. Using DPV under optimized conditions, the sensor demonstrated excellent linearity with correlation coefficients above 0.99. It also exhibited high sensitivity, with detection limits of 348 ng/L, 110.1 ng/L, 3.3 ng/L, and 7.2 μg/L for the respective compounds (S/N = 3) (Hayat et al., 2016). Very few studies have reported on the oxidation of BSPA using a magnetic-activated carbon-cobalt nanocomposite-modified screen-printed electrode (MAC-Co-SPE). In these regard, Fatemah et al. reported an electrochemical sensor of MAC-Co-SPE for sensitive quantification of BPA (Emambakhsh et al., 2022). The electrochemical measurements discovered that a blend of MAC-Co-SPE and screen-printed electrodes showed excellent voltammetric responses toward BPA oxidation with broad linear ranges between 0.1 and 350.0 μM and a lower limit of detection of 10.0 nM and good sensitivity of 218.1 μA mM^{-1} cm^{-2}. Further, BPA in various water and milk samples was identified by the standard addition procedure (Emambakhsh et al., 2022).

PFAS are characterized by highly stable carbon–fluorine bonds, contributing to their environmental persistence. Traditional separation techniques, such as membrane filtration or adsorption, often struggle with efficiency or demand substantial energy and chemical inputs for regeneration (Román Santiago et al., 2022). Electrochemical methods present a more sustainable and energy-efficient alternative, with advancements in electrode materials and electrochemical processes enabling effective separation and degradation of PFAS (Román Santiago et al., 2022). An inexpensive, commercially available granular activated carbon (GAC) was utilized to sorb and conduct electrical charge to achieve reductive defluorination of a 6-carbon (C6) PFAS, (*E*)-perfluoro(4-methylpent-2-enoic acid) (PFMeUPA), as well as perfluorooctane sulfonic acid (PFOS) (Ackerman Grunfeld et al., 2024). PFMeUPA is analogous to saturated, branched perfluorohexanoic acid. The results indicate PFMeUPA undergoes electrochemical reduction at an applied cell potential of 10 V without an electron-shuttling catalyst, such as vitamin B_{12}, which is typically required for reductive defluorination reactions (Ackerman Grunfeld et al., 2024). The rate of reduction was found to increase with decreasing reduction potential and increased temperature until −1.4 V vs. SHE. Less than 10% of the PFOS was reductively defluorinated, suggesting that more work is required to apply this technology for linear PFAS reduction. A recent study evaluated inexpensive granular activated carbon (GAC) for the reductive defluorination of a 6-carbon PFAS, PFMeUPA, and PFOS. PFMeUPA was electrochemically reduced at 10 V without requiring an electron-shuttling catalyst like vitamin B_{12}, with reduction rates increasing at lower potentials and higher temperatures up to −1.4 V vs. SHE (Ackerman Grunfeld et al., 2024). However, less than 10% of PFOS was defluorinated, indicating further optimization is needed for linear PFAS reduction.

6.2.4 Graphene

Graphene's unique atomically thin two-dimensional structure, along with its exceptional electrical and chemical properties, makes it highly effective for electrochemical sensing applications. This includes the detection of heavy metal ions, per- and polyfluoroalkyl substances (PFAS), and endocrine-disrupting compounds (Bobrinetskiy & Knezevic, 2018; Anegbe et al., 2024; Keramari et al., 2023; Duc La et al., 2024). Its high surface area allows adsorption of analytes, while its exceptional conductivity ensures efficient electron transfer for enhanced sensitivity (Yol Jeong et al., 2015; Ostertag et al., 2024; Kammarchedu et al., 2024; Goodrum et al., 2024; Fu & Gray, 2024). For heavy metal ions, functional groups or dopants facilitate adsorption and electrochemical reactions, such as reduction or oxidation, generating measurable signals (Navalon et al., 2017; Molina et al., 2016). In PFAS detection, graphene's hydrophobic nature promotes adsorption, and functionalized electrodes enhance reductive defluorination through electron transfer, breaking stable C-F bonds (Duinslaeger & Radjenovic, 2022; Navalon et al., 2017; Molina et al., 2016). For endocrine disruptors, π–π interactions with graphene's aromatic structure enable strong adsorption, while functionalized surfaces improve specificity (Alkhaldi et al., 2024; Anegbe et al., 2024; Tang et al., 2018; Ali et al., 2019b; Qureshi et al., 2020). Electrochemical techniques utilize these interactions to generate distinct current peaks, enabling precise detection (Kammarchedu et al., 2024; Keramari et al., 2023). The tunability of graphene with dopants or nanoparticles further increases its catalytic activity and selectivity, making it an excellent platform for sensitive, rapid, and specific electrochemical sensing in various environmental and analytical applications (Bobrinetskiy & Knezevic, 2018; Fu & Gray, 2024; Goodrum et al., 2024).

A colorimetric/electrochemical dual-mode Hg^{2+} sensing method was reported based on the analyte accelerating the peroxidase-mimetic activity of AuNPs loaded on graphene oxide (GO). The GO-AuNPs composite was readily fabricated via mechanically mixing AuNPs and GO with controllable proportions, exhibiting a weak peroxidase-like activity in catalyzing the oxidation reaction of 3,3′,5,5′-tetramethylbenzidine (TMB). Upon introducing Hg^{2+}, it combined with AuNPs rapidly, significantly enhancing the peroxidase-mimicking activity of GO-AuNPs. By capturing the color signal of the generated blue oxTMB and the electrooxidation response of residual TMB, a "turn-on" colorimetric and "turn-off" electrochemical bimodal method was established for Hg^{2+} measurement, providing a linear range of 10–60 μg/L in colorimetric detection and 0.001–20 μg/L in electrochemical analysis (Diao et al., 2025). In another approach, boron (B) and nitrogen (N) co-doped laser-induced porous graphene (LIGBN) nanostructured electrodes were developed for the individual and simultaneous electrochemical detection of Pb^{2+} and Cd^{2+} ions using a SWV technique (Chaudhary et al., 2024). The synergistic effect of B and N co-doping boosted the electrode's sensing performance with a low detection limit of 0.21 and 0.25 μM for Pb^{2+} and Cd^{2+} ions (Chaudhary et al., 2024). Yang et al. developed a sensor by assembling rGO and polyphosphate nanoparticles (polyP NPs) as a 3D porous structure (Yang et al., 2024). The functional groups present on the polyP NPs facilitated strong coordination with metal cations, enhancing the sensor's ability to detect heavy metal ions. This polyP NPs/rGO-based electrochemical sensor achieved sensitive detection of Pb^{2+} and Cu^{2+} ions, with detection limits of 1.6 and 0.9 nM, respectively, for both individual and simultaneous measurements (Yang et al., 2024). Nitrogen and bismuth co-doped, homogeneous porous petaloid graphene on polyimide films is used as a sensing platform for detecting Pb^{2+} and Cd^{2+} ions (Yuan et al., 2024). The mild surface ion exchange and self-metallization strategy enabled the in situ incorporation of bismuth elements with ultralow material consumption. A detection limit of 0.207 and 0.410 μg/L was achieved for Cd^{2+} and Pb^{2+}, respectively (Yuan et al., 2024).

The low electronic conductivity and poor cyclic stability of transition metal oxides present significant challenges. To address these issues, carbon-based frameworks are commonly used as supporting materials in heterostructures for catalytic and sensing applications (Sun et al., 2021). Among these, GO stands out due to its unique 2D structure, excellent biocompatibility, versatile surface chemistry, and extensive surface area featuring flat regions, defects, and wrinkles. GO is an ideal carrier, enhancing electron transfer efficiency in electrocatalysis, while its stable configuration minimizes catalyst agglomeration, extends lifespan, and promotes industrial applications (Maccaferri et al., 2019). Additionally, GO's abundant oxygen-containing functional groups enhance electron generation and impart hydrophilicity to the catalytic system (Kalluru et al., 2016). Ellipsoidal 3D Mn_2O_3 structures showed higher specific surface area, larger pore diameters, and enhanced electron transfer reactivity, leading to superior electrocatalytic activity for oxidizing phenolic endocrine disruptors, *o*-phenylphenol (OPP), and butylparaben (BP) (Sun et al., 2021). Wrapping these ellipsoids with GO further improved their performance by increasing active sites and surface-chemisorbed oxygen. The sensor achieved simultaneous detection of OPP and BP across wide linear ranges (0.002–20 and 0.003–24 μM) with low detection limits of 0.63 and 0.88 nM, respectively (Sun et al., 2021). Santos et al. developed a strategy for monitoring BPA using a screen-printed electrode modified with functionalized graphene, cadmium telluride quantum dots, and PEDOT:PSS polymer. The sensor demonstrated a detection limit of 3.4 nmol L^{-1} and achieved nearly 100% recovery efficiency (Santos et al., 2024). In another example, a Ni-doped ZnO/graphene nanoplatelet (Ni-ZnO/GNP) composite was employed for the electrochemical detection of commonly used drugs, acetaminophen and estradiol (Shen et al., 2022). The composite exhibited a significantly larger electrochemically active area and enhanced catalytic properties, improving the electrochemical responses of both drugs. This enabled their sensitive and reproducible detection with detection limits as low as 0.0072 μM for acetaminophen and 0.00056 μM for estradiol (Shen et al., 2022). A boron-doped graphene sponge anode was utilized for the electrochemical oxidation of C4-C8 per- and polyfluoroalkyl substances (Duinslaeger & Radjenovic, 2022). In a low-conductivity electrolyte (1 mS cm^{-1}) under a one-pass flow-through setup, removal efficiencies ranged from 16.7% to 67% at an anodic current density of 230 A m^{-2}, with energy consumption of 10.1 ± 0.7 kWh m^{-3}. The removal was achieved through electrosorption (7.4–35%) and electrooxidation (9.3–32%) (Duinslaeger & Radjenovic, 2022). Magnetic amine-functionalized graphene oxide (MAGO) was tested for its efficiency in adsorbing PFOA, PFOS, perfluorohexane sulfonate (PFHxS), and perfluorobutane sulfonate (PFBS) (Mahpishanian et al., 2024). It demonstrated exceptional removal rates of over 95% for long-chain PFAS and 85% for short-chain PFAS within 30 min, showcasing prompt kinetics and stability across a pH range of 4–7 (Mahpishanian et al., 2024). These outcomes highlight the synergistic role of graphene oxide and amine groups, utilizing both electrostatic and hydrophobic interactions for effective PFAS adsorption. Additionally, MAGO exhibits strong pollutant removal performance and excellent regeneration potential (Mahpishanian et al., 2024).

6.2.5 Carbon-Supported Single-Atom Catalysts (CSACs)

CSACs are crucial in detecting and removing heavy metals and organic pollutants, including emerging contaminants like endocrine disruptors and per- and polyfluoroalkyl substances (Rocha et al., 2023; Yuan & Ma, 2024; Liu et al., 2024; Zhang et al., 2023). The unique structure of CSACs, where individual metal atoms are dispersed on a carbon support, maximizes the surface area and active sites, resulting in enhanced catalytic activity and high selectivity for target pollutants (Rocha et al., 2023; Yuan & Ma, 2024; Liu et al., 2024; Zhang et al., 2023). This structure allows CSACs to efficiently adsorb heavy metals like lead, mercury, and cadmium, as well as organic contaminants such as pharmaceuticals, pesticides, and endocrine-disrupting chemicals, including PFAS. CSACs offer superior sensitivity for detecting low concentrations of these pollutants, with the metal centers facilitating the breakdown or transformation of organic contaminants through oxidation, reduction, or other catalytic processes (Li et al., 2023b). Additionally, their robust stability, reusability, and high surface-to-volume ratio make CSACs ideal for long-term environmental applications, such as in sensors for real-time monitoring or in water treatment systems. These properties allow CSACs to provide a cost-effective and efficient solution for detecting and degrading pollutants in water, air, and soil, making them an important tool in environmental remediation (Li et al., 2023b).

Nitrogen-doped carbon (N-C) is selected as the substrate due to its ability to disperse and securely anchor single transition metal atoms uniformly. Additionally, N-C is well-suited for modifying the coordination structure of metal atoms, enhancing their stability and performance (Wang et al., 2017; Yin et al., 2016). Li et al. reported the development of iron single-atom catalysts with distinct coordination structures, FeN_2C_2 and FeN_3P, designed for the selective detection of As(III) in the presence of Cu(II) (Li et al., 2023a). The FeN_3P catalyst demonstrated remarkable sensitivity, achieving 3.90 μA ppb^{-1} for As(III) detection in an $NH_4Cl/NH_3 \cdot H_2O$ electrolyte at pH 8.0, eliminating interference from Cu(II) (Li et al., 2023a). Experimental and theoretical electroanalysis models have been developed by Song et al. to explore the detection of Cu(II), Cd(II), As(III), and their mixtures using Ni single-atom catalysts with simple structures and stable coordination (Song et al., 2022). The combined use of thermodynamic modeling and Fourier transform infrared reflection identified the specific interactions between nanomaterials and heavy metal ions at the interface as the primary source of interference. This study provided a novel perspective on leveraging the interplay between nanomaterials and heavy metal ions to design anti-interference intelligent electrodes for stripping analysis (Song et al., 2022). Li et al. created a Pt single-atom catalyst supported on MoS_2 (Pt_1/MoS_2) for detecting As(III), achieving a high sensitivity of 3.31 μA ppb^{-1} (Li et al., 2021). Uric acid, a key biomarker for diseases like gout, kidney, and cardiovascular conditions, was detected by an electrochemical sensor developed by Hu et al. and achieved a detection limit of 33.3 nM (Hu et al., 2021). The sensor utilized a single-atom catalyst with Co(II) atoms anchored on an N-doped graphene matrix (A-Co-NG). Single-atom cobalt-anchored nitrogen-doped carbon catalysts (Co/N–C

SACs) showed a linear relationship with sulfamethoxazole in a wide range of concentration (0.08–66.5 μM) with a limit of detection of ~15 nM (Gu et al., 2022). A novel electrochemical sensor using Cu-N-C single-atom nanozymes (SAzymes) was developed, enabling simultaneous detection of multiple bisphenols with high sensitivity and selectivity. Cu-N-C SAzymes outperformed natural tyrosinase by catalyzing a wider range of bisphenols due to Cu^{+} and Cu^{2+} active centers anchored on graphene nanosheets through nitrogen coordination (Zhang et al., 2024). This method offers a reliable, rapid solution for detecting bisphenols in environmental and food samples.

6.3 Challenges and Future Perspective

Carbon-based nanomaterials offer great promise for detecting heavy metals and organic pollutants, including emerging contaminants like PFAS and endocrine disruptors, but face several challenges (Li et al., 2024). These include achieving high sensitivity and selectivity in complex environments, as interference from other substances can complicate detection, and ensuring the stability of functionalized materials under varying conditions (Li et al., 2024). Carbon materials like graphene, CNTs, and activated carbon have high surface areas. Still, their nonspecific adsorption can lead to interference from other substances, complicating the detection of low concentrations of emerging contaminants like PFAS and endocrine disruptors (Liao et al., 2022). Carbon nanomaterials often need to be functionalized with various chemical or biological group to improve the selectivity for specific contaminants. However, ensuring the stability of these functional groups under environmental conditions (e.g., varying pH, temperature, or ionic strength) remains a significant challenge. The functionalization process must also be efficient and reproducible, which can be difficult when scaling up for real-world applications. Achieving cost-effective mass production methods while maintaining the high performance of carbon nanomaterial sensors is a critical challenge for real-world applications, particularly for widespread environmental monitoring.

Future research could focus on developing new functionalization techniques that enhance the sensitivity and selectivity of carbon nanomaterials for specific contaminants. For example, incorporating molecular recognition sites (like antibodies or aptamers) on the surface of carbon nanomaterials could allow for highly selective detection of emerging pollutants like PFAS and endocrine disruptors, minimizing interference from other substances. Combining carbon nanomaterials with other advanced materials (e.g., metal-organic frameworks (MOFs), quantum dots, or 2D materials like MXenes) could improve the performance of sensors by enhancing their binding affinity for target pollutants. Integrating carbon nanomaterial-based sensors with innovative monitoring systems that provide real-time data could revolutionize environmental monitoring. Developing more sustainable and environmentally friendly methods for synthesizing carbon-based nanomaterials is a key area of future research. New approaches for sensor regeneration, such as electrochemical desorption or photochemical methods, could be explored to extend the lifetime of the sensors. Research efforts should make carbon nanomaterial-based sensors robust, portable, and user-friendly for widespread application in environmental monitoring, water quality control, and pollution management. As the use of carbon nanomaterial-based sensors becomes more widespread, regulatory bodies may begin to set guidelines for their use in environmental monitoring. This would help in standardizing detection methods for emerging pollutants like PFAS, endocrine disruptors, and heavy metals, ensuring that the sensors are consistent, reliable, and comparable across different regions.

6.4 Conclusions

In conclusion, carbon-based nanomaterials have emerged as powerful tools for enhancing electrochemical sensors used in environmental monitoring. Their unique physicochemical properties, including high surface area, excellent electrical conductivity, and catalytic potential, make them particularly effective in detecting various contaminants, from traditional heavy metals to emerging pollutants like PFAS and endocrine disruptors. This chapter has illustrated how various carbon nanomaterials, such as CDs, graphene, CNTs, activated carbon, and CSACs, improve sensor performance by increasing sensitivity, selectivity, and stability.

Despite the impressive progress made, several challenges remain, particularly regarding the reproducibility, long-term stability, and integration of these sensors into compact, portable devices suitable for field applications. These hurdles must be addressed to fully realize the potential of carbon-based nanomaterial sensors in practical environmental monitoring. By enabling rapid, on-site detection of environmental contaminants, these sensors could significantly mitigate pollution and protect public health. As research continues, interdisciplinary collaboration will be key to overcoming existing limitations and driving the next generation of sustainable, high-performance electrochemical sensors.

References

Ackerman Grunfeld, D., Jones, A. M., Sun, J., Le, S. T., Pickford, R., Huang, Q., Manefield, M., Kumar, N., Lee, M. J., & O'Carroll, D. M. (2024). Electrochemical degradation of a C6-perfluoroalkyl substance (PFAS) using a simple activated carbon cathode. *Environmental Science: Water Research & Technology, 10*(1), 272–287. https://doi.org/10.1039/D3EW00543G

Alam, A. U., & Deen, M. J. (2020). Bisphenol A electrochemical sensor using graphene oxide and β-cyclodextrin-functionalized multi-walled carbon nanotubes. *Analytical Chemistry, 92*(7), 5532–5539. https://doi.org/10.1021/acs.analchem.0c00402

Ali, I., Basheer, A. A., Mbianda, X. Y., Burakov, A., Galunin, E., Burakova, I., Mkrtchyan, E., Tkachev, A., & Grachev, V. (2019b, June). Graphene based adsorbents for remediation of noxious pollutants from wastewater. *Environment International, 127*, 160–180. https://doi.org/10.1016/j.envint.2019.03.029

Ali, H., Khan, E., & Ilahi, I. (2019a, March). Environmental chemistry and ecotoxicology of hazardous heavy metals: Environmental persistence, toxicity, and bioaccumulation. *Journal of Chemistry, 2019*, 1–14. https://doi.org/10.1155/2019/6730305

Alkhaldi, H., Alharthi, S., Alharthi, S., AlGhamdi, H. A., AlZahrani, Y. M., Mahmoud, S. A., Amin, L. G., et al. (2024). Sustainable polymeric adsorbents for adsorption-based water remediation and pathogen deactivation: A review. *RSC Advances, 14*(45), 33143–33190. https://doi.org/10.1039/D4RA05269B

Anegbe, B., Ifijen, I. H., Maliki, M., Uwidia, I. E., & Aigbodion, A. I. (2024). Graphene oxide synthesis and applications in emerging contaminant removal: A comprehensive review. *Environmental Sciences Europe, 36*(1), 15. https://doi.org/10.1186/s12302-023-00814-4

Arafa, F. N., Fahmy, H. M., Elshafiey, M., & Elshenawy, M. A. E. (2024). Nanostructured highly sensitive and selective electrochemical sensors and their applications. In *Handbook of nanosensors* (pp. 1529–1575). Springer. https://doi.org/10.1007/978-3-031-47180-3_19

Aravind, S. P., Anandhu, T. P., Jayakumar, K., Appukuttan, A., Rahul Krishna, B., Amala, J., & Bhuvaneshwari, S. (2024). Activated carbon based paste electrodes for the simultaneous and effective detection of divalent cadmium and lead ions in wastewater. *Environmental Technology, 45*(15), 3062–3075. https://doi.org/10.1080/09593330.2023.2206528

Aryal, P., Hefner, C., Martinez, B., & Henry, C. S. (2024). Microfluidics in environmental analysis: Advancements, challenges, and future prospects for rapid and efficient monitoring. *Lab on a Chip, 24*(5), 1175–1206. https://doi.org/10.1039/D3LC00871A

Asadian, E., Ghalkhani, M., & Shahrokhian, S. (2019, August). Electrochemical sensing based on carbon nanoparticles: A review. *Sensors & Actuators, B: Chemical, 293*, 183–209. https://doi.org/10.1016/j.snb.2019.04.075

Awewomom, J., Dzeble, F., Takyi, Y. D., Ashie, W. B., Ettey, E. N. Y. O., Afua, P. E., Sackey, L. N. A., Opoku, F., & Akoto, O. (2024). Addressing global environmental pollution using environmental control techniques: A focus on environmental policy and preventive environmental management. *Discover Environment, 2*(1), 8. https://doi.org/10.1007/s44274-024-00033-5

Barhoum, A., Hamimed, S., Slimi, H., Othmani, A., Abdel-Haleem, F. M., & Bechelany, M. (2023, June). Modern designs of electrochemical sensor platforms for environmental analyses: Principles, nanofabrication opportunities, and challenges. *TrEAC Trends in Environmental Analytical Chemistry, 38*, e00199. https://doi.org/10.1016/j.teac.2023.e00199

Beas-Bernuy, L. C., Cardenas-Riojas, A. A., Calderon-Zavaleta, S. L., Quiroz-Aguinaga, U., La Rosa-Toro, A., López, E. O., Asencios, Y. J. O., Baena-Moncada, A. M., & Muedas-Taipe, G. (2023). Cd^{2+} detection by an electrochemical electrode based on MWCNT–orange peel activated carbon. *ACS Omega, 8*(40), 37341–37352. https://doi.org/10.1021/acsomega.3c05154

Bhimaraya, K., Manjunatha, J. G., Nagarajappa, H., Tighezza, A. M., Albaqami, M. D., & Sillanpää, M. (2023). Development of a sensitive and inexpensive electrochemical sensor for indigotin using poly(valine) modified carbon paste electrode. *Heliyon, 9*(11), e20937. https://doi.org/10.1016/j.heliyon.2023.e20937

Bhuvaneswari, K., Radha, S., Sreeja, B. S., & Senthil Kumar, P. (2023, May). Development of in-situ electrochemical heavy metal ion sensor using integrated 1D/0D/1D hybrid by MWCNT and CQDs supported MnO_2 nanomaterial. *Environmental Research, 225*, 115570. https://doi.org/10.1016/j.envres.2023.115570

Bobrinetskiy, I. I., & Knezevic, N. Z. (2018). Graphene-based biosensors for on-site detection of contaminants in food. *Analytical Methods, 10*(42), 5061–5070. https://doi.org/10.1039/C8AY01913D

Cantoni, B., Turolla, A., Wellmitz, J., Ruhl, A. S., & Antonelli, M. (2021, November). Perfluoroalkyl substances (PFAS) adsorption in drinking water by granular activated carbon: Influence of activated carbon and PFAS characteristics. *Science of the Total Environment, 795*, 148821. https://doi.org/10.1016/j.scitotenv.2021.148821

Chaudhary, Y., Suman, S., Rakesh, B., Ojha, G. P., Deshpande, U., Pant, B., & Sankaran, K. J. (2024). Boron and nitrogen co-doped porous graphene nanostructures for the electrochemical detection of poisonous heavy metal ions. *Nanomaterials, 14*(9), 806. https://doi.org/10.3390/nano14090806

Cui, M., Xu, S., Song, W., Ye, H., Huang, J., Liu, B., Dong, B., & Xu, Z. (2022, February). Trace metals, polycyclic aromatic hydrocarbons and polychlorinated biphenyls in the surface sediments from Sanya River, China: Distribution, sources and ecological risk. *Environmental Pollution, 294*, 118614. https://doi.org/10.1016/j.envpol.2021.118614

Deshmukh, M. A., Bodkhe, G. A., Shirsat, S., Ramanavicius, A., & Shirsat, M. D. (2018b, October). Nanocomposite platform based on EDTA modified Ppy/SWNTs for the sensing of Pb(II) ions by electrochemical method. *Frontiers in Chemistry, 6*, 451. https://doi.org/10.3389/fchem.2018.00451

Deshmukh, M. A., Celiesiute, R., Ramanaviciene, A., Shirsat, M. D., & Ramanavicius, A. (2018a). EDTA_PANI/SWCNTs nanocomposite modified electrode for electrochemical determination of copper (II), lead (II) and mercury (II) ions. *Electrochimica Acta, 259*, 930–938.

Deshmukh, M. A., Shirsat, M. D., Ramanaviciene, A., & Ramanavicius, A. (2018c). Composites based on conducting polymers and carbon nanomaterials for heavy metal ion sensing (review). *Critical Reviews in Analytical Chemistry, 48*(4), 293–304. https://doi.org/10.1080/10408347.2017.1422966

Deshmukh, M. A., Kang, B.-C., & Ha, T.-J. (2020). Non-enzymatic electrochemical glucose sensors based on polyaniline/reduced-graphene-oxide nanocomposites functionalized with silver nanoparticles. *Journal of Materials Chemistry C, 8*, 5112–5123. https://pubs.rsc.org/en/content/articlelanding/2020/tc/c9tc06836h#!divAbstract

Diao, Q., Bu, Z., Feng, R., Chen, X., Liu, J., Tang, Z., Liang, H., Tian, Q., Li, S., & Niu, X. (2025, January). Performance-complementary colorimetric/electrochemical bimodal detection of Hg^{2+} based on analyte-accelerated peroxidase-mimicking activity of GO-AuNPs. *Sensors & Actuators, B: Chemical, 422*, 136598. https://doi.org/10.1016/j.snb.2024.136598

Duc La, D., Khong, H. M., Nguyen, X. Q., Dang, T.-D., Bui, X. T., Nguyen, M. K., Hao Ngo, H., & Duc Nguyen, D. (2024, September). A review on advances in graphene and porphyrin-based electrochemical sensors for pollutant detection. *Sustainable Chemistry One World, 3*, 100017. https://doi.org/10.1016/j.scowo.2024.100017

Duinslaeger, N., & Radjenovic, J. (2022, April). Electrochemical degradation of per- and polyfluoroalkyl substances (PFAS) using low-cost graphene sponge electrodes. *Water Research, 213*, 118148. https://doi.org/10.1016/j.watres.2022.118148

Emambakhsh, F., Asadollahzadeh, H., Rastakhiz, N., & Mohammadi, S. Z. (2022, August). Highly sensitive determination of bisphenol A in water and milk samples by using magnetic activated carbon—Cobalt nanocomposite-screen printed electrode. *Microchemical Journal, 179*, 107466. https://doi.org/10.1016/j.microc.2022.107466

Fu, H., & Gray, K. A. (2024, December). Graphene-encapsulated nanocomposites: Synthesis, environmental applications, and future prospects. *Science of the Total Environment, 955*, 176753. https://doi.org/10.1016/j.scitotenv.2024.176753

Gabrielle Sutanto, L., Sabilla, S., Wardhana, B. Y., Ramadani, A., Sari, A. P., Anjani, Q. K., Basirun, W. J., Amrillah, T., Amalina, I., & Jiwanti, P. K. (2024). Carbon nanomaterials as electrochemical sensors for theophylline: A review. *RSC Advances, 14*(39), 28927–28942. https://doi.org/10.1039/D4RA03585B

Goodrum, R., Weldekidan, H., Li, H., Mohanty, A. K., & Misra, M. (2024). Graphene-based nanostructures from green processes and their applications in biomedical sensors. *Advanced Industrial and Engineering Polymer Research, 7*(1), 37–53. https://doi.org/10.1016/j.aiepr.2023.03.001

Gu, Q., Wang, Z., Ding, Q., Li, H., Wu, P., & Cai, C. (2022). Electrochemical quantification of sulfamethoxazole antibiotic in environmental water using zeolitic imidazolate framework (ZIF)-derived single-atom cobalt catalyst in nitrogen-doped carbon nanostructures. *Sensors & Diagnostics, 1*(5), 1052–1062. https://doi.org/10.1039/D2SD00117A

Guan, J., Fang, Y., Zhang, T., Wang, L., Zhu, H., Mingliang, D., & Zhang, M. (2020). Kelp-derived activated porous carbon for the detection of heavy metal ions via square wave anodic stripping voltammetry. *Electrocatalysis, 11*(1), 59–67. https://doi.org/10.1007/s12678-019-00568-9

Gurban, A.-M., Rotariu, L., Baibarac, M., Baltog, I., & Bala, C. (2011). Sensitive detection of endocrine disrupters using ionic liquid—Single walled carbon nanotubes modified screen-printed based biosensors. *Talanta, 85*(4), 2007–2013. https://doi.org/10.1016/j.talanta.2011.07.045

Hassan, M. H., Khan, R., & Andreescu, S. (2022). Advances in electrochemical detection methods for measuring contaminants of emerging concerns. *Electrochemical Science Advances, 2*(6), e2100184. https://doi.org/10.1002/elsa.202100184

Hayat, A., Rhouati, A., Mishra, R. K., Alonso, G. A., Nasir, M., Istamboulie, G., & Marty, J. L. (2016). An electrochemical sensor based on TiO_2/activated carbon nanocomposite modified screen printed electrode and its performance for phenolic compounds detection in water samples. *International Journal of Environmental Analytical Chemistry, 96*(3), 237–246. https://doi.org/10.1080/03067319.2015.1137910

Hu, F. X., Hu, T., Chen, S., Wang, D., Rao, Q., Liu, Y., Dai, F., Guo, C., Yang, H. B., & Li, C. M. (2021). Single-atom cobalt-based electrochemical biomimetic uric acid sensor with wide linear range and ultralow detection limit. *Nano-Micro Letters, 13*(1), 7. https://doi.org/10.1007/s40820-020-00536-9

Hu, C.-Y., Qiao, J.-C., Gui, S.-Y., Xu, K.-X., Dzhambov, A. M., & Zhang, X.-J. (2023, August). Perfluoroalkyl and polyfluoroalkyl substances and hypertensive disorders of pregnancy: A systematic review and meta-analysis. *Environmental Research, 231*, 116064. https://doi.org/10.1016/j.envres.2023.116064

Huang, C.-W., Lin, C., Nguyen, M. K., Hussain, A., Bui, X.-T., & Ngo, H. H. (2023). A review of biosensor for environmental monitoring: Principle, application, and corresponding achievement of sustainable development goals. *Bioengineered, 14*(1), 58–80. https://doi.org/10.1080/21655979.2022.2095089

Ismail, N. S., Ariffin, A. S. K., Azman, N. I. A. M., Halim, N. H. A., Sabani, N., Juhari, N., & Shamsudin, S. A. (2024, June). Electrochemical detection of heavy metal ions using gold nanoparticles on carbon dots extracted from curry leaves. *IJNeaM, 17*, 195–202. https://doi.org/10.58915/ijneam.v17iJune.855

Jaffrezic-Renault, N., Kou, J., Tan, D., & Guo, Z. (2020). New trends in the electrochemical detection of endocrine disruptors in complex media. *Analytical and Bioanalytical Chemistry, 412*(24), 5913–5923. https://doi.org/10.1007/s00216-020-02516-9

Jimenez, M., Mawin, J., Avila, A., de Barros, A., Lopez, E. O., Alvarez, F., Riul, A., & Perez-Taborda, J. A. (2021). Polyethyleneimine-functionalized carbon nanotube/graphene oxide composite: A novel sensing platform for Pb(II) acetate in aqueous solution. *ACS Omega, 6*(28), 18190–18199. https://doi.org/10.1021/acsomega.1c02085

Kalluru, P., Vankayala, R., Chiang, C.-S., & Hwang, K. C. (2016, July). Nano-graphene oxide-mediated in vivo fluorescence imaging and bimodal photodynamic and photothermal destruction of tumors. *Biomaterials, 95*, 1–10. https://doi.org/10.1016/j.biomaterials.2016.04.006

Kamel, A. H., & Abd-Rabboh, H. S. M. (2024). Electrochemical sensors based on molecularly imprinted polymers for the detection of chlorophenols as emergent distributing chemicals (EDCs): A review. *Analytical Methods, 16*(25), 4024–4040. https://doi.org/10.1039/D4AY00386A

Kammarchedu, V., Asgharian, H., Zhou, K., Khamsi, P. S., & Ebrahimi, A. (2024). Recent advances in graphene-based electroanalytical devices for healthcare applications. *Nanoscale, 16*(27), 12857–12882. https://doi.org/10.1039/D3NR06137J

Keerthana, P., George, A., Bharath, M., Ghosh, M., & Varghese, A. (2024, January). Dual ion specific electrochemical sensor using aminothiazole-engineered carbon quantum dots. *Chemical Engineering Journal, 480*, 148018. https://doi.org/10.1016/j.cej.2023.148018

Keramari, V., Karastogianni, S., & Girousi, S. (2023). New prospects in the electroanalysis of heavy metal ions (Cd, Pb, Zn, Cu): Development and application of novel electrode surfaces. *Methods and Protocols, 6*(4), 60. https://doi.org/10.3390/mps6040060

Kumar, H., Kumari, R., Singh, D., & Mangla, B. (2024, December). Advances in nanomaterials based electrochemical sensors for rapid detection of food additives: A comprehensive review. *TrAC Trends in Analytical Chemistry, 181*, 118011. https://doi.org/10.1016/j.trac.2024.118011

Lekawa-Raus, A., Patmore, J., Kurzepa, L., Bulmer, J., & Koziol, K. (2014). Electrical properties of carbon nanotube based fibers and their future use in electrical wiring. *Advanced Functional Materials, 24*(24), 3661–3682. https://doi.org/10.1002/adfm.201303716

Li, X., Chen, S., Quan, X., & Zhang, Y. (2011). Enhanced adsorption of PFOA and PFOS on multiwalled carbon nanotubes under electrochemical assistance. *Environmental Science & Technology, 45*(19), 8498–8505. https://doi.org/10.1021/es202026v

Li, P.-H., Yang, M., Song, Z.-Y., Chen, S.-H., Xiao, X.-Y., Lin, C.-H., & Huang, X.-J. (2021). Highly sensitive and stable determination of As(III) under near-neutral conditions: Benefit from the synergetic catalysis of Pt single atoms and active S atoms over Pt1/MoS2. *Analytical Chemistry, 93*(45), 15115–15123. https://doi.org/10.1021/acs.analchem.1c03416

Li, P.-H., Song, Z.-Y., Xiao, X.-Y., Liang, B., Yang, M., Chen, S.-H., Liu, W.-Q., & Huang, X.-J. (2023a, January). Coordination engineering strategy of iron single-atom catalysts boosts anti Cu(II) interference detection of As(III) with a high sensitivity. *Journal of Hazardous Materials, 442*, 130122. https://doi.org/10.1016/j.jhazmat.2022.130122

Li, Z., Tian, E., Wang, S., Ye, M., Li, S., Wang, Z., Ma, Z., et al. (2023b). Single-atom catalysts: Promotors of highly sensitive and selective sensors. *Chemical Society Reviews, 52*(15), 5088–5134. https://doi.org/10.1039/D2CS00191H

Li, X., Shen, X., Jiang, W., Xi, Y., & Li, S. (2024, June). Comprehensive review of emerging contaminants: Detection technologies, environmental impact, and management strategies. *Ecotoxicology and Environmental Safety, 278*, 116420. https://doi.org/10.1016/j.ecoenv.2024.116420

Liao, Z., Zi, Y., Zhou, C., Zeng, W., Luo, W., Zeng, H., Xia, M., & Luo, Z. (2022). Recent advances in the synthesis, characterization, and application of carbon nanomaterials for the removal of endocrine-disrupting chemicals: A review. *The International Journal of Molecular Sciences, 23*(21), 13148. https://doi.org/10.3390/ijms232113148

Liu, L., Yung, K.-F., Yang, H., & Liu, B. (2024). Emerging single-atom catalysts in the detection and purification of contaminated gases. *Chemical Science, 15*(17), 6285–6313. https://doi.org/10.1039/D4SC01030B

Maccaferri, G., Terzi, F., Xia, Z., Vulcano, F., Liscio, A., Palermo, V., & Zanardi, C. (2019, February). Highly sensitive amperometric sensor for morphine detection based on electrochemically exfoliated graphene oxide. Application in screening tests of urine samples. *Sensors & Actuators, B: Chemical, 281*, 739–745. https://doi.org/10.1016/j.snb.2018.10.163

Maduraiveeran, G. (2022). Nanomaterials-based portable electrochemical sensing and biosensing systems for clinical and biomedical applications. *Journal of Analytical Science and Technology, 13*(1), 35. https://doi.org/10.1186/s40543-022-00344-3

Mahpishanian, S., Zhou, M., & Foudazi, R. (2024). Magnetic amino-functionalized graphene oxide nanocomposite for PFAS removal from water. *Environmental Science: Advances, 3*(12), 1698–1713. https://doi.org/10.1039/D4VA00171K

Manasa, G., & Rout, C. S. (2024). Versatile MXenes as electrochemical sensors for heavy metal ions and phenolic moiety-containing industrial chemicals: Recent development and prospects. *Materials Advances, 5*(1), 83–122. https://doi.org/10.1039/D3MA00362K

Marlatt, V. L., Bayen, S., Castaneda-Cortès, D., Delbès, G., Grigorova, P., Langlois, V. S., Martyniuk, C. J., et al. (2022, May). Impacts of endocrine disrupting chemicals on reproduction in wildlife and humans. *Environmental Research, 208*, 112584. https://doi.org/10.1016/j.envres.2021.112584

Molina, J., Cases, F., & Moretto, L. M. (2016, November). Graphene-based materials for the electrochemical determination of hazardous ions. *Analytica Chimica Acta, 946*, 9–39. https://doi.org/10.1016/j.aca.2016.10.019

Morton, J., Havens, N., Mugweru, A., & Wanekaya, A. K. (2009). Detection of trace heavy metal ions using carbon nanotube-modified electrodes. *Electroanalysis, 21*(14), 1597–1603. https://doi.org/10.1002/elan.200904588

Murtaza, G., Ahmed, Z., Valipour, M., Ali, I., Usman, M., Iqbal, R., Zulfiqar, U., et al. (2024). Recent trends and economic significance of modified/functionalized biochars for remediation of environmental pollutants. *Scientific Reports, 14*(1), 217. https://doi.org/10.1038/s41598-023-50623-1

Navalon, S., Dhakshinamoorthy, A., Alvaro, M., Antonietti, M., & García, H. (2017). Active sites on graphene-based materials as metal-free catalysts. *Chemical Society Reviews, 46*(15), 4501–4529. https://doi.org/10.1039/C7CS00156H

Olawade, D. B., Wada, O. Z., Ige, A. O., Egbewole, B. I., Olojo, A., & Oladapo, B. I. (2024, December). Artificial intelligence in environmental monitoring: Advancements, challenges, and future directions. *Hygiene and Environmental Health Advances, 12*, 100114. https://doi.org/10.1016/j.heha.2024.100114

Orou-Seko, A., Chirawurah, D., Ndago, J. A., Nkansah-Baido, M., Pwatirah, D., Kolekang, A. S., & Adokiya, M. N. (2024, December). Farmers' pesticide use and knowledge of aquatic ecosystem contamination with its perceived health risk from contaminated fish consumption in Northern Ghana. *Scientific African, 26*, e02351. https://doi.org/10.1016/j.sciaf.2024.e02351

Ostertag, B. J., Porshinsky, E. J., Nawarathne, C. P., & Ross, A. E. (2024). Surface-roughened graphene oxide microfibers enhance electrochemical reversibility. *Langmuir, 40*(23), 12124–12136. https://doi.org/10.1021/acs.langmuir.4c01004

Pandey, R. R., & Chusuei, C. C. (2021). Carbon nanotubes, graphene, and carbon dots as electrochemical biosensing composites. *Molecules, 26*(21), 6674. https://doi.org/10.3390/molecules26216674

Park, S.-J., Deshmukh, M. A., Kang, B.-C., Jeon, J.-Y., Chen, C., & Ha, T.-J. (2020). Review—A review of advanced electronic applications based on carbon nanomaterials. *ECS Journal of Solid State Science and Technology, 9*(7), 071002. https://doi.org/10.1149/2162-8777/abb035

Patil, H. K., Deshmukh, M. A., Bodkhe, G. A., & Shirsat, M. D. (2018a). Sensitive detection of heavy metal ions: An electrochemical approach. *International Journal of Modern Physics B, 32*(19), 1840042. https://doi.org/10.1142/S0217979218400428

Patil, H. K., Deshmukh, M. A., Bodkhe, G. A., Shirsat, S. M., Asokan, K., & Shirsat, M. D. (2018b). Dimethylglyoxime modified swift heavy oxygen ions irradiated polyaniline/single walled carbon nanotubes composite electrode for detection of cobalt ions. *Materials Research Express, 5*(6), 065048. https://doi.org/10.1088/2053-1591/aaccb3

Porto, L. S., Silva, D. N., de Oliveira, A. E. F., Pereira, A. C., & Borges, K. B. (2020). Carbon nanomaterials: Synthesis and applications to development of electrochemical sensors in determination of drugs and compounds of clinical interest. *Reviews in Analytical Chemistry, 38*(3), 1–16. https://doi.org/10.1515/revac-2019-0017

Qureshi, U. A., Hameed, B. H., & Ahmed, M. J. (2020, December). Adsorption of endocrine disrupting compounds and other emerging contaminants using lignocellulosic biomass-derived porous carbons: A review. *JWPE, 38*, 101380. https://doi.org/10.1016/j.jwpe.2020.101380

Rocha, G. F. S. R., da Silva, M. A. R., Rogolino, A., Diab, G. A. A., Noleto, L. F. G., Antonietti, M., & Teixeira, I. F. (2023). Carbon nitride based materials: More than just a support for single-atom catalysis. *Chemical Society Reviews, 52*(15), 4878–4932. https://doi.org/10.1039/D2CS00806H

Rodríguez-Hernández, J. A., Araújo, R. G., López-Pacheco, I. Y., Rodas-Zuluaga, L. I., González-González, R. B., Parra-Arroyo, L., Sosa-Hernández, J. E., et al. (2022). Environmental persistence, detection, and mitigation of endocrine disrupting contaminants in wastewater treatment plants—A review with a focus on tertiary treatment technologies. *Environmental Science: Advances, 1*(5), 680–704. https://doi.org/10.1039/D2VA00179A

Román Santiago, A., Medina, P. B., & Su, X. (2022, January). Electrochemical remediation of perfluoroalkyl substances from water. *Electrochimica Acta, 403*, 139635. https://doi.org/10.1016/j.electacta.2021.139635

Saguin, N. S. G., Maulik, G., Cao, X., Luo, X., Nag, A., Gao, J., Deng, S., & Wong, J. W. C. (2024, October). CNTs-based biosensors for enzyme detection. *Sensors and Actuators A: Physical, 377*, 115753. https://doi.org/10.1016/j.sna.2024.115753

Sahragard, A., Varanusupakul, P., & Miró, M. (2023, September). Nanomaterial decorated electrodes in flow-through electrochemical sensing of environmental pollutants: A critical review. *TrAC Trends in Analytical Chemistry, 39*, e00208. https://doi.org/10.1016/j.teac.2023.e00208

Saisree, S., Arya Nair, J. S., & Yesodha, S. K. (2023). Graphene quantum dots doped with sulfur and nitrogen as versatile electrochemical sensors for heavy metal ions Cd(II), Pb(II), and Hg(II). *ACS Applied Nano Materials, 6*(2), 1224–1234. https://doi.org/10.1021/acsanm.2c04804

Sampath, K., Thirupathi, B., Venishetty, S. K., Bharadwaj, D. L., & Kummari, S. (2024). The development process of point-of-care sensor-based detection modules for biomedical diagnoses. In *Biosensors for personalized healthcare* (pp. 297–313). Springer. https://doi.org/10.1007/978-981-97-5473-1_10

Sánchez-Romate, X. F., Jiménez-Suárez, A., & Ureña, A. (2022). Electrical properties of carbon nanotubes. In *Handbook of carbon nanotubes* (pp. 213–247). Springer. https://doi.org/10.1007/978-3-030-91346-5_60

Santana, E. R., & Spinelli, A. (2020). Electrode modified with graphene quantum dots supported in chitosan for electrochemical methods and non-linear deconvolution of spectra for spectrometric methods: Approaches for simultaneous determination of triclosan and methylparaben. *Microchimica Acta, 187*(4), 250. https://doi.org/10.1007/s00604-020-04225-7

Santos, A. M., Feitosa, M. H. A., Wong, A., Fatibello-Filho, O., Sotomayor, M. D. P. T., & Moraes, F. C. (2024, January). Functionalized graphene, quantum dots, and PEDOT:PSS based screen-printed electrode for the endocrine disruptor bisphenol a determination. *Sensors & Actuators, B: Chemical, 399*, 134745. https://doi.org/10.1016/j.snb.2023.134745

Satyam, S., & Patra, S. (2024). Innovations and challenges in adsorption-based wastewater remediation: A comprehensive review. *Heliyon, 10*(9), e29573. https://doi.org/10.1016/j.heliyon.2024.e29573

Schroeder, V., Savagatrup, S., He, M., Lin, S., & Swager, T. M. (2019). Carbon nanotube chemical sensors. *Chemical Reviews, 119*(1), 599–663. https://doi.org/10.1021/acs.chemrev.8b00340

Shanbhag, M. M., Manasa, G., Mascarenhas, R. J., Mondal, K., & Shetti, N. P. (2023, November). Fundamentals of bio-electrochemical sensing. *Chemical Engineering Journal Advances, 16*, 100516. https://doi.org/10.1016/j.ceja.2023.100516

Shen, Y., Yang, J., Zhang, Y., Zeng, T., & Wan, Q. (2022). Fabrication of a branch-like Ni-doped ZnO/graphene nanoplatelet composite for enhanced electrochemical determination of 17β-estradiol and acetaminophen. *Journal of Electronic Materials, 51*(9), 5310–5321. https://doi.org/10.1007/s11664-022-09763-9

Singh, B. J., Chakraborty, A., & Sehgal, R. (2023, December). A systematic review of industrial wastewater management: Evaluating challenges and enablers. *Journal of Environmental Management, 348*, 119230. https://doi.org/10.1016/j.jenvman.2023.119230

Song, Z.-Y., Xiao, X.-Y., Chen, S.-H., Li, Y.-Y., Yang, Y.-F., Huang, C.-C., Duan, W., Yang, M., Li, P.-H., & Huang, X.-J. (2022). Sensing material-dependent interference of multiple heavy metal ions: Experimental and simulated thermodynamics study on Cu(II), Cd(II), and As(III) electroanalysis. *Analytical Chemistry, 94*(16), 6225–6233. https://doi.org/10.1021/acs.analchem.1c05617

Srinivasan, B., & Tung, S. (2015). Development and applications of portable biosensors. *SLAS Technology, 20*(4), 365–389. https://doi.org/10.1177/2211068215581349

Srivastava, A., Singh, K. R. B., Ahmed, K., Singh, R. P., Pandey, M. D., Pandey, S. S., & Singh, J. (2024). Aloe vera gel (*Aloe barbadensis Miller*) derived self-assembled flower-shaped carbon quantum dot interface for ultra-sensitive electrochemical detection of the endocrine disruptor bisphenol-A. *New Journal of Chemistry, 48*(12), 5543–5552. https://doi.org/10.1039/D4NJ00061G

Sun, J., Xu, L., Shi, Z., Zhao, Q., Wang, H., & Gan, T. (2021, January). Morphology-tunable hollow Mn_2O_3 nanostructures: Highly efficient electrocatalysts and their electrochemical sensing for phenolic endocrine disruptors via toughening of graphene oxide. *Sensors & Actuators, B: Chemical, 327*, 128889. https://doi.org/10.1016/j.snb.2020.128889

Tang, H., Zhao, Y., Shan, S., Yang, X., Liu, D., Cui, F., & Xing, B. (2018). Theoretical insight into the adsorption of aromatic compounds on graphene oxide. *Environmental Science: Nano, 5*(10), 2357–2367. https://doi.org/10.1039/C8EN00384J

Tien, N. N., Dat, N. T., Manh, N. B., Hieu, N. D., Tu, V. C., Ngan, N. T. T., Ha, V. T. T., Decorse, P., Piro, B., & Thu, V. T. (2023). Cu,Zr-BTC/CNT composite for electrochemical detection of endocrine disruptor bisphenol A. *Journal of Materials Science, 58*(43), 16699–16713. https://doi.org/10.1007/s10853-023-09083-1

Venkateswara Raju, C., Cho, C. H., Rani, G. M., Manju, V., Umapathi, R., Huh, Y. S., & Park, J. P. (2023, February). Emerging insights into the use of carbon-based nanomaterials for the electrochemical detection of heavy metal ions. *Coordination Chemistry Reviews, 476*, 214920. https://doi.org/10.1016/j.ccr.2022.214920

Wang, Y.-C., Cokeliler, D., & Gunasekaran, S. (2015). Reduced graphene oxide/carbon nanotube/gold nanoparticles nanocomposite functionalized screen-printed electrode for sensitive electrochemical detection of endocrine disruptor bisphenol A. *Electroanalysis, 27*(11), 2527–2536. https://doi.org/10.1002/elan.201500120

Wang, J., Huang, Z., Liu, W., Chang, C., Tang, H., Li, Z., Chen, W., et al. (2017). Design of N-coordinated dual-metal sites: A stable and active Pt-free catalyst for acidic oxygen reduction reaction. *Journal of the American Chemical Society, 139*(48), 17281–17284. https://doi.org/10.1021/jacs.7b10385

Wang, F., Xiang, L., Leung, K. S.-Y., Elsner, M., Zhang, Y., Guo, Y., Pan, B., et al. (2024). Emerging contaminants: A one health perspective. *The Innovation, 5*(4), 100612. https://doi.org/10.1016/j.xinn.2024.100612

Windmiller, J. R., & Wang, J. (2013). Wearable electrochemical sensors and biosensors: A review. *Electroanalysis, 25*(1), 29–46. https://doi.org/10.1002/elan.201200349

Wu, W., Jia, M., Zhang, Z., Chen, X., Zhang, Q., Zhang, W., Li, P., & Chen, L. (2019, July). Sensitive, selective and simultaneous electrochemical detection of multiple heavy metals in environment and food using a low cost Fe_3O_4 nanoparticles/fluorinated multi-walled carbon nanotubes sensor. *Ecotoxicology and Environmental Safety, 175*, 243–250. https://doi.org/10.1016/j.ecoenv.2019.03.037

Yadav, R., Lahariya, V., Vikas, Singh, A. K., Das, A., Yadav, A., & Gupta, G. (2024, September). Fluorometric sensing and nanomolar level detection of heavy metal ions using nitrogen doped carbon dots. *Emergent Materials*. https://doi.org/10.1007/s42247-024-00825-8

Yáñez-Sedeño, P., Pingarrón, J. M., Riu, J., & Xavier Rius, F. (2010). Electrochemical sensing based on carbon nanotubes. *TrAC Trends in Analytical Chemistry, 29*(9), 939–953. https://doi.org/10.1016/j.trac.2010.06.006

Yang, C., Denno, M. E., Pyakurel, P., & Jill Venton, B. (2015, August). Recent trends in carbon nanomaterial-based electrochemical sensors for biomolecules: A review. *Analytica Chimica Acta, 887*, 17–37. https://doi.org/10.1016/j.aca.2015.05.049

Yang, M., He, D., Zheng, S., & Yang, L. (2024, March). In situ biosynthesized polyphosphate nanoparticles/reduced graphene oxide composite electrode for highly sensitive detection of heavy metal ions. *Environmental Research, 244*, 117966. https://doi.org/10.1016/j.envres.2023.117966

Yin, P., Yao, T., Wu, Y., Zheng, L., Lin, Y., Liu, W., Ju, H., et al. (2016). Single cobalt atoms with precise N-coordination as superior oxygen reduction reaction catalysts. *Angewandte Chemie, 128*(36), 10958–10963. https://doi.org/10.1002/ange.201604802

Yol Jeong, S., Jeong, S., Lee, S. W., Kim, S. T., Kim, D., Jeong, H. J., Han, J. T., et al. (2015). Enhanced response and sensitivity of self-corrugated graphene sensors with anisotropic charge distribution. *Scientific Reports, 5*(1), 11216. https://doi.org/10.1038/srep11216

Yuan, J., & Ma, X. (2024). Single atom catalysts for environmental remediation. In *Advanced materials for multidisciplinary applications* (pp. 267–282). Springer. https://doi.org/10.1007/978-3-031-39404-1_10

Yuan, X., Wu, X., Ling, Y., Li, S., Chen, J., & Zhang, Z. (2024). In situ bismuth ion exchange plating micro-electrochemical sensor based on laser-induced graphene for trace Cd^{2+} and Pb^{2+} detection. *Journal of Environmental Chemical Engineering, 12*(2), 112161. https://doi.org/10.1016/j.jece.2024.112161

Yuwen, T., Shu, D., Zou, H., Yang, X., Wang, S., Zhang, S., Liu, Q., et al. (2023). Carbon nanotubes: A powerful bridge for conductivity and flexibility in electrochemical glucose sensors. *Journal of Nanobiotechnology, 21*(1), 320. https://doi.org/10.1186/s12951-023-02088-7

Zhang, T., Jin, H., Fang, Y., Guan, J. B., Ma, S. J., Pan, Y., Zhang, M., Zhu, H., Liu, X. D., & Du, M. L. (2019, March). Detection of trace Cd^{2+}, Pb^{2+} and Cu^{2+} ions via porous activated carbon supported palladium nanoparticles modified electrodes using SWASV. *Materials Chemistry and Physics, 225*, 433–442. https://doi.org/10.1016/j.matchemphys.2019.01.010

Zhang, X., Sun, L., Sun, Y., Zhou, M., Wang, S., Cao, Z., Zhang, X., Wei, Y., & Xu, Y. (2022, January). Effect of CNTs concentration on the microstructure and the sensing behavior of UIO-66-NH2/CNTs towards Pb^{2+} detection. *Results in Chemistry, 4*, 100595. https://doi.org/10.1016/j.rechem.2022.100595

Zhang, J., Tang, X., Hong, Y., Chen, G., Chen, Y., Zhang, L., Gao, W., Zhou, Y., & Sun, B. (2023). Carbon-based single-atom catalysts in advanced oxidation reactions for water remediation: From materials to reaction pathways. *Eco-Environment and Health, 2*(2), 47–60. https://doi.org/10.1016/j.eehl.2023.04.002

Zhang, M., Wang, G., Chen, J., & Lu, X. (2024, June). Single atom nanozyme sensing platform for simultaneous rapid detection of multiple bisphenols. *Analytica Chimica Acta, 1307*, 342628. https://doi.org/10.1016/j.aca.2024.342628

Zhou, X. (2024). Electrochemical detection of heavy metal ions in water using MWCNT/ZnO nanocomposite. *International Journal of Electrochemical Science, 19*(5), 100559. https://doi.org/10.1016/j.ijoes.2024.100559

Zhu, C., Yang, G., Li, H., Du, D., & Lin, Y. (2015). Electrochemical sensors and biosensors based on nanomaterials and nanostructures. *Analytical Chemistry, 87*(1), 230–249. https://doi.org/10.1021/ac5039863

Chapter 7
Enhancing the Sensitivity of Contaminant Detection: Surface-Enhanced Raman Spectroscopy (SERS) with Carbon Nanomaterials

Rajesh Unnathpadi, Navami Sunil, and Biji Pullithadathil

Abstract The use of carbon-based nanomaterials to improve the sensitivity of surface-enhanced Raman spectroscopy (SERS) for the detection of contaminants has been reviewed in this chapter. Because of its high sensitivity, specificity, and potential for label-free detection of various chemical pollutants, SERS has become a potent analytical tool in response to the growing need for quick, sensitive environmental, and food safety monitoring detection techniques. Reduced graphene oxide (rGO), carbon nanotubes (CNTs), carbon quantum dots (CQDs), and carbon nanofibers (CNFs) are examples of carbon nanomaterials that are useful SERS substrates because of their special physicochemical characteristics, which include chemical stability, high surface area, and tunable electronic characteristics. Specifically, hybrid nanocomposites that combine metallic nanoparticles like gold (Au) or silver (Ag) with carbon nanomaterials provide synergistic enhancements through chemical and electromagnetic mechanisms, enabling detection limits for various environmental contaminants. This chapter also examines the potential of carbon nanomaterials in next-generation monitoring technologies for thorough contaminant profiling and future directions for their integration in scalable, environmentally friendly SERS platforms. Case studies showing the sensitive and selective detection of priority pollutants in complex matrices highlight the effectiveness of different carbon-based SERS platforms. With the promise of easily accessible applications in industrial and environmental settings, special attention is paid to developing portable SERS devices that use carbon-based substrates for on-site testing. These developments position carbon nanomaterial-enhanced SERS as a game-changing method for accurate, real-time contaminant detection in various fields.

R. Unnathpadi · N. Sunil · B. Pullithadathil (✉)
Nanosensors and Clean Energy Laboratory, Department of Chemistry and Nanoscience and Technology, PSG Institute of Advanced Studies, Coimbatore, India
e-mail: pbm@psgias.ac.in

A. M. Parambil et al. (eds.), *Carbon*,
https://doi.org/10.1007/978-3-031-90613-8_7

Keywords Surface-enhanced Raman spectroscopy · Surface plasmon resonance · Hybrid nanocomposites · Biosensing · Label-free detection · Green nanomaterials · Carbon-based nanomaterials

Abbreviations

Au/GO NCs	Gold/reduced graphene oxide nanocomposites
CE	Chemical enhancement
CNFs	Carbon nanofibers
CNTs	Carbon nanotubes
CQDs	Carbon quantum dots
CT	Charge transfer
CVD	Chemical vapor deposition
DFT	Density functional theory
DMF	Dimethyl formamide
EFs	Enhancement factors
EM	Electromagnetic
GO	Graphene oxide
G-SERS	Graphene oxide-mediated surface-enhanced Raman spectroscopy
HOMO	Highest occupied molecular orbital
LOD	Limit of detection
LSPR	Localized surface plasmon resonance
LUMO	Lowest unoccupied molecular orbital
MB	Methylene blue
MWCNT	Multi-walled carbon nanotube
NPs	Nanoparticles
NWs	Nanowires
PAN	Polyacrylonitrile
PNPs	Plasmonic nanoparticles
PVA	Polyvinyl alcohol
RGO	Reduced graphene oxide
SERS	Surface-enhanced Raman spectroscopy
SPR	Surface plasmon resonance
SWCNT	Single-walled carbon nanotube
TE	Textile effluents
VOCs	Volatile organic compounds

7.1 Introduction

Contaminants are substances that can be harmful and are present in the environment in amounts higher than the natural levels. These contaminants can be found in water, soil, and air. Contaminants emerged from industrialization, globalization, and civilization, which can impact/destroy the ecosystem nowadays. The proper handling of these contaminants needs major attention for a sustainable environment. Contaminants can be physical, chemical, and biological, harming animals and humans. For example, air contaminants (ozone), volatile organic compounds (VOCs), and soil contaminants include heavy metals (lead, mercury) and pesticides (Zamora-Ledezma et al., 2021). Nanomaterials are the class of compounds whose dimension is within 1–100 nm. Generally, nanomaterials are classified into natural nanomaterials: materials that are found in nature, for example, cellulose nanofibrils (plant cell walls), clay nanoparticles (kaolinite), keratin nanofibers (human hair, skin, nails), graphite (graphite ore), and starch (corn, potato), and have potential applications in food packaging and preservation, biomedical field for drug delivery, tissue engineering, biosensors, and in water treatment and purification techniques (Asha & Narain, 2020; Baig et al., 2021; Saleh, 2020). Synthetic nanomaterials are man-made materials like metal nanoparticles, metal oxides, quantum dots, carbon nanotubes, nanogels, nanocomposites, and dendrimers and have revolutionized various applications such as in biomedical, electronics, energy, and so on. Nowadays, environmental contamination is occurring repeatedly everywhere, including air, water, and soil contamination, which can be detected using various detection methods such as chromatographic techniques, electrochemical techniques, and colorimetric based and are time-consuming and cost-effective (Pena-Pereira et al., 2021). Therefore, developing accurate analytical methods for identifying and detecting contaminants is in great demanded (Zhang et al., 2021). Raman spectroscopy offers fast, sensitive, and high-throughput analysis of various environmental contaminants in this scenario.

Inelastic light scattering from a sample is the basis for the method known as Raman spectroscopy. Photons can experience inelastic or elastic scattering when they come into contact with matter. The energy of the incident and scattered photons stays constant in elastic scattering, also called Rayleigh scattering (Mulvaney & Keating, 2000; Kudelski, 2008; Das & Agrawal, 2011). In contrast, inelastic scattering causes photons to interact with the material and either gain or lose energy, resulting in scattered photons having a different frequency than incident ones. Stokes and anti-Stokes Raman scattering are terms used to describe how a photon interacts with a molecule in its ground state or first vibrationally excited state, respectively. The dispersed Raman rays give a unique fingerprint in the Raman spectrum, revealing details about the vibrational modes of the material or molecule being studied. A strong tool for identification and structural study, its fingerprint is specific to the material. Based on variables such as the sample's composition, symmetry, and surroundings, a Raman spectrum offers intrinsic vibrational information that represents the chemical components, molecule structure, conformation, and intermolecular interactions. Despite its benefits, Raman scattering's intrinsic low sensitivity, which produces weak Raman signals, restricts its wider use in the biomedical area, particularly concerning ultrasensitive detection. To overcome this drawback of Raman spectroscopy, surface-enhanced Raman spectroscopy (SERS) was developed (Cialla et al., 2012; Zhang et al., 2017; Halvorson & Vikesland, 2010).

7.2 Surface-Enhanced Raman Spectroscopy

Surface-enhanced Raman scattering is a powerful phenomenon that amplifies the Raman signals by several orders of magnitude when analytes are adsorbed on or near the surface of metallic nanostructures such as gold or silver. The enhancement of Raman signals is achieved by using noble metal nanostructures, such as gold and silver, which amplify the weak Raman scattering signals, often by factors ranging from 10^7 to 10^{14}. This remarkable enhancement capability has made SERS an attractive tool for identifying and detecting targets, especially due to its high efficiency, sensitivity, and stability in aqueous environments. By leveraging the strengths of both infrared and Raman spectroscopy, SERS allows for detecting samples in solution and offers heightened sensitivity and discrimination. The remarkable million-fold enhancement provided by SERS enables the detection of signals from monolayer species on metal surfaces, and more recently, it has demonstrated sensitivity at the single-molecule level (Camden et al., 2008). Over the past four decades, SERS has been applied across various fields, including chemistry, material sciences, analytical sciences, surface sciences, and biomedical research. SERS offers several distinct advantages in bioanalysis compared to traditional methods: (1) It possesses ultra-high surface sensitivity, capable of detecting even single molecules; (2) SERS signals capture the intrinsic molecular fingerprint of biological systems; (3) it is resistant to photobleaching and photodegradation, making it ideal for long-term monitoring; (4) SERS peaks are much narrower—10–100 times—than the fluorescence emission peaks from organic dyes or quantum dots (McNay et al., 2011; Schlücker, 2014); (5) it allows for easy multiplex detection using a single wavelength excitation; and (6) SERS-active nanostructures can be customized in size, shape, and coating for specific detection purposes. Additionally, SERS substrates can be optimized for near-infrared lasers to reduce autofluorescence from biological samples and minimize photodamage from visible lasers to living cells. Consequently, SERS has made great strides, particularly in biomedical and bioanalytical applications such as in vivo tumor imaging and the identification of viruses, cancer cells, and biomolecules (Cialla-May et al., 2017).

7.2.1 Mechanisms of SERS

Various factors, including the laser excitation parameters such as wavelength, polarization, and incident direction, influence the enhancement achieved through SERS. The detection setup, including the scattering configuration, polarizer, and collection angle, also plays a critical role. The intrinsic Raman cross-section of the probe molecules or materials, along with the surrounding environment's dielectric properties and the SERS substrates' optical characteristics, also significantly impacts the level of enhancement (Le Ru & Etchegoin, 2008). The Raman signal of a molecule on a plasmonic substrate can be amplified through two distinct mechanisms: electromagnetic enhancement and chemical enhancement depicted in Fig. 7.1. The overall SERS enhancement (G_{SERS}) is represented as the product of the

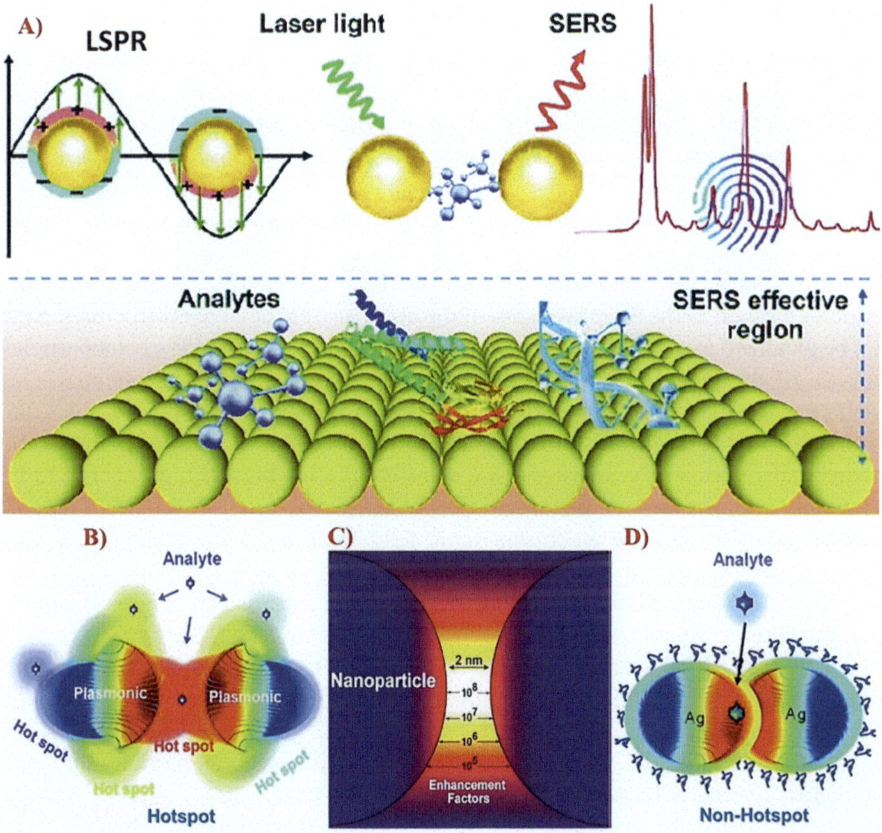

Fig. 7.1 (**a**) Schematic illustration of the mechanism of SERS for detection of analytes. (Reproduced with permission from Xie Liping et al. Copyright 2022, Springer). (**b**) Chemical enhancement (CM) "hot spot" formation in the gap of silver nanoparticles. (**c**) An illustration of a hot spot in the gap between the particles. (Reproduced with permission from Hakonen et al. Copyright 2015, Analytica Chimica Acta). (**d**) "Non-hot spots." (Reproduced with permission from Radziuk et al. Copyright 2014, Physical Chemistry Chemical Physics)

electromagnetic ($G_{\text{SERS}}^{\text{EM}}$) and chemical ($G_{\text{SERS}}^{\text{Chem}}$) contributions, expressed as follows (Pilot et al., 2017):

$$G_{\text{SERS}} = G_{\text{SERS}}^{\text{Em}} \cdot G_{\text{SERS}}^{\text{Chem}} \tag{7.1}$$

The electromagnetic (EM) mechanism is the primary contributor to the overall SERS enhancement, often resulting in amplifications of the Raman signal ranging from 4 to 11 orders of magnitude. Although the chemical enhancement (CE) is comparatively smaller, typically enhancing the signal by 10–100 times, it can still substantially impact the SERS signal's characteristics.

7.2.2 Electromagnetic Enhancement (EM)

Surface-enhanced Raman scattering (SERS) is a phenomenon that arises from the interplay between light–metal interactions (plasmonic processes) and light–molecule interactions (vibrational spectroscopy). When a laser strikes the interface between metal and dielectric materials, it can induce collective oscillation of the delocalized conduction electrons in metal nanostructures. Surface plasmon resonance (SPR) occurs when the frequency of the incident light aligns with the natural oscillation frequency of these free electrons. This resonance frequency is influenced by factors such as the size, shape, dielectric environment, and electron density of the particles, as well as the effective mass of the electrons. In metal nanostructures, SPR can be localized to a specific region, known as localized SPR (LSPR). Nanoparticles capable of producing a strong LSPR effect are termed plasmonic nanoparticles (PNPs), with silver (Ag), gold (Au), and copper (Cu) being common examples due to their strong SPR in the visible to near-infrared spectrum. LSPR results in the resonant absorption or scattering of incident light, allowing the light energy to be efficiently coupled into the metal nanoparticles, leading to a 2–5 orders of magnitude increase in the local electromagnetic field intensity at the nanoparticle surface—this is crucial for the substantial enhancement seen in SERS. When molecules are positioned near a plasmonic material, the Raman process can be greatly amplified, resulting in what is known as the SERS effect (Ding et al., 2016).

A very high SERS enhancement can be produced in the small space between two nanoparticles when positioned very close. A "hot spot" is a highly localized area that exhibits a markedly increased electric field and generates a powerful SERS signal shown in Fig. 7.2. Hot spots can appear at sites such as particle–substrate nanogaps, interparticle nanogaps, or nanotips (Xu et al., 2001) and are usually found in the spaces between nanoparticle aggregates. Most molecules that affect the observed SERS spectrum are found in these hot spots, which are usually 1–5 nm in size. Because of the high electromagnetic connection, the EM field is especially strong in the spaces between nanoparticle dimers and oligomers. The gap size significantly impacts the SERS enhancement factors (EFs) in the nanogaps and the

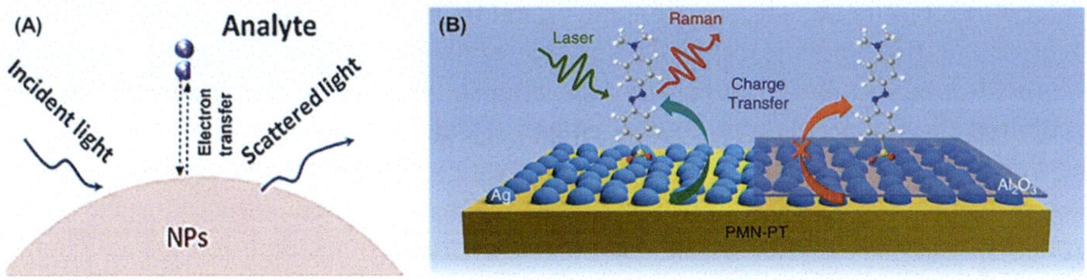

Fig. 7.2 (**a**) Electron transport between analytes and the surface of nanoparticles leads to chemical amplification. (Reproduced with permission from Wang et al. Copyright 2018, Nanomaterials). (**b**) Schematic illustration of the SERS experiment setup for understanding the SERS enhancement mechanism through charge transfer. (Reproduced with permission from Zhuo et al. Copyright 2024, Nanoscale Horizons)

extinction spectra of these dimers. For instance, reducing the gap size in a dimer from 10 to 2 nm can increase the SERS enhancement factor (Lee et al., 2013). However, the increase rapidly decreases by orders of magnitude just a few nanometers from the heated region, suggesting that the greater the distance between the nanoparticles, the lower the enhancement factor.

7.2.3 Chemical Enhancement (EM)

Chemical enhancement in surface-enhanced Raman scattering (SERS), which goes beyond the effects of electromagnetic enhancement alone, is the increase in Raman signal strength caused by specific molecule interactions with the metal surface. This type of improvement results from various chemical processes at the metal–molecule interface. Chemical enhancement results from a molecule's altered electronic polarizability upon interaction with a SERS substrate. This contact may occur directly with the metal surface or through a layer of specific receptors designed to raise the substrate's affinity for the target molecule. One important process is charge transfer, in which electrons flow from the molecule to the metal, changing its electronic structure and increasing its Raman scattering efficiency (Sharma et al., 2012; Guerrini et al., 2007; Morton & Jensen, 2009).

Another factor is the creation of chemical compounds, which can stabilize specific molecular vibrational modes and increase their Raman spectrum detection (Kiefer, 2003). This compound might show optical absorption linked to a charge transfer (CT) transition between the metal and the molecule, which could coincide with the wavelength of the excitation laser. Consequently, the resonance Raman effect can significantly amplify certain Raman bands (Lombardi & Birke, 2009). The metal's Fermi level needs to be near the molecule's lowest unoccupied molecular orbital (LUMO) or highest occupied molecular orbital (HOMO) in energy for a CT state to form. Since the Fermi level of metals frequently lies between the HOMO and LUMO of different organic molecules, this alignment is instead common (Yu et al., 2011). Additionally, functional groups on the metal surface or within the molecule can influence chemical enhancement by modifying the electronic environment or creating specific binding sites that enhance Raman scattering. Although chemical enhancement is generally less intense with EF in the order of 10^2–10^4 compared to electromagnetic enhancement, it can still substantially contribute to the overall SERS signal, particularly for molecules with strong interactions with the metal surface. The extent of this enhancement can vary based on factors such as the type of metal used, the nature of the molecular interactions, and the surface chemistry of the nanoparticles.

7.3 Carbon Nanomaterials in SERS

Applications for carbon-based materials are numerous and varied, spanning from biology to sensing to catalysis. Green chemistry has advanced rapidly with the discovery of environmentally benign processes for creating nanomaterials from carbon-based materials, such as graphene, carbon quantum dots (CQDs), fullerenes, carbon nanotubes (CNTs), and carbon nanofibers (CNFs). Carbon-based materials have a high surface area/volume ratio, making them suitable for environmental contamination detection. Carbon-based nanomaterials for sensing applications are shown in Fig. 7.3. These carbon materials have femto, pico, and nanosensitivity to a wide range of analytes and conditions. Carbon-based SERS substrates will allow for more developments in the flexible microelectronic area when combined with semiconductor chips and microelectronic processing technologies. They might be

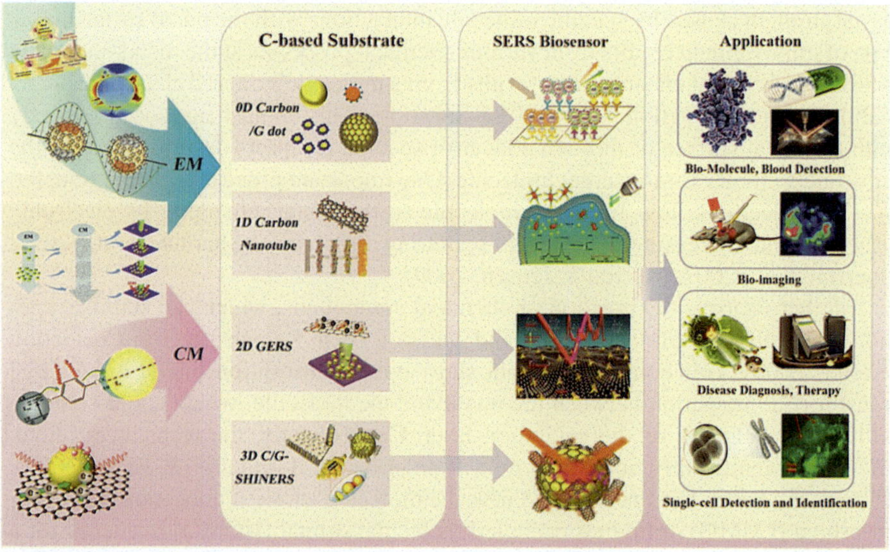

Fig. 7.3 Schematic illustration of the main content of carbon-based nanomaterials as SERS biosensors with two main enhancement mechanisms in different biological application fields. (Reproduced with permission from Liang et al. Copyright 2021, Nature)

employed as multipurpose sensors with quantitative and in situ analytical capabilities. Because carbon nanostructures have outstanding electrical characteristics, they can be used for real-time applications such as in vivo analysis. Because of the metal nanostructures' high surface area-to-volume ratio, analytes can easily be absorbed by their surfaces, increasing the sensitivity of SERS. Metals like Ag, Au, Cu, and other noble metals are frequently used as plasmonic active materials to increase signal enhancements. Furthermore, magnetic materials paired with noble metals are also referred to as SERS active substrates because of their distinct magnetic and optical characteristics, which offer them a wide range of possible SERS sensing applications. For a variety of biosensing applications, including illness diagnostics, bioimaging, therapy, and analytical chemistry, numerous researchers have reported magnetic materials functionalized with carbon components like Fe, Ni, Co, and Ti (Mauter & Elimelech, 2008; Díez-Pascual, 2021).

7.3.1 Carbon Nanotube (CNT)

The discovery of carbon nanotubes in 1991 by Ijima has generated great interest among researchers in exploring their unique electrical, physical, mechanical, and chemical properties to develop high-performance devices (Dai, 2002). With a large surface area and a very high aspect ratio, CNT is a sensitive and highly reproducible SERS substrate for detecting various analyte molecules (Lefrant et al., 2003). In addition to sensitivity, another parameter is specificity, which can usually be achieved by controlling the surface interaction of the metal and CNT structure. When the metal nanoparticles are in close contact, they generate hot sites to enhance the local electric fields and can enhance the Raman signals up to 10^{12}. Signal enhancement can be different in single-walled carbon nanotubes (SWCNT) and multi-walled carbon nanotubes (MWCNT); SWCNT offers high surface area and curvature to enable efficient molecule–nanotube interactions, and the effective functionalization and defects can improve the signal enhancement. The advantage of MWCNT is that the functionalization is much easier and more robust, and surface interaction is effective for generating the hot spot (Chen et al., 2007).

Currently, the SERS technique can be used to detect various kinds of pesticides, fungicides, chemical residues, and micro-organisms and monitor the quality and quantity of food products. Integrating CNT with metal nanostructures improves the uniformity of the substrate because of their large length-to-width ratio and provides a large surface area. In CNT/AuNP synthesized by the one-pot method, the CNT sheet provides abundant immobilization sites for gold nuclei as well as a spacious template for gold crystal growth (Fig. 7.4). Chloroauric acid is the gold precursor, and ascorbic acid is the reductant. KBr was used as the surfactant for gold anisotropic growth, and during the addition of Br^- ions, it was liberated in the solution and played as a surfactant. Carbon nanotube bundles are randomly entangled and cross-linked through the sheet surface, forming a porous network. Thus, CNT support provides a large surface area, namely the abundant anchoring sites for gold nuclei. The detection of melamine in water samples can be monitored by a flexible

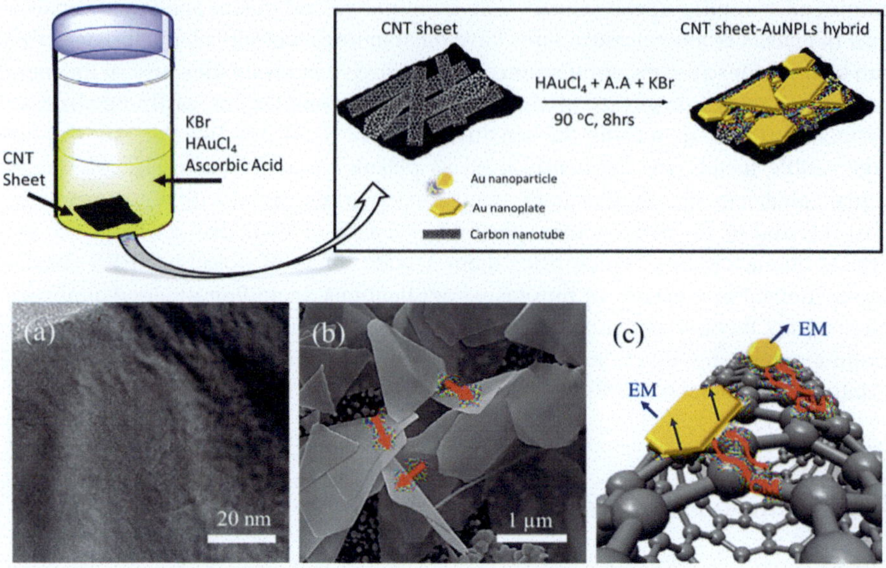

Fig. 7.4 Schematic illustration of the one-pot synthesis of AuNPLs on CNT sheet and characteristics of AuNPLs that favor the SERS performance. (**a**) Surface roughness. (**b**) AuNPLs with sharp edges and corners lie arbitrarily on the CNT sheet. (**c**) Schematic illustration of the mechanisms of Raman signal enhancement. (Reproduced with permission from Xin et al. Copyright 2015, Royal Society of Chemistry)

substrate developed from a CNT/AuNP-coated membrane. The EF was 2.4×10^6 for the CNT/AuNP membrane, exhibiting excellent sensitivity toward organic pollutants like melamine at a limit of detection (LOD) of 1 nM (0.126 ppb) (Zhang et al., 2015). Thus, the reported SERS substrate has great potential as a novel, flexible, and efficient SERS platform for real-life applications in food quality monitoring applications. Flexible SERS substrates are a promising alternative for routine application because of their flexibility, transparency, and high sensitivity.

The development of effective enrichment and sensitive detection approaches for environmental antibiotic monitoring is essential due to the increased antibiotic contamination of aquatic ecosystems. Due to the electrostatic interactions and π–π stacking, combining graphene oxide (GO), carbon nanotube intercalated graphene oxide membranes, and silver nanoparticles provides an efficient material for aquatic antibiotic detection. To make the AgNPs/CNT-GO, the membranes are formed into a positively charged GO and negatively charged CNT and AgNP on the other side of the membrane. Following the filtering of the antibiotic solution through AgNPs/CNT-GO membranes, the antibiotic molecules were enriched on the GO membranes by electrostatic and π–π stacking interactions. This technique significantly enhances the concentration of antibiotic molecules at the "hot spots"—zones of the AgNP electromagnetic field or interparticle junctions—by intensifying the Raman signals of the target molecules (Qu et al., 2016).

Carbon nanotubes (CNTs) have also been extensively investigated for their potential to eliminate hazardous pollutants from gas streams and aqueous solutions because of their large specific surface area, highly hollow and porous structure, light mass density, and strong contact with pollutant molecules. Very long helical-CNTs with a diameter of less than 20 nm were created using chemical vapor deposition (CVD) over the surface of PAA membranes functionalized with $CoFe_2O_4$ nanoparticles. Heavy metals Pb^{2+}, Hg^{2+}, and Cd^{2+} can be extracted from water residues by these CNTs at extremely low concentrations (ppm and ppb) (Shaban & Galaly, 2016). The generated blank PAA and CNTs/PAA membranes are used as ultra-high sensitive sensors to detect Hg^{2+}, Pb^{2+}, and Cd^{2+} ion traces in water. The sensing principle is based on in situ SERS spectroscopy. Through CVD, transition metal nanoparticles (NPs) and nanowires (NWs) placed inside the PAA membrane's pores can be converted into MWCNTs. The interpore distance is $D_{pp} = 100$ nm, the pore width is approximately $D_p = 70$ nm, and the pore density is approximately 1.142×101 cm^{-2}. The membrane's porosity, calculated using the formula $P = \frac{\pi}{2\sqrt{3}}\left(\frac{D_p}{D_{pp}}\right)^2$, is 44.4%. Important colored materials such as organic dyes are used extensively in many industrial applications, including textiles, coatings, detergents, and agriculture. Capillary electrophoresis and high-performance liquid chromatography are two time-consuming, costly sample pretreatment techniques that need intricate processes performed by a qualified specialist. SERS is a useful instrument for identifying organic contaminants in the environment. Multi-walled carbon nanotubes coated with functional nanocomposites provide a sensing instrument for methylene blue detection (Dinh et al., 2016). The MWCNT-Ag functional composite was created via a two-step chemical procedure. To create -OH/-COOH functionalized MWCNT, MWCNT was first thermally oxidized and acid refluxed in a strong acid medium (HNO_3:H_2SO_4) in 1:3 ratios at 90 °C. The photochemical method outlined in the second step is then used to decorate AgNPs on the OH- and/or COOH functionalized MWCNTs. AgNPs stuck to the inside and carbon nanotubes'exterior walls, creating AgNPs on the MWCNTs' surface. Figure 7.5 illustrates that the

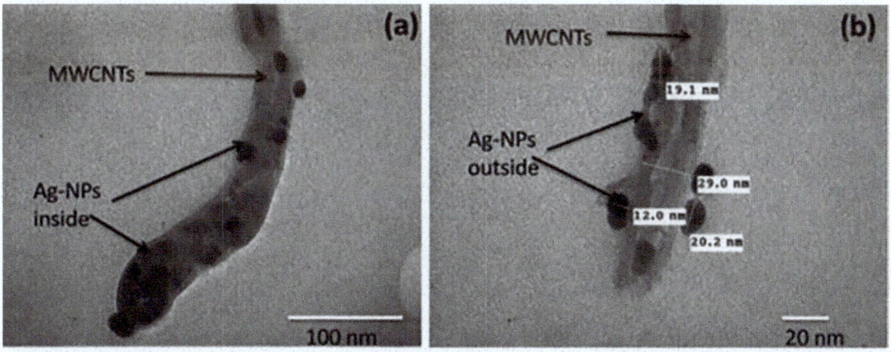

Fig. 7.5 TEM images of MWCNTs-Ag nanocomposite. Panel (**a**) displays multi-walled carbon nanotubes (MWCNTs) with silver nanoparticles (Ag-NPs) located inside. Panel (**b**) shows MWCNTs with Ag-NPs positioned outside. (Reproduced with permission from Dinh et al. Copyright 2016, Advanced Materials and Devices)

AgNPs connected to the MWCNTs are typically around 15–20 nm in size. Compared to AgNP alone, the MWCNT-Ag nanocomposite exhibits outstanding amplification up to the order of 4.68×10^6. Different conditions, including pure Ag and methylene blue (MB) molecules alone, were used to test the substrate's SERS activity. Raman bands at 1385 and 1621 cm^{-1} are present in pure metallic Ag. At 1621, 1385, and 445 cm^{-1}, MB exhibits vibrations that are attributed to C-C stretching, C-H stretching, and C-N-C bending, respectively. The SERS spectra of MB adsorbed at concentrations ranging from 1 to 70 ppm on the AgNPs-deposited SERS substrate. It is clear from all spectra that the three similar characteristic peaks for the locations of the MB dye are 450, 1390, and 1594 cm^{-1}. The shift in Raman spectra showed that the MB molecules were chemisorbed on the AgNP surface. It demonstrates a considerable quantity of MB dye adsorption on the metallic AgNP surface, and the correlation value was 0.99 at 449 cm^{-1}. The substrate as it is can be employed to identify MB in water residues; MB detection has been achieved at ppm quantities.

7.3.2 Carbon Quantum Dots (CQD)

Carbon dots are a new type of fluorescent zero-dimensional nanomaterials, also called carbon quantum dots (CQD), with a particle size <10 nm and can detect pesticides in the agricultural sectors. Atrazine is a restricted-use herbicide used in crops for increased yield and to enhance the quality of foods. These herbicides are used in corn, sugarcane, and sorghum production units. Various techniques are reported to monitor the presence of pesticides. SERS offers a simple, low-cost, effective sensing tool for detecting various contaminants in the agricultural sectors. Silver nanoparticles have been demonstrated as the most suitable substrate for SERS detection because of their unique properties and various nanostructures like rods, sheets, and nanoparticles. Integrating the silver structures with carbon dots can selectively detect the pesticides in the agricultural sectors. CD@Ag can selectively adsorb the atrazine molecule and can improve the signal enhancement 1×10^{-7} M using R6G as the Raman reporter molecule. This CD@Ag exhibits excellent analytical methods and offers highly sensitive and selective detection of the atrazine molecule in aqueous media (Tang et al., 2019). Thus, the as-developed CD@Ag substrate offers the LOD of around 5.0 and 10 nM, and this simple, rapid, label-free, cost-effective sensing system can pave the way for various practical applications.

Tuning the structural properties of the CQDs can improve the signal enhancement and offer a higher surface-to-volume ratio. The honeycomb structures are developed using a 2D or 3D silica colloid template on the gold film as substrate material. The CQD is synthesized by an electrochemical method using graphite as the working electrode, Pt wire, and Ag/AgCl as the counter and reference electrodes (Fan et al., 2012). After the electrochemical reaction, SiO_2 did a monolayer deposition onto the N-doped Si substrate, and then the formation of 3D colloidal crystals was accomplished by evaporation-induced multilayer assembly. CQD was uniformly distributed on the film's surface using the CVD deposition technique and morphologically characterized by SEM and TEM analysis. The Au nanoparticles are distributed on the honeycomb CQD skeleton's walls and size below 10 nm. Thus, the honeycomb architecture of the CQDs has more adsorption sites and exhibits stronger enhancement than the traditional Au nanoparticles on planar substrates.

7.3.3 Carbon Nanofibers (CNFs)

Fabiola et al. developed CNF-based SERS substrates using electrospinning techniques using polyacrylonitrile (PAN), polyvinyl alcohol (PVA), and dimethyl formamide (DMF) to form a 3D hierarchical structure (Armenta-Monzón et al., 2019). The nanofibrous mats provide a higher surface area for nanoparticle attachment and the generation of hot spots on the surface of nanofibers. They are embedded with Ag nanostructures directly onto the polymer matrix. The nanofibers are uniformly distributed by gold nanoparticles using a chemical reduction method, with the SPR band at 518 nm, respectively. The observed rough surface comprises hundreds of nanofibers; however, within these aggregates, the single nanofibers are distinguishable as individual pieces. The CNFs comprise an extensive range of diameters in the nanoscale from 38 to 412 nm. The SERS activity of the CNF-integrated AuNP was studied using Rhodamine 110 chloride molecule with concentrations ranging from 10^{-4} to 10^{-6} M. The vibrational frequencies of Rhodamine 110 were calculated using the density functional theory (DFT) method using the B3LYP/6-311++G (d,p) set. SERS substrates based on CNFs are easy to fabricate and offer a large surface area for molecule adsorption. They also resist corrosion and maintain structural integrity.

Wu et al. reported magnetic carbon nanofiber-based SERS substrates for the detection of Rhodamine 6G and amino acids using various concentrations of gold nanoparticles 20 uL , 40 uL , and 50 uL deposited on Ni@CNF surfaces and are denoted as Ni@CNFs@Au(20), Ni@CNFs@Au(40), Ni@CNFs@Au(50), respectively (Wu et al., 2018). The LOD was found to be 10^{-14} M and exhibited great potential for high sensitivity under a magnetic environment. The acid treatment method will remove the metal NPs from the surface of CNFs to avoid this developed CO_2 utilization process to produce carboxylic-functionalized CNFs integrated with magnetic Ni nanoparticles. The Ni@CNFs exhibit excellent magnetization and high surface area, increasing the hot spot density. The Ni@CNFs@Au substrate not only detects R6G but can also sense the phenylaniline and tyrosine at a concentration level of 10^{-11} and 10^{-13} M, respectively (Wu et al., 2018).

7.3.4 Graphene Oxide

A team of researchers demonstrated a highly sensitive and reliable SERS substrate that combines graphene oxide (GO) nanosheets with silver (Ag) nanocubes to detect organic contaminants (He et al., 2022). Using a portable Raman spectrometer, the substrate exhibits remarkable detection limits for malachite green, methylene blue, and crystal violet (10^{-9} to 10^{-10} M). It also detects complex pesticides, including tetramethylthiuram disulfide and diquat dibromide, at levels below U.S. EPA safety limits, which showed promise for sensitive, label-free SERS applications in environmental monitoring and food safety. Gold/reduced graphene oxide nanocomposites (Au/GO NCs) are used in a study by Parvathi et al. to investigate a graphene oxide-mediated SERS (G-SERS) sensor for the detection of different dye pollutants, such as monoazo, diazo, triarylmethane, thiazine, and xanthene dyes (Parvathi et al., 2019). The findings show that Au/GO NCs have the potential to be useful sensors for monitoring industrial effluents in water treatment and processing amenities. A work presented by Wang et al. developed a SERS platform that uses graphene to assemble PAH molecules and Ag nanoparticles for direct PAH detection (Wang et al., 2020). Pyrene, anthracene, and phenanthrene have remarkably low detection limits of 0.73, 1.1, and 0.57 ppb, respectively. The findings revealed that graphene effectively immobilizes PAHs, making it possible to create extremely sensitive sensors for aromatic pollutants with promising applications in environmental monitoring and food safety.

A study by a group of scientists uses a graphene-enhanced silver nanoparticle (Ag-rGO) SERS platform to detect toxic compounds in textile effluents (TE). The substrate's effectiveness was first validated on rhodamine 6G, Congo red, and malachite green dyes. In TE, carcinogenic azo dyes were identified via N=N and C–N bands in the SERS spectra. The results highlight Ag-rGO's potential for sensitive, scalable, and stable dye detection and removal, offering a promising solution for water recycling in textile dyeing industry effluent discharge (Parimaladevi et al., 2019). The development of nanostructured GO assemblies with Ag nanoprisms to improve SERS detection is described in another study by Shanta et al. The GO-Ag nanoprism substrates allow for multiplex SERS measurement of legacy aromatic pollutants. Three tetrachlorobiphenyl isomers can be identified using distinct molecular fingerprints, including PCB-77 trace detection (Shanta & Cheng, 2017). The LOD and EF of various carbon-based nanomaterials for SERS application are summarized in Table 7.1.

Table 7.1 Summarizing the key attributes of SERS substrates, analyte, LOD, and EF

Sl. no	Material	Analyte	LOD	EF	Reference
1	CNT/AuNP	Melamine	0.126 ppb	2.4×10^6	Zhang et al. (2015)
2	AgNP/CNT-GO membrane	Antibiotics	–	–	Qu et al. (2016)
3	CNT/PAA membrane	Pb^{2+}, Hg^{2+}, Cd^{2+}	–	–	Shaban and Galaly (2016)
4	MWCNT-Ag	Methylene blue	1 ppm	4.68×10^6	Dinh et al. (2016)
5	CD@Ag	Atrazine	5.0–10 nM	1×10^{-7} M	Tang et al. (2019)
6	Ni@CNFs@Au	R6G, amino acids	10^{-14} M	–	Wu et al. (2018)
7	AgNP-graphene	Polycyclic aromatic hydrocarbons	0.57 ppb	–	Wang et al. (2020)
8	Ag-rGO	Azo dyes	–	–	Parimaladevi et al. (2019)
9	ArGO	Polychlorinated biphenyls	100 nM	–	Shanta and Cheng (2017)

7.4 Future Perspectives

Sensors enabled by carbon nanomaterials are an interesting technology that offers exquisite detection of environmental contaminants, such as pesticides, heavy metals, and viruses, at the nanomolar to sub-picomolar level. SERS is a potent spectrum characterization method that has been utilized widely for the fast and trace detection of a range of analyte molecules. It has several benefits, such as high sensitivity, specificity, efficiency, and noninvasiveness. SERS will be more likely to be used in non-invasive clinical diagnostics, promoting human health and preventing disease in conjunction with advancements in optical technology, biochemical methods, nanotechnology, and data processing techniques. This will show how much better SERS will be in the upcoming generation. The primary challenge with SERS substrates are the requirement for intimate contact between the analyte and the enhancing surface and the slow substrate deterioration that decreases the signal, causes problems with the homogeneity and reproducibility of the SERS signal inside a substrate, and limits the reusability of the substrates. To solve these problems, SERS substrate tweaking is required, particularly to improve substrate stability, repeatability, and specificity—all of which are critical for quantitative detections.

7.5 Conclusion

Carbon-based nanomaterials'superior physical, chemical, and electrical characteristics for contamination detection have attracted interest. Carbon nanomaterials generally have a high surface-to-volume ratio, which ensures that most surface atoms are exposed. These atoms can then bind a sizable number of sensing components. Additionally, the *sp*^{2} hybridization structure and the exceptional surface-to-volume ratio guarantee surface modification by the sensing components and the conversion of the sensing signals of different environmental pollutants. Combining SERS and carbon nanoparticles provides quick, easy, and inexpensive environmental pollutant detection. These benefits encourage the quick development of biosensors for various uses, such as environmental monitoring, food safety, and illness detection.

Funding Sources and Acknowledgements This work was supported by the DST-INSPIRE Fellowship (No. DST/INSPIRE Fellowship/2020/IF200094), Government of India, Ministry of Science and Technology, Department of Science and Technology. The authors wish to acknowledge the PSG & Sons' Charities and the Department of Science and Technology, Government of India, for the financial support from grants, DST-INSPIRE (DST/INSPIRE Fellowship/2020/IF200094). The authors also acknowledge the facilities and support provided by the management of PSG Institute of Advanced Studies, Coimbatore.

Author Contributions The manuscript was written through contributions from all authors. All the authors have approved the final version of the manuscript.

References

Armenta-Monzón, F., Aguilar-Hernández, I., García-García, A., Holmberg, S., Pisano, E., Morales-Luna, G., García-Méndez, M., & Ornelas-Soto, N. (2019). Characterization of Rhodamine 110 adsorbed on carbon-based electrospun nanofibers decorated with gold nanoparticles by Raman spectroscopy and SERS. *Materials Research Express, 6*(12), 125012. https://doi.org/10.1088/2053-1591/ab53f0

Asha, A. B., & Narain, R. (2020). Nanomaterials properties. In *Polymer science and nanotechnology* (pp. 343–359). Elsevier. https://doi.org/10.1016/B978-0-12-816806-6.00015-7

Baig, N., Kammakakam, I., & Falath, W. (2021). Nanomaterials: A review of synthesis methods, properties, recent progress, and challenges. *Materials Advances, 2*(6), 1821–1871. https://doi.org/10.1039/D0MA00807A

Camden, J. P., Dieringer, J. A., Wang, Y., Masiello, D. J., Marks, L. D., Schatz, G. C., & Van Duyne, R. P. (2008). Probing the structure of single-molecule surface-enhanced Raman scattering hot spots. *Journal of the American Chemical Society, 130*(38), 12616–12617.

Chen, Y.-C., Young, R. J., Macpherson, J. V., & Wilson, N. R. (2007). Single-walled carbon nanotube networks decorated with silver nanoparticles: A novel graded SERS substrate. *The Journal of Physical Chemistry C, 111*(44), 16167–16173. https://doi.org/10.1021/jp073771z

Cialla, D., März, A., Böhme, R., Theil, F., Weber, K., Schmitt, M., & Popp, J. (2012). Surface-enhanced Raman spectroscopy (SERS): Progress and trends. *Analytical and Bioanalytical Chemistry, 403*, 27–54. https://doi.org/10.1007/s00216-011-5631-x

Cialla-May, D., Zheng, X.-S., Weber, K., & Popp, J. J. C. S. R. (2017). Recent progress in surface-enhanced Raman spectroscopy for biological and biomedical applications: From cells to clinics. *Chemical Society Reviews, 46*(13), 3945–3961.

Dai, H. (2002). Carbon nanotubes: Opportunities and challenges. *Surface Science, 500*(1–3), 218–241. https://doi.org/10.1016/S0039-6028(01)01558-8

Das, R. S., & Agrawal, Y. K. (2011). Raman spectroscopy: Recent advancements, techniques, and applications. *Vibrational Spectroscopy, 57*, 163–176. https://doi.org/10.1016/j.vibspec.2011.08.003

Díez-Pascual, A. M. (2021). Carbon-based nanomaterials. *International Journal of Molecular Sciences, 22*(14), 7726. https://doi.org/10.3390/ijms22147726

Ding, S.-Y., Yi, J., Li, J.-F., Ren, B., Wu, D.-Y., Panneerselvam, R., & Tian, Z.-Q. (2016). Nanostructure-based plasmon-enhanced Raman spectroscopy for surface analysis of materials. *Nature Reviews Materials, 1*(6), 1–16. https://doi.org/10.1038/natrevmats.2016.21

Dinh, N. X., Huy, T. Q., & Le, A.-T. (2016). Multi-walled carbon nanotubes/silver nanocomposite as effective SERS platform for detection of methylene blue dye in water. *Journal of Science: Advanced Materials and Devices, 1*(1), 84–89. https://doi.org/10.1016/j.jsamd.2016.04.007

Fan, Y., Cheng, H., Zhou, C., Xie, X., Liu, Y., Dai, L., Zhang, J., & Qu, L. (2012). Honeycomb architecture of carbon quantum dots: A new efficient substrate to support gold for stronger SERS. *Nanoscale, 4*(5), 1776–1781. https://doi.org/10.1039/C2NR12015A

Guerrini, L., Jurasekova, Z., Domingo, C., Pérez-Méndez, M., Leyton, P., Campos-Vallette, M., García-Ramos, J. V., & Sánchez-Cortés, S. (2007). Importance of metal–adsorbate interactions for the surface-enhanced Raman scattering of molecules adsorbed on plasmonic nanoparticles. *Plasmonics, 2*, 147–156. https://doi.org/10.1007/s11468-007-9044-9

Halvorson, R. A., & Vikesland, P. J. (2010). Surface-enhanced Raman spectroscopy (SERS) for environmental analyses. *Environmental Science & Technology, 44*, 7749–7755. https://pubs.acs.org/doi/10.1021/es101228z

He, J., Li, X., & Li, J. (2022). Facile construction of silver nanocubes/graphene oxide composites for highly sensitive SERS detection of multiple organic contaminants by a portable Raman spectrometer. *Journal of Environmental Chemical Engineering, 10*(5), 108278. https://doi.org/10.1016/j.jece.2022.108278

Kiefer, W. (2003). The Raman effect—A unified treatment of the theory of Raman scattering by molecules. Derek A. Long, John Wiley & Sons, Ltd., 2002, Pp 597. ISBN 0-471-49028-8. *Journal of Raman Spectroscopy, 34*(2), 180. https://doi.org/10.1002/jrs.987

Kudelski, A. (2008). Analytical applications of Raman spectroscopy. *Talanta, 76*(1), 1–8. https://doi.org/10.1016/j.talanta.2008.02.042

Le Ru, E., & Etchegoin, P. (2008). *Principles of surface-enhanced Raman spectroscopy: and related plasmonic effects*. Elsevier. https://doi.org/10.1016/B978-0-444-52779-0.X0001-3

Lee, H. M., Jin, S. M., Kim, H. M., & Suh, Y. D. (2013). Single-molecule surface-enhanced Raman spectroscopy: A perspective on the current status. *Physical Chemistry Chemical Physics, 15*(15), 5276–5287. https://doi.org/10.1039/C3CP44463E

Lefrant, S., Buisson, J. P., Schreiber, J., Chauvet, O., Baibarac, M., & Baltog, I. (2003). Study of interactions in carbon nanotube systems by using Raman and SERS spectroscopy. *Synthetic Metals, 139*(3), 783–785. https://doi.org/10.1016/S0379-6779(03)00264-9

Lombardi, J. R., & Birke, R. L. (2009). A unified view of surface-enhanced Raman scattering. *Accounts of Chemical Research, 42*(6), 734–742. https://doi.org/10.1021/ar800249y

Mauter, M. S., & Elimelech, M. (2008). Environmental applications of carbon-based nanomaterials. *Environmental Science & Technology, 42*(16), 5843–5859. https://doi.org/10.1021/es8006904

McNay, G., Eustace, D., Smith, W. E., Faulds, K., & Graham, D. (2011). Surface-enhanced Raman scattering (SERS) and surface-enhanced resonance Raman scattering (SERRS): A review of applications. *Applied Spectroscopy, 65*(8), 825–837.

Morton, S. M., & Jensen, L. (2009). Understanding the molecule–surface chemical coupling in SERS. *Journal of the American Chemical Society, 131*(11), 4090–4098. https://doi.org/10.1021/ja809143c

Mulvaney, S. P., & Keating, C. D. (2000). Raman spectroscopy. *Analytical Chemistry, 72*(12), 145–158. https://doi.org/10.1021/a10000155

Parimaladevi, R., Sathe, V., & Mahalingam, U. (2019). Graphene boosted silver nanoparticles as surface-enhanced Raman spectroscopic sensors and photocatalysts for removal of standard and industrial dye contaminants. *Sensors and Actuators B: Chemical, 281*, 679–688. https://doi.org/10.1016/j.snb.2018.11.007

Parvathi, V. P., Parimaladevi, R., Sathe, V., & Umadevi, M. (2019). Application of G-SERS for the efficient detection of toxic dye contaminants in textile effluents using gold/graphene oxide substrates. *Journal of Molecular Liquids, 273*, 203–214. https://doi.org/10.1016/j.molliq.2018.10.027

Pena-Pereira, F., Bendicho, C., Pavlović, D. M., Martín-Esteban, A., Díaz-Álvarez, M., Pan, Y., Cooper, J., et al. (2021). Miniaturized analytical methods for determination of environmental contaminants of emerging concern—A review. *Analytica Chimica Acta, 1158*, 238108. https://doi.org/10.1016/j.aca.2020.11.040

Pilot, R., Signorini, R., & Fabris, L. (2017). Surface-enhanced Raman spectroscopy: Principles, substrates, and applications. In *Metal nanoparticles and clusters: Advances in synthesis, properties, and applications* (pp. 89–164). https://doi.org/10.1007/978-3-319-68053-8_4

Qu, L.-L., Liu, Y.-Y., Liu, M.-K., Yang, G.-H., Li, D.-W., & Li, H.-T. (2016). Highly reproducible Ag NPs/CNT-intercalated GO membranes for enrichment and SERS detection of antibiotics. *ACS Applied Materials & Interfaces, 8*(41), 28180–28186. https://doi.org/10.1021/acsami.6b08790

Saleh, T. A. (2020). Nanomaterials: Classification, properties, and environmental toxicities. *Environmental Technology & Innovation, 20*, 101067. https://doi.org/10.1016/j.eti.2020.101067

Schlücker, S. (2014). Surface-enhanced Raman spectroscopy: Concepts and chemical applications. *Angewandte Chemie International Edition, 53*(19), 4756–4795.

Shaban, M., & Galaly, A. R. (2016). Highly sensitive and selective in-situ SERS detection of Pb^{2+}, Hg^{2+}, and Cd^{2+} using nanoporous membrane functionalized with CNTs. *Scientific Reports, 6*(1), 25307. https://doi.org/10.1038/srep25307

Shanta, P. V., & Cheng, Q. (2017). Graphene oxide nano prisms for sensitive detection of environmentally important aromatic compounds with SERS. *ACS Sensors, 2*(6), 817–827. https://doi.org/10.1021/acssensors.7b00182

Sharma, B., Frontiera, R. R., Henry, A.-I., Ringe, E., & Van Duyne, R. P. (2012). SERS: Materials, applications, and the future. *Materials Today, 15*(1–2), 16–25. https://doi.org/10.1016/S1369-7021(12)70017-2

Tang, J., Chen, W., & Ju, H. (2019). Sensitive surface-enhanced Raman scattering detection of atrazine based on aggregation of silver nanoparticles modified carbon dots. *Talanta, 201*, 46–51. https://doi.org/10.1016/j.talanta.2019.03.108

Wang, X., Xu, Q., Hu, X., Han, F., & Zhu, C. (2020). Silver-nanoparticles/graphene hybrids for effective enrichment and sensitive SERS detection of polycyclic aromatic hydrocarbons. *Spectrochimica Acta Part A: Molecular and Biomolecular Spectroscopy, 228*, 117783. https://doi.org/10.1016/j.saa.2019.117783

Wu, H.-C., Chen, T.-C., Tsai, H.-J., & Chen, C.-S. (2018). Au nanoparticles deposited on magnetic carbon nanofibers as the ultrahigh sensitive substrate for surface-enhanced Raman scattering: Detections of rhodamine 6G and aromatic amino acids. *Langmuir, 34*(47), 14158–14168. https://doi.org/10.1021/acs.langmuir.8b02488

Xu, H., Bjerneld, E. J., Aizpurua, J., Apell, P., Gunnarsson, L., Petronis, S., Kasemo, B., Larsson, C., Hook, F., & Kall, M. (2001). Interparticle coupling effects in surface-enhanced Raman scattering. In *Nanoparticles and nanostructured surfaces: Novel reporters with biological applications* (Vol. 4258, pp. 35–42). https://doi.org/10.1117/12.430771

Yu, X., Cai, H., Zhang, W., Li, X., Pan, N., Luo, Y., Wang, X., & Hou, J. G. (2011). Tuning chemical enhancement of SERS by controlling the chemical reduction of graphene oxide nanosheets. *ACS Nano, 5*(2), 952–958. https://doi.org/10.1021/nn102291j

Zamora-Ledezma, C., Negrete-Bolagay, D., Figueroa, F., Zamora-Ledezma, E., Ni, M., Alexis, F., & Guerrero, V. H. (2021). Heavy metal water pollution: A fresh look about hazards, novel and conventional remediation methods. *Environmental Technology & Innovation, 22*, 101504. https://doi.org/10.1016/j.eti.2021.101504

Zhang, K., Ji, J., Fang, X., Yan, L., & Liu, B. (2015). Carbon nanotube/gold nanoparticle composite-coated membrane as a facile plasmon-enhanced interface for sensitive SERS sensing. *Analyst, 140*(1), 134–139. https://doi.org/10.1039/C4AN01473A

Zhang, Y., Zhao, S., Zheng, J., & He, L. (2017). Surface-enhanced Raman spectroscopy (SERS) combined techniques for high-performance detection and characterization. *TrAC Trends in Analytical Chemistry, 90*, 1–13. https://doi.org/10.1016/j.trac.2017.02.006

Zhang, Y., Zhu, Y., Zeng, Z., Zeng, G., Xiao, R., Wang, Y., Hu, Y., Tang, L., & Feng, C. (2021). Sensors for the environmental pollutant detection: Are we already there? *Coordination Chemistry Reviews, 431*, 213681. https://doi.org/10.1016/j.ccr.2020.213681

Chapter 8
Carbon Adsorbents: A Remedy for Environmental Contaminants

Nishu Choudhary, Sanjay Yadav, and Alok Ranjan Paital

Abstract Environmental pollution caused by heavy metals, organic compounds, and emerging contaminants is one of the significant threats to ecosystems and human health. Carbon-based adsorbents, including activated carbon, biochar, carbon nanotubes (CNTs), graphene-based materials, and carbon dots (CDs), have been widely explored for their outstanding properties and high efficiency in pollutant removal. This chapter explores these carbon materials' synthesis, properties, performance, and application in environmental remediation, focusing on their use in short-bed adsorption columns for large-scale pollutant removal. The synthesis of carbon-based adsorbents includes physical and chemical activation (for activated carbon), pyrolysis (for biochar), and chemical vapor deposition or arc discharge (for CNTs). These processes impart unique characteristics, such as high surface area, tunable porosity, and functional groups, which enhance their adsorption capabilities. Carbon-based materials also exhibit superior mechanical strength, chemical stability, and reusability, making them ideal for continuous use in industrial-scale applications. The adsorption mechanisms in carbon-based adsorbents include physical adsorption (physisorption), chemical adsorption (chemisorption), and electrostatic interactions, all of which play a role in efficient pollutant removal. Packed with carbon-based materials, short-bed adsorption columns are an easy and effective means of water, air, and soil contamination treatment. Such systems are scalable, cost-effective, and can remove a wide range of pollutants, such as heavy metals, dyes, pharmaceuticals, and VOCs. The performance and scalability of carbon-based adsorbents make them suitable for sustainable environmental remediation. Regeneration methods enhance their long-term usability, reducing operational costs and waste generation. Carbon-based adsorbents are a promising solution for mitigating environmental contamination and ensuring a cleaner, safer future.

N. Choudhary (✉) · S. Yadav · A. R. Paital (✉)
Salt and Marine Chemicals Division, CSIR—Central Salt and Marine Chemicals Research Institute, Bhavnagar, Gujarat, India

Academy of Scientific and Innovative Research, Ghaziabad, India
e-mail: arpaital@csmcri.res.in

A. M. Parambil et al. (eds.), *Carbon*,
https://doi.org/10.1007/978-3-031-90613-8_8

Keywords Biochar · Carbon dots · Carbon nanotubes · Activated charcoal · Adsorption · Photocatalysis

Abbreviations

CAC	Commercial activated carbon
CDs	Carbon dots
CHKBC	Biochar co-modified with KOH and chitosan
CNTs	Carbon nanotubes
CVD	Chemical vapor deposition
DCC	Dichloroxyl carbodiimide
DFT	Density functional theory
GO	Graphene oxide
IBU	Ibuprofen
MT	Melamine thiourea
MT-AMC	Melamine thiourea modified activated carbon
MWCNTS	Multi-walled carbon nanotubes
OMP	Omeprazole
OTC	Oxytetracycline
rGO	Reduced graphene oxides
Si-CDs	Silica-carbon dot hybrid
SWCNTs	Single-walled carbon nanotubes
TAC	Tire-derived activated carbon
VOCs	Volatile organic compounds
XPS	X-ray photoelectron spectroscopy

8.1 Introduction

Environmental contamination from industrial, agricultural, and urban activities significantly endangers ecosystems and human health. Contaminants that penetrate water, air, and soil include heavy metals, pesticides, pharmaceuticals, dyes, and volatile organic compounds. They cause an imbalanced ecosystem at the general environmental level and risks for chronic diseases and toxicity to living organisms (Chen et al., 2021; Aguilar-Aguilar et al., 2023; Srivastava et al., 2017). Industrial discharge contributes to heavy metal pollution, while agricultural runoff introduces excess nutrients, pesticides, and herbicides into water bodies, resulting in eutrophication and biodiversity loss. Urbanization exacerbates the problem by generating solid waste, untreated sewage, and airborne pollutants. These challenges necessitate the development of robust and sustainable remediation strategies to mitigate environmental damage and safeguard public health. Common chemical treatment methods include precipitation, oxidation–reduction, and ion exchange, which have been applied to remove heavy metals and organic compounds. These processes usually result in secondary wastes such as toxic sludge, are usually selective, and are generally ineffective against complex mixtures of contaminants. The reagents' costs and the operations' complexities also make these methods less practical (Lee et al., 2024; Obaideen et al., 2022; Elbasiouny et al., 2021).

Biological treatments include bioremediation and phytoremediation, which use microorganisms or plants to degrade or immobilize pollutants. Although environmentally friendly, these methods are slow and highly dependent on external factors such as temperature, pH, and nutrient availability, making them less effective in harsh or variable conditions (Kafle et al., 2022; Ayilara & Babalola, 2023). Physical methods, which include membrane filtration, thermal desorption, and air stripping, demonstrate robust removal capability but often with high energy requirements, membrane fouling, and associated costly infrastructure; hence, there is a need to find alternative techniques that are also efficient and environmentally friendly (Obotey Ezugbe & Rathilal, 2020; Qasem et al., 2021; Chen et al., 2023). Due to its efficiency, versatility, and cost-effectiveness, adsorption presents an attractive opportunity for the treatment of environmental contaminants. This is an adsorption process wherein pollutants accumulate on the surface of an adsorbent, allowing contaminants in water, air, or soil to be removed. Its major advantage lies in the fact that it encompasses a large number of pollutants, including heavy metals, dyes, pharmaceuticals, and volatile organic compounds, all without requiring extensive chemical reactions or auxiliary systems (Pan & Xing, 2008; Chen et al., 2007; Dąbrowski, 2001). Key factors that made adsorption popular include its selectivity, which the surface functionalization of adsorbents can amplify, and fast pollutant capture, often showing high removal efficiencies at low concentrations. The design, operation, and scaling of adsorption systems are also relatively straightforward, making them useful for industrial-scale and decentralized applications (Zhang et al., 2024; Akhtar et al., 2024; Alkhaldi et al., 2024). In addition, many adsorbents have regeneration and reusability capabilities, reducing the long-term operation cost compared to chemical precipitation or membrane filtration techniques. More importantly, the ability of adsorption to maintain efficiency under varying environmental conditions like temperature and pH may limit biological or chemical treatments. Its suitability to different pollutant types and mixtures underscores its superiority as a remediation technique (Fouda-Mbanga et al., 2024; Kumar et al., 2019).

Nanomaterials have attracted immense interest as novel adsorbents for removing organic and inorganic pollutants because of their high surface area, nanoscale size, and tunable surface functionalities. The metal oxides, titanium dioxide (TiO_2) and zinc oxide (ZnO), have been proven to be efficient adsorbents for heavy metals and dyes through ion exchange and electrostatic attraction mechanisms (Naser et al., 2021; Chenab et al., 2020; Bushra et al., 2016). Similarly, silica-based functional nanomaterials are effective for the removal of organic and inorganic contaminants due to their hydrophilic surfaces and large pore volumes (Yadav et al., 2021, 2022a, 2022b, 2023a, 2023b, 2024a, 2024b). Magnetic nanoparticles, such as iron oxides, have gained popularity due to ease of separation after adsorption, making them ideal for water purification (Pinto et al., 2020; Keshta et al., 2024). However, these nanomaterials often have limitations, including aggregation tendencies, high synthesis costs, and poor chemical stability under varying environmental conditions.

Many inorganic nanomaterials lack versatility for addressing mixed contaminants, and their regeneration often involves complex processes, reducing long-term usability. Thus, when combined with advanced adsorbents such as carbon-based materials, adsorption offers a sustainable, high-performance solution to complex environmental contamination challenges. Carbon-based adsorbents have emerged as highly effective and versatile alternatives due to their exceptional properties, including high surface area, tunable porosity, and adaptable surface chemistry (Devi et al., 2023; Smith & Rodrigues, 2015; Yadav et al., 2024c). Unlike conventional methods, carbon adsorbents remove broad-spectrum contaminants without generating harmful by-products. Their capability to adsorb various pollutants such as heavy metals, dyes, pharmaceuticals, and volatile organic compounds makes them highly suitable for various environmental applications.

Furthermore, their functionalization ability and hybrid materials developed enhance the selectivity and reusability, thereby bringing an improvement in the performance with remediation efficiency addressing the deficiency of traditional strategies (Liu et al., 2022; Gusain et al., 2020; Cai et al., 2023). Activated carbon is considered one of the most widely applied adsorbents (Wang et al., 2022; Menéndez-Díaz & Martín-Gullón, 2006; Jeirani et al., 2017). Its excellent surface area and hierarchical pore structure afford ample active sites for binding contaminants. It is highly efficient in removing organic pollutants, heavy metals, and gases through the physical or chemical activation of carbon-rich precursors, such as coal, biomass, or polymers. The surface chemistry of activated carbon can be adjusted to enhance the adsorption processes through π–π stacking, van der Waals forces, and hydrogen bonding (Yin et al., 2007; Foo & Hameed, 2010; Bläker et al., 2019). Biochar is a sustainable, cost-effective alternative with significant environmental advantages, as it is obtained from biomass pyrolysis. The porous structure and the abundance of oxygen-rich functional groups, such as carboxyl, hydroxyl, and carbonyl groups, enhance its ability to adsorb a wide range of pollutants (Liang et al., 2021; Tan et al., 2015; Ambaye et al., 2021).

Modification techniques such as acid treatments or metal oxide impregnation improve carbon-based materials'adsorption efficiency and selectivity. Biochar, which can sequester carbon, is a sustainable adsorbent for environmental pollutants. Carbon nanotubes, especially single-walled (SWCNTs) and multi-walled (MWCNTs), have high surface areas, thermal and chemical stability, and strong adsorption capabilities through electrostatic forces and π–π interactions. Functionalization enhances their ability to remove heavy metals, hydrocarbons, and dyes (Sajid et al., 2022; Gao et al., 2023; Aslam et al., 2021). Graphene-based materials, for example, GO and rGO, possess a tremendous adsorption capacity due to their high surface area, oxygen-functional groups, and tunable chemistry. GO performs better in removing dyes, pharmaceuticals, and heavy metals, while rGO provides further hydrophobic interactions for organic pollutants. Graphene composites add to the extent of adsorption by combining functionalities. Small-sized and surface chemistry-tunable carbon dots have enhanced adsorption through fast uptake and high selectivity toward contaminants (Thakur & Kandasubramanian, 2019; Dayana Priyadharshini et al., 2022; Kong et al., 2023). Due to their physicochemical properties and reusability, carbon-based materials are a potential solution for effective, cost-efficient environmental pollution remediation, becoming one of the most crucial technologies for sustainable contamination treatment. This chapter focuses on the synthesis, properties, mechanisms, and application potential of carbon adsorbents, such as activated carbon, biochar, CNTs, graphene-based materials, and carbon dots, for environmental cleanup.

8.2 Activated Charcoal

Activated charcoal or activated carbon is a highly porous form of carbon made by heating materials, such as coal, coconut shells, or wood in a gas stream. Its surface area increases with activation and is an excellent adsorbent for gases, liquids, and dissolved solids. Activated charcoal, with a surface area ranging from 500 to 1500 m^2/g, effectively adsorbs organic compounds, heavy metals, and toxins. Activated charcoal is used for water purification, air filtration, medical treatments, and industrial processes. Adsorption captures impurities by sticking pollutants to their surface rather than absorption. Effectiveness depends on pore size, surface chemistry, and temperature (Spencer et al., 2024; Manocha, 2003; Ahmed et al., 2022).

8.2.1 Synthesis, Properties, and Adsorption Mechanisms

The synthesis of activated carbon has two major methods: physical and chemical activation, with different processes and results depending on the raw material and the application. In physical activation, carbon-rich materials such as coconut shells, coal, or wood are heated to high temperatures (700–1000 °C) in the presence of oxidizing gases, like steam or carbon dioxide (Sarathchandran et al., 2021; Rajak et al., 2018). This process creates a network of micropores and mesopores within the carbon structure, significantly increasing its surface area and adsorption capacity. For example, activated carbon derived from coconut shells is widely used in water purification due to its high surface area and effectiveness in adsorbing organic contaminants like volatile organic compounds (VOCs) and chlorine. The carbonization step removes volatile components, leaving behind a solid char that is then activated by exposure to steam or CO_2, producing a highly porous structure that can adsorb a variety of pollutants. After activation, the material is cooled and washed to remove any remaining impurities. Raw materials like wood, peat, or agriculture wastes are soaked with chemical activating agents like phosphoric acid, potassium hydroxide, and zinc chloride for chemical activation. This type of activation makes way for carbon structural openings through lower temperature carbonization at around 450–900 °C. For instance, activated carbon prepared with KOH is highly microporous and is, therefore, very suitable for gas-phase adsorption applications, such as air purification and industrial gas separation. Chemical activation provides better control over pore development. Usually, it leads to activated carbons with higher surface areas, sometimes even higher than 3000 m^2/g, compared to the surface areas (800–1500 m^2/g) that are usually obtained by physical activation. This activated carbon, following the process, is completely rinsed to neutralize any residues left behind with the chemicals for proper use in applications in the environment (Rodríguez-Reinoso & Molina-Sabio, 1992; Maciá-Agulló et al., 2004; Caturla et al., 1991; Hayashi et al., 2000).

Activated carbon, including activated charcoal, is an excellent adsorbent with versatility to remove both organic and inorganic pollutants through physical adsorption, electrostatic attraction, π–π stacking interactions, hydrogen bonding, and chemisorption. Physical adsorption is quite efficient for removing organic contaminants such as VOCs (benzene and toluene). Electrostatic attraction assists in the removal of heavy metals like Pb^{2+} and Hg^{2+} from aqueous solutions. Carboxyl and hydroxyl groups on the surface of AC allow for hydrogen bonding with polar organic compounds. The π–π stacking is significant for the adsorption of aromatic compounds, while chemisorption gives stronger bonds, especially for metals such as arsenic (Yang et al., 2019; Berber, 2020). Different isotherms follow the adsorption process depending on the nature of the contaminants. Langmuir isotherm applies to uniform adsorption sites in systems. It is more common in the removal of organic pollutants (Guo & Wang, 2019; Chang et al., 2020).

Conversely, Freundlich isotherm is used in heterogeneous systems, such as heavy metals or organic dyes. The adsorption may be exothermic, like for VOCs, or endothermic, which applies to more complex contaminants such as arsenic, for which specific conditions enhance spontaneity. Kinetic models describe the adsorption rate (Murphy et al., 2023; Dehmani et al., 2024). The pseudo-first-order model is applied to lighter organic molecules, while the pseudo-second-order model is used for heavy metals. The latter is indicative of stronger interactions between the adsorbent and contaminants. Understanding these adsorption mechanisms, thermodynamics, and kinetics is essential for optimizing the use of AC in environmental applications (Revellame et al., 2020).

In a study, Rahim Shahrokhi-Shahraki and co-authors compared the adsorption and desorption capabilities of tire-derived activated carbon (TAC) and commercial activated carbon (CAC) in removing selected heavy metals from aqueous solutions: Pb^{2+}, Cu^{2+}, and Zn^{2+} (Shahrokhi-Shahraki et al., 2021) (Fig. 8.1a). The main findings include the following: TAC was shown to be a more efficient sorbent than CAC. The maximum adsorption capacities of the ions Pb^{2+}, Cu^{2+}, and Zn^{2+} were found to be 322.5, 185.2, and 71.9 mg g^{-1}, respectively, while the capacities for CAC were lower. In both materials, the adsorption capacity was in the order Pb (II) > Cu (II) > Zn (II). TAC was also faster in achieving equilibrium, reaching 6 h, compared to CAC, which took 24 h. Pseudo-second-order kinetics best fitted both sorbents, which signifies chemisorption as a prevalent mechanism. Competitive adsorption experiments showed a noticeable decrease in maximum adsorption capacity for

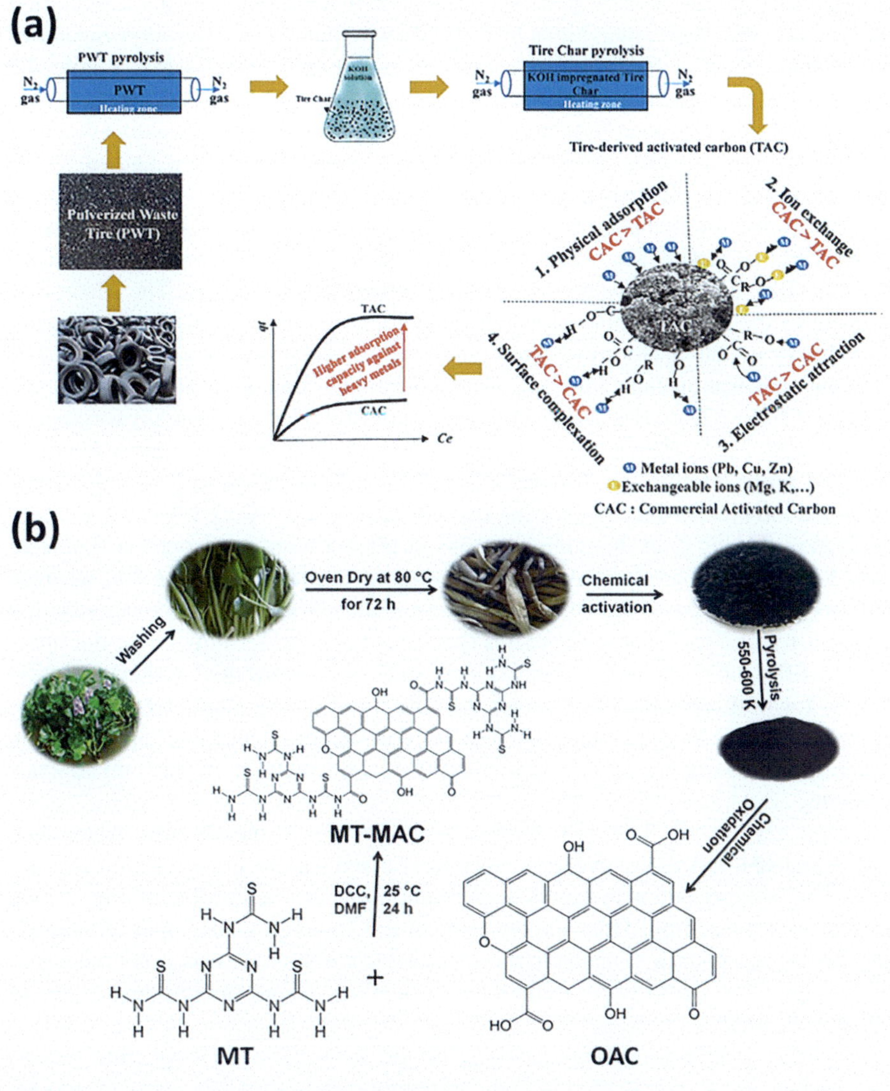

Fig. 8.1 (**a**) The synthesis and mechanistic insights of the activated carbon. (**b**) The representation of melamine thiourea modified activated carbon (MT-AMC). (Reproduced with permission from Rahim Shahrokhi-Shahraki et al. Copyright 2021, Elsevier)

both TAC and CAC in multicomponent compared to their single metal cases. The strongest impact on Zn was seen; therefore, it has lesser affinity than Pb and Cu. The leaching study established that metal release from TAC generally increased in highly acidic environments (pH 1.5), and zinc was the most leachable. However, under neutral and alkaline conditions, MCLs proposed by the US EPA were not exceeded for metal leaching from TAC; such applications can be safely used. The repeated adsorption–desorption cycles illustrated that neither TAC nor CAC could fully recover the loaded metal ions by 0.01 M HCl; thus, their adsorption capacities were reduced to about half after three adsorption–desorption cycles. Meanwhile, the pores in CAC were better for physical adsorption due to a bigger BET surface area and pore volume. However, TAC performed better than CAC and overcame the limitation where electrostatic attraction and surface complexation are known as the main adsorption mechanisms.

Conversely, ion exchange played a larger role in CAC, given the considerably higher CEC recorded. The adsorption behaviors of the heavy metals by the two sorbents, TAC and CAC, were analyzed by isotherms and kinetics. The adsorption behavior of the two sorbents followed the Langmuir isotherm model well, translating to heavy metals' adsorption onto homogeneous active sites in a monolayer. The maximum adsorption capacities (q_m) obtained from the Langmuir model were much higher for TAC (322.5, 185.2, and 71.9 mg g^{-1} for Pb^{2+}, Cu^{2+}, and Zn^{2+}, respectively) than for CAC (42.5, 15.0, and 14.0 mg g^{-1}). This indicates that TAC has more accessible and active binding sites for metal ions. The pseudo-second-order model better explained the kinetic data, suggesting that chemisorption, where valence forces are involved in sharing or exchanging electrons, is predominant in adsorption. TAC had faster kinetics compared to CAC, as TAC reached equilibrium within 6 h, whereas CAC required up to 24 h. This difference accounts for the better reactivity and accessibility of TAC's adsorption sites than those of CAC. Additionally, the intraparticle diffusion model suggested that pore diffusion contributed to the adsorption process for both sorbents but was not the sole rate-limiting step. Overall, the isotherm and kinetic analyses highlight TAC's enhanced efficiency and rapid sorption capabilities, making it a promising material for heavy metal removal from aqueous solutions.

Therefore, overall, TAC is better for removing heavy metals from water than CAC because it has more favorable electrostatic and surface complexation interactions. TAC's applications in highly acidic conditions need careful handling as they increase metal leaching. Thus, the outcome of this research gives important knowledge on the behavior and mechanisms of activated carbons, allowing rational selection of them for use in water treatment applications where the removal of heavy metals is desired.

Eva Díaz and her coworkers investigated the adsorption of omeprazole (OMP) and other pharmaceuticals that include amoxicillin, ciprofloxacin, carbamazepine, and ibuprofen onto activated carbon (Díaz et al., 2024). This work was focused on equilibrium, mass transfer, and OMP degradation. Adsorption isotherms of Langmuir, Temkin, Fowler–Guggenheim, and Jovanovic models were used to characterize these compounds' interactions with activated carbon. The research indicated that amoxicillin manifested irreversible endothermic adsorption, carbamazepine exhibited reversible adsorption, and ciprofloxacin, along with carbamazepine, possessed strong interactions at the molecular level. Ibuprofen's adsorption had a nature of mechanical contact with the adsorbent. In all instances, intraparticle diffusion was found to be the limiting step in mass transfer with coefficients of about 10^{-8} cm^2 h^{-1}. Mass transfer and diffusion coefficients decreased with increasing molecular size for ibuprofen (IBU), ciprofloxacin, and amoxicillin, but carbamazepine showed an unusual trend due to its pKa. The degradation of omeprazole over 77 h was modeled using first-order kinetics. The study provides insights into pharmaceutical pollutants' adsorption and degradation mechanisms, enhancing understanding of their removal and fate during water treatment.

Another emerging adsorbent is functionalized carbon, which can be tuned with the functional group to induce selectivity. For instance, Ahmad M. El-Wakil and his coworkers have recently developed a novel chelating adsorbent, melamine thiourea modified activated carbon (MT-MAC), through functionalizing activated carbon (AC) obtained from water hyacinth (WH) with melamine thiourea (MT) (El-Wakil et al., 2022) (Fig. 8.1b). An amidation reaction was applied to form carboxyl groups of the OAC and the amino groups of melamine thiourea, with a dicyclohexylcarbodiimide (DCC) as the coupling agent. The toxic metal ions like Hg^{2+}, Pb^{2+}, and Cd^{2+}, MT-MAC, displayed remarkable efficiencies in removing metal ions from an aqueous solution by adsorbing with the following maximum capacities of 292.6 mg/g for Hg^{2+}, 237.4 mg/g for Pb^{2+}, and 97.9 mg/g for Cd^{2+}. The adsorbent showed 100% removal efficiency of Hg^{2+} at pH 5.5 at low initial concentrations (25–1000 ppb). The adsorption data fit the Langmuir isotherm model well, indicating a monolayer adsorption process. Kinetic studies supported a chemisorption mechanism, with the adsorption process fitting the pseudo-second-order (PSO) model. In mixed metal ion solutions (50 ppm of each metal), MT-MAC indicated selective removals of Hg^{2+} (97.0%), Pb^{2+} (68%), and Cd^{2+} (45%), which was much lesser for other metals such as Cu^{2+}, Ni^{2+}, and Zn^{2+}, indicating its selectivity toward the Hg^{2+} was exceptional. Moreover, MT-MAC showed excellent stability, maintaining its performance at three adsorption–desorption cycles. These results suggest that MT-MAC is an efficient, eco-friendly, and highly selective adsorbent for removing Pb^{2+}, Hg^{2+}, and Cd^{2+} from polluted water, making it a promising material for environmental cleanup applications.

8.3 Biochar

Biochar is a carbon-rich, porous material produced through pyrolysis, which involves heating organic materials, such as agricultural residues, wood, or other biomass, in a low-oxygen environment. During pyrolysis, the biomass decomposes at high temperatures (typically between 300 and 700 °C), releasing volatile gases and leaving behind a solid, carbon-rich residue known as biochar. This material features a high surface area, porosity, and functional groups on its surface. It is a very effective material for adsorption of various contaminants: organic compounds, heavy metals, and greenhouse gases. Biochar has many diverse applications, including one of its most important—environmental remediation. Biochar can efficiently absorb pollutants from the air, water, or soil since it has a very high surface area combined with rich functional groups. Its major application scope is also in agriculture: biochar helps improve soil fertility by enhancing water-holding capacity, providing nutrients, and contributing to carbon sequestration due to climate change mitigation. Besides the environmental benefits, biochar is considered a sustainable material because it is produced from renewable feedstocks, often using agricultural waste or biomass that would otherwise be discarded. The carbon sequestered in biochar is stable for long periods, which may help to reduce the amount of carbon dioxide in the atmosphere, making it a promising tool for climate change mitigation.

8.3.1 Synthesis, Properties, and Adsorption Mechanisms

The synthesis of biochar begins with the selection of an appropriate feedstock, typically organic material like agricultural residues, which include rice husks, wheat straw, or corn stalks, forestry by-products such as wood chips or sawdust, animal manure, or even municipal waste. Choosing the right feedstock is significant because it defines the final product's characteristics in terms of carbon content, porosity, and minerals or nutrients. Feedstocks rich in cellulose, hemicellulose, and lignin tend to produce biochar with a higher carbon content and better adsorption properties. After the feedstock is chosen, it undergoes the pyrolysis process, a thermochemical conversion method in which the material is heated to high temperatures (usually between 300 and 700 °C) in an oxygen-limited environment. This low-oxygen condition prevents combustion, allowing the decomposition of the feedstock into biochar, volatile gases, and liquids. While performing pyrolysis, a three-phase material undergoes such stages: three primary phases with the initial or drying stage at the onset with possible removal of excess moisture in feedstock and with two last phases: breaking of the molecule in a small volatile gas with liquid devolatilization, then building up carbonaceous stable structure also called carbonizing. The temperature and heating rate of pyrolysis determine the properties of the biochar produced. Low temperatures during pyrolysis, usually between 300 and 450 °C, produce more reactive biochar with a higher surface area and more functional groups. Such characteristics make it ideal for contaminant adsorption and soil amendment applications. Higher temperatures, between 500 and 700 °C, result in more stable biochar with higher carbon content, which is appropriate for carbon sequestration and long-term environmental benefits. The type of pyrolysis reactor used, whether fixed-bed, fluidized-bed, or rotary kilns, also affects the final characteristics of the biochar. The residence time, the time the material is exposed to heat, and the presence of specific gases during pyrolysis can further influence the surface chemistry, pore structure, and adsorption capacity of the biochar. The produced biochar can be further modified or activated through various methods to enhance its performance for specific applications, such as removing heavy metals or organic contaminants from water or air (Guo et al., 2020; Mian & Liu, 2018; Amusat et al., 2021).

Biochar possesses a wide range of properties that make it very effective for removing contaminants from different environments. Its most significant property is its high surface area and porosity, ranging from 200 to 1200 m^2/g. This large surface area enables biochar to adsorb a wide range of contaminants, including organic compounds, such as pesticides and VOCs, and inorganic pollutants, such as heavy metals like lead, arsenic, and cadmium. The porous structure of biochar provides ample space for adsorption and facilitates the retention of these contaminants through physical trapping and surface interactions. Additionally, biochar contains a wide range of functional groups such as carboxyl, hydroxyl, and phenolic groups on its surface, which are important in binding and retaining positively charged ions and some organic molecules. These functional groups enhance its ability to adsorb heavy metals and organic contaminants, making it a versatile solution for various types of pollution. These chemical groups also make biochar attractive and capable of capturing various pollutants. Moreover, the hydrophobic nature of biochar allows it to remove non-polar organic compounds effectively, and the negative surface charge of biochar can facilitate the adsorption of positively charged metal ions (Dai et al., 2019; Jagadeesh & Sundaram, 2023). These unique properties make biochar an invaluable material for environmental remediation, effective and sustainable in removing soil, water, and air contaminants.

The adsorption mechanisms of biochar toward both organic and inorganic contaminants are driven by a combination of physical adsorption and chemical interactions, each playing a critical role in its effectiveness as an adsorbent. Physical adsorption, driven by Van der Waals forces, occurs when contaminants are attracted to the biochar surface due to its highly porous structure, providing an extensive surface area for the contaminants to adhere. This mechanism suits volatile organic compounds (VOCs) such as benzene and toluene or pesticides like atrazine and glyphosate. As these organic compounds are adsorbed through a weak interaction involving no chemical bonds, the operation is reversible with the possibility of regenerating the same biochar. Similarly, biochar can adsorb gaseous pollutants such as ammonia and chlorine by physical interactions within its porous network, thus lowering the concentration of these harmful substances in air or water.

Chemical adsorption is utilized to remove inorganic contaminants, especially heavy metals, in addition to physical adsorption. The biochar surface is characterized by functional groups such as carboxyl, hydroxyl, and phenolic groups that can form strong chemical bonds with contaminants. For instance, lead (Pb^{2+}), cadmium (Cd^{2+}), and arsenic (As^{3+}) can be well adsorbed through ion exchange and chelation. Here, the surface groups of biochar interact with the metal ions to form stable complexes. Carboxyl groups can bind with metals, and hydroxyl groups can engage in hydrogen bonding. The negative surface charge of biochar also promotes the adsorption of positively charged metal ions such as copper (Cu^{2+}) and zinc (Zn^{2+}) through electrostatic attraction. This dual mechanism of physical and chemical adsorption makes biochar a versatile and efficient adsorbent for a wide range of environmental contaminants, including both organic and inorganic substances, thereby offering a sustainable solution for remediation applications (Gwenzi et al., 2021; Zhao et al., 2022).

For instance, Xu Su et al. prepared a biochar material co-modified with KOH and chitosan (CHKBC), which improved its industrial dye pollutant adsorption capacities by 7.03-fold with excellent performances at different temperature and

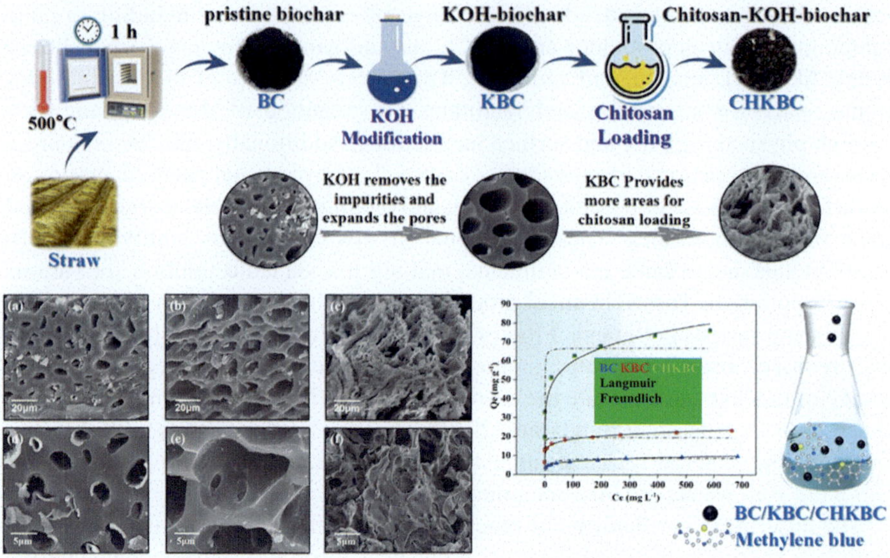

Fig. 8.2 (**a**) The schematic representation of the possible adsorption mechanism. (**b**) The FeSEM images of the synthesized biochar. (**c**) The adsorption of methylene blue in the presence of co-existing ions. (Reproduced with permission from Xu Su et al. Copyright 2024, Elsevier)

pH levels (Su et al., 2024) (Fig. 8.2). The CHKBC showed excellent improvement in maximum adsorption capacities for methylene blue (MB) from the original 8.83 mg g^{-1} up to 62.04 mg g^{-1}. This corresponds to a 7.03-fold improvement, which means that the modifications enhance the biochar's ability to remove dye pollutants from water. The CHKBC showed a consistent adsorption efficiency for MB at various temperatures (288–318 K) and pH levels (5–11). This stability implies that the material can be effectively used in different environmental conditions, making it a versatile option for wastewater treatment. Several key mechanisms controlling the removal of MB by CHKBC, namely ion exchange, pore filling, electrostatic attraction, π–π interactions, and hydrogen bonding, were determined (Fig. 8.2a). Knowing these mechanisms can be crucial for optimizing a material for practical applications.

8.3.1.1 Mechanisms of Adsorption

This chapter identifies several key mechanisms that contribute to the effective removal of methylene blue (MB) by the CHKBC material. These mechanisms include:

- *Ion Exchange*: This process involves the exchange of ions between the dye molecules and the functional groups on the biochar surface, enhancing the adsorption capacity.
- *Pore Filling*: The biochar pores can be filled with the dye molecules, enhancing the surface area for adsorption.

- *Electrostatic Attraction*: There might be oppositely charged functional groups in CHKBC, which would attract the oppositely charged dye molecules to adsorb on it.
- π–π *Interactions*: The methylene blue aromatic structure can also interact with the aromatic groups within the biochar. This increases the adsorption as a result of π–π stacking interactions.
- *Hydrogen Bonding*: The functional groups such as –OH and $-NH_2$ on the CHKBC can form hydrogen bonds with the dye molecules, further promoting adsorption efficiency.

This approach is holistic, contributing to the advancement of knowledge and the application of biochar-based solutions to overcome the challenges that emerging pollutants pose in aquatic ecosystems.

Similarly, biochars can be tailored or engineered for specific applications. For example, Hong Nam Nguyen and others studied biochars from five by-products of agricultural materials: macadamia nut shells, rice husks, durian peel, coir, and straw, synthesized from a combination of hydrothermal treatment and pyrolysis to address oxytetracycline (OTC) contamination. These biochars' surface elements and functional groups were very similar, yet there were obvious differences in the porous structures among them. The batch and continuous adsorption tests gave insights into the dynamic behavior of biochar materials in removing OTC, which showed diversified performances of different biochars with a preference for long-term sustained efficacy rather than one-time efficacy for practical implementation. Coir biochar showed the highest OTC adsorption capacity at 93% with batch adsorption and showed full treatment for eight cyclic reuses. Both durian peel and straw biochars also demonstrated high performance in batch and continuous adsorption. Rice husk and macadamia nut shell biochars were less effective for the OTC treatment. Pore filling appears to be the major mechanism for adsorption by the van der Waals forces between OTC molecules and micro-mesopore walls at a 1.2–10 nm range. This work has findings that are key to innovation, modeling, and designing efficient systems for removing emerging organic pollutants from water.

8.4 Carbon Nanotubes

Carbon nanotubes (CNTs) are cylindrical structures formed by rolled-up graphene sheets. They can be broadly divided into two types based on their structure: Single-walled carbon nanotubes (SWCNTs) and multi-walled carbon nanotubes (MWCNTs). The only difference is the number of graphene layers forming the nanotube (Prasek et al., 2011).

8.4.1 Single-Walled Carbon Nanotubes (SWCNTs)

The name indicates that SWCNTs are made up of a single layer of graphene rolled into a cylindrical shape. These nanotubes have diameters between 0.4 and 2 nm, characterized by their high surface area, flexibility, high electrical conductivity, and mechanical strength. The SWCNTs are quite effective at adsorbing organic pollutants like dyes, pesticides, and pharmaceutical residues because of their small diameter and high surface area. The functionalization of SWCNTs with different chemical groups, such as carboxyl or amine groups, can improve their adsorptive properties and make them effective for heavy metal removal. This allows SWCNTs to adsorb both hydrophobic and hydrophilic contaminants, including arsenic (As) and cadmium (Cd) (Niyogi et al., 2002; McEuen, 2000).

8.4.2 Multi-Walled Carbon Nanotubes (MWCNTs)

MWCNTs are made up of multiple concentric graphene cylinders, each one nested inside the other. The diameters of these nanotubes are larger compared to SWCNTs, measuring from 2 to 100 nm, and the structural nature of these makes them mechanically stronger and thermally stable. Compared to MWCNTs, with their large surface area and presence of more than one layer of graphene, they have multiple adsorption sites for many different contaminants. Because of their higher surface area and hydrophobic character, they are highly efficient in adsorbing hydrophobic organic contaminants, such as oils, grease, and solvents. In addition, MWCNTs can be functionalized to increase the adsorption of inorganic contaminants such as heavy metals (Pb^{2+}, Hg^{2+}, and Cr^{3+}). The multi-layers also enhance the mechanical strength of MWCNTs, making them more capable of withstanding high-performance adsorption processes without breaking their structure (Andrews et al., 2002; Lehman et al., 2011).

In conclusion, SWCNTs are the best adsorbent for small, organic pollutants due to their small diameter and high surface area. In contrast, MWCNTs are conducive to large contaminants and have better mechanical properties due to their multi-layered structure. CNTs are highly versatile in applications like environmental remediation due to the unique merits of removing a wide scope of pollutants.

8.4.3 Synthesis, Properties, and Adsorption Mechanisms

Carbon nanotubes'synthesis involves different methods, but each has certain advantages depending upon the properties in which the CNTs need to be tailored. One of the earliest approaches is the arc-discharge method, wherein a high-temperature arc is established between two electrodes made of graphite in an inert gas such as helium or argon. In this process, the arc attained 3000–4000 °C temperatures, thereby vaporizing carbon from the cathode, giving rise to CNTs. This method has a problem with yield, although it is possible to produce single-walled as well as multi-walled CNTs. In addition, amorphous carbon impurity often limits the application of the method. Other popular methods include laser ablation that vaporizes, with the high-powered laser, the graphite target, along with a metal catalyst. It is highly reputed for yielding excellent-quality single-walled CNTs because it can accurately control the vaporization process with the laser. It is often expensive and not scalable to larger synthesis for practical applications. The preferred method for the large-scale production of CNTs is through chemical vapor deposition (CVD). In this method, a carbon-containing gas such as methane or acetylene decomposes at high temperatures in the presence of a metal catalyst that is usually iron, nickel, or cobalt deposited on a substrate. This method is excellent for precisely controlling the growth of CNTs and is applicable to producing both single-walled and multi-walled CNTs with definite diameters and lengths. CVD is scalable and, therefore, can be applied industrially, although the synthesis products often require purification post-synthesis to remove the remaining catalyst particles. Another method is flame synthesis, a variant of CVD in which the flame acts as the heat source for the growth of CNTs. This is a relatively simple and inexpensive method, but it tends to produce more defective CNTs. Each synthesis method has disadvantages and merits, but the CVD method is among the most versatile and scalable options to produce high-quality CNTs for most applications (Ando et al., 1999; Odom et al., 2000; Mintmire & White, 1995).

Increasingly, there is growing interest in developing sustainable and environmentally friendly approaches to regenerating spent nanomaterials for resource efficiency and reducing the costs involved in environmental remediation processes. This study by Mohamed Ateia and coworkers examines the potential of using ozone to regenerate magnetic carbon nanotubes after they are applied to remove organic pollutants such as atrazine from water (Ateia et al., 2018). Regeneration involves repeated cycles, including magnetic collection, ozone treatment, and sequential washing with ethanol and water. Experimental results showed that the adsorption capacity of MCNTs for atrazine decreased from 57.8 to 27.6 mg/g after three cycles when only ozone treatment was used. However, adding ethanol washing significantly improved the performance, allowing the MCNTs to retain 85–93% of their initial adsorption capacity even after ten consecutive regeneration cycles. DFT calculations were carried out using a three-layer graphite slab as a model for CNTs to understand the mechanism of this enhanced regeneration. These calculations estimated the free energy of adsorption and solvation of atrazine and its degradation by-products in water and ethanol. Complementary X-ray photoelectron spectroscopy (XPS) measurements and density functional theory (DFT) analysis confirmed that the π–π interactions critical for MCNT functionality remained intact during the ozonation and ethanol wash. With the combined regeneration strategy, the degradation by-products adhering to the surface of the material were removed successfully without affecting its adsorption efficacy. This work represents the first

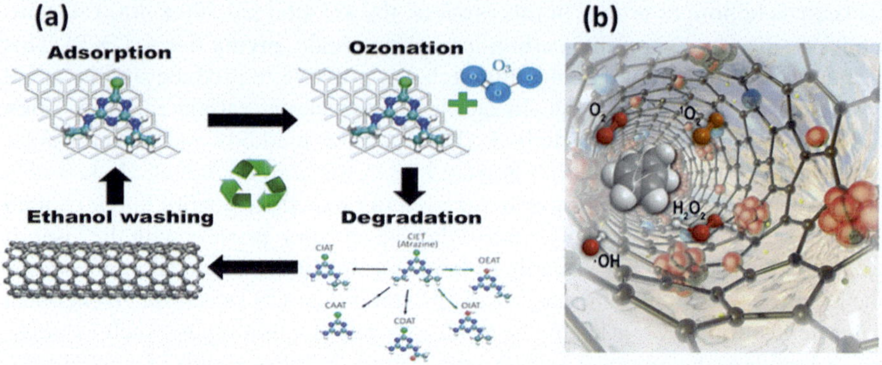

Fig. 8.3 (**a**) The adsorption, ozonation, degradation, and regeneration mechanism. (Reproduced with permission from Mohamed Ateia et al. Copyright 2018, Elsevier). (**b**) The schematic representation of the catalytic mechanism of the composite material Mo@CNT. (Reproduced with permission from Chencheng Dong et al. Copyright 2023, Elsevier)

demonstration of ozone-assisted regeneration of MCNTs augmented by ethanol washing as an efficient, practical, and sustainable method for maintaining the adsorption performance of MCNTs in water treatment applications. The study provides a promising framework for advancing the reuse of nanomaterials in environmental remediation and systematically identifies the primary cause of reduced adsorption capacity after ozonation. Atrazine, a commonly used and toxic s-triazine herbicide, was chosen as a model organic pollutant to target adsorption and degradation mechanisms in aquatic environments (Fig. 8.3).

8.5 Graphene Oxide and Derivatives

Graphene oxide (GO) is a graphene derivative created by the oxidation of graphite, which introduces oxygen-containing functional groups like hydroxyl, epoxide, and carboxyl groups onto its surface. These groups make GO highly hydrophilic, enabling it to disperse easily in water and other polar solvents. However, the presence of these oxygenated groups disrupts the graphene's π-conjugation, making GO an electrical insulator with significantly reduced conductivity compared to pure graphene. Despite this, GO has a large surface area and high chemical reactivity, making it suitable for water filtration, sensors, and drug delivery applications. Reduced graphene oxide (rGO) is the chemical or thermally reduced GO product that aims to restore some of the electrical properties of graphene. Reduction removes some of the oxygen groups, improving electrical conductivity and mechanical properties. However, rGO still retains some oxygen functionalities, which can benefit functionalization and further chemical modifications. The extent of reduction and the remaining oxygen groups in rGO can vary depending on the reduction method used. rGO finds applications in energy storage devices, such as supercapacitors and batteries, as well as composites, sensors, and conductive inks, where high conductivity is essential. The trade-off between improved conductivity and the retention of functional groups often determines the specific applications of rGO (Zhu et al., 2010; Marcano et al., 2010; Jiříčková et al., 2022; Gutiérrez-Cruz et al., 2022).

8.5.1 Synthesis, Properties, and Adsorption Mechanisms

8.5.1.1 Top-Down Approaches

Graphene sheets can be obtained by exfoliating natural graphite into single or few-layer graphene sheets. This can be achieved via mechanical, chemical, or liquid-phase exfoliation methods (Farjadian et al., 2020). Mechanical exfoliation, known as the Scotch tape method, involves peeling graphene layers from graphite using adhesive tapes. While effective for small-scale production, it is not suitable for large-scale applications. In chemical exfoliation, graphene oxide is produced by oxidizing graphite using strong oxidizing agents like potassium permanganate ($KMnO_4$) in concentrated sulfuric acid. The resulting graphene oxide sheets can be reduced to graphene oxide or reduced graphene oxide (rGO) using reducing agents, such as hydrazine. In liquid-phase exfoliation. This method involves dispersing graphite in a solvent and applying ultrasonic waves to separate the layers of graphite into graphene sheets.

8.5.1.2 Bottom-Up Approaches

In the chemical vapor deposition (CVD) method, carbon-rich gases (such as methane) are decomposed at high temperatures on a copper or nickel substrate to produce single-layer graphene. CVD offers high-quality, large-area graphene but is more expensive than other methods. The solvothermal synthesis method involves using a high-temperature, high-pressure solution to promote chemical reactions that synthesize graphene sheets from precursor compounds.

Graphene oxide (GO) is a highly effective material for the adsorption of pollutants due to its unique structural and chemical properties. The high surface area of GO provides a large number of active sites for pollutant interactions, making it ideal for adsorbing both organic and inorganic pollutants. The material contains various oxygenated functional groups, including hydroxyl, carboxyl, carbonyl, and epoxide groups, which significantly enhance its ability to interact with various contaminants. These functional groups can form hydrogen bonds, electrostatic interactions, and covalent bonds with pollutants, increasing their adsorption capacity. GO's hydrophilicity, a result of the presence of these oxygenated groups, allows it to easily disperse in water and interact with water-soluble pollutants, such as metal ions, dyes, and polar organic molecules. Additionally, the layered structure of GO enables the intercalation of molecules between its sheets, allowing the adsorption of both small and large pollutants. The negatively charged surface of GO at neutral pH further promotes the adsorption of positively charged metal ions and cationic organic molecules. This combination of high surface area, functional versatility, and surface charge makes GO an excellent adsorbent for a variety of environmental pollutants, offering an efficient and sustainable solution for water treatment and pollution remediation (Chowdhury & Balasubramanian, 2014; Yan et al., 2016).

For instance, A. P. Melchor-Durán et al. explore, in-depth, the adsorption properties of GO for heavy metal ion removal in the magnetic-assisted system (Melchor-Durán et al., 2024). Overall, it emerges that the surface chemistry of GO plays a major role in affecting the efficacy of the adsorption process when any specific type of heavy metal is under the action of magnetic fields. Adding a magnetic field proved to significantly enhance the removal of Pb^{2+} ions by GO while showing relatively slight effects on the adsorption of Cu^{2+} ions. Characterization of the GO samples showed that oxygenated functional groups on the graphene oxide surface played a crucial role in the interaction with and separation of metallic cations, and these groups were more actively involved in removing Pb^{2+} ions when subjected to a magnetic field. Although the GO samples are more sensitive to magnetic fields, they exhibited higher adsorption capacity for Cu^{2+} than Pb^{2+}, indicating a multi-ionic monolayer adsorption process. This process involves the interaction of oxygen-containing functional groups with both cations, contributing to removing these pollutants from water. The study shows that magnetic-assisted adsorption systems are promising in cleaning water contaminated by toxic heavy metals using sintered graphene oxide materials. This study emphasizes the necessity of understanding the effects of magnetic field exposure on the adsorption properties of graphene-based adsorbents. Optimization of this intensified adsorption process could enhance the performance of water treatment systems significantly. The results of this research have critical implications for designing and improving heavy metal removal technologies, where magnetic field applications can serve as an effective strategy to enhance the separation and removal efficiency of carbonaceous adsorbents in water purification processes.

Sahil Thakur and coworkers investigated the synthesis of graphene oxide (GO), focusing on tailoring its surface and optical properties to enhance its effectiveness as both a superior adsorbent and a visible light-active photocatalyst (Thakur et al., 2024) (Fig. 8.4). GO is one of the most promising materials for environmental remediation, particularly for removing organic pollutants from water. Scanning and transmission electron microscopy confirm that the oxidation process disrupts the ordered graphite structure, forming crumpled nanosheets. GO showed excellent adsorption properties, especially for MB dye, with a maximum adsorption capacity of 454.54 mg/g, superior to many other materials. Adsorption equilibrium was achieved within 60 min, and optimal conditions were observed under alkaline pH, with GO's surface chemistry facilitating strong interactions with organic pollutants. Besides being an efficient adsorbent, GO is a robust photocatalyst under sunlight, removing over 99% of organic pollutants. This broad UV absorption makes the

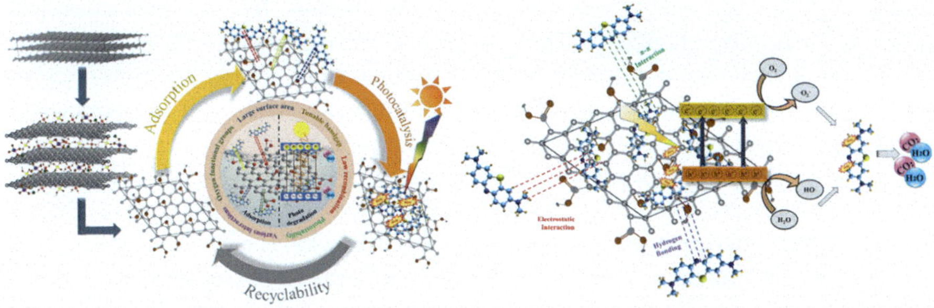

Fig. 8.4 The schematic representation of the composite material's adsorption, photocatalysis, and regeneration toward dye. (Reproduced with permission from Sahil Thakur et al. Copyright 2024, American Chemical Society)

material suitable for sunlight activation. Scavenger studies assured the involvement of reactive species in pollutant degradation. Moreover, recyclability and stability characterize GO, maintaining high performance even after regeneration cycles. The self-cleaning ability of GO in sunlight further increases its sustainability. In summary, GO is an effective, multi-functional material with superior adsorption and photocatalytic properties. It is recyclable and stable, and surface chemistry can be tailored, which puts it as one of the hopeful candidates for future sustainable water treatment and environmental cleanup technologies.

8.6 Carbon Dots

8.6.1 Synthesis, Properties, and Adsorption Mechanisms

Carbon dots (CDs) are one of the families of carbon nanomaterials that have recently attracted the attention of numerous researchers due to their unique optical properties, biocompatibility, and potential applications in various fields, such as environmental pollution remediation. CDs are usually made of small carbon nanoparticles, which are less than 10 nm in size, and have a core–shell structure. The core comprises sp^2 hybridized carbon atoms, while the surface is usually functionalized with oxygenated or nitrogenated groups (Liu et al., 2019; Xia et al., 2019). The synthesis of CDs can be broadly classified into two categories: top-down (Shi et al., 2022; Pillar-Little et al., 2018) and bottom-up approaches(Choi et al., 2018; Han et al., 2018; Bian et al., 2023). Each method brings its own advantages and possibilities for tuning the size, surface functionalization, and properties of the CDs.

CDs are hydrophilic due to the presence of surface groups such as hydroxyl, carboxyl, and amine and can be easily dispersed in water, which improves their interaction with pollutants through electrostatic interactions, hydrogen bonding, and coordination. Due to their high surface area, small size, and functional groups, CDs allow efficient adsorption of a wide range of contaminants, including heavy metals, organic pollutants, and emerging contaminants like microplastics and endocrine-disrupting chemicals. The fluorescence properties of CDs allow for the real-time monitoring of pollutant removal. Excellent stability and eco-friendly synthesis make CDs promising adsorbents for water treatment and environmental remediation (Ajith et al., 2022; Parambil & Rajamani, 2024).

For instance, Ajith et al. reported the synthesis of highly stable and photoluminescent CDs derived from the leaf extract of *Ficus benghalensis* (Ajith et al., 2020). These CDs exhibited exceptional selectivity toward nickel (Ni^{2+}) and iron (Fe^{3+}) in a dose-dependent manner, making them highly effective for targeted heavy metal removal. To enhance their adsorption capacity, silica-carbon dot (Si-CD) hybrid conjugates were developed and integrated into a column purification system for industrial wastewater treatment. The Si-CD conjugates demonstrated a threefold increase in Fe^{3+} adsorption and a twofold increase in Ni^{2+} adsorption, significantly improving their removal efficiency. The enhanced adsorption performance is attributed to the strong electrostatic interactions between Fe^{3+}/Ni^{2+} ions and functional groups on the CD surface, leading to immobilization within the Si-CD column bed. These findings highlight the potential of Si-CD hybrid materials in establishing scalable continuous bed filtration systems for industrial water purification, offering a sustainable and high-efficiency approach for heavy metal remediation.

8.7 Carbon Nanomaterials and Short-Bed Adsorption Columns

Short-bed adsorption columns are conventional for removing pollutants from water and wastewater treatment systems. These columns ensure the efficient removal of contaminants from fluid streams based on the passage through an adsorbent material. The key advantages of short-bed columns encompass their ability to remove contaminants at a high flow rate, compactness, and the optimal contact time between pollutants and absorbers, making them scalable to large industrial sites (Nithya et al., 2020; Debebe et al., 2024).

8.7.1 Working Principle

In a short-bed adsorption column, a fluid containing pollutants like heavy metals, organic contaminants, or dyes is passed through a bed of adsorbent material. In this process, the adsorbent captures contaminants as the fluid passes through the column. Factors affecting the process include characteristics of the adsorbent, concentration of pollutants, flow rate, and column dimensions. Shorter columns are usually used for smaller volumes of water or higher concentrations of pollutants to ensure that the contact time between the fluid and the adsorbent is sufficient for effective removal (Parambil et al., 2024; Hethnawi et al., 2018). Carbon nanomaterials, including activated carbon and CNTs, play a crucial role in short-bed adsorption columns due to their high surface area and well-developed porosity, enhancing their ability to adsorb pollutants efficiently. These materials have a large surface area that provides ample active sites to capture contaminants, ensuring rapid adsorption even at short contact times. Carbon nanomaterials can also be chemically functionalized with amino, carboxyl, or hydroxyl groups to enhance their selectivity toward specific pollutants, such as heavy metals, organic dyes, and chemicals. This functionalization enhances the adsorption capacity and efficiency in targeting a wide range of contaminants. The mechanical strength and stability of carbon-based materials such as CNTs and graphene make them ideal for industrial applications, offering durability and long-lasting performance in pollutant removal processes (Gupta et al., 2011; Liu et al., 2009). Moreover, carbon nanomaterials'superior conductivity and structure result in enhanced adsorption kinetics, enabling faster pollutant removal even at high flow rates. Finally, these materials have very good regeneration and reusability properties that can be used in several cycles with minimum degradation, making them an appropriate solution for large-scale or emergency treatments.

8.7.2 *Advantages of Carbon Nanomaterials in Short-Bed Adsorption Columns*

8.7.2.1 Rapid Adsorption Kinetics

Carbon nanomaterials allow for the fast removal of pollutants, thus reducing the contact time required in short-bed columns and enhancing treatment efficiency.

8.7.2.2 Scalability

Carbon nanomaterials can be produced in large quantities and are versatile for both small-scale and large-scale adsorption systems, thus suitable for various industrial applications.

8.7.2.3 Economy of Operation

Although carbon nanomaterials are costlier than conventional adsorbents in the initial stage, their recyclability, strength, and excellent performance often pay back in the long term.

8.7.2.4 Environmental Sustainability

Carbon-based materials, especially bio-derived carbon nanomaterials, are eco-friendly and environmentally friendly, responding to the emerging need for green, sustainable technologies against the traditional use of conventional adsorbents.

8.7.3 Challenges and Future Directions

8.7.3.1 Column Fouling

Saturation of the adsorbent in the column occurs after some time, decreasing efficiency. Regeneration or replacement of the adsorbent is essential to ensure long-term performance.

8.7.3.2 Scale-Up Issues

Although it is possible to obtain a high level of efficiency in laboratory or pilot-scale short-bed columns, scaling them up for applications on a larger scale often meets operational requirements for uniform fluid flow distribution and mass transfer limitations.

8.7.3.3 Optimization

Further research into optimizing column design, adsorbent selection, and operational conditions can enhance the performance and longevity of these systems.

Conclusively, short-bed adsorption columns offer a simple yet effective method for removing pollutants from water and air, with the potential for broad applications in environmental cleanup and industrial processes.

8.8 Conclusion

Carbon-based adsorbents, such as activated carbon, biochar, CNTs, graphene-based materials, and CDs, have been identified as highly effective for removing environmental pollutants. Their high surface area, tunable porosity, and functionalization capabilities allow them to target various contaminants, including heavy metals, organic pollutants, and emerging contaminants. Short-bed adsorption columns, utilizing these materials, provide a simple yet highly efficient method for pollutant removal, with the potential for large-scale industrial applications. The cost-effectiveness of these carbon-based materials also facilitates its further promise in the area of sustainable environmental remediation. Due to continuous research and development, carbon-based adsorbents are likely to fill the growing demand for the future challenges of contamination in the environment.

References

Aguilar-Aguilar, A., de León-Martínez, L. D., Forgionny, A., Acelas Soto, N. Y., Mendoza, S. R., & Zárate-Guzmán, A. I. (2023, December). A systematic review on the current situation of emerging pollutants in Mexico: A perspective on policies, regulation, detection, and elimination in water and wastewater. *Science of the Total Environment, 905*, 167426. https://doi.org/10.1016/j.scitotenv.2023.167426

Ahmed, A. S., Alsultan, M., Hameed, R. T., Assim, Y. F., & Swiegers, G. F. (2022). High surface area activated charcoal for water purification. *Journal Of Composites Science, 6*(10), 311. https://doi.org/10.3390/jcs6100311

Ajith, M. P., Priyadarshini, E., & Rajamani, P. (2020). Effective and selective removal of heavy metals from industrial effluents using sustainable Si–CD conjugate based column chromatography. *Bioresource Technology, 314*, 123786. https://doi.org/10.1016/j.biortech.2020.123786

Ajith, M. P., Pardhiya, S., & Rajamani, P. (2022). Carbon dots: An excellent fluorescent probe for contaminant sensing and remediation. *Small, 18*(15), e2105579. https://doi.org/10.1002/smll.202105579

Akhtar, M. S., Ali, S., & Zaman, W. (2024). Innovative adsorbents for pollutant removal: Exploring the latest research and applications. *Molecules, 29*(18), 4317. https://doi.org/10.3390/molecules29184317

Alkhaldi, H., Alharthi, S., Alharthi, S., AlGhamdi, H. A., AlZahrani, Y. M., Mahmoud, S. A., Amin, L. G., et al. (2024). Sustainable polymeric adsorbents for adsorption-based water remediation and pathogen deactivation: A review. *RSC Advances, 14*(45), 33143–33190. https://doi.org/10.1039/D4RA05269B

Ambaye, T. G., Vaccari, M., van Hullebusch, E. D., Amrane, A., & Rtimi, S. (2021). Mechanisms and adsorption capacities of biochar for the removal of organic and inorganic pollutants from industrial wastewater. *International journal of Environmental Science and Technology, 18*(10), 3273–3294. https://doi.org/10.1007/s13762-020-03060-w

Amusat, S.O., Temesgen G.K., Simiso D., Mathew M.N. 2021. "Ball-Milling Synthesis of Biochar and Biochar–Based Nanocomposites and Prospects for Removal of Emerging Contaminants: A Review." Journal of Water Process Engineering 41 (June):101993. https://doi.org/10.1016/j.jwpe.2021.101993.

Ando, Y., Zhao, X., Shimoyama, H., Sakai, G., & Kaneto, K. (1999). Physical properties of multiwalled carbon nanotubes. *International Journal of Inorganic Materials, 1*(1), 77–82. https://doi.org/10.1016/S1463-0176(99)00012-5

Andrews, R., Jacques, D., Qian, D., & Rantell, T. (2002). Multiwall carbon nanotubes: Synthesis and application. *Accounts of Chemical Research, 35*(12), 1008–1017. https://doi.org/10.1021/ar010151m

Aslam, M. M.-A., Kuo, H.-W., Den, W., Usman, M., Sultan, M., & Ashraf, H. (2021). Functionalized carbon nanotubes (CNTs) for water and wastewater treatment: Preparation to application. *Sustainability, 13*(10), 5717. https://doi.org/10.3390/su13105717

Ateia, M., Ceccato, M., Budi, A., Ataman, E., Yoshimura, C., & Johnson, M. S. (2018, March). Ozone-assisted regeneration of magnetic carbon nanotubes for removing organic water pollutants. *Chemical Engineering Journal, 335*, 384–391. https://doi.org/10.1016/j.cej.2017.10.166

Ayilara, M. S., & Babalola, O. O. (2023, May). Bioremediation of environmental wastes: The role of microorganisms. *Frontiers in Agronomy, 5*. https://doi.org/10.3389/fagro.2023.1183691

Berber, M. R. (2020). Surface-functionalization of activated carbon with polyglucosamine polymer for efficient removal of cadmium ions. *Polymer Composites, 41*(8), 3074–3086. https://doi.org/10.1002/pc.25599

Bian, Z., Wallum, A., Mehmood, A., Gomez, E., Wang, Z., Pandit, S., Nie, S., Link, S., Levine, B. G., & Gruebele, M. (2023). Properties of carbon dots versus small molecules from 'bottom-up' synthesis. *ACS Nano, 17*(22), 22788–22799. https://doi.org/10.1021/acsnano.3c07486

Bläker, C., Muthmann, J., Pasel, C., & Bathen, D. (2019). Characterization of activated carbon adsorbents—State of the art and novel approaches. *ChemBioEng Reviews, 6*(4), 119–138. https://doi.org/10.1002/cben.201900008

Bushra, R., Ahmed, A., & Shahadat, M. (2016). Mechanism of adsorption on nanomaterials. In *Advanced environmental analysis* (pp. 90–111). The Royal Society of Chemistry. https://doi.org/10.1039/9781782623625-00090

Cai, Y., Chen, Z., Wang, S., Chen, J., Hu, B., Shen, C., & Wang, X. (2023, March). Carbon-based nanocomposites for the elimination of inorganic and organic pollutants through sorption and catalysis strategies. *Separation and Purification Technology, 308*, 122862. https://doi.org/10.1016/j.seppur.2022.122862

Caturla, F., Molina-Sabio, M., & Rodríguez-Reinoso, F. (1991). Preparation of activated carbon by chemical activation with $ZnCl_2$. *Carbon, 29*(7), 999–1007. https://doi.org/10.1016/0008-6223(91)90179-M

Chang, C.-K., Tun, H., & Chen, C.-C. (2020). An activity-based formulation for Langmuir adsorption isotherm. *Adsorption, 26*(3), 375–386. https://doi.org/10.1007/s10450-019-00185-4

Chen, H., Zhiying L., Yangming C., Enrico D., Zhaohui W., Feng Z., Zhaoliang C. 2023. "Development and Emerging Application of Membrane Degassing Technology." Advanced Membranes 3:100076. https://doi.org/10.1016/j.advmem.2023.100076.

Chen, W., Duan, L., & Zhu, D. (2007). Adsorption of polar and nonpolar organic chemicals to carbon nanotubes. *Environmental Science & Technology, 41*(24), 8295–8300. https://doi.org/10.1021/es071230h

Chen, Y.-G., He, X.-L.-S., Huang, J.-H., Luo, R., Ge, H.-Z., Wołowicz, A., Wawrzkiewicz, M., et al. (2021, August). Impacts of heavy metals and medicinal crops on ecological systems, environmental pollution, cultivation, and production processes in China. *Ecotoxicology and Environmental Safety, 219*, 112336. https://doi.org/10.1016/j.ecoenv.2021.112336

Chenab, K. K., Sohrabi, B., Jafari, A., & Ramakrishna, S. (2020, June). Water treatment: Functional nanomaterials and applications from adsorption to photodegradation. *Materials Today Chemistry, 16*, 100262. https://doi.org/10.1016/j.mtchem.2020.100262

Choi, Y., Choi, Y., Kwon, O.-H., & Kim, B.-S. (2018). Carbon dots: Bottom-up syntheses, properties, and light-harvesting applications. *Chemistry—An Asian Journal, 13*(6), 586–598. https://doi.org/10.1002/asia.201701736

Chowdhury, S., & Balasubramanian, R. (2014, February). Recent advances in the use of graphene-family nanoadsorbents for removal of toxic pollutants from wastewater. *Advances in Colloid and Interface Science, 204*, 35–56. https://doi.org/10.1016/j.cis.2013.12.005

Dąbrowski, A. (2001). Adsorption—From theory to practice. *Advances in Colloid and Interface Science, 93*(1–3), 135–224. https://doi.org/10.1016/S0001-8686(00)00082-8

Dai, Y., Zhang, N., Xing, C., Cui, Q., & Sun, Q. (2019, May). The adsorption, regeneration and engineering applications of biochar for removal organic pollutants: A review. *Chemosphere, 223*, 12–27. https://doi.org/10.1016/j.chemosphere.2019.01.161

Dayana Priyadharshini, S., Manikandan, S., Kiruthiga, R., Rednam, U., Babu, P. S., Subbaiya, R., Karmegam, N., Kim, W., & Govarthanan, M. (2022, August). Graphene oxide-based nanomaterials for the treatment of pollutants in the aquatic environment: Recent trends and perspectives—A review. *Environmental Pollution, 306*, 119377. https://doi.org/10.1016/j.envpol.2022.119377

Debebe, Y., Worku, Z., & Alemayehu, E. (2024). Fixed bed column adsorption system for the removal of 2,4-dichlorophenoxyacetic acid herbicide from an aqueous solution. *Journal of Engineering, 2024*(1), 1629004. https://doi.org/10.1155/2024/1629004

Devi, R., Vinay K., Sunil K., Mamta B., Anushree J., Renu R., Ajay K.M., Paul S. 2023. "Recent Advancement in Biomass-Derived Activated Carbon for Waste Water Treatment, Energy Storage, and Gas Purification: A Review." Journal of Materials Science 58 (30): 12119–42. https://doi.org/10.1007/s10853-023-08773-0.

Dehmani, Y., Mohammed, B. B., Oukhrib, R., Dehbi, A., Lamhasni, T., Brahmi, Y., El-Kordy, A., et al. (2024). Adsorption of various inorganic and organic pollutants by natural and synthetic zeolites: A critical review. *Arabian Journal of Chemistry, 17*(1), 105474. https://doi.org/10.1016/j.arabjc.2023.105474

Díaz, E., García, L., & Ordóñez, S. (2024). Adsorption of stable and labile emerging pollutants on activated carbon: Degradation and mass transfer kinetic study. *Applied Water Science, 14*(4), 79. https://doi.org/10.1007/s13201-023-02087-x

Elbasiouny, H., Darwesh, M., Elbeltagy, H., Abo-alhamd, F. G., Amer, A. A., Elsegaiy, M. A., Khattab, I. A., et al. (2021). Ecofriendly remediation technologies for wastewater contaminated with heavy metals with special focus on using water hyacinth and black tea wastes: A review. *Environmental Monitoring and Assessment, 193*(7), 449. https://doi.org/10.1007/s10661-021-09236-2

El-Wakil, A. M., Waly, S. M., Abou El-Maaty, W. M., Waly, M. M., Yılmaz, M., & Awad, F. S. (2022). Triazine-based functionalized activated carbon prepared from water hyacinth for the removal of Hg^{2+}, Pb^{2+}, and Cd^{2+} ions from water. *ACS Omega, 7*(7), 6058–6069. https://doi.org/10.1021/acsomega.1c06441

Farjadian, F., Abbaspour, S., Sadatlu, M. A. A., Mirkiani, S., Ghasemi, A., Hoseini-Ghahfarokhi, M., Mozaffari, N., Karimi, M., & Hamblin, M. R. (2020). Recent developments in graphene and graphene oxide: Properties, synthesis, and modifications: A review. *ChemistrySelect, 5*(33), 10200–10219. https://doi.org/10.1002/slct.202002501

Foo, K. Y., & Hameed, B. H. (2010). Detoxification of pesticide waste via activated carbon adsorption process. *Journal of Hazardous Materials, 175*(1–3), 1–11. https://doi.org/10.1016/j.jhazmat.2009.10.014

Fouda-Mbanga, B. G., Onotu, O. P., & Tywabi-Ngeva, Z. (2024, December). Advantages of the reuse of spent adsorbents and potential applications in environmental remediation: A review. *Green Analytical Chemistry, 11*, 100156. https://doi.org/10.1016/j.greeac.2024.100156

Gao, C., Guo, M., Liu, Y., Zhang, D., Gao, F., Sun, L., Li, J., Chen, X., Terrones, M., & Wang, Y. (2023, August). Surface modification methods and mechanisms in carbon nanotubes dispersion. *Carbon, 212*, 118133. https://doi.org/10.1016/j.carbon.2023.118133

Guo, X., & Wang, J. (2019, December). Comparison of linearization methods for modeling the Langmuir adsorption isotherm. *Journal of Molecular Liquids, 296*, 111850. https://doi.org/10.1016/j.molliq.2019.111850

Guo, X., Menghong L., Aijv L., Man J., Xiaoyin N., Xinpeng L. 2020. "Adsorption Mechanisms and Characteristics of Hg2+ Removal by Different Fractions of Biochar." Water 12 (8): 2105. https://doi.org/10.3390/w12082105.

Gupta, V. K., Agarwal, S., & Saleh, T. A. (2011). Synthesis and characterization of alumina-coated carbon nanotubes and their application for lead removal. *Journal of Hazardous Materials, 185*(1), 17–23. https://doi.org/10.1016/j.jhazmat.2010.08.053

Gusain, R., Kumar, N., & Ray, S. S. (2020, February). Recent advances in carbon nanomaterial-based adsorbents for water purification. *Coordination Chemistry Reviews, 405*, 213111. https://doi.org/10.1016/j.ccr.2019.213111

Gutiérrez-Cruz, A., Ruiz-Hernández, A. R., Vega-Clemente, J. F., Luna-Gazcón, D. G., & Campos-Delgado, J. (2022). A review of top-down and bottom-up synthesis methods for the production of graphene, graphene oxide and reduced graphene oxide. *Journal of Materials Science, 57*(31), 14543–14578. https://doi.org/10.1007/s10853-022-07514-z

Gwenzi, W., Chaukura, N., Wenga, T., & Mtisi, M. (2021, January). Biochars as media for air pollution control systems: Contaminant removal, applications and future research directions. *Science of the Total Environment, 753*, 142249. https://doi.org/10.1016/j.scitotenv.2020.142249

Han, M., Zhu, S., Lu, S., Song, Y., Feng, T., Tao, S., Liu, J., & Yang, B. (2018, April). Recent progress on the photocatalysis of carbon dots: Classification, mechanism and applications. *Nano Today, 19*, 201–218. https://doi.org/10.1016/j.nantod.2018.02.008

Hayashi, J.'i., Kazehaya, A., Muroyama, K., & Watkinson, A. P. (2000). Preparation of activated carbon from lignin by chemical activation. *Carbon, 38*(13), 1873–1878. https://doi.org/10.1016/S0008-6223(00)00027-0

Hethnawi, A., Manasrah, A. D., Vitale, G., & Nassar, N. N. (2018, March). Fixed-bed column studies of total organic carbon removal from industrial wastewater by use of diatomite decorated with polyethylenimine-functionalized pyroxene nanoparticles. *Journal of Colloid and Interface Science, 513*, 28–42. https://doi.org/10.1016/j.jcis.2017.10.078

Jagadeesh, N., & Sundaram, B. (2023, February). Adsorption of pollutants from wastewater by biochar: A review. *Journal of Hazardous Materials Advances, 9*, 100226. https://doi.org/10.1016/j.hazadv.2022.100226

Jeirani, Z., Niu, C. H., & Soltan, J. (2017). Adsorption of emerging pollutants on activated carbon. *Reviews in Chemical Engineering, 33*(5), 491–522. https://doi.org/10.1515/revce-2016-0027

Jiříčková, A., Jankovský, O., Sofer, Z., & Sedmidubský, D. (2022). Synthesis and applications of graphene oxide. *Materials, 15*(3), 920. https://doi.org/10.3390/ma15030920

Kafle, A., Timilsina, A., Gautam, A., Adhikari, K., Bhattarai, A., & Aryal, N. (2022, July). Phytoremediation: Mechanisms, plant selection and enhancement by natural and synthetic agents. *Environmental Advances, 8*, 100203. https://doi.org/10.1016/j.envadv.2022.100203

Keshta, B. E., Gemeay, A. H., Sinha, D. K., Elsharkawy, S., Hassan, F., Rai, N., & Arora, C. (2024, January). State of the art on the magnetic iron oxide nanoparticles: Synthesis, functionalization, and applications in wastewater treatment. *Results in Chemistry, 7*, 101388. https://doi.org/10.1016/j.rechem.2024.101388

Kong, Q., Zhang, H., Lan, Y., Shi, X., Fang, Z., Chang, Q., Liu, J., & Wei, C. (2023). Functional graphene oxide for organic pollutants removal from wastewater: A mini review. *Environmental Technology, 44*(21), 3183–3195. https://doi.org/10.1080/09593330.2022.2053754

Kumar, P. S., Korving, L., van Loosdrecht, M. C. M., & Witkamp, G.-J. (2019, August). Adsorption as a technology to achieve ultra-low concentrations of phosphate: Research gaps and economic analysis. *Water Research X, 4*, 100029. https://doi.org/10.1016/j.wroa.2019.100029

Lee, H., Sam, K., Coulon, F., De Gisi, S., Notarnicola, M., & Labianca, C. (2024, February). Recent developments and prospects of sustainable remediation treatments for major contaminants in soil: A review. *Science of the Total Environment, 912*, 168769. https://doi.org/10.1016/j.scitotenv.2023.168769

Lehman, J. H., Terrones, M., Mansfield, E., Hurst, K. E., & Meunier, V. (2011). Evaluating the characteristics of multiwall carbon nanotubes. *Carbon, 49*(8), 2581–2602. https://doi.org/10.1016/j.carbon.2011.03.028

Liang, L., Xi, F., Tan, W., Meng, X., Hu, B., & Wang, X. (2021). Review of organic and inorganic pollutants removal by biochar and biochar-based composites. *Biochar, 3*(3), 255–281. https://doi.org/10.1007/s42773-021-00101-6

Liu, X., O'Carroll, D. M., Petersen, E. J., Huang, Q., & Lindsay Anderson, C. (2009). Mobility of multi-walled carbon nanotubes in porous media. *Environmental Science & Technology, 43*(21), 8153–8158. https://doi.org/10.1021/es901340d

Liu, M. L., Chen, B. B., Li, C. M., & Huang, C. Z. (2019). Carbon dots: Synthesis, formation mechanism, fluorescence origin and sensing applications. *Green Chemistry, 21*(3), 449–471. https://doi.org/10.1039/C8GC02736F

Liu, Z., Ling, Q., Cai, Y., Xu, L., Su, J., Yu, K., Wu, X., Xu, J., Hu, B., & Wang, X. (2022). Synthesis of carbon-based nanomaterials and their application in pollution management. *Nanoscale Advances, 4*(5), 1246–1262. https://doi.org/10.1039/D1NA00843A

Maciá-Agulló, J. A., Moore, B. C., Cazorla-Amorós, D., & Linares-Solano, A. (2004). Activation of coal tar pitch carbon fibres: Physical activation vs. chemical activation. *Carbon, 42*(7), 1367–1370. https://doi.org/10.1016/j.carbon.2004.01.013

Manocha, S. M. (2003). Porous carbons. *Sadhana, 28*(1–2), 335–348. https://doi.org/10.1007/BF02717142

Marcano, D. C., Kosynkin, D. V., Berlin, J. M., Sinitskii, A., Sun, Z., Slesarev, A., Alemany, L. B., Lu, W., & Tour, J. M. (2010). Improved synthesis of graphene oxide. *ACS Nano, 4*(8), 4806–4814. https://doi.org/10.1021/nn1006368

McEuen, P. L. (2000). Single-wall carbon nanotubes. *Physics World, 13*(6), 31–36. https://doi.org/10.1088/2058-7058/13/6/26

Melchor-Durán, A. P., Moreno-Virgen, M. R., Bonilla-Petriciolet, A., Reynel-Ávila, H. E., Lucio Ortiz, E., & González-Vázquez, O. F. (2024). Heavy metal removal from water using graphene oxide in magnetic-assisted adsorption systems: Characterization, adsorption properties, and modelling. *Separations, 11*(10), 294. https://doi.org/10.3390/separations11100294

Menéndez-Díaz, J. A., & Martín-Gullón, I. (2006). Chapter 1—Types of carbon adsorbents and their production. In *Interface science and technology* (pp. 1–47). https://doi.org/10.1016/S1573-4285(06)80010-4

Mian, Md M., Guijian L., Balal Y., Biao F., Habib U., Muhammad U.A., Qumber A., Mehr A.M.M., Liu R. 2018. "Simultaneous Functionalization and Magnetization of Biochar via NH3 Ambiance Pyrolysis for Efficient Removal of Cr (VI)." Chemosphere 208 (October):712–21. https://doi.org/10.1016/j.chemosphere.2018.06.021.

Mintmire, J. W., & White, C. T. (1995). Electronic and structural properties of carbon nanotubes. *Carbon, 33*(7), 893–902. https://doi.org/10.1016/0008-6223(95)00018-9

Murphy, O. P., Vashishtha, M., Palanisamy, P., & Vasanth Kumar, K. (2023). A review on the adsorption isotherms and design calculations for the optimization of adsorbent mass and contact time. *ACS Omega, 8*(20), 17407–17430. https://doi.org/10.1021/acsomega.2c08155

Naser, J. A., Ahmed, Z. W., & Ali, E. H. (2021). Nanomaterials usage as adsorbents for the pollutants removal from wastewater; a review. *Materials Today: Proceedings, 42*, 2590–2595. https://doi.org/10.1016/j.matpr.2020.12.584

Nithya, K., Sathish, A., & Senthil Kumar, P. (2020, February). Packed bed column optimization and modeling studies for removal of chromium ions using chemically modified Lantana camara adsorbent. *Journal of Water Process Engineering, 33*, 101069. https://doi.org/10.1016/j.jwpe.2019.101069

Niyogi, S., Hamon, M. A., Hu, H., Zhao, B., Bhowmik, P., Sen, R., Itkis, M. E., & Haddon, R. C. (2002). Chemistry of single-walled carbon nanotubes. *Accounts of Chemical Research, 35*(12), 1105–1113. https://doi.org/10.1021/ar010155r

Obaideen, K., Shehata, N., Sayed, E. T., Abdelkareem, M. A., Mahmoud, M. S., & Olabi, A. G. (2022, September). The role of wastewater treatment in achieving sustainable development goals (SDGs) and sustainability guideline. *Energy Nexus, 7*, 100112. https://doi.org/10.1016/j.nexus.2022.100112

Obotey E.E., Sudesh R. 2020. "Membrane Technologies in Wastewater Treatment: A Review." Membranes 10 (5): 89. https://doi.org/10.3390/membranes10050089.

Odom, T. W., Huang, J.-L., Kim, P., & Lieber, C. M. (2000). Structure and electronic properties of carbon nanotubes. *The Journal of Physical Chemistry B, 104*(13), 2794–2809. https://doi.org/10.1021/jp993592k

Pan, B., & Xing, B. (2008). Adsorption mechanisms of organic chemicals on carbon nanotubes. *Environmental Science & Technology, 42*(24), 9005–9013. https://doi.org/10.1021/es801777n

Parambil, A. M., & Rajamani, P. (2024). Carbon dots: A promising path towards environmental sustainability. *Environmental Science: Advances, 3*(11), 1513–1523. https://doi.org/10.1039/D4VA00273C

Parambil, A. M., Priyadarshini, E., Goutam, R., Tsai, P.-C., Huang, P.-C., Rajamani, P., Lin, Y.-C., & Ponnusamy, V. K. (2024, November). Self-assembled mesoporous silica decorated with biogenic carbon dot nanospheres hybrid nanomaterial for efficient removal of aqueous methoxy-DDT via a short-bed adsorption column technique. *Environmental Research, 260*, 119653. https://doi.org/10.1016/j.envres.2024.119653

Pillar-Little, T. J., Wanninayake, N., Nease, L., Heidary, D. K., Glazer, E. C., & Kim, D. Y. (2018, December). Superior photodynamic effect of carbon quantum dots through both type I and type II pathways: Detailed comparison study of top-down-synthesized and bottom-up-synthesized carbon quantum dots. *Carbon, 140*, 616–623. https://doi.org/10.1016/j.carbon.2018.09.004

Pinto, M., Ramalho, P. S. F., Moreira, N. F. F., Gonçalves, A. G., Nunes, O. C., Pereira, M. F. R., & Soares, O. S. G. P. (2020, December). Application of magnetic nanoparticles for water purification. *Environmental Advances, 2*, 100010. https://doi.org/10.1016/j.envadv.2020.100010

Prasek, J., Drbohlavova, J., Chomoucka, J., Hubalek, J., Jasek, O., Adam, V., & Kizek, R. (2011). Methods for carbon nanotubes synthesis—Review. *Journal of Materials Chemistry, 21*(40), 15872. https://doi.org/10.1039/c1jm12254a

Qasem, N.A.A., Ramy H.M., Dahiru U.L. 2021. "Removal of Heavy Metal Ions from Wastewater: A Comprehensive and Critical Review." Npj Clean Water 4 (1): 36. https://doi.org/10.1038/s41545-021-00127-0.

Rajak, V. K., Kumar, S., Thombre, N. V., & Mandal, A. (2018). Synthesis of activated charcoal from saw-dust and characterization for adsorptive separation of oil from oil-in-water emulsion. *Chemical Engineering Communications, 205*(7), 897–913. https://doi.org/10.1080/00986445.2017.1423288

Revellame, E.D., Dhan L.F., Wayne S., Rafael H., Mark E.Z. 2020. "Adsorption Kinetic Modeling Using Pseudo-First Order and Pseudo-Second Order Rate Laws: A Review." Cleaner Engineering and Technology 1 (December):100032. https://doi.org/10.1016/j.clet.2020.100032.

Rodríguez-Reinoso, F., & Molina-Sabio, M. (1992). Activated carbons from lignocellulosic materials by chemical and/or physical activation: An overview. *Carbon, 30*(7), 1111–1118. https://doi.org/10.1016/0008-6223(92)90143-K

Sajid, M., Asif, M., Baig, N., Kabeer, M., Ihsanullah, I., & Mohammad, A. W. (2022, June). Carbon nanotubes-based adsorbents: Properties, functionalization, interaction mechanisms, and applications in water purification. *Journal of Water Process Engineering, 47*, 102815. https://doi.org/10.1016/j.jwpe.2022.102815

Sarathchandran, C., Devika, M. R., Prakash, S., Sujatha, S., & Ilangovan, S. A. (2021). Activated carbon: Synthesis, properties, and applications. In *Handbook of carbon-based nanomaterials* (pp. 783–827). Elsevier. https://doi.org/10.1016/B978-0-12-821996-6.00008-7

Shi, W., Han, Q., Wu, J., Ji, C., Zhou, Y., Li, S., Gao, L., Leblanc, R. M., & Peng, Z. (2022). Synthesis mechanisms, structural models, and photothermal therapy applications of top-down carbon dots from carbon powder, graphite, graphene, and carbon nanotubes. *International Journal of Molecular Sciences, 23*(3), 1456. https://doi.org/10.3390/ijms23031456

Shahrokhi-Shahraki, R., Chelsea B., Mohamed G. El-D., Junboum P. 2021. "High Efficiency Removal of Heavy Metals Using Tire-Derived Activated Carbon vs Commercial Activated Carbon: Insights into the Adsorption Mechanisms." Chemosphere 264 (February):128455. https://doi.org/10.1016/j.chemosphere.2020.128455.

Smith, S.C., Debora F.R. 2015. "Carbon-Based Nanomaterials for Removal of Chemical and Biological Contaminants from Water: A Review of Mechanisms and Applications." Carbon 91 (September):122–43. https://doi.org/10.1016/j.carbon.2015.04.043.

Spencer, W., Senanayake, G., Altarawneh, M., Ibana, D., & Nikoloski, A. N. (2024, July). Review of the effects of coal properties and activation parameters on activated carbon production and quality. *Minerals Engineering, 212*, 108712. https://doi.org/10.1016/j.mineng.2024.108712

Srivastava, V., Sarkar, A., Singh, S., Singh, P., de Araujo, A. S. F., & Singh, R. P. (2017, October). Agroecological responses of heavy metal pollution with special emphasis on soil health and plant performances. *Frontiers in Environmental Science, 5*. https://doi.org/10.3389/fenvs.2017.00064

Su, X., Wang, X., Ge, Z., Bao, Z., Lin, L., Chen, Y., Dai, W., et al. (2024, April). KOH-activated biochar and chitosan composites for efficient adsorption of industrial dye pollutants. *Chemical Engineering Journal, 486*, 150387. https://doi.org/10.1016/j.cej.2024.150387

Tan, X., Liu, Y., Zeng, G., Wang, X., Hu, X., Gu, Y., & Yang, Z. (2015, April). Application of biochar for the removal of pollutants from aqueous solutions. *Chemosphere, 125*, 70–85. https://doi.org/10.1016/j.chemosphere.2014.12.058

Thakur, K., & Kandasubramanian, B. (2019). Graphene and graphene oxide-based composites for removal of organic pollutants: A review. *Journal of Chemical & Engineering Data, 64*(3), 833–867. https://doi.org/10.1021/acs.jced.8b01057

Thakur, S., Badoni, A., Samriti, Sharma, P., Ojha, A., Swart, H. C., Kuznetsov, A. Y., & Prakash, J. (2024). Standalone highly efficient graphene oxide as an emerging visible light-driven photocatalyst and recyclable adsorbent for the sustainable removal of organic pollutants. *Langmuir, 40*(35), 18486–18502. https://doi.org/10.1021/acs.langmuir.4c01727

Wang, X., Cheng, H., Ye, G., Fan, J., Yao, F., Wang, Y., Jiao, Y., Zhu, W., Huang, H., & Ye, D. (2022, January). Key factors and primary modification methods of activated carbon and their application in adsorption of carbon-based gases: A review. *Chemosphere, 287*, 131995. https://doi.org/10.1016/j.chemosphere.2021.131995

Xia, C., Zhu, S., Feng, T., Yang, M., & Yang, B. (2019). Evolution and synthesis of carbon dots: From carbon dots to carbonized polymer dots. *Advanced Science, 6*(23). https://doi.org/10.1002/advs.201901316

Yadav, S., Choudhary, N., Bhai, S., Bhojani, G., Chatterjee, S., Ganguly, B., & Paital, A. R. (2021). Recyclable functionalized material for sensitive detection and exceptional sorption of hexava lent chromium and permanganate ions with biosensing applications. *ACS Applied Bio Materials, 4*(8), 6430–6440. https://doi.org/10.1021/acsabm.1c00609

Yadav, S., Choudhary, N., Dash, M. R., & Paital, A. R. (2022a, December). High surface area dendritic silica pairing with anthraquinone derivative: A promising single platform for dual applications of detection and remediation of nitroaromatics and copper ion. *Chemical Engineering Journal, 450*, 138042. https://doi.org/10.1016/j.cej.2022.138042

Yadav, S., Rajpurohit, D., Dash, S. R., Bhojani, G., Chatterjee, S., & Paital, A. R. (2022b, June). Hybrid material for ferric ion detection & remediation: Exceptional selectivity & adsorption capacity with biological applications. *Microporous and Mesoporous Materials, 338*, 111945. https://doi.org/10.1016/j.micromeso.2022.111945

Yadav, S., Choudhary, N., Sonpal, V., & Paital, A. R. (2023a, September). Engineering excitation-independent turn-on fluorescent probe for mercury: Functionalized dendritic silica doped with red-emissive carbon dots towards simultaneous detection and remediation with biosensing application. *Chemical Engineering Journal, 471*, 144715. https://doi.org/10.1016/j.cej.2023.144715

Yadav, S., Choudhary, N., & Paital, A. R. (2023b, March). Tailored graphene/silica functional composite as signaling and adsorbent material: A sequential excimer probe with a fluoro-switching response with zinc and adenosine triphosphate. *Carbon, 205*, 527–539. https://doi.org/10.1016/j.carbon.2023.01.063

Yadav, S., Choudhary, N., Sonpal, V., & Paital, A. R. (2024a). Carbon dots-embedded silica tubes: An excitation-independent yellow-emitting turn-on probe for simultaneous detection and removal of inorganic arsenic with in vivo tracking in living organisms. *Small, 20*(11). https://doi.org/10.1002/smll.202307491

Yadav, S., Choudhary, N., Vasave, A. T., Sonpal, V., & Paital, A. R. (2024b). Perylene diimide functionalized nano-silica: Green emissive material for selective probing and remediation of 4-nitrocatechol, Ru^{3+}, and Cu^{2+} with biosensing applications. *Materials Advances, 5*(22), 8937–8952. https://doi.org/10.1039/D4MA00862F

Yadav, S., Choudhary, N., & Paital, A. R. (2024c). Carbon within carbon: Growth of excitation-independent CDs within functional mesoporous carbon towards detection and adsorption of a specific nitrofuran class of antibiotics. *Journal of Materials Chemistry A, 12*(1), 202–213. https://doi.org/10.1039/D3TA05273G

Yadav, S., Choudhary, N., Sonpal, V., Vyas, B. G., & Paital, A. R. (2025). Sunlight-mediated photocatalytic uranium extraction from seawater: A bioresistant heterojunction for highly selective uranium sensing and extraction with in vivo application. *Journal of Materials Chemistry A, 13*, 3858. https://doi.org/10.1039/D4TA08292C

Yan, H., Yang, H., Li, A., & Cheng, R. (2016, January). pH-tunable surface charge of chitosan/graphene oxide composite adsorbent for efficient removal of multiple pollutants from water. *Chemical Engineering Journal, 284*, 1397–1405. https://doi.org/10.1016/j.cej.2015.06.030

Yang, X., Wan, Y., Zheng, Y., He, F., Yu, Z., Huang, J., Wang, H., Ok, Y. S., Jiang, Y., & Gao, B. (2019, June). Surface functional groups of carbon-based adsorbents and their roles in the removal of heavy metals from aqueous solutions: A critical review. *Chemical Engineering Journal, 366*, 608–621. https://doi.org/10.1016/j.cej.2019.02.119

Yin, C., Aroua, M., & Daud, W. (2007). Review of modifications of activated carbon for enhancing contaminant uptakes from aqueous solutions. *Separation and Purification Technology, 52*(3), 403–415. https://doi.org/10.1016/j.seppur.2006.06.009

Zhang, X., Zhang, K., Shi, Y., Xiang, H., Yang, W., & Zhao, F. (2024, April). Surface engineering of multifunctional nanostructured adsorbents for enhanced wastewater treatment: A review. *Science of the Total Environment, 920*, 170951. https://doi.org/10.1016/j.scitotenv.2024.170951

Zhao, Z., Wang, B., Theng, B. K. G., Lee, X., Zhang, X., Chen, M., & Xu, P. (2022). Removal performance, mechanisms, and influencing factors of biochar for air pollutants: A critical review. *Biochar, 4*(1), 30. https://doi.org/10.1007/s42773-022-00156-z

Zhu, Y., Murali, S., Cai, W., Li, X., Suk, J. W., Potts, J. R., & Ruoff, R. S. (2010). Graphene and graphene oxide: Synthesis, properties, and applications. *Advanced Materials, 22*(35), 3906–3924. https://doi.org/10.1002/adma.201001068

Chapter 9
Filtering Solutions for Environmental Pollution: Membrane Filtration with Carbon Nanomaterials

Po-Jung Huang, Ju-Chun Kao, and Tina Jen

Abstract Pure water and clean air are the most fundamental necessities of life. The demand for safe drinking water and fresh air is high in many regions worldwide. The scarcity of these resources is mainly driven by population growth, which leads to pollution and environmental issues like droughts and poor air quality. Carbon nanomaterials are being utilized in a wide range of applications, and their unique properties make them ideal for ecological solutions, such as water purification, heavy metal removal, and gas separation. Advancements in membrane synthesis, a cutting-edge area of nanotechnology research, have resulted in both natural and synthetic membranes. This discussion highlights the exceptional potential of carbon-based nanomaterials, such as graphene, carbon nanotubes (CNTs), and their derivatives, which have been highly favored since their discovery. Graphene and CNTs are being extensively studied for environmental applications like filtration and adsorption, demonstrating remarkable efficiency in removing contaminants, such as organic matter, heavy metal ions, bacteria, and particulate matter (PMs). The separation mechanisms employed by carbon-based nanomaterials largely depend on Van der Waals forces and π–π interactions. Moreover, the functionalization of carbon nanomaterials can enhance molecular interactions, such as hydrogen bonding and chelation, further improving separation and filtration efficiency.

P.-J. Huang (✉)
Department of Chemical and Materials Engineering, National Central University, Taoyuan, Taiwan

Graduate College of Sustainability and Green Energy, National Central University, Taoyuan, Taiwan
e-mail: pjhuang@cc.ncu.edu.tw

J.-C. Kao
Department of Chemical and Materials Engineering, National Central University, Taoyuan, Taiwan

T. Jen
Graduate College of Sustainability and Green Energy, National Central University, Taoyuan, Taiwan

A. M. Parambil et al. (eds.), *Carbon*,
https://doi.org/10.1007/978-3-031-90613-8_9

Functionalized carbon nanomaterials also offer selective separation capabilities, making them even more effective in environmental applications.

Keywords Graphene · Carbon nanotubes · Carbon dots · Adsorption and purification · Membrane separation mechanism · Functionalization

Abbreviations

2-NO	2-Naphthol
AAO	Acid aluminum oxide
AAP	Acetaminophen
AB	Acid black 1
ACT	Acetaminophen
AF	Acid fuchsin
AG	Alizarin yellow GG
AMX	Amoxicillin
APT	Attapulgite
ATT HCl	Amitriptyline HCl
BPA	Bisphenol-A
BZF	Bezafibrate
CAP	Chloramphenicol
CBD	Carbendazim
CBZ	Carbamazepine
CFN	Caffeine
ChNCs	Chitin nanocrystals
COF-1	Covalent organic framework-01
CR	Congo red
CV	Crystal violet
DB	Disperse black 9
DBP	Dibutyl phthalate
DCF	Diclofenac
DIC	Diclofenac sodium
DMAEMA	Dimethylaminoethyl methacrylate
DR	Direct Red
DR23	Direct Red 23
E2	17β-estradiol
EB	Evan Blue
EDA	Ethylenediamine
ERY	Erythromycin
ET	Eriochrome Black T
GE	Ginger plant extractive
IBF	Ibuprofen
IOP	Iopromide

MB	Methylene blue
MWCNT	Multi-walled carbon nanotube
NIPS	Non-solvent phase separation
NP	4-Nonylphenol
ONCT	Oxidized carbon nanotubes
PAN	Polyacrylonitrile
PC	Polycarbonate
PES	Polyethersulfone
PMIA	Poly(m-phenylene isophthalamide)
PNC	Penicillin
p-NP	p-Nitrophenol
PSF	Polysulfone
RB	Rose Bengal
RB5	Reactive Black 5
RhB	Rhodamine B
RR 195	Reactive Red 195
SA	Sodium alginate
SMX	Sulfamethoxazole
ST	Safranin T
SWCNT	Single-walled carbon nanotube
TC	Tetracycline
TCS	Triclosan

9.1 Introduction

9.1.1 *Membrane Filtration for Environmental Remediation*

Membrane filtration is an advanced, efficient method for wastewater and air remediation that is widely used in environmental applications. This technique employs semi-permeable membranes to separate particles (Tanis-Kanbur et al., 2022), microorganisms (Cian et al., 2020), and other impurities from water, relying on selective permeability to achieve high-quality filtration. Membrane filtration is crucial in environmental sectors for addressing pollution, ensuring safe drinking water, and supporting sustainable water management practices (Othman et al., 2022). Membrane filtration processes are categorized according to the particle size they filter, typically ranging from macro- to nanoscale. They include microfiltration (MF), which removes larger particles, such as suspended solids, bacteria, and certain protozoa. MF typically operates at lower pressures and is often a pre-treatment step in more advanced filtration systems; ultrafiltration (UF) filters out smaller particles like viruses, proteins, and colloidal materials. UF membranes exhibit smaller pore sizes than MF and effectively provide high water clarity levels. Nanofiltration (NF) is primarily used for softening water and removing smaller contaminants, including some organic molecules and divalent ions. NF operates at higher pressures and is commonly used in wastewater remediation facilities. Reverse osmosis (RO) utilizes the smallest pore sizes to filter out even soluble salts, heavy metals, and other ions, making it ideal for desalination and advanced water purification applications (Othman et al., 2022). With advances in materials science, membranes are becoming more robust, selective, and efficient, enhancing their role in environmental protection. For example, nanotechnology developments enable the creation of membranes with anti-fouling (Zhao et al., 2018), antimicrobial (Epand et al., 2016), and higher-selectivity properties. These improvements will likely expand membrane filtration's application in resource recovery, environmental monitoring, and circular water management solutions. Carbon material membranes are advanced filtration membranes composed of carbon-based nanomaterials, like graphene oxide (GO), carbon nanotubes (CNTs), and carbon dots (CDs). These materials offer remarkable properties such as high permeability, selectivity, mechanical strength, and chemical stability, making them highly suitable for wastewater remediation, gas separation, desalination, and pollutant removal applications.

9.1.2 Graphene

Graphene is a highly promising 2D carbon material composed of sp^2 carbon atoms arranged in hexagonal rings, forming a structure similar to a honeycomb. Graphene features exceptional charge carrier mobility, a vast specific surface area, and outstanding electrical conductivity, making it frequently used in electrode materials and the semiconductor industry. The structure of graphene extends to graphene oxide (GO), which differs by incorporating both sp^2 and sp^3 carbon atoms due to oxygen-containing functional groups on its edges and surfaces. Reduced graphene oxide (rGO) is a form of graphene oxide that has undergone reduction to eliminate most of these oxygen-containing functional groups, increasing its porosity for pollutant adsorption. Modified graphene oxide can be synthesized into sulfonated graphene oxide (sGO), where the sulfonic acid groups form more stable hydrogen bonds, providing enhanced flux and pollution prevention benefits.

9.1.3 Graphene Oxide (GO)

Graphene oxide (GO) forms through the oxidation of graphite, creating carbon-hydrogen bonds and a layered structure. A common preparation method is the modified Hummers method, which uses $KMnO_4$, $NaNO_3$, and H_2SO_4 to oxidize graphite (Shahriary and Athawale 2014). This process results in carboxyl and carbonyl groups at the edges of GO, along with epoxy and hydroxyl groups on the surface. These oxygen-containing functional groups contribute hydrophilicity and increase dispersion in water or organic solvents. GO is characterized by negatively charged oxygen functional groups on its surface, providing ample sites for reactions and adsorption with pollutants (Li et al., 2013). GO-containing structural benzene rings exhibit π–π stacking interactions with aromatic pollutants (Björk et al., 2010).

9.1.4 Reduced Graphene Oxide (rGO)

Reduced graphene oxide (rGO) has a structure similar to graphene and is produced through a simpler, cost-effective method that involves removing most of the oxygen functional groups from GO. Several common green reducing agents can be used, such as ascorbic acid, green tea, and carrots. After chemical reduction, the removal of oxygen functional groups creates vacancies, resulting in characteristics akin to graphene's excellent conductivity. rGO surfaces carry negative charges, interacting with pollutants primarily through electrostatic and van der Waals forces (Li et al., 2014). The vacancies and large specific surface area of rGO make it suitable as a potential adsorbent for pollutants (Jiang et al., 2024). GO can be reduced through chemical (Stankovich et al., 2007; Pei et al., 2010; De Silva et al., 2018; Vatandost et al., 2020; Uchôa Teixeira et al., 2023), thermal (Zhu et al., 2010; Sengupta et al., 2018), electrochemical (Zhou et al., 2022), and hydrothermal treatment (Huang et al., 2018).

9.1.5 Carbon Nanotubes

Carbon nanotubes (CNTs) are cylindrical nanostructures made of carbon atoms arranged in a hexagonal lattice, similar to a rolled-up graphene sheet. Discovered in the early 1990s, CNTs have garnered significant interest due to their extraordinary physical, chemical, and mechanical properties. They come in two main forms: Single-walled carbon nanotubes (SWCNTs), a single graphene layer rolled into a tube, with diameters typically ranging from 0.4 to 3 nm, and multi-walled carbon nanotubes (MWCNTs), multiple concentric graphene layers rolled into a larger tube, with diameters spanning a few nanometers to several hundred nanometers.

Electric arc discharge is a widely used technique for the synthesis of CNTs. The method involves establishing a high-current electric arc between two graphite electrodes within an inert atmosphere (Ribeiro et al., 2021). The resulting high temperature causes the graphite to sublime, and the carbon vapor's subsequent condensation leads to carbon nanotubes'formation. Laser ablation is a technique applied for the production of CNTs. A high-powered laser beam is focused onto a graphite target in an inert gas environment. The intense laser energy vaporizes the graphite, creating a plume of carbon atoms and ions. These carbon species then condense and rearrange into the nanotubes (Chrzanowska et al., 2015). Solar energy can be harnessed to produce CNTs through a novel process known as solar-assisted chemical vapor deposition (CVD) or solar-driven pyrolysis. These methods utilize concentrated sunlight to achieve the high temperatures necessary for CNT synthesis, offering an environmentally friendly and sustainable approach to CNT production (Arribas et al., 2017). Electrolysis involves passing an electric current through the graphite electrodes that are immersed in a molten salt solution, resulting in the formation of carbon nanotubes and other nanomaterials on the electrode surfaces (Novoselova et al., 2008).

9.1.6 *Carbon Dots*

Carbon dots (CDs) are carbon-based nanoparticles, typically less than 10 nm in diameter, renowned for their distinctive optical, chemical, and biological properties. First identified in the early 2000s as fluorescent byproducts during carbon nanomaterials research, CDs have since been extensively developed for various applications due to their high stability, low toxicity, biocompatibility, and straightforward synthesis. CDs can be synthesized through various top-down and bottom-up methods. Top-down approaches include laser ablation (Kaczmarek et al., 2021), arc discharge (Chao-Mujica et al., 2021), or chemical oxidation of larger carbon structures (like graphite or carbon nanotubes) (González-González et al., 2022). While bottom-up approaches include pyrolysis (Wang et al., 2022), hydrothermal (Nammahachak et al., 2022), or microwave-assisted synthesis of organic precursors (such as citric acid, glucose, or polyethylene glycol) (George et al., 2024) to form carbon nanoparticles directly. These methods allow control over particle size, shape, and functionalization, influencing the carbon dots'resulting optical and chemical properties.

9.2 Membrane Fabrication

9.2.1 *Vacuum Filtration Method*

Figure 9.1a is a schematic diagram of the vacuum filtration process for manufacturing membranes with carbon nanomaterials. Vacuum filtration is a powerful and widely used technique for fabricating carbon nanotube-based (CNT) and graphene oxide-based (GO) membranes, known for creating uniform and highly ordered structures. In this process, CNTs or GO suspensions are filtered under strong vacuum pressure, rapidly removing water and preventing GO flakes' agglomeration. This results in the formation of laminated membranes with tunable properties, such as thickness, pore size, and interlayer spacing (Latorrata et al., 2021).

This method also supports the fabrication of reduced graphene oxide (rGO) membranes with enhanced functionalities, such as improved ion rejection and biofouling resistance. Notably, recent advancements in vacuum filtration have led to the development of rGO composite membranes with exceptional water permeability, achieving rates above 10,000 L m^2 h^{-1} bar^{-1}, making vacuum filtration a key

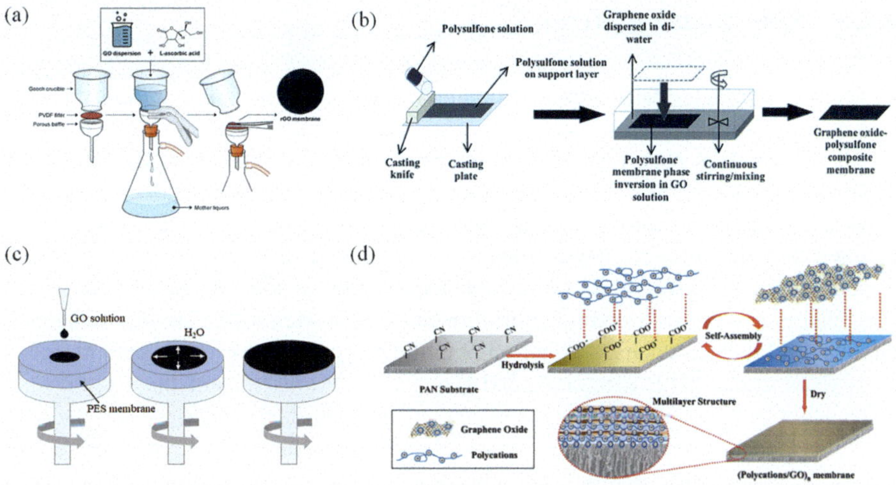

Fig. 9.1 Fabrication of membrane with carbon nanomaterials: (**a**) vacuum filtration. (Reproduced with permission from Latorrata et al. Copyright 2021 MDPI). (**b**) Solvent casting. (Reproduced with permission from Mokkapati et al. Copyright 2017 Royal Society of Chemistry). (**c**) Spin coating. (Reproduced with permission from Kim et al. Copyright 2014 Royal Society of Chemistry), and (**d**) layer-by-layer self-assembly. (Reproduced with permission from Wang et al. Copyright 2016 Elsevier)

method for high-performance membrane fabrication in water treatment and other filtration processes.

9.2.2 Solvent Casting Method

The solvent casting method shown in Fig. 9.1b is a straightforward and widely used technique for fabricating thin films or membranes, particularly in polymer science and membrane technology. A polymer or material of interest is dissolved in a suitable solvent, forming a homogeneous solution. This solution is poured or cast onto a flat glass plate or mold. The membrane formed after solvent exchange at room temperature or in suitable conditions. One of the key advantages of the solvent casting method is its simplicity and versatility. It allows for easy control over the film's thickness by adjusting the solution's concentration or the amount of solution cast. Additionally, the method can be used to incorporate various additives or fillers, such as nanoparticles or functional materials, to enhance the properties of the resulting film (Mokkapati et al., 2017). Solvent casting is commonly used to prepare membranes for filtration, biomedical applications, and coatings. However, one limitation of this method is the potential for solvent residues in the final product, which may require additional processing steps to remove. Despite this, the solvent casting method remains popular for producing films and membranes due to its ease of use and adaptability to different materials.

9.2.3 Spin Coating Method

Figure 9.1c shows the fabrication of membranes with carbon nanomaterials through spin coating methods. In this process, a suspension of GO or rGO is dropped onto a substrate, which is then rapidly rotated. The centrifugal force spreads the liquid uniformly, creating a thin film. This method offers precise control over the membrane's thickness and uniformity by adjusting factors such as the amount of GO solution and the rotation speed. Research has shown that spin coating produces thinner and more uniform GO films than drop-coating or solvent-induced precipitation. Nair et al. demonstrated that ultrathin GO membranes with a laminar structure, having interlayer distances of about 10 Å, could be synthesized with excellent selectivity (Nair et al., 2012; Kim et al., 2014).

9.2.4 Layer-by-Layer (LBL) Self-Assembly Method

The layer-by-layer (LBL) self-assembly method is highly versatile and precise for fabricating multilayered thin films or membranes. Figure 9.1d presents the working mechanism of LBL methods. This method involves depositing alternating layers of oppositely charged materials onto a substrate, such as polyelectrolytes, nanoparticles, or other functional substances. The process is driven by electrostatic interactions, hydrogen bonding, or van der Waals forces, which form highly organized and uniform films. The LBL self-assembly method offers several advantages, including excellent control over film thickness at the nanometer scale by adjusting the number of deposited layers (Wang et al., 2016). It also allows for incorporating various functional materials into each layer, enabling the development of membranes with tailored properties, such as enhanced selectivity, permeability, or mechanical strength. Moreover, the method can be applied to a wide range of substrates, including polymers, ceramics, and metals. This technique is commonly used in applications like nanofiltration, drug delivery, biosensing, and surface coatings, where precise control over film structure and composition is crucial. One limitation of the LBL method is that it can be time-consuming, as each layer must be deposited individually. However, its ability to create highly controlled and multifunctional films makes it an attractive choice for advanced material design and membrane fabrication.

9.3 Filtration Mechanism of Membrane with Carbon Materials

The filtration mechanism of membranes containing carbon materials shown in Fig. 9.2 includes physical filtration, chemical adsorption, and electrical effects (Khanzada et al., 2020).

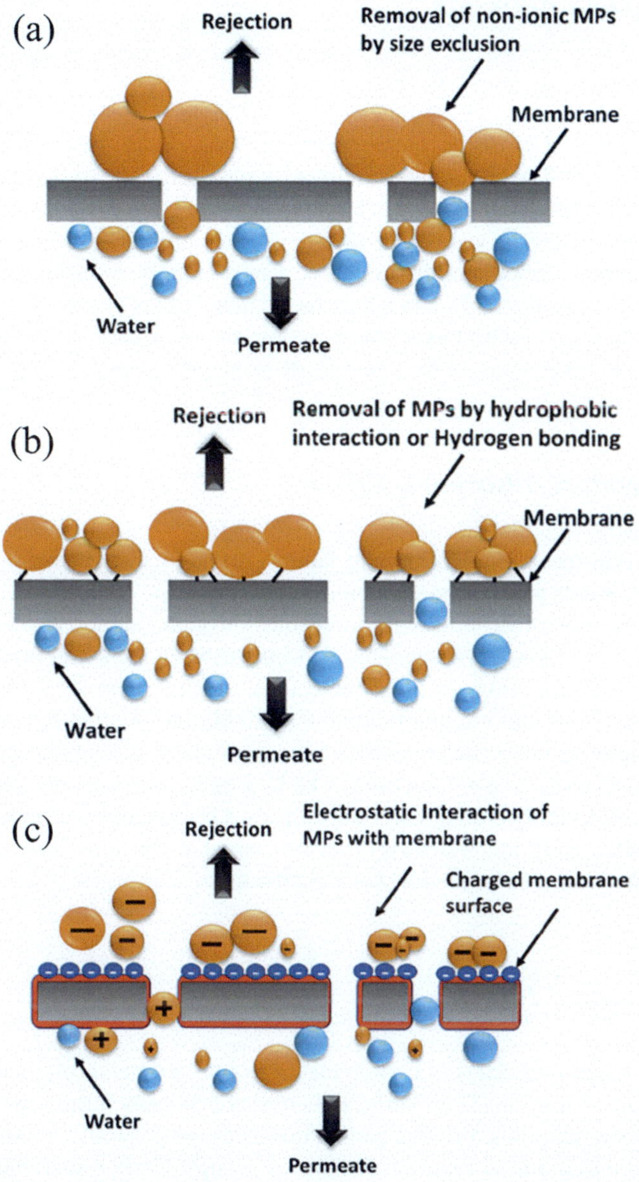

Fig. 9.2 Filtration mechanism of membrane with carbon materials: (**a**) Physical filtration, (**b**) chemical adsorption, and (**c**) electrical effect. (Reproduced with permission from Khanzada et al. Copyright 2020 Elsevier)

9.3.1 Physical Filtration

Carbon materials have nanoscale diameters, typically ranging from a few nanometers to tens of nanometers, which enables them to filter particles based on size exclusion shown in Fig. 9.2a. Molecules or particles larger than the carbon material's pore size are blocked, while smaller molecules pass through. By selecting carbon materials with specific diameters, it is possible to target contaminants in water or air filtration, such as bacteria, viruses, or specific organic molecules. Moreover, carbon materials construct smooth and hydrophobic nanochannels, resulting in high water permeability. This makes carbon material-based membranes highly efficient, allowing for high throughput with lower energy requirements, which is advantageous for desalination and other large-scale water purification processes.

9.3.2 Chemical Adsorption Effect

Figure 9.2b presents the filtration mechanism through the chemical adsorption effect. Van der Waals force is mainly contributed by pristine carbon materials, which allows them to interact strongly with molecules through van der Waals forces and π–π stacking interactions. π–π interactions occur between the π-electrons of carbon materials and aromatic rings in other molecules, making them ideal for adsorbing aromatic compounds and organic pollutants. In addition, pristine carbon materials exhibit hydrophobic properties to enhance the adsorption of nonpolar organic molecules. Stronger molecular interactions, like hydrogen bonding and chelation, can be generated by functionalizing carbon materials. Oxygen and nitrogen-containing groups can effectively and selectively adsorb target metal ions from an aqueous environment, which is valuable for water purification.

9.3.3 Electrical Effect

The surface charge properties of carbon materials play a crucial role in their behavior in membrane filtration. Their surface chemistry primarily influences the surface charge of carbon materials. Pristine carbon materials are generally hydrophobic and have limited charge due to minimal functional groups. Carbon materials are often functionalized to improve dispersibility in solvents or compatibility in composites. Functionalization typically involves the addition of acidic groups, like carboxyl (–COOH), hydroxyl (–OH), and sulfonic acid (–SO_3H) through chemical treatment. These groups impart a negative charge in aqueous environments. The surface charge influences the electrostatic interactions between carbon materials and environmental pollutants. Negatively charged carbon materials can attract and absorb positively charged ions and molecules. In addition, the modified acidic groups on carbon materials exhibit pH-dependent properties due to the ionization of functional groups. Acidic functional groups are protonated at lower pH values, leading to a lower or even positive surface charge. As pH increases, these groups deprotonate, creating a more negative surface charge. Consequently, environmental remediation through carbon materials would also be adjusted to optimize conditions. Figure 9.2c arranges the filtration mechanism via an electrical effect.

9.4 Carbon Material Membrane for Environmental Pollution Remediation

9.4.1 Heavy Metal Removal

The efficiency of an adsorption process heavily relies on the surface area and pore size of the adsorbents. Specifically, the adsorption of metal ions occurs through chemical interactions at designated sites, meaning that an adsorbent's capacity improves with more functional groups. Graphene is particularly advantageous because it contains several active functional groups, such as carboxylic, hydroxyl, and epoxy groups, which serve as effective adsorption sites (Stafiej & Pyrzynska, 2007). For example, EDFA decorated GO contributes not only chelating properties resulting in EDTA-Pb ion complexes formation but also performs ion-exchange properties toward Pb(II) ions (Madadrang et al., 2012); Huang et al. designed polyethyleneimine-graphene oxide (PEI-GO), which demonstrates exceptional uranium adsorption capabilities due to its abundant amino and oxygen-containing functional groups. The GO-PEI structure utilizes these functional groups to effectively coordinate with uranium, achieving an adsorption capacity of 215.6 mg/g at pH 5 (Huang et al., 2018). Other natural materials, like carboxymethyl chitosan (CMS) and sodium alginate (SA) (Wu et al., 2019), and ceramic materials, like silica and MnO_2-Fe_3O_4 (Liu et al., 2015), are also introduced as a modifier to enhance the metal ion removal from wastewater.

Erçarikci et al. prepared rGO/ZnO-NPs EDTA paper for the adsorption of Cd^{2+}, Pb^{2+}, Cu^{2+}, and Hg^{2+} and their adsorption performances are 2963, 8056, 600, and 1753 mg/g, respectively. This high adsorption performance was associated with EDTA, attributing tertiary amine, carbonyl, and hydroxyl groups. Meanwhile, water permeability also achieves 53 L m^{-2} L^{-1} by introducing ZnO-NPs (Erçarikci & Alanyalioglu, 2021). Zhang et al. demonstrated that the ion radius influences the rejection of metal ions. Tunable interlayer spacing of graphene oxide membranes was controlled by different cations, like K^+, Mg^{2+}, Ca^{2+}, and Ba^{2+}, resulting in performing different rejection rates of Na^+, Cu^{2+}, Ni^{2+}, Pb^{2+}, and Mg^{2+}. The rejection rate of these ions was in the order $Mg^{2+} > Cu^{2+} > Ni^{2+} > Pb^{2+} > Na^+$, followed by the order of their hydrated radius (Ghaffar et al., 2019). Additionally, indirect adsorption of heavy metal is another strategy for water remediation. Luan et al. designed a Cu/CNT integrated PVDF membrane for As(III) removal through direct and

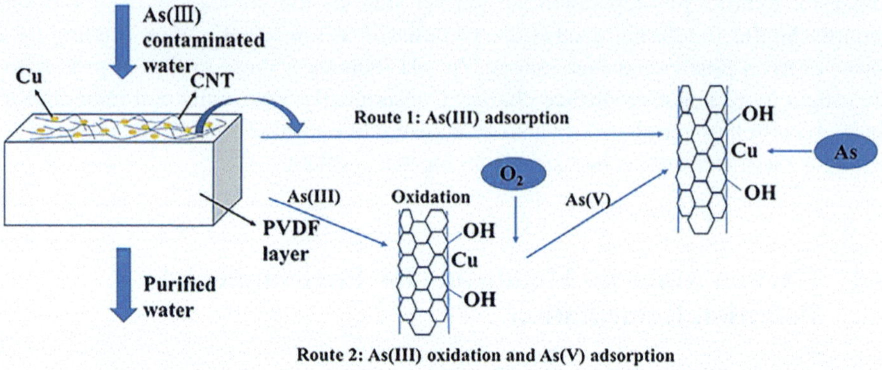

Fig. 9.3 The removal mechanism of As(III) contaminated water using a PVDF membrane embedded with Cu/CNT nanocomposites: Route 1 involves the direct adsorption of As(III). At the same time, Route 2 entails the oxidation of As(III) to As(V), followed by its adsorption onto the Cu/CNT nanocomposites. (Reproduced with permission from Luan et al. Copyright 2019 Elsevier)

indirect adsorption (oxidation), as shown in Fig. 9.3. The removal rate of As(III) is over 95% (Luan et al., 2019). Table 9.1 summarizes the arrangement of graphene oxide-based and CNT-based membranes for heavy metal ion removal.

9.4.2 Dye Removal

Numerous industries have discharged a significant amount of colored wastewater, particularly in fields like printing, textiles, leather, dyes, plastics, and paper. The dyes used in these processes contribute to colored wastewater, which is currently one of the most common organic pollutants and includes aromatic organic dyes in industrial wastewater. Therefore, removing dyes from water is crucial. The visibility of dyes in water is considerable; even a small amount of dye can result in high visibility, blocking sunlight and reducing dissolved oxygen levels in water, which is harmful to the environment. Most dyes belong to organic compounds that have teratogenic and carcinogenic properties. Dyes are generally classified according to their application or chemical structure. When classified by chemical structure, dyes can be divided into cationic, anionic, and neutral dyes. Cationic dyes, such as methylene blue (MB), are commonly used to dye silk, linen, and paper and can also be applied in biology for cell staining. Anionic dyes, such as methyl orange (MO), are often used as acid–base titration indicators in analytical chemistry; MO also belongs to azo dyes and is used for dyeing textiles; neutral dyes, such as alizarin (AR), can be used for dyeing and printing cotton and serve as a raw material for synthetic mordant dyes with similar structures. Figure 9.4 presents the schematic diagram of the dye removal mechanism through membrane materials (Xu et al., 2020). Table 9.2 presents the arrangement of graphene oxide-based, CNT-based, and CDs-based membranes for dye removal.

Table 9.1 Carbon material-based membrane for heavy metal removal

Polymer	Carbon type	Method	Rejection rate	Reference
Nylon	GO/onion extract	Vacuum filtration	$R(Cr^{6+}) = 87.5\%$ $R(As^{3+}) = 89.5\%$ $R(Cd^{2+}) = 93.0\%$ $R(Pb^{2+}) = 99.5\%$ $F(H_2O) = 150\ L\ m^{-2}\ h^{-1}\ bar^{-1}$	Janwery et al. (2023)
AAO	Thermally reduced GO	Vacuum filtration	$R(Na^{+}) = 99.37\%$ $F(H_2O) = 0.42\ L\ m^{-2}\ h^{-1}$	Zhao et al. (2019)
PVDF	GO-K^{+}/GO-Ba^{2+}/GO-Ca^{2+}/GO-Mg^{2+}/GO-Ca^{2+}-R	NIPS	$R(Na^{+}) = 72.3–88.6\%$ $R(Pb^{2+}) = 91.2–96.0\%$ $R(Ni^{2+}) = 92.4–97.0\%$ $R(Cu^{2+}) = 95.7–98.4\%$ $R(Mg^{2+}) = 95.6–99.5\%$ $F(H_2O) = 12.26–106.35\ L\ m^{-2}\ h^{-1}$	Ghaffar et al. (2019)
PES/PEEA	GO-TMC	Phase inversion	$R(Mg^{2+}) = 95.14\%$ $R(Li^{+}) = 20.93\%$ $F(H_2O) = 11.15\ L\ m^{-2}\ h^{-1}\ bar^{-1}$	Xu et al. (2020)
PES/PA	GO	Vacuum filtration	$R(Na^{+}) = 93.56\%$ $F(H_2O) = 35.14\ L\ m^{-2}\ h^{-1}\ bar^{-1}$	Kang et al. (2020)
Jatropha PU	GO-Fe_3O_4	Thermal curing	$R(Cu^{2+}) = 71.60\%$ $F(H_2O) = 523.33\ L\ m^{-2}\ h^{-1}$	Harun et al. (2020)
AAO	N-GO	Vacuum filtration	$R(Na^{2+}) = 99.0\%$ $F(H_2O) = 120\ L\ m^{-2}\ h^{-1}$	Qian et al. (2021)
Polyphenylsulfone	MWCNTs	Phase inversion	$R(Pb^{2+}) = 98\%$ $R(Hg^{2+}) = 76\%$ $R(Cd^{2+}) = 72\%$ $F(H_2O) > 185\ L\ m^{-2}\ h^{-1}$	Chandrashekhar Nayak et al. (2020)

(continued)

Table 9.1 (continued)

Polymer	Carbon type	Method	Rejection rate	Reference
PVDF	Cu/CNT	Vacuum filtration	$R(As^{3+}) = 95\%$ $F(H_2O) = 4639–4854\ L\ m^{-2}\ h^{-1}\ bar^{-1}$	Luan et al. (2019)
Nitrocellulose	GO/CNTs	Vacuum filtration	$R(Co^{2+}) = 99.52\%$ $R(Ni^{2+}) = 99.43\%$ $R(Cu^{2+}) = 99.29\%$ $R(Zn^{2+}) = 98.65\%$ $R(Cd^{2+}) = 99.85\%$ $R(Pb^{2+}) = 99.62\%$ $F(H_2O) = 4639–4854\ L\ m^{-2}\ h^{-1}\ bar^{-1}$	Musielak et al. (2019)
PAN	MWCNT-PEI	Electrospinning	$R(Cu^{2+}) = 93.6\%$ $R(Pb^{2+}) = 97.4\%$ $F(H_2O) = 145.8\ L\ m^{-2}\ h^{-1}\ bar^{-1}$	Deng et al. (2019)

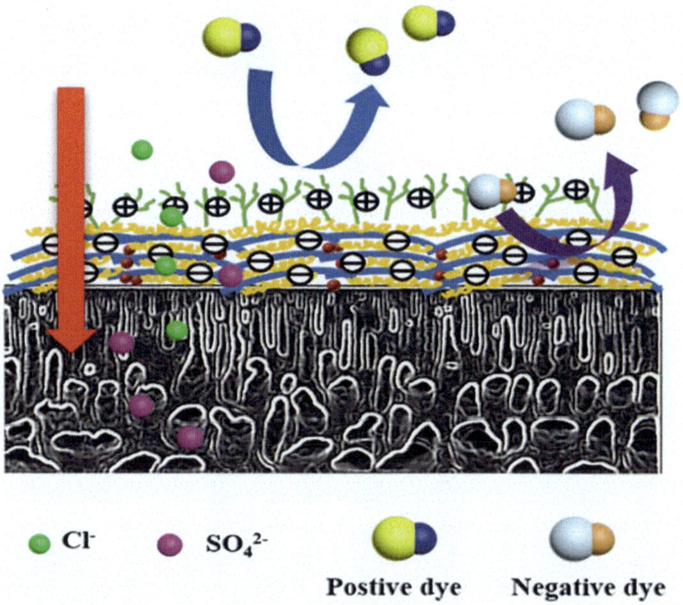

Fig. 9.4 Dye removal mechanism for using GO/DTiO_2-PDA-PEI composite membrane. Positive dye: Crystal violet (CV) and safranin T (ST); Negative dye: Eriochrome Black T (ET) and Alizarin yellow GG (AG). (Reproduced with permission from Xu et al. Copyright 2020 Elsevier)

9.4.3 PPCPs Removal

Pharmaceuticals and personal care products (PPCPs) are a diverse group of chemicals widely used in everyday products like medications, cosmetics, fragrances, sunscreens, and cleaning agents. While beneficial for health and hygiene, these compounds have become emerging contaminants in the environment due to their persistence, bioactivity, and potential ecological impacts. The strategy for PPCPs removal through carbon materials membrane occurs in hydrogen bonds and electrostatic and π–π interactions. Tables 9.3, 9.4, and 9.5 present the arrangement of graphene oxide-based, CNT-based, and CDs-based membranes for PPCP removal.

Table 9.2 Carbon material-based membrane for dye removal

Polymer	Carbon type	Method	Rejection rate	Reference
PAN	GO-g-psf	Vacuum filtration	R(AB) = 90.3%; R(CR) = 99.5%; R(MB) = 99.8% $F(H_2O) = 88.0\ L\ m^{-2}\ h^{-1}$	Li et al. (2020)
PES	GO-DES	Vacuum filtration	R(CR) = 99.4%; R(MB) = 96.7%; R(EB) = 98.5%; R(DR) = 96.7% $F(H_2O) = 17.7\ L\ m^{-2}\ h^{-1}$	Mehrabi et al. (2021)
PA	GO-ONCTs	Layer-by-layer	R(MB) = 99.3% $F(H_2O) = 7.2\ L\ m^{-2}\ h^{-1}$	Kang et al. (2018)
PES	GO	Solution casting	R(AB) = 99.7%; R(RB) = 99.0% $F(H_2O) = 116.5\ L\ m^{-2}\ h^{-1}$	Kadhim et al. (2020)
PSF	GO/DTiO_2-PDA-PEI	Vacuum filtration	R(ET) = 99.9%; R(CV) = 99.9%; R(AG) = 92.8%; R(ST) = 97.0% $F(H_2O) = 41.6\ L\ m^{-2}\ h^{-1}$	Xu et al. (2020)
Cellulose	GO-MB	Vacuum filtration	R(MO) = 96.37%; R(DB) = 99.78%; R(RhB) = 82.57% $F(H_2O) = 7.67\ L\ m^{-2}\ h^{-1}$	Hou et al. (2020)
PAN	GO-COF 1	Vacuum filtration	R(ET) = 100%; R(CR) = 100%; R(MB) = 99%; R(RB5) = 100%; R(DR23) = 99% $F(H_2O) = 310\ L\ m^{-2}\ h^{-1}\ MPa^{-1}$	Zhang et al. (2019)
PVDF	rGO	Vacuum filtration	R(MB) = 99%; R(CR) = 99%; R(RhB) = 90%; R(CV) = 99% $F(H_2O) = 11.9–16\ L\ m^{-2}\ h^{-1}$	Fan et al. (2020)
Nylon	GO/PEI-GO	Vacuum filtration	GO R(DR) > 99%; R(MB) > 99% $F(H_2O) = 17.4\ L\ m^{-2}\ h^{-1}$ PGO R(DR) > 99%; R(MB) > 99% $F(H_2O) = 131.1\ L\ m^{-2}\ h^{-1}$	Lu et al. (2019)

Cellulose acetate	GO/rGO/GO-PAM	Vacuum filtration	GO R(RhB) = 85.03% $F(H_2O) = 399.04\ L\ m^{-2}\ h^{-1}$ rGO R(RhB) = 95.43% $F(H_2O) = 85.85\ L\ m^{-2}\ h^{-1}$ GO-PAM R(RhB) = 97.06% $F(H_2O) = 188.89\ L\ m^{-2}\ h^{-1}$	Cheng et al. (2019)
PES	GO-GE	Vacuum filtration	R(RhB) = 95–99%; R(MB) = 95–99% $F(H_2O) = 80–320\ L\ m^{-2}\ h^{-1}$	Janjhi et al. (2021)
PES	Nitrogen-doped GO	NIPS	R(RR 195) = 96% $F(H_2O) = 190\ L\ m^{-2}\ h^{-1}$	Vatanpour et al. (2021)
Cellulose acetate	GO@cubic HKUST-1	Vacuum filtration	R(MB) = 99.5%; R(CR) = 71.2% $F(H_2O) = 214.7\ L\ m^{-2}\ h^{-1}$	Zhu et al. (2020)
PVDF	Ca/GO-SAx	Vacuum filtration	R(CR) > 99%; R(CV) > 99%; R(MB) > 99% $F(H_2O) = 38.9\ L\ m^{-2}\ h^{-1}\ bar^{-1}$	Yu et al. (2021)
Cellulose acetate	D-ChNCs/GO	Vacuum filtration	R(MB) = 99.3%; R(CR) = 98.3% $F(H_2O) = 135.6\ L\ m^{-2}\ h^{-1}$	Ou et al. (2019)
PC	GO/APT	Vacuum filtration	R(RhB) ~ 100% $F(H_2O) = 13.3\ L\ m^{-2}\ h^{-1}$	Wang et al. (2019)
PVDF	GO/NH_2-Fe_3O_4	Vacuum filtration	R(CR) = 94% $F(H_2O) = 78\ L\ m^{-2}\ h^{-1}$	Dong et al. (2020)
PDA/PES	GO-EDA	Vacuum filtration	R(MB) = 99.98%; R(MO) = 99.0% $F(H_2O) = 2.3\ L\ m^{-2}\ h^{-1}$	Lin et al. (2020)
Nylon	GO-COF-TpPa	Vacuum filtration	R(MB) = 97.1%; R(MO) = 68.6% $F(H_2O) = 166.8\ L\ m^{-2}\ h^{-1}$	Kong et al. (2019)
Nylon	GO-MXene	Vacuum filtration	R(MB) = 98.56% $F(H_2O) = 21.02\ L\ m^{-2}\ h^{-1}$	Wei et al. (2019)

(continued)

Table 9.2 (continued)

Polymer	Carbon type	Method	Rejection rate	Reference
Cellulose acetate	GO-CNTs	Vacuum filtration	R(CR) = 98.7%; R(MB) = 94.1% $F(H_2O)$ = 26.3 L m^{-2} h^{-1}	Huang et al. (2021)
PES	GO-asparagine	Vacuum filtration	R(MB) = 99.8%; R(RB) = 99.9% $F(H_2O)$ = 1740 L m^{-2} h^{-1}	Jatoi et al. (2024)
PAN	PEI/GO	Electrospinning	R(DR) = 99.7% $F(H_2O)$ = 59.1 L m^{-2} h^{-1}	Xing et al. (2023)
PES	GO-serine	Vacuum filtration	R(RhB) = 99%; R(MB) = 98% F(RhB) = 930 L m^{-2} h^{-1} bar^{-1}; F(RhB) = 880 L m^{-2} h^{-1} bar^{-1}	Chandio et al. (2021)
Cellulose acetate	GO-g-C_3N_4	Vacuum filtration	R(CR) = 94.6%; R(MB) = 99.8% $F(H_2O)$ = 152.1–274.7 L m^{-2} h^{-1}	Yang et al. (2023)
PES	PEBAX®1657/chitosan-wrapped MWCNTs	Pouring	R(MG) = 98.7% $F(H_2O)$ = 13.85 L m^{-2} h^{-1}	Mousavi et al. (2020)
PVDF	MWCNTs-OH and MWCNTs-COOH	NIPS	R(RG19) = 97.3% $F(H_2O)$ = 138 L m^{-2} h^{-1}	Gholami et al. (2022)
PAN	OMWCNT	Electrospinning	R(indigo) = 98.73% $F(H_2O)$ = 3891.85 L m^{-2} h^{-1}	Xu et al. (2017)
PVC	TETA-MWCNTs	Vacuum filtration	R(Lanasol blue 3R) = 86.3%; $F(H_2O)$ = 46 L m^{-2} h^{-1}	Vatanpour and Haghighat (2019)
PSf	Acid-MWCNTs	Pouring	R(MB) =99.29%; R(CV) = 99.54% R(RhB) = 99.04% $F(H_2O)$ = 38.9 L m^{-2} h^{-1} bar^{-1}	Goswami et al. (2021)
PS	Lignin-MWCNTs	Interface polymerization	R(CR) =99.2–99.7% $F(H_2O)$ = 45.5 L m^{-2} h^{-1} bar^{-1}	He et al. (2024)
PES	MOS_2-OMWCNT	NIPS	R(RB 19) = 93.5%; R(CV) = 99.54%; R(RR 195) = 98.4% $F(H_2O)$ = 38.9 L m^{-2} h^{-1} bar^{-1}	AReferencei-Oskoui et al. (2022)
PVDF	β-CD/TMC-MWCNTs)	Vacuum filtration	R(BG) = 97.41%; R(CR) = 96.39% $F(H_2O)$ = 179.93 L m^{-2} h^{-1} bar^{-1}	Li et al. (2021)
PVDF	MWCNTs-COOH	Vacuum filtration	R(MO) = 97.79% $F(H_2O)$ = 9.32 L m^{-2} h^{-1} bar^{-1}	Li et al. (2023)

PES	SWCNTs	Interface polymerization	R(DR 23) = 99.3% $F(H_2O) = 208\ L\ m^{-2}\ h^{-1}\ bar^{-1}$	Lu et al. (2020)
PAN-CF	MWCNT/FeOOH	VIPS	R(MB) = 97.6% $F(H_2O) = 17.2\ L\ m^{-2}\ h^{-1}\ bar^{-1}$	Yar and Parlayici (2022)
PVDF/SMA	CNT-COOH	NIPS	R(CR) = 99.9%; R(MB) = 96.7%; R(AF) = 96.1% $F(H_2O) = 251.6\ L\ m^{-2}\ h^{-1}\ bar^{-1}$	Kang et al. (2021)
PVDF	CS@CNTs-COOH	NIPS	R(MeB) > 99%; R(SY) > 99%; R(CV) > 99%; R(RB) > 99%; R(CR) > 99%; R(MB) > 99% $F(H_2O) = 85.25\ L\ m^{-2}\ h^{-1}\ bar^{-1}$	Zhao et al. (2021)
PES	QCDs-TMC	NIPS	R(Alcian blue) = 92.9% R(CR) = 98.8% $F(H_2O) = 244.7\ L/m^2\ h$	Bi et al. (2019)
PES	QCDs-β-CDs	NIPS	R(CR) > 93.0% $F(H_2O) = 474.7\ L/m^2\ h$	Xue et al. (2020)
PVC	QCD	Phase inversion	R(RB 19) = 97.6% $F(H_2O) = 19.1\ L/m^2\ h$	Vatanpour et al. (2020)
PES	Nitrogen-doped carbon dots (NCDs)	Phase inversion	R(RB 19) = 99.2% $F(H_2O) = 44.6\ L/m^2\ h$	Koulivand et al. (2020)
PAN fibers	Amine-rich carbon dots (CDs)	Electron spinning	R(RB 19) = 95%	Wang et al. (2020)
TFC PA	Graphene oxide quantum dots (GO QDs)	Physical coating	R(MO) = 97%; R(MB) = 71%; F(MO) = 124.4 L/m^2 h; F(MB) = 75.4 L/m^2 h	Zhao et al. (2019)
PVP-COOH	Amine-rich carbon dots (CDs)	Cross-linking	R(EY) = 29.2%; R(MG) =19.2%; R(CV) = 17.7%	Nayak et al. (2020)

Table 9.3 Graphene oxide-based membrane for PPCPs removal

Polymer	Graphene type	Method	Rejection rate	Reference
PVDF	EDA-GO	Vacuum filtration	R(CAP) = 57%; R(PNC) = 92%; R(ERY) = 81% $F(H_2O) = 7.5\ L\ m^{-2}\ h^{-1}$	Liang et al. (2023)
PSf	TEA-GO	Vacuum filtration	R(ATT HCl) ~ 90%; R(BPA) ~ 100%; R(CFN) ~ 90%; R(ACT) ~ 80% $F(H_2O) = 22.5\ L\ m^{-2}\ h^{-1}$	Baig and Matin (2024)
PSf	GO	Phase inversion	R(AMX) = 87.33%; R(DIC) = 66.03%; R(CFN) = 80.61% $F(H_2O) = 37.81\ L\ m^{-2}\ h^{-1}$	Cadore et al. (2024)
PES	GO	Phase inversion	R(ACT) = 97.7%; R(DCF) = 99.13%; R(IBF) = 95.78% $F(H_2O) = 7.2\ L\ m^{-2}\ h^{-1}\ bar^{-1}$	Ntone et al. (2023)
PP	DMAEMA/GO	Electrospinning	R(BPA) = 89%; R(2-NO) = 82%; R(E2) = 88%; R(DBP) = 92%; R(NP) = 91%; R(p-NP) = 85%; R(TC) = 90%	Tian et al. (2019)
PAN	$GO-Fe^{3+}$	Vacuum filtration	R(BPA) = 81% $F(H_2O) = 4\ L\ m^{-2}\ h^{-1}$	Li et al. (2021)

Table 9.4 CNT-based membrane for PPCPs removal

Polymer	CNT type	Method	Rejection rate	Reference
PSf	Functionalized CNTs	Phase inversion	R(phenol) = 65% $F(H_2O) = 28.24\ L\ m^{-2}\ h^{-1}\ bar^{-1}$	Daramola et al. (2019)
PMIA	MWCNT-COOH	Phase inversion	R(IBF) = 45.5%; R(SMX) = 13%; R(CBZ) = 7.5% $F(H_2O) = 165\ L\ m^{-2}\ h^{-1}$	Ren et al. (2023)
PAN	SWCNT	Electron spinning	R(ACT) = 73–76%; R(BZF) = 53–81%; R(IOP) = 64–74%; R(DCF) = 65–82%; R(CBZ) = 83–88%; R(SMX) = 74–77% $F(H_2O) = 330\ L\ m^{-2}\ h^{-1}$	Soberman et al. (2020)
PVC	MWCNT	Phase inversion	R(AAP) = 29%; R(CFN) = 7%; R(TCS) = 44%; R(CBD) = 69%	Wang et al. (2018)
PVC	MWCNT	Phase inversion	R(AAP) = 73%; R(CAF) = 85%; R(TCS) = 87%; R(CBD) = 93% $F(H_2O) = 72.5\ L\ m^{-2}\ h^{-1}$	Wang et al. (2018)

Table 9.5 CDs-based membrane for gas separation

Polymer	CNT type	Method	Species	Rejection rate	Reference
Amino-modified cellulose	Pyrene-boronic acid-based CDs (CDs-PyB)	Pouring	F ion	$R(F^-) = 90.2\%$	Fan et al. (2020)
Cellulose acetate	Zwitterionic carbon quantum dots (ZQDs)	Dry-wet spinning	High pure water and diesel oil–water emulsion	R(diesel oil) = 98.9% R(dodecane) = 94.2% R(canola oil) = 99.1% R(engine oil) = 99.2% $F(H_2O) = 583.1$ L/ m^2 h bar F(diesel oil / H_2O) = 350.5 L/ m^2 h bar	Carballo et al. (2024)
NCM	Graphene oxide quantum dots (GO QDs)	Vacuum filtration	Pharmaceuticals and personal care products (PPCPs)	R(negatively charged PPCPs) >80% R(neutral PPCPs) = 20–80% R(positively charged PPCPs) >56%	Oikawa et al. (2023)

9.4.4 Gas Separation

In gas separation, membranes act as selective barriers that allow certain gas molecules to permeate faster than others based on molecular size, solubility/diffusivity, and polarity/chemical affinity differences. Carbon material-based membranes can be designed to have specific molecular sizes through different membrane fabrication processes. For example, membrane formation through phase inversion would generate a dense surface layer and asymmetric structure in the membranes. In addition, the porosity of the skin layer would also be adjusted by adding a pore generator, like polyethylene glycol (PEG) or polyvinylpyrrolidone (PVP). Table 9.6 arranges graphene oxide-based and CNT-based membranes for gas separation.

Table 9.6 Graphene oxide-based membrane for gas separation

Polymer	Carbon type	Method	Permeability/selectivity	Reference
Borate	GO	Vacuum filtration	$P(CO_2)$ = 100 Barrer $S(CO_2/N_2$ = 91	Wang et al. (2016)
PEBA	GO	Film casting	$P(CO_2)$ = 650 GPU $S(CO_2/CH_4)$ = 75	Shen et al. (2015)
PS	PEI-GO	Film casting	$P(CO_2)$ = 36 GPU $S(CO_2/N_2)$ = 107	Shen et al. (2015)
PSf	PEG-GO	Film casting	$P(CO_2)$ = 28.2 GPU $S(CO_2/CH_4)$ = 22.39	Raouf et al. (2020)
Nylon 66	ZIF-8@GO	Vacuum filtration	$P(CO_2) = 1.64 \times 10^4$ Barrer; $P(CH_4) = 0.11 \times 10^4$ Barrer; $P(He) = 57.8 \times 10^4$ Barrer; $P(H_2) = 6.5 \times 10^4$ Barrer $S(H_2/CH_4)$ = 43	Liu et al. (2019)
AAO	GO	Spin coating	$P(CO_2)$ = 4 L/(m² bar h); $P(CH_4)$ = 13.9 L/(m² bar h) $S(H_2O/N_2)$ = 13,000	Petukhov et al. (2019)
PDMS	GO	Spin coating	$P(CO_2)$ = 183 GPU $S(CO_2/N_2)$ = 259	Zhou et al. (2020)
Pebax® 1657	Aminosilane-GO	Film casting	$P(CO_2)$ = 943 GPU $S(CO_2/N_2)$ = 71.1; $S(CO_2/CH_4)$ = 40.9	Zhang et al. (2019)
Nylon	GO/COF	Vacuum filtration	$P(H_2) = 1.067 \times 10^6$ Barrer $S(H_2/CO_2)$ = 25.57	Tang et al. (2019)
PEBAX	ZIF-8/rGO	Casting	$S(O_2/N_2)$ = 8; $S(CO_2/CH_4)$ = 35	Jamil et al. (2019)
PES	EDA-GO	Coating	$P(CO_2)$ = 660 Barrer $S(CO_2/N_2) > 500$	Zhou et al. (2019)
PSf/PES/ PVDF	PVA-GO	Coating	$P(CO_2)$ = 27.8 GPU $S(CO_2/H_2)$ = 22	Xu et al. (2023)
PEBA	MWCNTs-COOH; MWCNTs-NCO; MWCNTs-NH_2	Film casting	$P(CO_2)$ = 148.86 Barrer; $P(CH_4)$ = 5.14 Barrer; $P(N_2)$ = 1.42 Barrer $S(CO_2/N_2)$ = 104.92; $S(CO_2/CH_4)$ = 28.95	Amirkhani et al. (2020)
PSf	PANI@CNTs	Coating	$P(CO_2)$ = 264 GPU $S(CO_2/N_2)$ = 149.8	Wang et al. (2019)
PES	CNTs	Phase inversion	$P(CO_2)$ = 6720 Barrer; $P(CH_4)$ = 11,358 Barrer; $P(N_2)$ = 7540 Barrer; $P(H_2)$ = 28,553 Barrer $S(CO_2/N_2)$ = 1.2; $S(CH_4/N_2)$ = 1.62; $S(H_2/N_2)$ = 3.95	Yousef et al. (2021)
PI	β-CD-MWCNTs	Phase inversion	$P(CO_2)$ = 11 GPU $S(CO_2/CH_4)$ = 8	Sanip et al. (2011)

(continued)

Table 9.6 (continued)

Polymer	Carbon type	Method	Permeability/selectivity	Reference
PES	APTES-MWCNTs	Phase inversion	$P(CO_2) = 6.79$ GFU; $P(CH_4) = 0.028$ GPU; $P(N_2) = 0.29$ GPU; $P(O_2) = 3.19$ GPU $S(CO_2/CH_4) = 250.13$; $S(O_2/N_2) = 10.6$	Ismail et al. (2011)
PES	MWCNTs	Phase inversion	$P(CO_2) = 4.5$ Barrer $S(CO_2/N_2) = 22$	Ge et al. (2011)
Matrimid matrix	GO-MWCNTs	Casting	$P(CO_2) = 38.07$ Barrer $S(CO_2/CH_4) = 84.6$; $S(CO_2/N_2) = 81.0$	Li et al. (2015)
PVA-NH_2	MWCNTs	Phase inversion	$P(CO_2) = 836$ Barrer $S(CO_2/H_2) = 43$	Zhao et al. (2014)
Cellulose acetate	β-CD-MWCNTs	Casting	$P(CO_2) = 733.90–741.67$ GPU; $P(N_2) = 18.46–18.56$ GPU $S(CO_2/N_2) = 39.54–40.18$	Ahmad et al. (2014)
PVA-POS	NH_2-MWCNTs	Cross-linking	$P(CO_2) = 957$ Barrer $S(CO_2/N_2) = 384$; $S(CO_2/H_2) = 56$; $S(CO_2/CH_4) = 264$	Ansaloni et al. (2015)
Pebax-1657	SWCNTs	Casting	$P(CO_2) = 105$ Barrer; $P(N_2) = 1.45$ Barrer; $P(H_2) = 9.11$ Barrer $S(CO_2/N_2) = 72.4$; $S(CO_2/H_2) = 11.5$	Yu et al. (2013)
Pebaxs	MWNTs, MWNTs-OH, MWNTs-NH_2	Casting	$P(CO_2) = 1400$ Barrer; $P(N_2) = 40$ Barrer; $P(H_2) = 110$ Barrer; $P(CH_4) = 100$ Barrer $S(CO_2/N_2) = 53$; $S(CO_2/CH_4) = 16$; $S(CO_2/H_2) = 4$	Zhao et al. (2017)
PAN	Polymer matrix-MWCNTs	Casting	$P(CO_2) = 2.9 \times 10^4$ Barrer $S(CO_2/N_2) = 24.2$	Sun et al. (2020)

1 GPU = 10^{-6} cm^3 and 1-Barrer = 10^{-10} cm^3 and 10^{-12} mol m^{-2} s^{-1} Pa^{-1}

9.5 Conclusion

Membrane technology stands at the forefront of addressing the most pressing global challenges, from ensuring clean water and air to enabling sustainable energy solutions. Membranes incorporating carbon nanomaterials can construct precise microstructures through various manufacturing processes and introduce diverse intramolecular interactions, enabling advanced pollutant separation. Integrating membrane with carbon nanomaterials technology represents a transformative leap forward, enhancing performance, efficiency, and adaptability. The continuous evolution of carbon nanomaterial-based membranes, coupled with breakthroughs in process integration and AI-driven optimization techniques, promises unprecedented efficiency and versatility.

References

Ahmad, A. L., Jawad, Z. A., Low, S. C., & Zein, S. H. S. (2014). A cellulose acetate/multi-walled carbon nanotube mixed matrix membrane for CO/N separation. *Journal of Membrane Science, 451*, 55–66. https://doi.org/10.1016/j.memsci.2013.09.043

Amirkhani, F., Mosadegh, M., Asghari, M., & Parnian, M. J. (2020). The beneficial impacts of functional groups of CNT on structure and gas separation properties of PEBA mixed matrix membranes. *Polymer Testing, 82*, 106285. https://doi.org/10.1016/j.polymertesting.2019.106285

Ansaloni, L., Zhao, Y. N., Jung, B. T., Ramasubramanian, K., Baschetti, M. G., & Ho, W. S. W. (2015). Facilitated transport membranes containing amino-functionalized multi-walled carbon nanotubes for high-pressure CO separations. *Journal of Membrane Science, 490*, 18–28. https://doi.org/10.1016/j.memsci.2015.03.097

AReferencei-Oskoui, S., Khataee, A., Behrouz, S. J., Vatanpour, V., Gharamaleki, S. H., Orooji, Y., & Safarpour, M. (2022). Development of MoS2/O-MWCNTs/PES blended membrane for efficient removal of dyes, antibiotic, and protein. *Separation and Purification Technology, 280*, 119822. https://doi.org/10.1016/j.seppur.2021.119822

Arribas, L., Arconada, N., González-Fernández, C., Löhrl, C., González-Aguilar, J., Kaltschmitt, M., & Romero, M. (2017). Solar-driven pyrolysis and gasification of low-grade carbonaceous materials. *International Journal of Hydrogen Energy, 42*(19), 13598–13606. https://doi.org/10.1016/j.ijhydene.2017.02.026

Baig, N., & Matin, A. (2024). Incorporating functionalized graphene oxide into diethylene triamine-based nanofiltration membranes can improve the removal of emerging organic micropollutants. *Journal of Colloid and Interface Science, 676*, 657–669. https://doi.org/10.1016/j.jcis.2024.06.180

Bi, R., Zhang, R. N., Shen, J. L., Liu, Y. N., He, M. R., You, X. D., Su, Y. L., & Jiang, Z. Y. (2019). Graphene quantum dots engineered nanofiltration membrane for ultrafast molecular separation. *Journal of Membrane Science, 572*, 504–511. https://doi.org/10.1016/j.memsci.2018.11.044

Björk, J., Hanke, F., Palma, C.-A., Samori, P., Cecchini, M., & Persson, M. (2010). Adsorption of aromatic and anti-aromatic systems on graphene through π−π stacking. *The Journal of Physical Chemistry Letters, 1*(23), 3407–3412. https://doi.org/10.1021/jz101360k

Cadore, J. S., Fabro, L. F., Giubel, G. O. M., Costa, L. F. S., Pires, M. J. R., & Brião, V. B. (2024). Synthesis of composite membranes modified with graphene oxide and polyethylene glycol for the removal of emerging contaminants from water. *Colloids and Surfaces A: Physicochemical and Engineering Aspects, 696*, 134348. https://doi.org/10.1016/j.colsurfa.2024.134348

Carballo, G. V., Yang, H. L., Hsu, Y. X., Leron, R. B., Tsai, H. A., & Lee, K. R. (2024). Incorporation of zwitterionic carbon quantum dots in cellulose acetate tubular membrane for oil/water separation. *Separation and Purification Technology, 337*, 126301. https://doi.org/10.1016/j.seppur.2024.126301

Chandio, I., Janjhi, F. A., Memon, A. A., Memon, S., Ali, Z., Thebo, K. H., Pirzado, A. A. A., Hakro, A. A., & Khan, W. S. (2021). Ultrafast ionic and molecular sieving through graphene oxide based composite membranes. *Desalination, 500*, 114848. https://doi.org/10.1016/j.desal.2020.114848

Chandrashekhar Nayak, M., Isloor, A. M., Inamuddin, Lakshmi, B., Marwani, H. M., & Khan, I. (2020). Polyphenylsulfone/multiwalled carbon nanotubes mixed ultrafiltration membranes: Fabrication, characterization and removal of heavy metals Pb^{2+}, Hg^{2+}, and Cd^{2+} from aqueous solutions. *Arabian Journal of Chemistry, 13*(3), 4661–4672. https://doi.org/10.1016/j.arabjc.2019.10.007

Chao-Mujica, F., Garcia-Hernández, L., Camacho-López, S., Camacho-López, M., Camacho-López, M., Reyes Contreras, D., Pérez-Rodríguez, A., Peña-Caravaca, J., Páez-Rodríguez, A., & Darias-Gonzalez, J. J. J. A. P. (2021). Carbon quantum dots by submerged arc discharge in water: Synthesis, characterization, and mechanism of formation. *Journal of Applied Physics, 129*(16), 163301. https://doi.org/10.1063/5.0040322

Cheng, M. M., Huang, L. J., Wang, Y. X., Zhao, Y. C., Tang, J. G., Wang, Y., Zhang, Y., Hedayati, M., Kipper, M. J., & Wickramasinghe, S. R. (2019). Synthesis of graphene oxide/polyacrylamide composite membranes for organic dyes/water separation in water purification. *Journal of Materials Science, 54*(1), 252–264. https://doi.org/10.1007/s10853-018-2828-9

Chrzanowska, J., Hoffman, J., Małolepszy, A., Mazurkiewicz, M., Kowalewski, T. A., Szymanski, Z., & Stobinski, L. (2015). Synthesis of carbon nanotubes by the laser ablation method: Effect of laser wavelength. *Physica Status Solidi (B), 252*(8), 1860–1867. https://doi.org/10.1002/pssb.201451614

Cian, M. B., Giordano, N. P., Mettlach, J. A., Minor, K. E., Dalebroux, Z. D., & Minor, K. E. (2020). Separation of the cell envelope for Gram-negative bacteria into inner and outer membrane fractions with technical adjustments for Acinetobacter baumannii. *JoVE, 158*, e60517. https://doi.org/10.3791/60517

Daramola, M., Sadare, O., Oluwasina, O., & Iyuke, S. (2019). Synthesis and application of functionalized carbon nanotube infused polymer membrane (fCNT/PSF/PVA) for treatment of phenol-containing wastewater. *Journal of Membrane Science and Research, 5*(4), 310–316. https://doi.org/10.22079/jmsr.2019.98343.1235

De Silva, K. K. H., Huang, H.-H., & Yoshimura, M. (2018). Progress of reduction of graphene oxide by ascorbic acid. *Applied Surface Science, 447*, 338–346. https://doi.org/10.1016/j.apsusc.2018.03.243

Deng, S., Liu, X., Liao, J., Lin, H., & Liu, F. (2019). PEI modified multiwalled carbon nanotube as a novel additive in PAN nanofiber membrane for enhanced removal of heavy metal ions. *Chemical Engineering Journal, 375*, 122086. https://doi.org/10.1016/j.cej.2019.122086

Dong, L., Li, M., Zhang, S., Si, X., Bai, Y., & Zhang, C. (2020). NH_2-Fe_3O_4-regulated graphene oxide membranes with well-defined laminar nanochannels for desalination of dye solutions. *Desalination, 476*, 114227. https://doi.org/10.1016/j.desal.2019.114227

Epand, R. M., Walker, C., Epand, R. F., & Magarvey, N. A. (2016). Molecular mechanisms of membrane targeting antibiotics. *Biochimica et Biophysica Acta (BBA)—Biomembranes, 1858*(5), 980–987. https://doi.org/10.1016/j.bbamem.2015.10.018

Erçarikci, E., & Alanyalioglu, M. (2021). Dual-functional graphene-based flexible material for membrane filtration and electrochemical sensing of heavy metal ions. *IEEE Sensors Journal, 21*(3), 2468–2475. https://doi.org/10.1109/JSEN.2020.3021988

Fan, X. T., Cai, C. B., Gao, J., Han, X. L., & Li, J. D. (2020). Hydrothermal reduced graphene oxide membranes for dyes removing. *Separation and Purification Technology, 241*, 116730. https://doi.org/10.1016/j.seppur.2020.116730

Ge, L., Zhu, Z. H., & Rudolph, V. (2011). Enhanced gas permeability by fabricating functionalized multi-walled carbon nanotubes and polyethersulfone nanocomposite membrane. *Separation and Purification Technology, 78*(1), 76–82. https://doi.org/10.1016/j.seppur.2011.01.024

George, H. S., Selvaraj, H., Ilangovan, A., Almutairi, B. O., Sheit, M. K., & Sathishkumar, K. (2024). Microwave-assisted one-pot synthesis of carbon dots from biomass: Fe(III) ion sensing and in-situ methylene blue reduction catalytic activity. *Desalination and Water Treatment, 320*, 100624. https://doi.org/10.1016/j.dwt.2024.100624

Ghaffar, A., Zhang, L. N., Zhu, X. Y., & Chen, B. L. (2019). Scalable graphene oxide membranes with tunable water channels and stability for ion rejection. *Environmental Science-Nano, 6*(3), 904–915. https://doi.org/10.1039/C8EN01273C

Gholami, S., Llacuna, J. L., Vatanpour, V., Dehqan, A., Paziresh, S., & Cortina, J. L. (2022). Impact of a new functionalization of multiwalled carbon nanotubes on antifouling and permeability of PVDF nanocomposite membranes for dye wastewater treatment. *Chemosphere, 294*, 133699. https://doi.org/10.1016/j.chemosphere.2022.133699

González-González, R. B., González, L. T., Madou, M., Leyva-Porras, C., Martinez-Chapa, S. O., & Mendoza, A. (2022). Synthesis, purification, and characterization of carbon dots from non-activated and activated pyrolytic carbon black. *Nanomaterials, 12*(3), 298. https://doi.org/10.3390/nano12030298

Goswami, R., Gogoi, M., Borah, A., Sarmah, H., Ingole, P. G., & Hazarika, S. (2021). Functionalized activated carbon and carbon nanotube hybrid membrane with enhanced antifouling activity for removal of cationic dyes from aqueous solution. *Environmental Nanotechnology, Monitoring & Management, 16*, 100492. https://doi.org/10.1016/j.enmm.2021.100492

Harun, N. H., Abidin, Z. Z., Abdullah, A. H., & Othaman, R. (2020). Sustainable oil-based membrane with graphene oxide for potential application in Cu(II) ion removal from aqueous solution. *Processes, 8*(2). https://doi.org/10.3390/pr8020230

He, Z., Zhang, N., Zhao, L., Li, Z., Cui, W., Zhu, Y., Wang, Y., Deng, H., Si, C., Bo, W., Zhou, J., Xu, S., & Li, Q. (2024). Carbon nanotube incorporated quaternized lignin-based loose nanofiltration membrane toward dye/salt separation. *Journal of Environmental Chemical Engineering, 12*(3), 112834. https://doi.org/10.1016/j.jece.2024.112834

Hou, J. K., Chen, Y. B., Shi, W. X., Bao, C. L., & Hu, X. Y. (2020). Graphene oxide/methylene blue composite membrane for dyes separation: Formation mechanism and separation performance. *Applied Surface Science, 505*, 144145. https://doi.org/10.1016/j.apsusc.2019.144145

Huang, H.-H., De Silva, K. K. H., Kumara, G. R. A., & Yoshimura, M. (2018). Structural evolution of hydrothermally derived reduced graphene oxide. *Scientific Reports, 8*(1), 6849. https://doi.org/10.1038/s41598-018-25194-1

Huang, Z. W., Li, Z. J., Wu, Q. Y., Zheng, L. R., Zhou, L. M., Chai, Z. F., Wang, X. L., & Shi, W. Q. (2018). Simultaneous elimination of cationic uranium(VI) and anionic rhenium(VII) by graphene oxide-poly(ethyleneimine) macrostructures: A batch, XPS, EXAFS, and DFT combined study. *Environmental Science-Nano, 5*(9), 2077–2087. https://doi.org/10.1039/C8EN00677F

Huang, L., Li, Z., Luo, Y., Zhang, N., Qi, W., Jiang, E., Bao, J., Zhang, X., Zheng, W., An, B., & He, G. (2021). Low-pressure loose GO composite membrane intercalated by CNT for effective dye/salt separation. *Separation and Purification Technology, 256*, 117839. https://doi.org/10.1016/j.seppur.2020.117839

Ismail, A. F., Rahim, N. H., Mustafa, A., Matsuura, T., Ng, B. C., Abdullah, S., & Hashemifard, S. A. (2011). Gas separation performance of polyethersulfone/multi-walled carbon nanotubes mixed matrix membranes. *Separation and Purification Technology, 80*(1), 20–31. https://doi.org/10.1016/j.seppur.2011.03.031

Jamil, N., Othman, N. H., Alias, N. H., Shahruddin, M. Z., Roslan, R. A., Lau, W. J., & Ismail, A. F. (2019). Mixed matrix membranes incorporated with reduced graphene oxide (rGO) and zeolitic imidazole framework-8 (ZIF-8) nanofillers for gas separation. *Journal of Solid State Chemistry, 270*, 419–427. https://doi.org/10.1016/j.jssc.2018.11.028

Janjhi, F. A., Chandio, I., Memon, A. A., Ahmed, Z., Thebo, K. H., Pirzado, A. A. A., Hakro, A. A., & Iqbal, M. (2021). Functionalized graphene oxide based membranes for ultrafast molecular separation. *Separation and Purification Technology, 274*, 117969. https://doi.org/10.1016/j.seppur.2020.117969

Janwery, D., Memon, F. H., Memon, A. A., Iqbal, M., Memon, F. N., Ali, W., Choi, K. H., & Thebo, K. H. (2023). Lamellar graphene oxide-based composite membranes for efficient separation of heavy metal ions and desalination of water. *ACS Omega, 8*(8), 7648–7656. https://doi.org/10.1021/acsomega.2c07243

Jatoi, A. H., Ali, A., Nadeem, A., Phulpoto, S. N., Iqbal, M., Memon, A. A., Yang, J., & Thebo, K. H. (2024). High-performance asparagine-modified graphene oxide membranes for organic dyes and heavy metal ion separation. *New Journal of Chemistry, 48*(4), 1715–1723. https://doi.org/10.1039/d3nj04552h

Jiang, S., Do, H., Yusuf, A., Xiao, Z., Wang, C., Li, J., Sun, Y., Ren, Y., & He, J. (2024). Camellia oleifera shell–reduced graphene oxide for adsorption of copper(II). *Materials Chemistry and Physics, 314*, 128818. https://doi.org/10.1016/j.matchemphys.2023.128818

Kaczmarek, A., Hoffman, J., Morgiel, J., Mościcki, T., Stobiński, L., Szymański, Z., & Małolepszy, A. (2021). Luminescent carbon dots synthesized by the laser ablation of graphite in polyethylenimine and ethylenediamine. *Materials, 14*(4), 729. https://doi.org/10.3390/ma14040729

Kadhim, R. J., Al-Ani, F. H., Al-shaeli, M., Alsalhy, Q. F., & Figoli, A. (2020). Removal of dyes using graphene oxide (GO) mixed matrix membranes. *Membranes, 10*(12), 366. https://doi.org/10.3390/membranes10120366

Kang, H., Shi, J., Liu, L. Y., Shan, M. J., Xu, Z. W., Li, N., Li, J., Lv, H. M., Qian, X. M., & Zhao, L. H. (2018). Sandwich morphology and superior dye-removal performances for nanofiltration membranes self-assembled via graphene oxide and carbon nanotubes. *Applied Surface Science, 428*, 990–999. https://doi.org/10.1016/j.apsusc.2017.09.212

Kang, X., Liu, X., Liu, J. H., Wen, Y., Qi, J. Y., & Li, X. (2020). Spin-assisted interfacial polymerization strategy for graphene oxide-polyamide composite nanofiltration membrane with high performance. *Applied Surface Science, 508*, 145198. https://doi.org/10.1016/j.apsusc.2019.145198

Kang, D., Shao, H., Chen, G., Dong, X., & Qin, S. (2021). Fabrication of highly permeable PVDF loose nanofiltration composite membranes for the effective separation of dye/salt mixtures. *Journal of Membrane Science, 621*, 118951. https://doi.org/10.1016/j.memsci.2020.118951

Khanzada, N. K., Farid, M. U., Kharraz, J. A., Choi, J., Tang, C. Y., Nghiem, L. D., Jang, A., & An, A. K. (2020). Removal of organic micropollutants using advanced membrane-based water and wastewater treatment: A review. *Journal of Membrane Science, 598*, 117672. https://doi.org/10.1016/j.memsci.2019.117672

Kim, H. W., Yoon, H. W., Yoo, B. M., Park, J. S., Gleason, K. L., Freeman, B. D., & Park, H. B. (2014). High-performance CO_2-philic graphene oxide membranes under wet-conditions. *Chemical Communications, 50*(88), 13563–13566. https://doi.org/10.1039/C4CC06207H

Kong, G., Pang, J., Tang, Y., Fan, L., Sun, H., Wang, R., Feng, S., Feng, Y., Fan, W., Kang, W., Guo, H., Kang, Z., & Sun, D. (2019). Efficient dye nanofiltration of a graphene oxide membrane via combination with a covalent organic framework by hot pressing. *Journal of Materials Chemistry A, 7*(42), 24301–24310. https://doi.org/10.1039/C9TA07684K

Koulivand, H., Shahbazi, A., Vatanpour, V., & Rahmandoost, M. (2020). Novel antifouling and antibacterial polyethersulfone membrane prepared by embedding nitrogen-doped carbon dots for efficient salt and dye rejection. *Materials Science & Engineering C-Materials for Biological Applications, 111*, 110787. https://doi.org/10.1016/j.msec.2020.110787

Latorrata, S., Cristiani, C., Basso Peressut, A., Brambilla, L., Bellotto, M., Dotelli, G., Finocchio, E., Gallo Stampino, P., & Ramis, G. (2021). Reduced graphene oxide membranes as potential self-assembling filter for wastewater treatment. *Minerals, 11*(1), 15. https://doi.org/10.3390/min11010015

Li, M. J., Liu, C. M., Cao, H. B., & Zhang, Y. (2013). Surface charge research of graphene oxide, chemically reduced graphene oxide and thermally exfoliated graphene oxide. *Advanced Materials Research, 716*, 127–131. https://doi.org/10.4028/www.scientific.net/AMR.716.127

Li, M.-J., Liu, C.-M., Xie, Y.-B., Cao, H.-B., Zhao, H., & Zhang, Y. (2014). The evolution of surface charge on graphene oxide during the reduction and its application in electroanalysis. *Carbon, 66*, 302–311. https://doi.org/10.1016/j.carbon.2013.09.004

Li, X. Q., Ma, L., Zhang, H. Y., Wang, S. F., Jiang, Z. Y., Guo, R. L., Wu, H., Cao, X. Z., Yang, J., & Wang, B. Y. (2015). Synergistic effect of combining carbon nanotubes and graphene oxide in mixed matrix membranes for efficient CO separation. *Journal of Membrane Science, 479*, 1–10. https://doi.org/10.1016/j.memsci.2015.01.014

Li, J., Hu, M. Y., Pei, H. C., Ma, X. H., Yan, F., Dlamini, D. S., Cui, Z. Y., He, B. Q., Li, J. X., & Matsuyama, H. (2020). Improved water permeability and structural stability in a polysulfone-grafted graphene oxide composite membrane used for dye separation. *Journal of Membrane Science, 595*, 117547. https://doi.org/10.1016/j.memsci.2019.117547

Li, J., Gong, J.-L., Zeng, G.-M., Song, B., Cao, W.-C., Fang, S.-Y., Tang, S.-Q., Guan, Y., Tan, Z.-K., Chen, Z.-P., Mao, X.-Q., & Zhu, R.-L. (2021). Thin-film composite polyester nanofiltration membrane with high flux and efficient dye/salts separation fabricated from precise molecular sieving structure of β-cyclodextrin. *Separation and Purification Technology, 276*, 119352. https://doi.org/10.1016/j.seppur.2021.119352

Li, S., Wan, Y., Guo, S., & Luo, J. (2021). Ferric ions mediated defects narrowing of graphene oxide nanofiltration membrane for robust removal of organic micropollutants. *Chemical Engineering Journal, 411*, 128587. https://doi.org/10.1016/j.cej.2021.128587

Li, J., Gong, J.-L., Fang, S.-Y., Cao, W.-C., Tang, S.-Q., Qin, M., Wang, Y.-W., Zhou, H.-Y., & Long, M. (2023). Antifouling and chemical-resistant nanofiltration membrane modulated by a functionalized MWCNT interlayer for efficient dye/salt separation at high salinity. *ACS Applied Materials & Interfaces, 15*(28), 34272–34289. https://doi.org/10.1021/acsami.3c05677

Liang, Z., Zhao, X., Huang, W., Qi, H., & Wang, C. (2023). Removal of antibiotics with different charges in water by graphene oxide membranes. *Water Reuse, 13*(2), 220–232. https://doi.org/10.2166/wrd.2023.084

Lin, H., Li, Y., & Zhu, J. (2020). Cross-linked GO membranes assembled with GO nanosheets of differently sized lateral dimensions for organic dye and chromium separation. *Journal of Membrane Science, 598*, 117789. https://doi.org/10.1016/j.memsci.2019.117789

Liu, Y., Luo, C., Cui, G. J., & Yan, S. Q. (2015). Synthesis of manganese dioxide/iron oxide/graphene oxide magnetic nanocomposites for hexavalent chromium removal. *RSC Advances, 5*(67), 54156–54164. https://doi.org/10.1039/C5RA06455D

Liu, D., Pang, G., Tang, Z., & Feng, S. (2019). Interfacial engineering of metal–organic frameworks/graphene oxide composite membrane by polyethyleneimine for efficient H_2/CH_4 gas separation. *Inorganic Chemistry Frontiers, 6*(8), 2043–2049. https://doi.org/10.1039/C9QI00455F

Lu, J. J., Gu, Y. H., Chen, Y., Yan, X., Guo, Y. J., & Lang, W. Z. (2019). Ultrahigh permeability of graphene-based membranes by adjusting D-spacing with poly (ethylene imine) for the separation of dye wastewater. *Separation and Purification Technology, 210*, 737–745. https://doi.org/10.1016/j.seppur.2018.08.065

Lu, Y., Fang, W., Kong, J., Zhang, F., Wang, Z., Teng, X., Zhu, Y., & Jin, J. (2020). A microporous polymer ultrathin membrane for the highly efficient removal of dyes from acidic saline solutions. *Journal of Membrane Science, 603*, 118027. https://doi.org/10.1016/j.memsci.2020.118027

Luan, H., Teychene, B., & Huang, H. (2019). Efficient removal of As(III) by Cu nanoparticles intercalated in carbon nanotube membranes for drinking water treatment. *Chemical Engineering Journal, 355*, 341–350. https://doi.org/10.1016/j.cej.2018.08.104

Madadrang, C. J., Kim, H. Y., Gao, G. H., Wang, N., Zhu, J., Feng, H., Gorring, M., Kasner, M. L., & Hou, S. F. (2012). Adsorption behavior of EDTA-graphene oxide for Pb (II) removal. *ACS Applied Materials & Interfaces, 4*(3), 1186–1193. https://doi.org/10.1021/am201645g

Mehrabi, N., Lin, H. Q., & Aich, N. (2021). Deep eutectic solvent functionalized graphene oxide nanofiltration membranes with superior water permeance and dye desalination performance. *Chemical Engineering Journal, 412*, 128577. https://doi.org/10.1016/j.cej.2021.128577

Mokkapati, V. R. S. S., Koseoglu-Imer, D. Y., Yilmaz-Deveci, N., Mijakovic, I., & Koyuncu, I. (2017). Membrane properties and anti-bacterial/anti-biofouling activity of polysulfone–graphene oxide composite membranes phase inversed in graphene oxide non-solvent. *RSC Advances, 7*(8), 4378–4386. https://doi.org/10.1039/C6RA25015G

Mousavi, S. R., Asghari, M., & Mahmoodi, N. M. (2020). Chitosan-wrapped multiwalled carbon nanotube as filler within PEBA thin film nanocomposite (TFN) membrane to improve dye removal. *Carbohydrate Polymers, 237*, 116128. https://doi.org/10.1016/j.carbpol.2020.116128

Musielak, M., Gagor, A., Zawisza, B., Talik, E., & Sitko, R. (2019). Graphene oxide/carbon nanotube membranes for highly efficient removal of metal ions from water. *ACS Applied Materials & Interfaces, 11*(31), 28582–28590. https://doi.org/10.1021/acsami.9b11214

Nair, R. R., Wu, H. A., Jayaram, P. N., Grigorieva, I. V., & Geim, A. K. (2012). Unimpeded permeation of water through helium-leak–tight graphene-based membranes. *Science, 335*(6067), 442–444. https://doi.org/10.1126/science.1211694

Nammahachak, N., Aup-Ngoen, K. K., Asanithi, P., Horpratum, M., Chuangchote, S., Ratanaphan, S., & Surareungchai, W. (2022). Hydrothermal synthesis of carbon quantum dots with size tunability via heterogeneous nucleation. *RSC Advances, 12*(49), 31729–31733. https://doi.org/10.1039/D2RA05989D

Nayak, S., Prasad, S. R., Mandal, D., & Das, P. (2020). Carbon dot cross-linked polyvinylpyrrolidone hybrid hydrogel for simultaneous dye adsorption, photodegradation and bacterial elimination from waste water. *Journal of Hazardous Materials, 392*, 122287. https://doi.org/10.1016/j.jhazmat.2020.122287

Novoselova, I. A., Oliinyk, N. F., Volkov, S. V., Konchits, A. A., Yanchuk, I. B., Yefanov, V. S., Kolesnik, S. P., & Karpets, M. V. (2008). Electrolytic synthesis of carbon nanotubes from carbon dioxide in molten salts and their characterization. *Physica E: Low-dimensional Systems and Nanostructures, 40*(7), 2231–2237. https://doi.org/10.1016/j.physe.2007.10.069

Ntone, E. P. N., Rahman, S. A., Samah, R. A., Wahab, M. S. A., Jawad, Z. A., & Hasbullah, H. (2023). A comparison study on performance of thin film composite membrane embedded with graphene oxide for acetaminophen, diclofenac and ibuprofen separation from waste water. *Chemical Engineering Research and Design, 195*, 28–37. https://doi.org/10.1016/j.cherd.2023.05.023

Oikawa, M., Takeuchi, H., Koide, R., Yoshizawa, N., Wang, Z. M., & Koura, S. (2023). Achieving overall rejection of pharmaceuticals and personal care products of different polarities by controlling interlayer charging environment of graphene oxide membrane using carbon quantum dots. *Chemical Engineering Journal, 472*, 144811. https://doi.org/10.1016/j.cej.2023.144811

Othman, N. H., Alias, N. H., Fuzil, N. S., Marpani, F., Shahruddin, M. Z., Chew, C. M., David Ng, K. M., Lau, W. J., & Ismail, A. F. (2022). A review on the use of membrane technology systems in developing countries. *Membranes, 12*(1), 30. https://doi.org/10.3390/membranes12010030

Ou, X., Yang, X., Zheng, J., & Liu, M. (2019). Free-standing graphene oxide–chitin nanocrystal composite membrane for dye adsorption and oil/water separation. *ACS Sustainable Chemistry & Engineering, 7*(15), 13379–13390. https://doi.org/10.1021/acssuschemeng.9b02619

Pei, S., Zhao, J., Du, J., Ren, W., & Cheng, H.-M. (2010). Direct reduction of graphene oxide films into highly conductive and flexible graphene films by hydrohalic acids. *Carbon, 48*(15), 4466–4474. https://doi.org/10.1016/j.carbon.2010.08.006

Petukhov, D. I., Chernova, E. A., Kapitanova, O. O., Boytsova, O. V., Valeev, R. G., Chumakov, A. P., Konovalov, O. V., & Eliseev, A. A. (2019). Thin graphene oxide membranes for gas dehumidification. *Journal of Membrane Science, 577*, 184–194. https://doi.org/10.1016/j.memsci.2019.01.041

Qian, Y. J., Shang, J., Liu, D., Yang, G. L., Wang, X. G., Chen, C., Kou, L. Z., & Lei, W. W. (2021). Enhanced ion sieving of graphene oxide membranes via surface amine functionalization. *Journal of the American Chemical Society, 143*(13), 5080–5090. https://doi.org/10.1021/jacs.1c00575

Raouf, M., Abedini, R., Omidkhah, M., & Nezhadmoghadam, E. (2020). A favored CO separation over light gases using mixed matrix membrane comprising polysulfone/polyethylene glycol and graphene hydroxyl nanoparticles. *Process Safety and Environmental Protection, 133*, 394–407. https://doi.org/10.1016/j.psep.2019.11.002

Ren, X., Shi, L., He, P., Qian, Y., Wu, Y., Zhang, Z., & Gong, C. (2023). Effects of carboxylated MWCNTs additives on the structure and performance of PMIA membrane: Size, content and PPCPs removal specificity. *Applied Surface Science, 629*, 157454. https://doi.org/10.1016/j.apsusc.2023.157454

Ribeiro, H., Schnitzler, M. C., da Silva, W. M., & Santos, A. P. (2021). Purification of carbon nanotubes produced by the electric arc-discharge method. *Surfaces and Interfaces, 26*, 101389. https://doi.org/10.1016/j.surfin.2021.101389

Sanip, S. M., Ismail, A. F., Goh, P. S., Soga, T., Tanemura, M., & Yasuhiko, H. (2011). Gas separation properties of functionalized carbon nanotubes mixed matrix membranes. *Separation and Purification Technology, 78*(2), 208–213. https://doi.org/10.1016/j.seppur.2011.02.003

Sengupta, I., Chakraborty, S., Talukdar, M., Pal, S. K., & Chakraborty, S. (2018). Thermal reduction of graphene oxide: How temperature influences purity. *Journal of Materials Research, 33*(23), 4113–4122. https://doi.org/10.1557/jmr.2018.338

Shahriary, L., & Athawale, A. A.(2014). Graphene oxide synthesized by using modified hummers approach. International Journal Renewable Energy and Environment Engineering, 2, 58–63

Shen, J., Liu, G. P., Huang, K., Jin, W. Q., Lee, K. R., & Xu, N. P. (2015). Membranes with fast and selective gas-transport channels of laminar graphene oxide for efficient CO capture. *Angewandte Chemie-International Edition, 54*(2), 578–582. https://doi.org/10.1002/ange.201409563

Shen, Y. J., Wang, H. X., Liu, J. D., & Zhang, Y. T. (2015). Enhanced performance of a novel polyvinyl amine/chitosan/graphene oxide mixed matrix membrane for CO capture. *ACS Sustainable Chemistry & Engineering, 3*(8), 1819–1829. https://doi.org/10.1021/acssuschemeng.5b00409

Soberman, M. J., Farnood, R. R., & Tabe, S. (2020). A low pressure SWCNT-ENM sandwich membrane system for the removal of PPCPs from water. *The Canadian Journal of Chemical Engineering, 98*(5), 1047–1058. https://doi.org/10.1002/cjce.23703

Stafiej, A., & Pyrzynska, K. (2007). Adsorption of heavy metal ions with carbon nanotubes. *Separation and Purification Technology, 58*(1), 49–52. https://doi.org/10.1016/j.seppur.2007.07.008

Stankovich, S., Dikin, D. A., Piner, R. D., Kohlhaas, K. A., Kleinhammes, A., Jia, Y., Wu, Y., Nguyen, S. T., & Ruoff, R. S. (2007). Synthesis of graphene-based nanosheets via chemical reduction of exfoliated graphite oxide. *Carbon, 45*(7), 1558–1565. https://doi.org/10.1016/j.carbon.2007.02.034

Sun, H. X., Gao, W., Zhang, Y. W., Cao, X. Z., Bao, S. S., Li, P., Kang, Z. X., & Niu, Q. J. (2020). Bis(phenyl)fluorene-based polymer of intrinsic microporosity/functionalized multi-walled carbon nanotubes mixed matrix membranes for enhanced CO separation performance. *Reactive & Functional Polymers, 147*, 104465. https://doi.org/10.1016/j.reactfunctpolym.2019.104465

Tang, Y. C., Feng, S., Fan, L. L., Pang, J., Fan, W. D., Kong, G. D., Kang, Z. X., & Sun, D. F. (2019). Covalent organic frameworks combined with graphene oxide to fabricate membranes for H/CO separation. *Separation and Purification Technology, 223*, 10–16. https://doi.org/10.1016/j.seppur.2019.04.069

Tanis-Kanbur, M. B., Tamilselvam, N. R., Lai, H. Y., & Chew, J. W. (2022). Impact of particle shape and surface group on membrane fouling. *Membranes, 12*(4), 403. https://doi.org/10.3390/membranes12040403

Tian, J., Wei, J., Zhang, H., Kong, Z., Zhu, Y., & Qin, Z. (2019). Graphene oxide-functionalized dual-scale channels architecture for high-throughput removal of organic pollutants from water. *Chemical Engineering Journal, 359*, 852–862. https://doi.org/10.1016/j.cej.2018.12.048

Uchôa Teixeira, J. V., Cavalcante Pimentel, G. J., Santos, A. A., Gonçalves Dias, L. F., Mastelaro, V. R., & Lisboa-Filho, P. N. (2023). Partially reduced graphene oxide produced by glucose in alkaline conditions using probe sonication: The role of the base in reduction. *Journal of Materials Research and Technology, 26*, 1785–1797. https://doi.org/10.1016/j.jmrt.2023.07.211

Vatandost, E., Ghorbani-HasanSaraei, A., Chekin, F., Naghizadeh Raeisi, S., & Shahidi, S.-A. (2020). Green tea extract assisted green synthesis of reduced graphene oxide: Application for highly sensitive electrochemical detection of sunset yellow in food products. *Food Chemistry: X, 6*, 100085. https://doi.org/10.1016/j.fochx.2020.100085

Vatanpour, V., & Haghighat, N. (2019). Improvement of polyvinyl chloride nanofiltration membranes by incorporation of multiwalled carbon nanotubes modified with triethylenetetramine to use in treatment of dye wastewater. *Journal of Environmental Management, 242*, 90–97. https://doi.org/10.1016/j.jenvman.2019.04.060

Vatanpour, V., Khadem, S. S. M., Masteri-Farahani, M., Mosleh, N., Ganjali, M. R., Badiei, A., Pourbashir, E., Mashhadzadeh, A. H., Munir, M. T., Mahmodi, G., Zarrintaj, P., Ramsey, J. D., Kim, S. J., & Saeb, M. R. (2020). Anti-fouling and permeable polyvinyl chloride nanofiltration membranes embedded by hydrophilic graphene quantum dots for dye wastewater treatment. *Journal of Water Process Engineering, 38*, 101652. https://doi.org/10.1016/j.jwpe.2020.101652

Vatanpour, V., Mousavi Khadem, S. S., Dehqan, A., Al-Naqshabandi, M. A., Ganjali, M. R., Sadegh Hassani, S., Rashid, M. R., Saeb, M. R., & Dizge, N. (2021). Efficient removal of dyes and proteins by nitrogen-doped porous graphene blended polyethersulfone nanocomposite membranes. *Chemosphere, 263*, 127892. https://doi.org/10.1016/j.chemosphere.2020.127892

Wang, L., Wang, N., Li, J., Li, J., Bian, W., & Ji, S. (2016). Layer-by-layer self-assembly of polycation/GO nanofiltration membrane with enhanced stability and fouling resistance. *Separation and Purification Technology, 160*, 123–131. https://doi.org/10.1016/j.seppur.2016.01.024

Wang, S. F., Wu, Y. Z., Zhang, N., He, G. W., Xin, Q. P., Wu, X. Y., Wu, H., Cao, X. Z., Guiver, M. D., & Jiang, Z. Y. (2016). A highly permeable graphene oxide membrane with fast and selective transport nanochannels for efficient carbon capture. *Energy & Environmental Science, 9*(10), 3107–3112. https://doi.org/10.1039/C3NR05380F

Wang, Y., Huang, H., & Wei, X. (2018). Influence of wastewater precoagulation on adsorptive filtration of pharmaceutical and personal care products by carbon nanotube membranes. *Chemical Engineering Journal, 333*, 66–75. https://doi.org/10.1016/j.cej.2017.09.149

Wang, Y., Yang, Q., Dong, J., & Huang, H. (2018). Competitive adsorption of PPCP and humic substances by carbon nanotube membranes: Effects of coagulation and PPCP properties. *Science of the Total Environment, 619-620*, 352–359. https://doi.org/10.1016/j.scitotenv.2017.11.117

Wang, C.-Y., Zeng, W.-J., Jiang, T.-T., Chen, X., & Zhang, X.-L. (2019). Incorporating attapulgite nanorods into graphene oxide nanofiltration membranes for efficient dyes wastewater treatment. *Separation and Purification Technology, 214*, 21–30. https://doi.org/10.1016/j.seppur.2018.04.079

Wang, Y. H., Li, L., Zhang, X. R., Li, J. P., Liu, C. C., Li, N. W., & Xie, Z. L. (2019). Polyvinylamine/graphene oxide/PANI@CNTs mixed matrix composite membranes with enhanced CO/N separation performance. *Journal of Membrane Science, 589*, 117246. https://doi.org/10.1016/j.memsci.2019.117246

Wang, B., Chen, P. Y., Zhao, R. X., Zhang, L., Chen, Y., & Yu, L. P. (2020). Carbon-dot modified polyacrylonitrile fibers: Recyclable materials capable of selectively and reversibly adsorbing small-sized anionic dyes. *Chemical Engineering Journal, 391*, 123484. https://doi.org/10.1016/j.cej.2019.123484

Wang, C., Yang, M., Shi, H., Yao, Z., Liu, E., Hu, X., Guo, P., Xue, W., & Fan, J. (2022). Carbon quantum dots prepared by pyrolysis: Investigation of the luminescence mechanism and application as fluorescent probes. *Dyes and Pigments, 204*, 110431. https://doi.org/10.1016/j.dyepig.2022.110431

Wei, S., Xie, Y., Xing, Y., Wang, L., Ye, H., Xiong, X., Wang, S., & Han, K. (2019). Two-dimensional graphene Oxide/MXene composite lamellar membranes for efficient solvent permeation and molecular separation. *Journal of Membrane Science, 582*, 414–422. https://doi.org/10.1016/j.memsci.2019.03.085

Wu, Z. G., Deng, W. J., Zhou, W., & Luo, J. W. (2019). Novel magnetic polysaccharide/graphene oxide @FeO gel beads for adsorbing heavy metal ions. *Carbohydrate Polymers, 216*, 119–128. https://doi.org/10.1016/j.carbpol.2019.04.020

Xing, C., Liu, C.-Y., Lai, C., & Zhang, S.-Q. (2023). Tuning d-spacing of graphene oxide nanofiltration membrane for effective dye/salt separation. *Rare Metals, 42*(2), 418–429. https://doi.org/10.1007/s12598-022-02153-4

Xu, Z., Li, X., Teng, K., Zhou, B., Ma, M., Shan, M., Jiao, K., Qian, X., & Fan, J. (2017). High flux and rejection of hierarchical composite membranes based on carbon nanotube network and ultrathin electrospun nanofibrous layer for dye removal. *Journal of Membrane Science, 535*, 94–102. https://doi.org/10.1016/j.memsci.2017.04.029

Xu, P., Hong, J., Qian, X. M., Xu, Z. Z., Xia, H., & Ni, Q. Q. (2020). "Bridge" graphene oxide modified positive charged nanofiltration thin membrane with high efficiency for Mg/Li separation. *Desalination, 488*, 114522. https://doi.org/10.1016/j.desal.2020.114522

Xu, Y. Q., Peng, G. B., Liao, J. B., Shen, J. N., & Gao, C. J. (2020). Preparation of molecular selective GO/DTiO-PDA-PEI composite nanofiltration membrane for highly pure dye separation. *Journal of Membrane Science, 601*, 117727. https://doi.org/10.1016/j.memsci.2019.117727

Xu, W. Q., Lindbråthen, A., Janakiram, S., Ansaloni, L., & Deng, L. Y. (2023). Enhanced CO_2/H_2 separation by GO and PVA-GO embedded PVAm nanocomposite membranes. *Journal of Membrane Science, 671*, 121397. https://doi.org/10.1016/j.memsci.2023.121397

Xue, J., Shen, J. L., Zhang, R. N., Wang, F., Liang, S. W., You, X. D., Yu, Q. Q., Hao, Y. F., Su, Y. L., & Jiang, Z. Y. (2020). High-flux nanofiltration membranes prepared with β-cyclodextrin and graphene quantum dots. *Journal of Membrane Science, 612*, 118465. https://doi.org/10.1016/j.memsci.2020.118465

Yang, L., Jia, F., Juan, Z., Yu, D., Sun, L., Song, Y., Wang, Y., Huang, L., & Tang, J. (2023). High-permeable graphene oxide/graphitic carbon nitride composite nanofiltration membrane for selective separation of dye and desalination. *Journal of Environmental Chemical Engineering, 11*(2), 109306. https://doi.org/10.1016/j.jece.2023.109306

Yar, A., & Parlayici, Ş. (2022). Carbon nanotubes/ polyacrylonitrile composite nanofiber mats for highly efficient dye adsorption. *Colloids and Surfaces A: Physicochemical and Engineering Aspects, 651*, 129703. https://doi.org/10.1016/j.colsurfa.2022.129703

Yousef, S., Tuckute, S., Tonkonogovas, A., Stankevicius, A., & Mohamed, A. (2021). Ultra-permeable CNTs/PES membranes with a very low CNTs content and high H_2/N_2 and CH_4/N_2 selectivity for clean energy extraction applications. *Journal of Materials Research and Technology-JMR&T, 15*, 5114–5127. https://doi.org/10.1016/j.jmrt.2021.10.125

Yu, B., Cong, H. L., Li, Z. J., Tang, J. G., & Zhao, X. S. (2013). Pebax-1657 nanocomposite membranes incorporated with nanoparticles/colloids/carbon nanotubes for CO/N and CO/H separation. *Journal of Applied Polymer Science, 130*(4), 2867–2876. https://doi.org/10.1002/app.39500

Yu, J., Wang, Y., He, Y., Gao, Y., Hou, R., Ma, J., Zhang, L., Guo, X., & Chen, L. (2021). Calcium ion-sodium alginate double cross-linked graphene oxide nanofiltration membrane with enhanced stability for efficient separation of dyes. *Separation and Purification Technology, 276*, 119348. https://doi.org/10.1016/j.seppur.2021.119348

Zhang, J. H., Xin, Q. P., Li, X., Yun, M. Y., Xu, R., Wang, S. F., Li, Y. F., Lin, L. G., Ding, X. L., Ye, H., & Zhang, Y. Z. (2019). Mixed matrix membranes comprising aminosilane-functionalized graphene oxide for enhanced CO separation. *Journal of Membrane Science, 570*, 343–354. https://doi.org/10.1016/j.memsci.2018.10.075

Zhang, X. K., Li, H., Wang, J., Peng, D. L., Liu, J. D., & Zhang, Y. T. (2019). In-situ grown covalent organic framework nanosheets on graphene for membrane-based dye/salt separation. *Journal of Membrane Science, 581*, 321–330. https://doi.org/10.1016/j.memsci.2019.03.070

Zhao, Y. N., Jung, B. T., Ansaloni, L., & Ho, W. S. W. (2014). Multiwalled carbon nanotube mixed matrix membranes containing amines for high pressure CO/H separation. *Journal of Membrane Science, 459*, 233–243. https://doi.org/10.1016/j.memsci.2014.02.022

Zhao, D., Ren, J., Wang, Y., Qiu, Y., Li, H., Hua, K., Li, X., Ji, J., & Deng, M. (2017). High CO_2 separation performance of Pebax®/CNTs/GTA mixed matrix membranes. *Journal of Membrane Science, 521*, 104–113. https://doi.org/10.1016/j.memsci.2016.08.061

Zhao, X., Zhang, R., Liu, Y., He, M., Su, Y., Gao, C., & Jiang, Z. (2018). Antifouling membrane surface construction: Chemistry plays a critical role. *Journal of Membrane Science, 551*, 145–171. https://doi.org/10.1016/j.memsci.2018.01.039

Zhao, G. K., Hu, R. R., He, Y. J., & Zhu, H. W. (2019). Physically coating nanofiltration membranes with graphene oxide quantum dots for simultaneously improved water permeability and salt/dye rejection. *Advanced Materials Interfaces, 6*(5). https://doi.org/10.1002/admi.201801742

Zhao, Z. Y., Ni, S. N., Su, X., Gao, Y., & Sun, X. Q. (2019). Thermally reduced graphene oxide membrane with ultrahigh rejection of metal ions' separation from water. *ACS Sustainable Chemistry & Engineering, 7*(17), 14874–14882. https://doi.org/10.1021/acssuschemeng.9b02972

Zhao, J., Liu, H., Xue, P., Tian, S., Sun, S., & Lv, X. (2021). Highly-efficient PVDF adsorptive membrane filtration based on chitosan@CNTs-COOH simultaneous removal of anionic and cationic dyes. *Carbohydrate Polymers, 274*, 118664. https://doi.org/10.1016/j.carbpol.2021.118664

Zhou, F. L., Tien, H. N., Dong, Q. B., Xu, W. W. L., Li, H. Z., Li, S. G., & Yu, M. (2019). Ultrathin, ethylenediamine-functionalized graphene oxide membranes on hollow fibers for CO capture. *Journal of Membrane Science, 573*, 184–191. https://doi.org/10.1016/j.memsci.2018.11.080

Zhou, Y., Yan, K. L., Yang, Z., Bauer, R. A., Hong, N. N., & Verweij, H. (2020). Nanometer-thick supported graphene oxide membrane for CO capture. *ACS Applied Nano Materials, 3*(7), 6654–6663. https://doi.org/10.1021/acsanm.0c01081

Zhou, A., Bai, J., Hong, W., & Bai, H. (2022). Electrochemically reduced graphene oxide: Preparation, composites, and applications. *Carbon, 191*, 301–332. https://doi.org/10.1016/j.carbon.2022.01.056

Zhu, Y., Murali, S., Stoller, M. D., Velamakanni, A., Piner, R. D., & Ruoff, R. S. (2010). Microwave assisted exfoliation and reduction of graphite oxide for ultracapacitors. *Carbon, 48*(7), 2118–2122. https://doi.org/10.1016/j.carbon.2010.02.001

Zhu, M., Liu, Y., Chen, M., Gan, D., Wang, M., Zeng, H., Liao, M., Chen, J., Tu, W., & Niu, W. (2020). Ultrahigh flux of graphene oxide membrane modified with orientated growth of MOFs for rejection of dyes and oil-water separation. *Chinese Chemical Letters, 31*(10), 2683–2688. https://doi.org/10.1016/j.cclet.2020.04.011

Chapter 10
Carbon-Based Catalyst for Degradation of Environmental Contaminants

Shounik Paul and Ajith Manayil Parambil

Abstract The increasing scarcity of drinkable water and rising pollution levels have led to the rapid development of effective methods for treating water contaminants. POPs are particularly concerning among various pollutants due to their robust chemical nature, high toxicity, and aquatic endurance. AOPs have emerged as a promising solution for degrading such pollutants by mineralizing them into CO_2, H_2O, and inorganic ions. Key AOPs include photocatalysis, sonocatalysis, electrochemical oxidation, and Fenton and Fenton-like processes. However, challenges such as high capital costs and significant energy demands limit their broader application. AOPs depend not only on transition metal-based catalysts but also on carbon-based materials, reducing the application cost. Carbon-based materials in pristine form or as hybrids have high surface area, providing anchoring sites for transition metals and tolerating harsh chemical environments, which lowers toxicity while increasing biocompatibility. Thus, this chapter offers insights into the mechanism of action of different carbon-based materials to facilitate the development of new carbon-based materials and hybrids. This leads to advancements in water treatment technologies, emphasizing the potential of EAOPs for large-scale implementation and commercialization.

Keywords Water pollution · Fenton oxidation · Advanced oxidation process · Photocatalysis · Electrocatalysis · Sonocatalysis

Abbreviations

AC	Activated carbon
AOPs	Advanced oxidation processes
CDs	Carbon dots

S. Paul (✉) · A. M. Parambil (✉)
Nanotechnology Centre, Centre for Energy and Environmental Technologies, VŠB–Technical University of Ostrava, Ostrava-Poruba, Czech Republic

A. M. Parambil et al. (eds.), *Carbon*,
https://doi.org/10.1007/978-3-031-90613-8_10

CNTs	Carbon nanotubes
EAOPs	Electrochemical advanced oxidation processes
e^-–h^+	Electron–hole
eF	Electro-Fenton
FOP	Fenton oxidation process
g-C_3N_4	Graphitic carbon nitride
GO	Graphene oxide
H_2O_2	Hydrogen peroxide
$HO_2^{\bullet}$	Hydroperoxyl radicals
MG	Multilayer graphene
ML	Machine learning
MTZ	Metronidazole
MWCNTs	Multi-walled carbon nanotubes
$O_2^{\bullet-}$	Superoxide radicals
$^{\bullet}OH$	Hydroxyl radicals
PDS	Peroxydisulfate
pF	Photo-Fenton
PMS	Peroxymonosulfate
POPs	Persistent organic pollutants
ROS	Reactive oxygen species
SACs	Single-atom catalysts
$SO_4^{\bullet-}$	Sulfate radicals
spF	Sono-photo-Fenton
SWCNTs	Single-walled carbon nanotubes
US	Ultrasonic
UV	Ultraviolet

10.1 Introduction

Water pollution has become a critical global challenge, affecting all life forms, including humans. The rapid growth of industries, increasing populations, and shifting lifestyles have turned water pollution into a serious threat to human health and the ecosystems that support life on Earth. The contamination of natural resources, especially water, has reached alarming levels, affecting countries around the world, both developed and developing (Mallick et al., 2024). As cities grow and industries expand, organic and inorganic pollutants are increasingly released into freshwater sources. This harms the environment and poses serious health risks to communities. Though water covers about 70% of the Earth's surface, less than 3% is freshwater, and much of it is either unsafe to drink or challenging to access. The World Health Organization (WHO) warns humankind that billions of people do not have access to safe drinking water, and the prime cause related to it is quality (Moradi et al., 2020). Toxic substances like industrial waste, pesticides, and heavy metals continue to contaminate water sources, making it even harder to secure safe drinking water (Parambil et al., 2024).

Water pollution comes from many different sources, which can be divided into organic and inorganic pollutants. Inorganic pollutants, such as nutrients, sediments, and heavy metals, often enter water through mining, agriculture, and industrial processes. While some of these pollutants naturally exist in the environment, human activities have significantly increased their concentrations, often making them harmful. For example, heavy metals such as cadmium, lead, mercury, and arsenic are non-biodegradable and can accumulate in the human body. They are the prime cause of serious health issues. These metals, along with other inorganic contaminants like nitrates and fluorides, pose significant risks to both the environment and human health.

Organic pollutants, however, include substances like pharmaceuticals, dyes, pesticides, and industrial chemicals. These often enter water systems through agricultural runoff, sewage, and industrial waste. These pollutants can remain in the environment for long periods and accumulate in ecosystems, causing lasting damage to biodiversity and public health. The accumulation of organic and inorganic pollutants has made water pollution one of the most pressing environmental challenges, affecting all life forms. With clean water in such short supply, developing practical solutions has become more urgent (Lu & Astruc, 2020).

Over the last few decades, many methods have been developed to tackle water pollution. Traditional methods like oxidation, solvent extraction, filtration, and reverse osmosis have been implemented to de-contaminate polluted water bodies. However, these methods have limitations, such as high costs, excessive energy use, and the creation of secondary pollutants. One promising area of research is AOPs, which use radical-based chemical reactions to break down stubborn pollutants. AOPs typically rely on H_2O_2 to generate $^{\bullet}OH$, powerful oxidants that can degrade organic pollutants. While effective, these methods can be complex and may not remove all types of toxins (Mukherjee et al., 2023).

Despite the variety of existing techniques, many still face practical challenges, such as long treatment times and low efficiency in removing POPs. To address these limitations, nanotechnology-based solutions are gaining attention. Techniques like nano-adsorption and photocatalysis are at the forefront of water purification research. Nanomaterials are particularly effective at removing organic and inorganic pollutants due to their high surface area, low cost, and eco-friendly nature. These technologies offer higher removal efficiencies, reduce energy consumption, and minimize secondary waste production. Photocatalysis, for example, uses light-activated materials to break down pollutants, making it a highly efficient and sustainable option, as it relies on solar energy to drive chemical reactions that degrade harmful substances (Ahmad et al., 2020).

The growing water pollution crisis highlights the need for preventative and corrective actions. The rising energy demands, rapid industrialization, and global population growth have intensified the pollution problem, making it crucial for governments, industries, and researchers to work together on innovative and scalable solutions. A multifaceted approach is needed to mitigate the long-term effects of pollution, including policy reforms to impose stricter regulations on industrial discharges, investments in clean technologies to reduce pollutant generation, and efforts to promote public awareness of sustainable practices. As science and technology continue to advance, developing cost-effective, efficient, and environmentally sustainable methods will be key to addressing the water pollution crisis. Nanotechnology and AOP represent just a few innovative tools that could help restore water resources and protect ecosystems for future generations. However, the success of these efforts will ultimately depend on global cooperation and a commitment to sustainable development. While progress has been made in water pollution remediation, the complexity and scale of the issue require continued innovation, investment, and collaboration. Addressing this urgent environmental challenge is essential to preserve biodiversity and ensure a healthy and secure future for human populations worldwide.

Carbon-based materials have garnered significant attention in recent years for their role in catalytic processes, particularly catalytic aromatization. They are explored both as a support and as catalyst material owing to their high surface area, excellent electron conductivity, chemical stability, and porous structure, which make them highly effective catalysts (Mallick et al., 2021). Additionally, these materials can be derived from coal and renewable biomass, offering advantages such as low cost and a reduced carbon footprint. Various oxygen functionalities on carbon materials enhance their acid–base properties, further improving their performance as carbon-supported catalysts. Moreover, these materials can be activated at high temperatures and functionalized or decorated with metallic nanoparticles to enhance catalytic activity.

AOPs have emerged as a highly efficient and attractive solution in wastewater treatment, bridging the gap between conventional physicochemical and biological methods. Industrial effluents from the chemical, mining, petrochemical, and pharmaceutical sectors are major contributors to water pollution. While physical methods like coagulation, filtration, and sedimentation help separate waste materials from water, they do not degrade contaminants, making complete mineralization of pollutants necessary. Similarly, biological methods use bacteria to break down organic waste but require optimal temperature, pH, and aeration conditions for effective performance. Furthermore, large amounts of organic matter and highly acidic conditions often make biological treatment challenging (Kumari & Kumar, 2023). Although traditional pollutant treatment methods remain widely used, they often demonstrate limited efficiency and fail to eliminate contaminants. As a result, researchers are exploring environmentally friendly alternatives, with studies highlighting the potential of AOPs as an effective solution to address these challenges.

10.2 Methods and Mechanisms of Catalytic Degradation of Contaminants

10.2.1 Fenton

The FOP is an AOP that uses H_2O_2 and ferrous ions (Fe^{2+}) to generate $^{\bullet}OH$ that break down organic pollutants (Eq. 10.1). This process relies on Fe^{2+}/Fe^{3+} redox cycling with H_2O_2 to produce both $^{\bullet}OH$ and $HO_2^{\bullet}$ radicals (Eq. 10.2), enabling the degradation of persistent contaminants such as phenols, pesticides, and pharmaceuticals due to the high oxidation potential of $^{\bullet}OH$ (2.0–2.8 V). While the goal is complete mineralization of pollutants into CO_2 and H_2O, efficiency is often hampered by the slow regeneration of Fe^{2+} from Fe^{3+} (Eq. 10.3) and competitive side reactions (Eqs. 10.4–10.9). However, the Fenton process also faces challenges, including narrow pH requirements (pH 2–4) and iron leaching. The slow regeneration of Fe^{2+} constrains $^{\bullet}OH$ production, leading to research on electron-donating promoters like hydroxylamine and ascorbic acid to accelerate Fe^{3+} reduction. Yet, secondary pollution from these additives raises concerns about sustainability.

A promising solution is carbon-based catalysis, especially MG, which accelerates the Fe^{3+}/Fe^{2+} cycle and enhances Fe^{3+} reactivity, boosting $^{\bullet}OH$ generation without sacrificial electron donors. Kinetic studies show that MG increases the Fe^{3+}-H_2O_2 reaction rate by 100 times compared to aqueous solutions, positioning carbon catalysis as an efficient, eco-friendly enhancement to traditional FOP (Sun et al., 2022).

Chain reactions of the classical Fenton system.

$$Fe^{2+} + H_2O_2 \rightarrow {}^{\bullet}OH + Fe^{3+} + OH^- \tag{10.1}$$

$$Fe^{3+} + H_2O_2 \rightarrow Fe^{2+} + H^+ + HO_2^{\bullet} \tag{10.2}$$

$$Fe^{3+} + HO_2^{\bullet} \rightarrow Fe^{2+} + O_2 + H^+ \tag{10.3}$$

$$Fe^{2+} + {}^{\bullet}OH \rightarrow Fe^{3+} + OH^- \tag{10.4}$$

$$Fe^{2+} + HO_2^{\bullet} \rightarrow Fe^{3+} + HO_2^- \tag{10.5}$$

$${}^{\bullet}OH + {}^{\bullet}OH \rightarrow H_2O_2 \tag{10.6}$$

$${}^{\bullet}OH + H_2O_2 \rightarrow HO_2^{\bullet} + H_2O \tag{10.7}$$

$$HO_2^{\bullet} + HO_2^{\bullet} \rightarrow H_2O_2 + O_2 \tag{10.8}$$

$${}^{\bullet}OH + HO_2^{\bullet} \rightarrow H_2O + O_2 \tag{10.9}$$

10.2.2 Drawback of Fenton

Antibiotics make up 15% of global drug production, frequently entering ecosystems through contaminated water from households, hospitals, and pharmaceutical facilities. MTZ, a nitroimidazole antibiotic for anaerobic bacterial and protozoal infections, is particularly problematic due to its high water solubility and low degradability, which enable it to persist in the environment and contribute to antimicrobial resistance. This resistance diminishes the effectiveness of biological treatments, causing antibiotic buildup in natural water sources (Usman et al., 2023).

AOPs offer a promising chemical alternative for degrading organic pollutants like MTZ, with the Fenton process standing out due to its rapid degradation rate, low toxicity, cost-effectiveness, and environmental safety. The Fenton process generates ROS such as $^{\bullet}OH$, $O_2^{\bullet-}$, and singlet oxygen, breaking organic compounds into water and carbon dioxide. However, the homogeneous Fenton process has limitations: it operates efficiently only in a narrow acidic pH range (2.5–3.0), requiring substantial acid and post-treatment neutralization. Iron recycling and Fe^{3+} to Fe^{2+} conversion are challenging at neutral pH (5–9), where $Fe(OH)_3$ or Fe_2O_3 precipitates form. High oxidant (H_2O_2) consumption and metal hydroxide sludge generation also create secondary pollution, reducing the process's cost-effectiveness and environmental sustainability. Lastly, catalyst recovery and reusability challenges further limit the use of the homogeneous Fenton process, as these factors increase treatment costs (Ziembowicz & Kida, 2022).

10.2.3 Electro-Fenton

$^{\bullet}OH$, with a high oxidation potential (2.8 V) and an extremely short half-life (~10^{-9} s), enables rapid, non-selective oxidation of various environmental pollutants. The development of electrochemical processes for in situ Fenton reagent generation at selected cathodes minimizes the use of large reagent quantities. It reduces iron-containing sludge (Eq. 10.10). Specifically, H_2O_2 is generated directly on carbon electrodes via oxygen reduction (Eq. 10.11), circumventing issues of transport, storage, and handling of H_2O_2. Studies indicate that electrogenerated H_2O_2, due to synergistic radical species from oxygen reduction, facilitates faster and more effective oxidation than stabilized commercial H_2O_2, which is prone to catalytic decomposition from trace metal contaminants.

An added benefit of eF processes is continuous iron regeneration through cathodic reduction, allowing the use of small catalytic iron quantities and reducing sludge production, which lowers costs. While Fe^{3+} can also promote a Fenton-like reaction (Eq. 10.12), the resulting $HO_2^{\bullet}$ radical has a weaker oxidizing potential (E_0 = 1.65 V vs SHE), making it ineffective against organic molecules. Therefore, effective oxidation depends on the rapid reduction of Fe^{3+} to Fe^{2+}, which is typically slower than the reverse process (Eq. 10.13). Thus, cathode materials for eF must efficiently reduce both oxygen to H_2O_2 and Fe^{3+} to Fe^{2+}, promoting $^{\bullet}OH$ production. Carbon-based cathodes are ideal for this purpose due to their high overpotential toward water discharge (Petrucci et al., 2016).

$$Fe^{2+} + H_2O_2 \rightarrow {}^{\bullet}OH + Fe^{3+} + OH^- \tag{10.10}$$

$$O_2 + 2H_3O^+ + 2e^- \rightarrow H_2O_2 + 2H_2O \tag{10.11}$$

$$Fe^{3+} + e^- \rightarrow Fe^{2+} \; E^{o}\left(Fe^{3+}/Fe^{2+}\right) = 0.77\,\text{V vs S.H.E.} \tag{10.12}$$

$$H_2O_2 + Fe^{3+} \rightarrow HO_2^{\bullet} + Fe^{2+} + H^+ \tag{10.13}$$

For instance, N, O-codoped biomass-derived porous carbon was modified with Fe/Co bimetallic sites. This served as an inexpensive and efficient cathode material for the electrochemical oxidation of dimethyl phthalate (DMP), which was facilitated by the cathodic production of $^{\bullet}OH$, demonstrating the eF mechanism (Qu et al., 2024) (Fig. 10.1a).

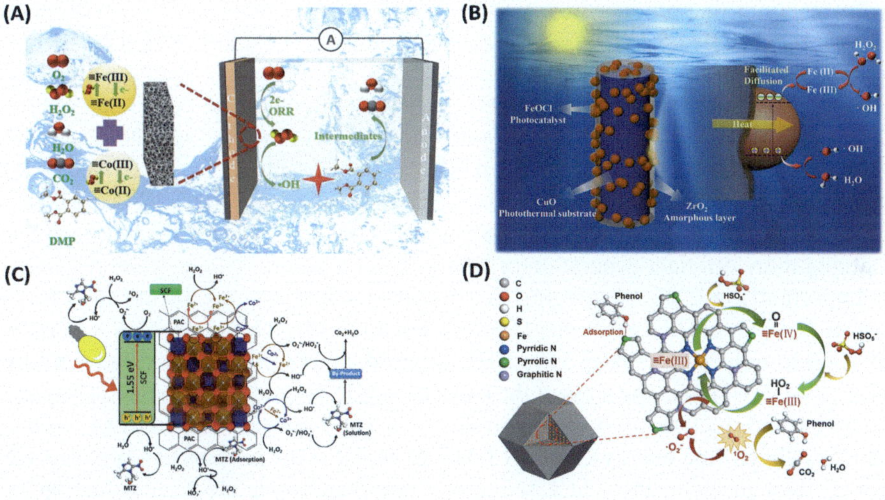

Fig. 10.1 Schematic representation of (**a**) mechanism of dimethyl phthalate (DMP) oxidation via eF process with FeCo@NOPC cathode. (Reproduced with permission from Qu et al. Copyright 2024, Elsevier). (**b**) CuO nanowire array-coated amorphous ZrO_2-induced pF process at FeOCl for degradation of ciprofloxacin. (Reproduced with permission from Yang et al. Copyright 2025, Elsevier). (**c**) Proposed mechanistic pathways of H_2O_2 decomposition and metronidazole (MTZ) degradation over spinel cobalt ferrite nanoparticles on porous AC. (Reproduced with permission from Kakavandi et al. Copyright 2024, Elsevier). (**d**) Phenol removal by 1.2Fe–N–C/PMS system via PMS (peroxymonosulfate) activation. (Reproduced with permission from He et al. Copyright 2021, Elsevier)

10.2.4 Photo-Fenton

For efficient pF-like catalysis, various semiconductor (SC)-based photocatalysts have been extensively explored, such as metal oxides, metal sulfides, metal-organic frameworks (MOFs), and g-C_3N_4. Among these, metal-free g-C_3N_4 has garnered increasing attention due to its low cost, chemical stability, strong light-harvesting capabilities, nontoxicity, and easy synthesis. These advantages make g-C_3N_4 an ideal candidate for integrating interfacial heterogeneous Fenton catalysis with photocatalysis, opening up new possibilities for advanced wastewater treatment and environmental remediation. g-C_3N_4 is a layered polymer semiconductor, where van der Waals forces hold the layers together. The tri-s-triazine unit, a common structural motif in g-C_3N_4, forms an π-conjugated system with sp^2-hybridized carbon and nitrogen atoms. With a moderate bandgap of 2.7–2.8 eV, g-C_3N_4 absorbs visible light around 450–460 nm. It also shows excellent thermal stability (<600 °C) and chemical robustness in organic solvents and under acidic or basic conditions, making it a durable photocatalyst.

In the context of a g-C_3N_4-driven pF system, the process involves three key steps: (i) absorption of solar energy to generate photoexcited e^-–h^+ pairs (Eq. 10.14), (ii) separation and migration of charge carriers to the g-C_3N_4 surface, and (iii) surface redox reactions that produce highly ROS. The photogenerated electrons on the conduction band (CB) of g-C_3N_4 can directly decompose H_2O_2 into $^{\bullet}OH$ (Eq. 10.15). Additionally, since the CB position of g-C_3N_4 is more negative than the standard potential for O_2/ $O_2^{\bullet -}$, −0.33 eV vs NHE, electrons migrating to the CB can reduce O_2 to form $O_2^{\bullet -}$ (Eq. 10.16). Furthermore, singlet oxygen ($^1O_2$) can be generated through reactions between $O_2^{\bullet -}$ and $^{\bullet}OH$ (Eq. 10.17). While the valence band (VB) of g-C_3N_4 is not sufficiently positive to oxidize H_2O or OH^- to form $^{\bullet}OH$ (Eq. 10.18), tuning the electronic structure and bandgap of g-C_3N_4 can improve the VB position, enabling more efficient generation of $^{\bullet}OH$.

Importantly, g-C_3N_4 is emerging as a promising candidate for the in situ generation of H_2O_2 in pF-like reactions. The CB potential of g-C_3N_4 (−1.27 V vs NHE) allows it to reduce O_2 (−0.28 V vs NHE) to produce H_2O_2, while its relatively low VB (1.4 V vs NHE) helps prevent the oxidation of H_2O_2 under irradiation. The photocatalytic production of H_2O_2 involves a reaction between O_2 and H_2O, with a standard Gibbs free energy change of 117 kJ mol^{-1}. This process can proceed via two distinct mechanisms: (1) a sequential two-step single-electron reduction of molecular oxygen ($O_2 \rightarrow O_2^{\bullet -} \rightarrow H_2O_2$), or (2) a direct two-electron reduction of O_2 to hydrogen peroxide ($O_2 \rightarrow H_2O_2$). In the single-electron reduction pathway, photoexcited electrons in the CB of g-C_3N_4 interact with adsorbed O_2 molecules to generate $O_2^{\bullet -}$ (Eq. 10.19). Simultaneously, O_2 can react with H^+ to produce $HO_2^{\bullet -}$ (Eq. 10.20), which is then reduced to HO_2^{2-} through another single-electron reduction (Eq. 10.21). Finally, H_2O_2 is generated from the reaction between HO_2^- and H^+ (Eq. 10.22). Alternatively, H_2O_2 can also be produced via a direct two-electron reduction of O_2 (Lin et al., 2022).

$$g-C_3N_4 + hv \rightarrow g-C_3N_4\left(h^+ + e^-\right) \quad (10.14)$$

$$e^- + H_2O_2 \rightarrow 2^{\bullet}OH \quad (10.15)$$

$$e^- + O_2 \rightarrow O_2^{\bullet -} \quad (10.16)$$

$$^{\bullet}OH + O_2^{\bullet -} \rightarrow {}^1O_2 + OH^- \quad (10.17)$$

$$h^+ + 2H_2O / OH^- \rightarrow {}^{\bullet}OH \quad (10.18)$$

$$O_2 + 2H_2O + hv \rightarrow 2H_2O_2 \quad (10.19)$$

$$O_2 + H^+ \rightarrow HO_2^{\bullet -} \quad (10.20)$$

$$HO_2^{\bullet -} + e^- \rightarrow H_2O_2 \quad (10.21)$$

$$O_2 + 2e^- + 2H^+ \rightarrow H_2O_2 \quad (10.22)$$

Materials like $FeOCl/ZrO_2/CuO$, which participate in pF-like processes, enhance ciprofloxacin degradation. ZrO_2/CuO are subjected to photothermal heating, which increases the diffusion length of the charge carriers, thus effectively reducing the charge-carrier recombination process. The carrier concentration at FeOCl is highest at 60 °C, suppressing the carrier recombination and enhancing ciprofloxacin's degradation (Yang et al., 2025) (Fig. 10.1b).

10.2.5 Sono-Photo-Fenton

The spF process combines US waves, UV light, and Fenton chemistry to enhance $^{\bullet}OH$ production, pollutant degradation, and Fe^{2+} regeneration in aqueous systems. US initiates water sonolysis, producing $^{\bullet}OH$ and $^{\bullet}H$ atoms (Eq. 10.27). However, recombination reduces these species. UV light addresses this by converting H_2O_2 formed from $^{\bullet}OH$ recombination back into $^{\bullet}OH$ (Eq. 10.23), boosting overall radical availability and contaminant removal efficiency. A key challenge with the US alone is low contaminant removal efficiency due to radical recombination. Combining US with UV, amplifies radical production, aided by the acoustic cavitation effect (Eqs. 10.24–10.28). Cavitation bubbles form under intense US, reaching local pressures of 500 atm and temperatures up to 5200 K within the bubble (1900 K in the surrounding area). These conditions drive the production of radicals like $^{\bullet}OH$, $^{\bullet}H$, $^{\bullet}OOH$, and O, intensifying the degradation reactions (Kakavandi et al., 2024).

Pollutant degradation occurs via two primary pathways: (i) pyrolysis inside cavitation bubbles for volatile compounds and (ii) oxidation of non-volatile pollutants by radicals in the liquid phase. The cavitation collapse also generates shear forces and shock waves, enhancing mass transfer and turbulence, thereby improving the overall reaction efficiency. The spF process is highly effective, energy-efficient, and cost-efficient, gaining significant attention for its pollutant degradation capability. The enhanced •OH production and cavitation-induced turbulence ensure a high oxidation rate, making spF a powerful approach for environmental remediation (Van Nguyen et al., 2024). One example is the spinel cobalt ferrite nanoparticles over porous AC, which activates H_2O_2 for Fenton degradation of metronidazole. A US source was introduced into the system to enhance the effectiveness of UV irradiation. Introducing the US to this oxidation process increased the efficiency of the overall process by increasing cavitation and the transformation of Fe^{2+} to Fe^{3+}. Enhancement in the production of Fe^{2+} further catalyzes the Fenton reaction to produce radical oxygen species (Kakavandi et al., 2024) (Fig. 10.1c).

Fenton reaction:

$$Fe^{2+} + H_2O_2 + H^+ \rightarrow Fe^{3+} + OH^{\bullet} + H_2O \quad (10.23)$$

Under UV irradiation:

$$Fe^{3+} + H_2O \rightarrow Fe(OH)^{2+} + OH^+ \quad (10.24)$$

$$Fe(OH)^{2+} \quad Fe^{3+} + OH^- \quad (10.25)$$

$$Fe(OH)^{2+} + h\nu \rightarrow Fe^{2+} + OH^{\bullet} \quad (10.26)$$

Under US treatment:

$$H_2O \rightarrow H^{\bullet} + OH^{\bullet} \quad (10.27)$$

$$FeOOH^{2+} + US \rightarrow Fe^{2+} + {}^{\bullet}HO_2 \quad (10.28)$$

10.2.6 Sulfate Radical-Based AOP

This technique shows promise in breaking down organic contaminants found in air, soil, and water. It works by generating reactive oxygen species, specifically $SO_4^{\bullet-}$, from powerful oxidants like PDS and PMS. $SO_4^{\bullet-}$ are particularly advantageous due to their high oxidizing power (E^0 = 2.60 V), stability, efficiency, selectivity over •OH, and relatively low cost. The sulfate radical advanced oxidation process (SR-AOP) shares some similarities with •OH-based methods, such as high oxidative power and short lifespan, but the reaction patterns of $SO_4^{\bullet-}$ differ significantly (Gaur et al., 2022).

The activation of peroxymonosulfate (PMS) or peroxydisulfate (PDS) can occur through various methods, including heating, UV irradiation, and transition metal activation (e.g., with Fe^{2+}, Co^{2+}, or Ag^{+} ions). These $SO_4^{\bullet-}$ can be generated through physical means, such as heat, UV light, and US waves, and chemical methods, like activation via transition metals, alkaline conditions, strong oxidizers, or electrochemical processes. The distinct characteristics of PMS and PDS, such as the bond length and energy of the O–O bond (1.453 Å for PMS and 1.497 Å for PDS), also influence their activation. PMS is composed of potassium hydrogen sulfate, potassium sulfate, and potassium peroxymonosulfate in a 1:1:2 ratio, while PDS contains sulfate and positive ions, giving them different properties for activation.

Carbon-based materials like AC, graphene, CNT, and biochar have recently been explored as effective PS and PMS activation catalysts. These materials act as electron donors, facilitating the generation of $SO_4^{\bullet-}$. Functional groups in these materials, such as quinones, carbonyl, and carboxyl, contribute to the process by forming delocalized π-electrons, enhancing PDS and PMS activation (Yan et al., 2023). This activation process involves electron transfer mechanisms, as depicted in relevant chemical equations, highlighting the roles of oxygen functional groups in generating $SO_4^{\bullet-}$. On the contrary, phenol was reportedly oxidized by activating PMS. The catalysts involved in the process were carbon nanocubes having well-dispersed FeN_x sites where the electron-deficient Fe centers receive electrons from PMS, generating highly reactive singlet oxygen $^1O_2$. The rapid phenol oxidation is an attribute of the nonradical pathway involving $^1O_2$ over the conventional radical pathway (He et al., 2021). Moreover, $PDS/FeCl_2 \cdot 4H_2O$ based AOPs serve to recover precious metals like Pd and Au from used catalysts (Ding et al., 2024).

10.3 Carbonaceous Materials in Catalyst Design

See Fig. 10.2.

10.3.1 Activated Carbon

AC is renowned for its exceptional adsorption properties, owing to its high surface area and porosity, allowing it to effectively capture a wide range of contaminants. However, over time, AC's adsorption capacity decreases as it becomes saturated with pollutants, necessitating regeneration or disposal. Various methods exist for regenerating AC, including thermal treatment, solvent extraction, biological degradation, and AOPs (Liu et al., 2024). Recent research has increasingly focused on AOPs for AC regeneration, highlighting a variety of techniques such as electrochemical, Fenton-based, persulfate-based, ozone-based, and photocatalytic regeneration. The effectiveness of these regeneration processes is influenced by the intrinsic properties of AC, such as particle size, pore structure, and surface

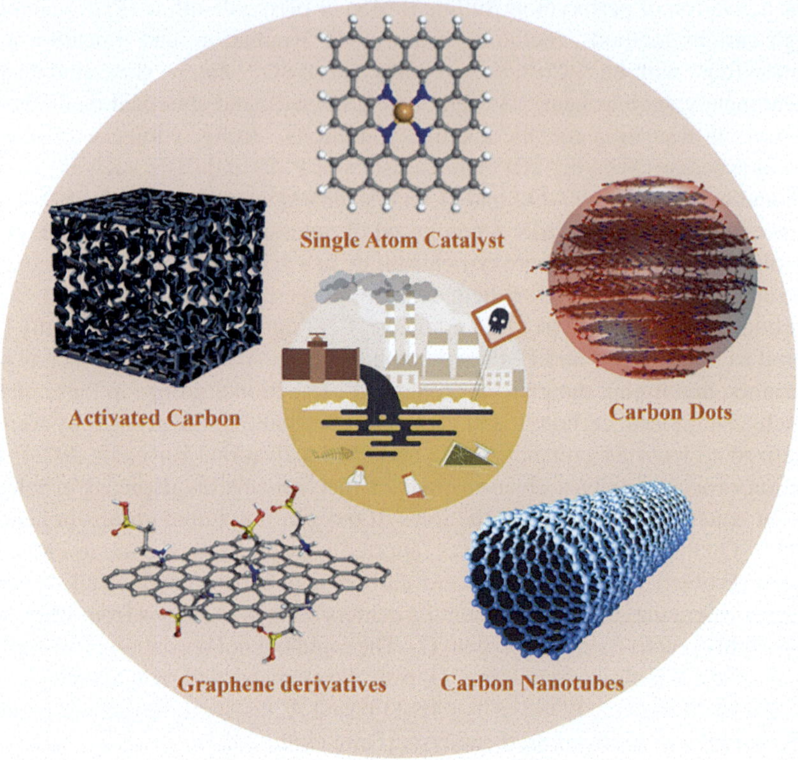

Fig. 10.2 Schematic representation of different classes of materials: Single-atom catalyst, CDs, CNTs, graphene derivatives, and AC used for water remediation. (Reproduced with permission from Xu et al. Copyright 2022, Springer; Fung et al. Copyright 2020, American Chemical Society and Ozyurt et al. Copyright 2023, Elsevier)

chemistry, which affect its interaction with contaminants and oxidants. Additionally, pH regulation is crucial in optimizing the balance between desorption and ROS degradation during regeneration.

10.3.2 Single-Atom Catalysts

SACs have emerged as transformative materials in AOPs and environmental catalysis due to their ability to maximize metal atom utilization, unique electronic properties, and ultralow metal loading requirements. By isolating single-atom (SA) active sites, SACs ensure that every metal atom is directly involved in catalytic activity, achieving optimal efficiency. These atomically dispersed metal sites also exhibit distinctive electronic characteristics that enhance reactant interactions, making them highly effective catalysts (Parambil et al., 2025).

A notable example is cobalt-based SACs, where atomically dispersed Co sites anchored on nitrogen-doped graphene (N-doped graphene) form Co–N_4 coordination structures (Zhu et al., 2025). These structures provide optimal binding energies for PMS, facilitating its activation and leading to remarkable catalytic performance. This enhanced activity stems from the tailored interaction between PMS and Co–N_4 sites, resulting in the efficient generation of $SO_4^{\bullet-}$ and $^{\bullet}OH$, powerful oxidants for degrading organic pollutants.

Similarly, carbon-supported iron-based SACs (Fe-SACs) have gained attention due to iron's abundance, affordability, and low toxicity compared to other transition metals like cobalt, nickel, and manganese. Fe-SACs have demonstrated a dramatic 33.2-fold improvement in PMS activation for phenol degradation compared to Fe nanoparticles on porous biochar. This extraordinary catalytic performance is attributed to the superior generation of reactive radicals at Fe single-atom sites compared to nanoparticle counterparts (Zhang et al., 2023).

The effectiveness of Fe-SACs, particularly those anchored on nitrogen-doped graphene (Fe-SA-NG), is tied to the Fe–N_4 active sites. These sites enhance radical generation through several mechanisms. First, the increased density of states near the Fermi level at Fe–N_4 sites facilitates electron transfer from the Fe single atom to PMS. Second, the Fe–N_4 sites are electron-rich, favoring the dissociation of PMS through O–O bond cleavage. Finally, the one-electron transfer process at Fe^{2+}-pyridinic N_4 sites is thermodynamically favorable, enabling the efficient production of $SO_4^{\bullet-}$ and $^{\bullet}OH$ radicals (Zhang et al., 2023).

These findings highlight the immense potential of SACs, particularly Fe- and Co-based systems, to revolutionize persulfate-based AOPs. Their atomic-level precision, efficient electron dynamics, and enhanced radical generation capabilities pave the way for innovative and sustainable solutions in water treatment and pollutant degradation technologies.

10.3.3 Carbon Dots

CDs are carbon-based nanoparticles with sizes below 10 nm, characterized by remarkable tunable photoluminescence, high quantum yield, and excellent biocompatibility. These features have made CDs versatile materials, leading to their integration into advanced semiconductor systems for applications such as medical imaging, environmental monitoring, and catalyst development. CDs exhibit significant potential as catalysts due to their highly adjustable surface functional groups, superior electron transfer and storage capabilities, and environmental friendliness (Parambil et al., 2024).

CDs are particularly effective in photocatalytic applications due to their strong sensitivity to visible light (Parambil & Rajamani, 2024). They enhance catalytic activity by functioning as photosensitizers or spectral converters, facilitating photoelectron transfer as both receptors and active catalysts. The principle of photocatalysis mirrors natural photosynthesis, where visible light photons (energy, *hν*) are absorbed to generate e^-–h^+. These photogenerated charge carriers lead to the formation of ROS, such as $^{\bullet}OH$ and $O_2^{\bullet-}$. These ROS are highly reactive, driving the degradation and mineralization of complex organic pollutants. Such capabilities make CDs promising materials in AOP for environmental remediation and sustainable chemistry (Hutton et al., 2016).

10.3.4 Carbon Nanotubes

CNTs, co-axial circular pipes formed from single or multiple layers of graphene sheets, play a crucial role in AOPs. Their carbon atoms are sp^2-hybridized, resulting in a large sp^2 orbital composition that grants CNTs remarkable mechanical properties, such as high modulus and strength. CNTs are categorized as SWCNTs or MWCNTs based on the number of layers stacked. SWCNTs are characterized by high chemical inertness and a simpler chemical structure, while MWCNTs exhibit greater stability, larger specific surface area (SSA), and more defect structures. These features enhance MWCNTs' ability to inhibit e^-–h^+ recombination, a critical aspect in catalytic processes. CNTs'unique properties have led to extensive use as green and efficient catalysts in AOPs. During catalytic ozonation, CNTs facilitate the sustained formation of the interphase "HO zone" and act as initiators in converting ozone (O_3) into $O_2^{\bullet-}$. Additionally, CNTs are effective in PDS activation, functioning as electron shuttles. In this context, CNTs transfer electrons from pollutants (acting as electron donors) to PDS (acting as electron acceptors), enabling organic pollutant degradation primarily through nonradical pathways without generating ROS (Peng et al., 2021).

The electrical conductivity of CNTs, attributed to their sp^2 hybridization and delocalized π-electron system, further enhances their catalytic performance. This high conductivity enables efficient electron transfer between electron donors (pollutants) and acceptors (PDS), simplifying organic pollutant degradation (Mallick & Roy, 2018). In nonradical processes, CNTs stabilize active species and catalyzed electron transfer. In radical processes, they act as electron donors and intermediaries, facilitating PDS activation. Structural defects on the CNT surface also serve as critical sites for PDS activation. These defects enhance adsorption and catalysis by altering the energy band structure and increasing the spin densities of carbon atoms. As a result, the adsorption capacity and electron transfer ability of CNTs are significantly improved, further underscoring their importance as catalysts in AOPs (Cruz et al., 2025).

10.3.5 Graphene Derivatives

Graphene derivatives, particularly GO, have shown significant promise for applications in AOP due to their unique structural and chemical properties. Graphene's inherent features, such as high conductivity, large specific surface area, and exceptional electron mobility, make it an excellent platform for carrier migration in photocatalytic reactions. However, the limitations posed by graphene's regular carbon atom arrangement necessitate modifications to tailor its properties for effective use in AOP. GO, a graphene derivative, has emerged as a versatile photocatalysis material offering several advantages (Cheruvathoor Poulose et al., 2023). It is synthesized with high yield and features abundant oxygenated functional groups, providing flexible solution processability. GO's surface chemistry enables it to act as a macromolecular surfactant, improving the dispersion of insoluble materials. Its two-dimensional structure and oxygen-rich functional groups serve as growth templates for composite materials with diverse morphologies. These properties enhance its compatibility with other materials in AOP applications.

GO's oxygen-containing groups (e.g., epoxy, carboxyl, hydroxyl) play a critical role in its photocatalytic performance. These groups render GO hydrophilic and dispersible in polar solvents, facilitating composite synthesis and interaction with pollutants. GO acts as an electron acceptor, mitigating electron–hole recombination and increasing the adsorption of organic pollutants through π–π interactions with the sp^2 regions of graphene. Additionally, its ability to create oxygen vacancies further enhances photocatalytic efficiency. GO's functional groups also allow for chemical modifications, including the immobilization of nanoparticles to form inorganic complexes, improving its AOP performance (Duan et al., 2018). By accelerating electron mobility at the interface and increasing pollutant adsorption rates, GO stands out as a highly effective and adaptable material for advancing the efficacy of AOP systems. Table 10.1 shows a comparison of different carbon-based materials in the degradation of contaminants.

Table 10.1 Comparison of different carbon-based materials in the degradation of contaminants with attributes, such as efficiency, mechanisms of action, targeted contaminants, and active species involved

Catalyst	Carbon support	Target reactants	Mechanism of action	Active species	Degradation efficiency (K)	Reference
A copolymer of dicyandiamide and L-cysteine	Carbon nitrides	Sulfamethoxazole	Visible light photocatalysis	$^{\bullet}O_2^-$ and h^+	0.106 min^{-1}	Zhou et al. (2019)
MIL-53(Fe)	MOF	Sulfamethoxazole	pF process	$^{\bullet}OH$, $^{\bullet}O_2^-$	0.022 min^{-1}	Ortega-Moreno et al. (2022)
N-doped graphene	Graphene derivative	Phenol	PMS activation	$SO_4^{\bullet-}$ and $^{\bullet}OH$	0.044 min^{-1}	Wang et al. (2016)
TiO_2/carbon dots	Carbon dots	Gemfibrozil	pF process	$^{\bullet}OH$, $^{\bullet}O_2^-$	0.101 min^{-1}	Chen et al. (2017)
Fe_3O_4@SiO_2/PAEDTC@MIL-101	MOF	Acid red 14 dye	spF process	h^+ and $^{\bullet}OH$	0.0327 min^{-1}	Al-Hawary et al. (2023)
FeCo@NOPC biochar	AC	Dimethyl phthalates	eF process	$^{\bullet}OH$	0.40 h^{-1}	Qu et al. (2024)

10.4 Potential Environmental and Industrial Applications

Based on the above, the primary challenges for advancing AOPs in wastewater treatment include developing efficient and cost-effective catalytic materials for large-scale treatment, powering the process with renewable energy sources while targeting emerging pollutants, and translating lab-scale processes to a commercial level. Despite advancements, several practical barriers hinder large-scale applications. These include the need for consistent monitoring of water quality parameters such as pH, chemical composition, and organic matter content, together with the difficulty in removing dissolved heavy metals; challenges in maintaining precise dosages and operational standards; high initial setup costs for machinery; and the need for skilled engineers to manage the excess $^{\bullet}OH$ required in the process. Additionally, removing residual H_2O_2 from treated water remains a complex task (Liu et al., 2024). Addressing these issues requires the development of innovative and effective solutions.

Future research on EAOPs should prioritize the integration of coupling technologies, the development of innovative materials, the reduction of energy consumption and costs, the minimization of toxicity, and the adoption of renewable energy sources for large-scale applications in pilot wastewater treatment plants. Advancements in these areas will enhance the sustainability of industrial wastewater treatment while supporting environmental protection and scientific and technological progress. Key research directions include: (i) integrating EAOPs with other wastewater treatment methods, such as bio-eF and EAOP-adsorption systems (Li et al., 2018), to create comprehensive solutions for degrading diverse pollutants; (ii) designing advanced catalyst materials with high catalytic activity and low environmental toxicity, particularly carbon-based materials, to improve efficiency and performance; (iii) developing electrode materials that optimize effectiveness, stability, and economic viability; (iv) reducing energy consumption and operational costs through improved reactor design, optimized process parameters, and pollutant-specific EAOP applications; and (v) assessing and mitigating the toxicity of byproducts while leveraging renewable energy sources to enhance sustainability (Olvera-Vargas et al., 2024). These advancements will ensure that EAOP systems are scalable, efficient, and viable for broader wastewater treatment applications, paving the way for future technological innovations.

Three-dimensional (3D) carbon-based electrodes have demonstrated significant efficiency in removing organic pollutants through the electro-Fenton (eF) mechanism. Traditional eF systems rely on two-dimensional (2D) electrodes, consisting of an anode and a cathode. Still, they suffer from several limitations, including poor mass transfer efficiency, low current utilization, and high energy consumption. These drawbacks hinder the overall effectiveness of pollutant degradation in wastewater treatment. The introduction of 3D electrode/eF technology addresses these issues by incorporating a particle electrode into the system, creating a more advanced electrochemical environment. This addition enhances the contact area between reactive species and pollutants, facilitating more efficient electrochemical reactions.

Specifically, the presence of the particle electrode accelerates the generation of hydrogen peroxide (H_2O_2) and ferrous ions (Fe^{2+}) during electrolysis. As a result, these intermediates react more rapidly to produce hydroxyl radicals ($^{\bullet}OH$), which are highly reactive and play a crucial role in the degradation of organic contaminants. This improvement significantly increases the production rate and overall yield of $^{\bullet}OH$, making the 3D electrode/eF system a more effective and energy-efficient solution for wastewater treatment (Li et al., 2021).

10.5 Conclusion, Challenges, and Future Perspective

Effectively mitigating environmental pollution from water bodies heavily depends on optimizing catalysts for stability and efficiency, particularly for large-scale industrial applications. Traditional modeling approaches such as density functional theory (DFT), microkinetic modeling, and quantum mechanics simulations have provided valuable insights into catalytic behavior and reaction mechanisms. However, their computational intensity and difficulty in accurately simulating complex catalytic systems present significant challenges. In contrast, artificial intelligence (AI), particularly ML, has emerged as a powerful tool to predict the catalytic degradation of carbon-based materials and unravel their underlying mechanisms (Aklilu & Bounahmidi, 2024). ML techniques offer unparalleled advantages, including rapid computation, adaptability, and the ability to process intricate, nonlinear interactions to design catalytic systems. By leveraging advanced models such as Gaussian process regression, random forests, gradient-boosting decision trees, and support vector machines, AI-driven approaches have demonstrated remarkable accuracy in predicting key performance indicators, optimizing catalytic processes, and performing computations many times faster. This transformative capability positions AI as an indispensable tool in advancing catalyst design and improving environmental remediation strategies.

References

Ahmad, K., Ghatak, H. R., & Ahuja, S. M. (2020). A review on photocatalytic remediation of environmental pollutants and H_2 production through water splitting: A sustainable approach. *Environmental Technology & Innovation, 19*, 100893. https://doi.org/10.1016/j.eti.2020.100893

Aklilu, E. G., & Bounahmidi, T. (2024). Machine learning applications in catalytic hydrogenation of carbon dioxide to methanol: A comprehensive review. *International Journal of Hydrogen Energy, 61*, 578–602. https://doi.org/10.1016/j.ijhydene.2024.02.309

Al-Hawary, S. I. S., Rahimpoor, R., Rahmani, A., Romero-Parra, R. M., Ramírez-Coronel, A. A., Alhachami, F. R., Mengelizadeh, N., & Balarak, D. (2023). Enhanced sonophotocatalytic degradation of acid red 14 using Fe_3O_4@ SiO_2/PAEDTC@ MIL-101 (Fe) based on metal-organic framework. *Catalysts, 13*(2), 411. https://doi.org/10.3390/catal13020411

Chen, P., Wang, F., Chen, Z.-F., Zhang, Q., Su, Y., Shen, L., Yao, K., et al. (2017). Study on the photocatalytic mechanism and detoxicity of gemfibrozil by a sunlight-driven TiO_2/carbon dots photocatalyst: The significant roles of reactive oxygen species. *Applied Catalysis B: Environmental, 204*, 250–259. https://doi.org/10.1016/j.apcatb.2016.11.040

Cheruvathoor Poulose, A., Medveď, M., Bakuru, V. R., Sharma, A., Singh, D., Kalidindi, S. B., Bares, H., et al. (2023). Acidic graphene organocatalyst for the superior transformation of wastes into high-added-value chemicals. *Nature Communications, 14*(1), 1373. https://doi.org/10.1038/s41467-023-36602-0

Cruz, D., Żółtowska, S., Savateev, O., Antonietti, M., & Giusto, P. (2025). Carbon nitride caught in the act of artificial photosynthesis. *Nature Communications, 16*(1), 374. https://doi.org/10.1038/s41467-024-55518-x

Ding, A., Li, M., Liu, C., Chee, T.-S., Yan, Q., Lei, L., & Xiao, C. (2024). Recovering palladium and gold by peroxydisulfate-based advanced oxidation process. *Science Advances, 10*(21), eadm9311. https://doi.org/10.1126/sciadv.adm9311

Duan, X., Sun, H., & Wang, S. (2018). Metal-free carbocatalysis in advanced oxidation reactions. *Accounts of Chemical Research, 51*(3), 678–687. https://doi.org/10.1021/acs.accounts.7b00535

Gaur, N., Dutta, D., Singh, A., Dubey, R., & Kamboj, D. V. (2022). Recent advances in the elimination of persistent organic pollutants by photocatalysis. *Frontiers in Environmental Science, 10*, 872514. https://doi.org/10.3389/fenvs.2022.872514

He, J., Wan, Y., & Zhou, W. (2021). ZIF-8 derived Fe–N coordination moieties anchored carbon nanocubes for efficient peroxymonosulfate activation via non-radical pathways: Role of FeNx sites. *Journal of Hazardous Materials, 405*, 124199. https://doi.org/10.1016/j.jhazmat.2020.124199

Hutton, G. A. M., Reuillard, B., Martindale, B. C. M., Caputo, C. A., Lockwood, C. W. J., Butt, J. N., & Reisner, E. (2016). Carbon dots as versatile photosensitizers for solar-driven catalysis with redox enzymes. *Journal of the American Chemical Society, 138*(51), 16722–16730. https://doi.org/10.1021/jacs.6b10146

Kakavandi, B., Ahmadi, M., Bedia, J., Hashamfirooz, M., Naderi, A., Oskoei, V., Yousefian, H., Kalantary, R. R., Pelalak, R., & Dewil, R. (2024). Metronidazole degradation mechanism by sono-photo-Fenton processes using a spinel ferrite cobalt on activated carbon catalyst. *Chemosphere, 358*, 142102. https://doi.org/10.1016/j.chemosphere.2024.142102

Kumari, P., & Kumar, A. (2023). Advanced oxidation process: A remediation technique for organic and non-biodegradable pollutant. *Results in Surfaces and Interfaces, 11*, 100122. https://doi.org/10.1016/j.rsurfi.2023.100122

Li, X., Chen, S., Angelidaki, I., & Zhang, Y. (2018). Bio-electro-Fenton processes for wastewater treatment: Advances and prospects. *Chemical Engineering Journal, 354*, 492–506. https://doi.org/10.1016/j.cej.2018.08.052

Li, H., Yang, H., Cheng, J., Chaoquan, H., Yang, Z., & Wu, C. (2021). Three-dimensional particle electrode system treatment of organic wastewater: A general review based on patents. *Journal of Cleaner Production, 308*, 127324. https://doi.org/10.1016/j.jclepro.2021.127324

Lin, J., Tian, W., Guan, Z., Zhang, H., Duan, X., Wang, H., Sun, H., Fang, Y., Huang, Y., & Wang, S. (2022). Functional carbon nitride materials in photo-Fenton-like catalysis for environmental remediation. *Advanced Functional Materials, 32*(24), 2201743. https://doi.org/10.1002/adfm.202201743

Liu, J., Chen, S., Ye, X., Chen, J., & Huang, J. (2024). Advanced oxidation processes for activated carbon regeneration: From fundamental to application. *ACS ES&T Water, 4*(8), 3119–3130. https://doi.org/10.1021/acsestwater.4c00280

Lu, F., & Astruc, D. (2020). Nanocatalysts and other nanomaterials for water remediation from organic pollutants. *Coordination Chemistry Reviews, 408*, 213180. https://doi.org/10.1016/j.ccr.2020.213180

Mallick, A., & Roy, S. (2018). Visible light driven catalytic gold decorated soft-oxometalate (SOM) based nanomotors for organic pollutant remediation. *Nanoscale, 10*(26), 12713–12722. https://doi.org/10.1039/C8NR03534B

Mallick, A., Kim, J., & Pumera, M. (2024). Magnetically propelled microrobots toward photosynthesis of green ammonia from nitrates. *Small, 21*, 2407050. https://doi.org/10.1002/smll.202407050

Mallick, A., Paul, S., Ben, T., Qiu, S., Verpoort, F., & Roy, S. (2021). Direct realization of an Operando Systems Chemistry Algorithm (OSCAL) for powering nanomotors. *Nanoscale*, 13(6), 3543–3551. https://doi.org/10.1039/D0NR06849G

Moradi, M., Elahinia, A., Vasseghian, Y., Dragoi, E.-N., Omidi, F., & Khaneghah, A. M. (2020). A review on pollutants removal by Sono-photo-Fenton processes. *Journal of Environmental Chemical Engineering, 8*(5), 104330. https://doi.org/10.1016/j.jece.2020.104330

Mukherjee, J., Lodh, B. K., Sharma, R., Mahata, N., Shah, M. P., Mandal, S., Ghanta, S., & Bhunia, B. (2023). Advanced oxidation process for the treatment of industrial wastewater: A review on strategies, mechanisms, bottlenecks and prospects. *Chemosphere, 345*, 140473. https://doi.org/10.1016/j.chemosphere.2023.140473

Olvera-Vargas, H., Trellu, C., Nidheesh, P. V., Mousset, E., Ganiyu, S. O., Martínez-Huitle, C. A., Zhou, M., & Oturan, M. A. (2024). Challenges and opportunities for large-scale applications of the electro-Fenton process. *Water Research, 266*, 122430. https://doi.org/10.1016/j.watres.2024.122430

Ortega-Moreno, G. A., Ayala-Duran, S. C., Barbero, B. P., Narda, G. E., Bernini, M. C., & Pupo Nogueira, R. F. (2022). Photo-Fenton degradation of sulfamethoxazole using MIL-53 (Fe) under UVA LED irradiation and natural sunlight. *Journal of Environmental Chemical Engineering, 10*(3), 107678. https://doi.org/10.1016/j.jece.2022.107678

Parambil, A. M., & Rajamani, P. (2024). Carbon dots: A promising path towards environmental sustainability. *Environmental Science: Advances, 3*(11), 1513–1523. https://doi.org/10.1039/D4VA00273C

Parambil, A. M., Priyadarshini, E., Goutam, R., Tsai, P.-C., Huang, P.-C., Rajamani, P., Lin, Y.-C., & Ponnusamy, V. K. (2024). Self-assembled mesoporous silica decorated with biogenic carbon dot nanospheres hybrid nanomaterial for efficient removal of aqueous Methoxy-DDT via a Short-Bed Adsorption column technique. *Environmental Research, 260*, 119653. https://doi.org/10.1016/j.envres.2024.119653

Parambil, A. M., Priyadarshini, E., Paul, S., Bakandritsos, A., Sharma, V. K., & Zboril, R. (2025). Emerging Nanomaterials for Per-and Polyfluorinated Substances Detection–A Review. Journal of Materials Chemistry A., https://doi.org/10.1039/D4TA07045C

Peng, J., He, Y., Zhou, C., Su, S., & Lai, B. (2021). The carbon nanotubes-based materials and their applications for organic pollutant removal: A critical review. *Chinese Chemical Letters, 32*(5), 1626–1636. https://doi.org/10.1016/j.cclet.2020.10.026

Petrucci, E., Da Pozzo, A., & Di Palma, L. (2016). On the ability to electrogenerate hydrogen peroxide and to regenerate ferrous ions of three selected carbon-based cathodes for electro-Fenton processes. *Chemical Engineering Journal, 283*, 750–758. https://doi.org/10.1016/j.cej.2015.08.030

Qu, S., Guo, W., Mou, H. D., Wang, R., Wei, M., & Hu, X. (2024). Improvement of heterogeneous electro-Fenton oxidation of dimethyl phthalate by embedding Fe/Co bimetallic active sites into waste wine lees biochar. *Journal of Water Process Engineering, 66*, 105966. https://doi.org/10.1016/j.jwpe.2024.105966

Sun, Y., Zhou, P., Zhang, P., Meng, S., Zhou, C., Liu, Y., Zhang, H., Xiong, Z., Duan, X., & Lai, B. (2022). New insight into carbon materials enhanced Fenton oxidation: A strategy for green iron (III)/iron (II) cycles. *Chemical Engineering Journal, 450*, 138423. https://doi.org/10.1016/j.cej.2022.138423

Usman, M., Monfort, O., Gowrisankaran, S., Hameed, B. H., Hanna, K., & Al-Abri, M. (2023). Dual functional materials capable of integrating adsorption and Fenton-based oxidation processes for highly efficient removal of pharmaceutical contaminants. *Journal of Water Process Engineering, 52*, 103566. https://doi.org/10.1016/j.jwpe.2023.103566

Van Nguyen, H., Pham, S. T., Vu, T. N., Van Nguyen, H., & La, D. D. (2024). Effective treatment of 2, 4, 6-trinitrotoluene from aqueous media using a sono–photo-Fenton-like process with a zero-valent iron nanoparticle (nZVI) catalyst. *RSC Advances, 14*(33), 23720–23729. https://doi.org/10.1039/d4ra03907f

Wang, C., Kang, J., Sun, H., Ang, H. M., Tadé, M. O., & Wang, S. (2016). One-pot synthesis of N-doped graphene for metal-free advanced oxidation processes. *Carbon, 102*, 279–287. https://doi.org/10.1016/j.carbon.2016.02.048

Yan, Y., Wei, Z., Duan, X., Long, M., Spinney, R., Dionysiou, D. D., Xiao, R., & Alvarez, P. J. J. (2023). Merits and limitations of radical vs. nonradical pathways in persulfate-based advanced oxidation processes. *Environmental Science & Technology, 57*(33), 12153–12179. https://doi.org/10.1021/acs.est.3c05153

Yang, Z., Xie, Y., Zhang, Z., Gu, R., & Xie, M. (2025). CuO nanowire array induced thermal effect from long wavelength light enhances photo-carrier transport in FeOCl for improved photo-Fenton reaction. *Materials Letters, 382*, 137831. https://doi.org/10.1016/j.matlet.2024.137831

Zhang, J., Tang, X., Hong, Y., Chen, G., Chen, Y., Zhang, L., Gao, W., Zhou, Y., & Sun, B. (2023). Carbon-based single-atom catalysts in advanced oxidation reactions for water remediation: From materials to reaction pathways. *Eco-Environment & Health, 2*(2), 47–60. https://doi.org/10.1016/j.eehl.2023.04.002

Zhou, C., Zeng, Z., Zeng, G., Huang, D., Xiao, R., Cheng, M., Zhang, C., et al. (2019). Visible-light-driven photocatalytic degradation of sulfamethazine by surface engineering of carbon nitride: Properties, degradation pathway and mechanisms. *Journal of Hazardous Materials, 380*, 120815. https://doi.org/10.1016/j.jhazmat.2019.120815

Zhu, Z.-S., Wang, Y., Wang, P., Zhong, S., Hu, K., Ren, S., Vongsvivut, J. P., Sun, H., Duan, X., & Wang, S. (2025). Multidimensional engineering of single-atom cobalt catalysts for ultrafast Fenton-like reactions. *Nature Water, 3*, 1–11. https://doi.org/10.1038/s44221-024-00382-8

Ziembowicz, S., & Kida, M. (2022). Limitations and future directions of application of the Fenton-like process in micropollutants degradation in water and wastewater treatment: A critical review. *Chemosphere, 296*, 134041. https://doi.org/10.1016/j.chemosphere.2022.134041

Chapter 11
Carbon Nanomaterials as Efficient Anti-microbial Agent Against Pathogens Including MDR Strains

Paridhi Sharma, Saksham Baliyan, and Nisha Gaur

Abstract The rapid rise in anti-microbial resistance (AMR) among pathogens from excessive antibiotic use presents a danger to global public health. Over 79% of known pathogens have been suggested to have developed resistance toward one or more anti-microbial agents, causing researchers to explore innovative solutions to this problem. Carbon nanomaterials such as carbon nanotubes, quantum dots, graphene, and fullerenes ranging between 1 and 100 nm in dimensions have been presented as promising avenues for developing new strategies to combat a broad spectrum of pathogens, including multi-drug resistance strains. The small size, substantially large surface-to-volume ratio, and higher interior volume as other diverse physicochemical features of these nanomaterials have been studied to disrupt microbial membranes physically, generate oxidative stress, and cause biofilm penetration, thus reducing the likelihood of resistance development. Carbon nanoparticles like fullerene have also been suggested to inhibit bacterial growth by impairing the respiratory chain and inhibiting metabolic processes. In addition to this, these nanomaterials also tend to act synergistically with antibiotics and other poly-indole nanocomposites to enhance anti-microbial efficiency. Despite their advantages over conventional anti-microbials, nanomaterials have been suggested to cause cellular toxicity and ecotoxicity. This chapter will explore the current landscape of anti-microbial resistance and how carbon nanomaterials could augment the efficacy of traditional antibiotics. We will also explore the mechanistic insights into the anti-microbial properties of these nanomaterials, their design, their applications, and their associated challenges.

Keywords Carbon nanoparticles · Graphene dots · Multi-drug resistance (MDR) · Anti-microbial resistance (AMR) · Reactive oxygen species (ROS) · Biofilm disruption · Nanotoxicity

---

P. Sharma · S. Baliyan · N. Gaur (✉)
School of Biotechnology, Gautam Buddha University, Greater Noida, India

A. M. Parambil et al. (eds.), *Carbon*,
https://doi.org/10.1007/978-3-031-90613-8_11

Abbreviations

AMR	Anti-microbial resistance
BSA	Bovine serum albumin
CBNs	Carbon-based nanoparticles
CD	Carbon nanodots
CNT	Carbon nanotubes
GO	Graphene oxide
MDR	Multi-drug resistance
MWCNT	Multi-walled carbon nanotubes
PEG	Polyethylene glycol
ROS	Reactive oxygen species
SAE	Single-atom advanced engineering
SEM	Scanning electron microscopy
SWCNT	Single-walled carbon nanotubes
TEM	Transmission electron microscope

11.1 Introduction

Since their discovery in 1928 by Alexander Fleming and commercialization, anti-microbial agents have revolutionized modern medicine and altered the therapeutic paradigm. However, this medicine's increased usage and ease of availability have given rise to resistance in many microorganisms (Fig. 11.1). Thus, it threatens the effective prevention and treatment of a wide spectrum of infections caused by bacteria, fungi, viruses, or parasites. Anti-microbial resistance (AMR) today has become a global problem, contributing to 9% of all mortalities (Ho et al., 2024; Prestinaci et al., 2015). AMR is the phenomenon by which different microorganisms evolve to develop special mechanisms to resist the actions of these drugs. Among the anti-microbial-resistant microorganisms, multi- and pan-drug-resistant pathogens are becoming more prevalent. This has been more commonly observed in the case of "gram-positive bacteria like *S. aureus, S. pneumoniae, E. faecium,* and *E. faecalis* as well as in gram-negative bacteria viz. *A. baumannii, E. coli, P. aeruginosa,* and *K. pneumonia*." This has been facilitated by horizontal gene transfer through plasmids in the environment. Many microbes have designed specific features to restrain themselves from the effects of antibiotics, like efflux pumps, reduced permeability of the LPS layer, secretion of degrading enzymes, and alteration of receptors (Bharadwaj et al., 2022).

The swift spread of multi-drug resistance (MDR) requires a pressing need to craft innovative approaches to control infectious diseases. The ineffectiveness of traditional antibiotics against these evolved microbes has led to treatment failure besides long-term disease. In this crisis, a multi-target strategy is needed to develop new anti-microbials to overcome existing resistance mechanisms. One promising

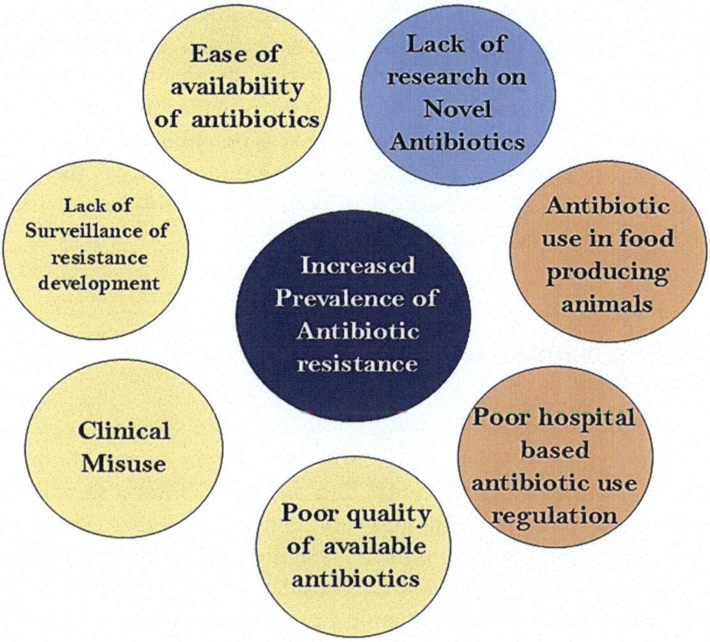

Fig. 11.1 Different factors contributing to an increased prevalence of antibiotic resistance

approach is to explore novel compounds obtained from natural sources such as plants, fungi, and marine organisms that may have unique properties that make them effective against resistant strains. In addition, advances in biotechnology and genomics can accelerate the development of specific anti-microbials that target bacterial virulence factors rather than simply kill bacteria, thereby preserving the beneficial microbiome and reducing selective pressure for resistance. Integrating these strategies with innovative delivery systems, such as nanotechnology-based carriers, can improve the pharmacokinetics and bioavailability of new anti-microbial, making them more effective against resistant organisms (Salesa et al., 2019).

Recent studies have indicated the potential of nanomaterials as anti-microbial agents. The small size of these particles (1–100 nm) allows them to pass through cellular barriers easily and disrupt various biological processes. Depending on the "size, place of origin, structural configuration, pore size, and potential toxicity, there could be a variety of nanoparticles such as carbon nanoparticles, metal oxides, liposomes, composites, etc." Nanomaterials find diverse usage due to their diverse properties. Studies have suggested nanomaterials to help drug and gene delivery improve therapeutic efficacy, fertilizer development, adjuvants in vaccine development, cancer diagnosis, medical device coatings, etc. Among these groups, carbon nanoparticles are the most diverse class of nanomaterials, owing to the remarkable properties and characteristics of atomic bonding shown by carbon atoms. Carbon nanomaterials like carbon nanotubes (CNTs) and nanographene have been found to aid in gene delivery, drug delivery, and carbon therapy, while fullerenes have been shown to have antiviral properties. However, carbon nanomaterials have not yet been fully explored for their antibacterial properties, especially against antibiotic-resistant microbes (Mekuye & Abera, 2023). This chapter provides a comprehensive overview of the effectiveness of carbon nanoparticles as potential therapeutics and adjuvants against anti-microbial-resistant microbes. We also discuss the mechanistic insights into carbon nanomaterial-mediated targeting of multi-drug resistant organisms and touch upon the relevant clinical works.

11.2 Carbon Nanomaterials

Carbon is the most profuse element on Earth, with extraordinary properties (Maduraiveeran & Jin, 2021; Rao et al., 2021). It exists in diverse allotropic forms. Its peculiar electronic configuration ($1s^2$, $2s^2$, and $2p^2$) allows it to undertake thermodynamically favorable conformations varying from sp^3 to sp^2 to sp, depending on the environmental conditions, to produce crystalline as well as amorphous entities such as nanodiamonds and graphene sheets besides allowing it to interact with other non-metallic atoms to form varied chemical compounds (Georgakilas et al., 2015; Sangnim et al., 2024). Carbon-based nanoparticles (CBNs) have gained interest across various academic fields because of their distinct physicochemical and structural characteristics. Carbon-based nanomaterials are gaining popularity because of their small size and potential characteristics, including mechanical, chemical, thermal, magnetic, electronic, catalytic, and optical properties. In addition, they have high surface area and high carrier mobility. These properties make them useful in various nanotechnology fields and can interact with biological systems (Makhado et al., 2024). A wide range of carbon nanoparticles exists today with varying dimensions, such as "fullerenes, nanodots, carbon nanotubes (CNTs), nanofibers, C_{60}, graphene, amorphous carbons, and nanocomposites like chitosan" (Fig. 11.2)

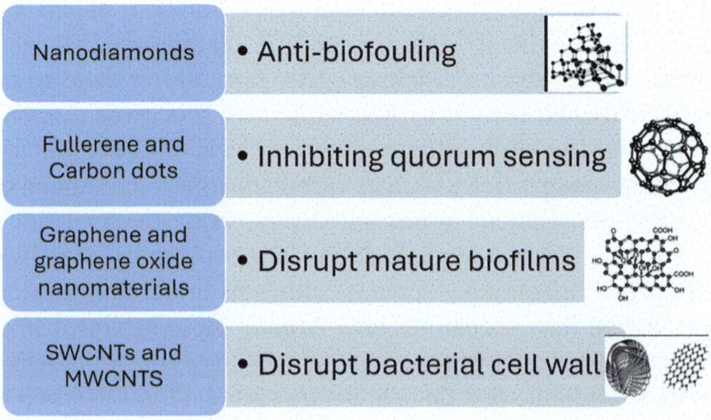

Fig. 11.2 Various types of carbon-based nanomaterials and their anti-microbial properties

Table 11.1 Different types of carbon nanomaterials

Carbon nanomaterial	Structure	Anti-microbial mechanism	Key characteristics	References
Carbon nanotubes (CNTs)	Cylindrical, rolled graphene sheets	Direct membrane disruption, cell viability reduction	Diameter 1–100 nm, types: SWCNT, MWCNT, etc.	Coville et al. (2011)
Carbon nanofibers (CNFs)	Cylindrical, stacked graphene layers	Large surface area interaction with microbial cells	Wider than CNTs (>100 nm), synthesized at lower temperatures	Johnston et al. (2010)
Fullerenes	Spherical, soccer ball-like structure	Reactive oxygen species (ROS) generation	60 carbon atoms, 12 pentagons, 20 hexagons	Al-Jumaili et al. (2017)
Carbon spheres	Spherical, single or multi-layer	Surface adsorption and functional group interactions	Diameter 10 nm to 10 μm, high surface area	Yang et al. (2014)
Graphene dots	sp^2 hybridized carbon atoms arranged in two-dimensional structure	Membrane disruption, ROS, adsorption of bacterial cell, biofilm disruption	Diameter >10 nm, high surface area, photoluminescence	Minsu et al. (2020)
Graphene oxide	Single-atomic layer	Membrane disruption, ROS, adsorption of bacteria, inhibition of biofilm formation	Diameter ~14 nm per carbon atom, high surface area, water solubility	Minsu et al. (2020)
Nanobiochar	Carbon rich, porous	Adsorption, cell membrane disruption, ROS	Diameter 1–100 nm, several atomic layers, high surface area	Chausali et al. (2021)

(Maduraiveeran & Jin, 2021). The important carbon nanomaterials are described in Table 11.1. "Nanotubes are the most common and exist in different sizes and characteristics, including single-walled carbon nanotubes (SWCNTs) and multi-walled carbon nanotubes (MWCNTs), among others. Both nanotubes and nanocomposites exhibit various anti-microbial properties" (Sangnim et al., 2024). "Kang et al. (2007) were the first to indicate that SWCNTs could serve as effective anti-microbial agents, capable of damaging cell membranes via direct contact and reinstating cell viability by 80% (Kang et al., 2007)." "Another study in 2008 indicated that the main way bactericidal CNTs operate is by damaging cell membranes. The author discovered that pathogens treated with CNTs showed signs of oxidative stress, leading to damaged cell membranes and the release of intracellular components (Kang et al., 2008)." Aslan et al. noted that the toxicity of shorter SWCNTs is increased due to the higher density at the open ends of the tubes, and smaller diameter nanotubes are more likely to interact with the cell surface, disrupting the underlying cell membrane (Johnston et al., 2010).

Fullerenes are known to have high anti-microbial activity due to their unique structure. They appear as closed-cell structures that communicate via electrons.

When irradiated with photons, C60 is elevated from its ground state to a transient (around 1.3 ns) excited state, rapidly transitioning to a lower-level triplet state. In the presence of molecular oxygen ($3O_2$), fullerenes are capable of producing singlet oxygen ($1O_2$) through the photochemical energy transfer route and reactive oxygen species (ROS), such as superoxide anion (O_2^-) through the electron transfer route. ROS are typically viewed as accountable for lipid peroxidation and membrane deterioration in eukaryotic cells. Elevated amounts of ROS are lethal to microorganisms, harming cellular components like lipids, proteins, and nucleic acids (Al-Jumaili et al., 2017). Several studies have reported the potential of environmentally friendly and steam-activated colloidal carbon spheres. "The surface of activated carbon spheres is abundant in functional groups and has porous structures, facilitating the adsorption of organic substances onto the carbon spheres." The carbon spheres' porous structure contains dodecyltributylphosphonium bromide. "Phosphonium carbon spheres demonstrate remarkable thermal stability and show significant antibacterial properties, with MICs of 150 and 100 mg/L against *E. coli* and *S. aureus*, respectively" (Yang et al., 2014). The antibacterial effectiveness of CNTs is influenced not only by their physicochemical characteristics but also by various factors, including the quantity of bacteria employed, duration of treatment, and its type (Haung et al., 2013). CNTs have been widely used to grow or coat antibiotics, with various factors such as length, size, composition, graphene layer count, dispersion, and agglomeration significantly influencing their antibacterial qualities (Ayanda et al., 2024; Xin et al., 2019). Carbon nanodots (CDs) are also emerging as a new direction in nanomaterials research. CDs are semiconductor nanoparticles of 1–10 nm, which are much more efficient than quantum dots (QDs) and provide optical and electrical activities. "CDs are superior in terms of chemical inertness, photon-induced electron transfer, and photoluminescence behavior, as well as improved biocompatibility and reduced toxicity." In addition, CDs are made from environmentally friendly raw materials compared to chemicals such as cadmium (Cd) and selenium (Se). Therefore, CDs are profitable, environmentally friendly, and produce less waste (Ajith et al., 2022). Moreover, the CDs can be functionalized by adding different organic groups, thus enhancing their activity (Li et al., 2022). The optical activity of CDs provides safe and effective anti-microbial activity, especially through direct interaction. Due to gene expression changes, the direct interaction of CDs with bacterial cells can result in ROS production, cell structure degradation, and cytoplasmic content leakage (Varghese & Balachandran, 2021).

In addition, biological CDs have attracted attention as a promising direction in nanomaterial research. Parambil et al. (2024) presented a unique method to produce CDs from *F. benghalensis* L. extract. "Biogenic CD demonstrated remarkable broad-spectrum anti-microbial activity against a variety of pathogens, including MDR strains." They observed that the regulation of ROS in bacterial cells was disturbed, resulting in lipid peroxidation of membranes, loss of membrane potential, and membrane breakage. This resulted in cytoplasmic leakage. Advanced imaging and microscopy uncovered membrane, structure, and morphology alterations in *Pseudomonas aeruginosa* and *Enterobacter cloacae* (Parambil et al., 2024). Graphene oxide (GO) is another interesting nanomaterial that has attracted considerable attention from researchers. "It is produced through the oxidation of graphite, resulting in multifunctional two-dimensional sheets rich in oxygen-containing functional groups like hydroxyl, epoxy, and carboxyl." This chemical alteration increases the solubility of GO in water and other polar solvents, making it suitable for numerous applications in medicine, electronics, and environmental science. The anti-microbial properties of GO are attributed to several interrelated mechanisms that can exert lethal effects on various microorganisms, such as bacteria, fungi, and viruses (Liu et al., 2011). Recently, graphene quantum nanodots (GQDs) or graphene dots have also attracted considerable attention as anti-microbial agents. "This is primarily because of their distinct characteristics: small size (<10 nm), spatially organized sp^2-hybridized carbon atoms in a two-dimensional configuration, and electrical, mechanical, and thermal attributes. One of the interesting features of GQDs is their large surface area and high permeability, which makes them highly reactive and efficient." In addition, graphene dots are richly composed of various oxygen-containing functional groups, which can enhance their solubility in aqueous media and facilitate their functionalization for specific applications. Various studies have shown that GQDs exert their anti-microbial effects by disrupting and lysing cell membranes and generating ROS upon exposure to light, which induces oxidative stress and consequently impairs cell viability. The increased surface area of GQDs allows them to be strongly adsorbed onto the bacterial surface, inhibiting their metabolic activity and enhancing their antibacterial efficacy (Minsu et al., 2020). Studies have also shown that GO exhibits some antibacterial activity while scavenging metal ions (single-atom nanomaterials).

Single-atom advanced engineering (SAE) holds tremendous potential to enhance materials with previously unattainable properties and has wide-ranging applications in catalysis, energy storage, and biomedicine. One of the most important aspects of SAE is the discovery of dynamic sites with rich and sophisticated metal coordination sites for single-atom attachment (SA). The location and other properties, such as conductivity and biocompatibility, determine and control SA's ultimate nature and function (tight binding, oxidation state, space density, multi-level cooperativity, etc.). Substrates fabricated from graphene-based nanomaterials are generally attractive. This is because it is composed of light, diverse, non-toxic elements and has a large surface area, semiconductor properties, and good stability. Graphene also provides efficient charge transfer between metal atoms and substrates, stabilizing rare oxidation states. SAE has so far allowed anti-microbial agents to function only through intermediate pathways, which involve the production of highly reactive and toxic species (viz. hydroxyl radicals or superoxide) and application of laser light for photothermally enhanced antibacterial activity (Panáček et al., 2024). In addition, nanobiochar has also shown remarkable antibacterial properties. It is obtained through biomass pyrolysis and has a large surface area and unique structural properties. This has been shown to enhance interaction with microbial cells. This antibacterial effect is thought to be due to cell membrane and metabolic disturbances. Studies have shown that the cell viability of various known pathogens, such as *E. coli*, is significantly reduced. "Exposure to nanobiochar causes the growth of *Escherichia coli, Staphylococcus aureus,* and *Candida albicans*, making this nanomaterial a promising candidate for use in a variety of fields, including soil remediation, wastewater treatment, and the development of anti-microbial coatings" (Chausali et al., 2021).

11.3 Mechanisms of Anti-microbial Action of CNTs

Carbon nanotubes (CNTs) exhibit anti-microbial properties via multiple mechanisms: (1) Interaction with membranes and physical disintegration:

(a) CNTs can engage with microbial membranes, leading to physical disruption and destabilization. These interactions compromise the integrity of the cell membrane, ultimately resulting in cell death.
(b) Oxidative stress via producing reactive oxygen species (ROS): CNTs trigger oxidative stress in microbial cells by producing ROS. This oxidative stress damages DNA, proteins, and lipids, leading to the demise of microbial cells.
(c) Biofilm Penetration: CNTs enhance the effectiveness of anti-microbial action by penetrating microbial biofilms and disrupting their structure.
(d) Interference with Microbial Metabolic Pathways: CNTs hinder microbial metabolic pathways, interfering with essential processes and resulting in the death of microbial cells.

These diverse anti-microbial properties make CNTs promising agents for combating microbial infections (Saleemi et al., 2021). Table 11.2 describes how different CBNs can help in anti-microbial activity. CNTs attach to the bacterial cell surface and facilitate electron transfer across the cell membrane, damaging the cell wall and cell membrane, leading to protein malfunction, and producing DNA damage and ROS as by-products during distribution into the bacterial cell. Research has demonstrated that damaging the microbial cell membrane causes the release of intracellular components, resulting in bacterial cell death. Kang et al. (2007) provide the initial proof of damage to bacterial cell membranes, as mentioned earlier. They observed the DNA, RNA, and protein leakage caused by direct interaction between SWCNTs and pathogens. Another research also indicated lateral interactions involving SWCNTs and *Salmonella*.

They found alterations in cell structure, permeability, and cell membrane because of CNT observation, which often leads to distortion of cell morphology. Therefore, "the demise of bacterial cells resulting from the antibacterial properties of CNTs begins with distortion and destruction of the cell membrane, causing the release of intracellular substances." Nevertheless, while various other studies have noted that 60 min of interaction between SWCNTs and bacteria damage the cell membrane, some studies have indicated that extended contact durations are necessary to achieve similar outcomes, as Ayanda et al. (2024) highlighted. The exact mechanisms and macromolecule-level interaction are not yet deciphered. An illustration of different mechanisms is provided in Fig. 11.3.

Table 11.2 Anti-microbial activity of different carbon-based nanomaterials

Carbon nanomaterial	Against microbes	MIC	Mechanism of action	Toxicity and environmental stress	References
Graphene oxide (GO)	*E. coli, S. aureus*	5–10 μg mL^{-1}	Physical contact with the cell membrane promotes cell membrane rupture and results in the creation of cell-GO aggregates	Low toxicity; potential accumulation to the organism	Berry et al. (2017), Maleki Dizaj et al. (2015)
Fullerene	*E. coli, Salmonella,* and *Streptococcus* spp.	1–50 μg mL^{-1}	Hinder the respiratory chain to stop the growth of microorganisms and inhibit energy metabolism	Low cytotoxicity: Lack of toxicity to anaerobic microbes but significantly impacts microorganisms in soils	Berry et al. (2017), Maleki Dizaj et al. (2015)
SWTNs	*E. coli*	50 μg mL^{-1} 200 μg mL^{-1}	Physical contact with the cell membrane creates cell CNT aggregates and stimulates cell membrane disruption	Potential cytotoxicity: Environmental stress depends on dispersion and functionalization	Arias and Yang (2009), Maleki Dizaj et al. (2015), Saleemi et al. (2021)
Carbon nanotubes (CNTs)	*E. coli*	10 μg mL^{-1} 20 μg mL^{-1}	Disruption of the cell membrane, induction of oxidative stress by the generation of ROS and hazardous compounds	Their low toxicity to 1321N1 human astrocytoma cells allows them to be used in biological applications, but they also have toxicological impacts on the environment	Azizi-Lalabadi et al. (2020), Liu et al. (2018)
Fullerene C60	*E. coli, S. aureus*	1–50 μg mL^{-1}	The role of electrostatic attraction is significant	Moderate toxicity; risk to aquatic organism	Snow et al. (2012)

(continued)

Table 11.2 (continued)

Carbon nanomaterial	Against microbes	MIC	Mechanism of action	Toxicity and environmental stress	References
Fullerene C70	MS2 bacteriophage viruses	1–50 μg mL^{-1}	AgNPs and C70 target bacterial cells that produce more ROS from photosynthesis	Low toxicity; potential bioaccumulation risk to the organism	Snow et al. (2012)
Multi-walled carbon nanotubes (MWCNTs)	*E. coli, S. aureus*	Less than 500 μg mL^{-1}	Bacterial adhesion or deposition onto bacterial cell	Indicating low microbial toxicity; persistent environment stress concern	Arias and Yang (2009)
SN-CNPs	*E. coli* and *L. lactis*	47 μg mL^{-1} and 188 μg mL^{-1}	Interaction with SN-CNPs drastically changed the cell's outer surface, making the impacted cells much rougher on the outside	Further research is needed to assess the environmental stress	Nurtay et al. (2023)
F-MWNTs	*Escherichia coli Staphylococcus aureus*	80 μg/mL and 60 μg/mL	Injury of the cell membrane, inactivation of enzymatic activity, and inhibition of metabolic performance	Oxidative stress leading to cell death	Abo-Neima et al. (2020)
Graphene oxide	*Staphylococcus aureus*	16 mg/mL	Inhibit metabolic activity, multiplication, and colonization	Not yet studied	Gilsanz-Muñoz et al. (2024)
Silver-coordinated fullerenes (Ag-C_{60}-Cl and Cu-C_{60}-Cl)	*E. coli* and methicillin-resistant *Staphylococcus aureus* (MRSA)	3.906 μg/mL	ROS production and disruption of cell membrane	Not yet studied	Ibrahim et al. (2024)
Fullerenes on nanodiamond	*E. coli* and *Staphylococcus aureus*	175 μg/mL	Biofilm disruption	Low toxicity	Bolshakova et al. (2023)

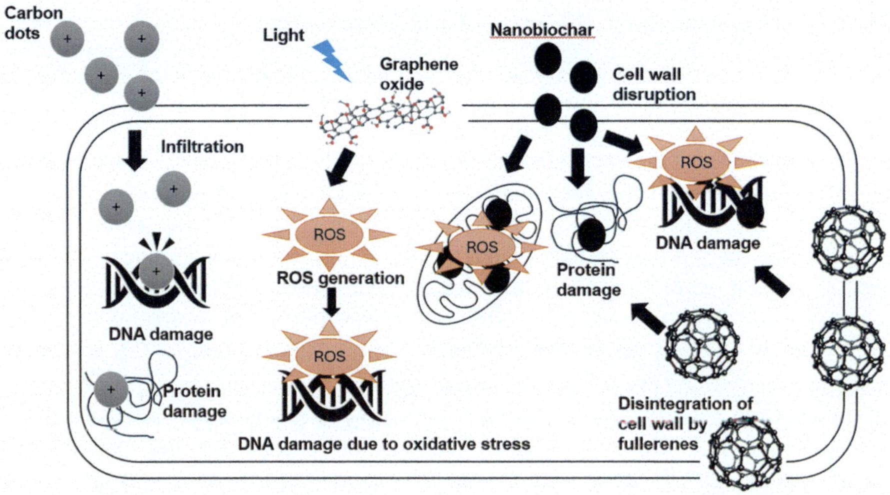

Fig. 11.3 Different anti-microbial actions of carbon nanomaterials: Physical disruption of cell membranes, oxidation of proteins, inhibition of biofilm formation, DNA damage, interruption of electron transport, and production of reactive oxygen species (ROS)

Carbon nanodots (CDs) have recently been studied for their antibacterial properties. This is mainly confirmed regardless of whether they are photoexcited. "In the absence of photoexcitation, CDs adhere to bacterial and fungal cell walls via diffusion and electrostatic interactions, altering the permeability of the cell wall and membrane." Cytoplasm escapes when CDs infiltrate the cell membranes of bacterial and fungal cells, and the binding of DNA and RNA by CDs damages the nucleic acid structure, thereby hindering bacterial growth. However, these CDs produce ROS during photoexcitation, leading to oxidative harm to bacterial cells. This alters the expression pattern of vital genes, leading to the death of bacterial cells. When photoexcitation occurs, these CDs generate ROS, which causes oxidative damage to bacterial cells. Electroplating CDs containing antibiotics as precursors using the detected light has enhanced antibacterial activity, providing a more capable alternative to combat multidrug-resistant bacteria. Similarly, "graphene oxide (GO), an isomorph of carbon, also interacts with bacterial cell membranes to propagate antibacterial activity, inducing ROS production, resulting in oxidative stress, DNA fragmentation, and subsequent microbial cell death" (Anand et al., 2019). It has also been noted that a cell membrane damage mechanism was created by directly trimming the sharp edges of GO nanosheets. The mechanical coating of bacterial cells results in cell lysis and the removal of phospholipids from the cell membrane. Therefore, the harmonious results of these mechanisms suggest that GO also acts as an anti-microbial agent similar to other CBNs, namely CNTs and fullerenes, which are likely toxic to humans but also have insolubility limitations (Nishshankage et al., 2024).

11.3.1 Disruption of Bacterial Cell Membrane

CBN tends to disrupt and destabilize cell membranes. This has been made clear through extensive microscopic analysis using SEM/TEM. The size, structure, and shape of these CBNs are very important for their antibacterial activity. Several studies have demonstrated the very sharp edges of GO under a microscope. "These sharp edges allow them to interact with bacteria such as *Staphylococcus aureus* and *E. coli*." The sharp edges of nanoplatelets and GO nanoplatelets can penetrate and destroy the cell membrane. Modeling studies have shown that GO membrane degradation is caused by phospholipid extraction. "This may be due to the hydrophobic sp^2 region of GO and the hydrophobic interaction of phospholipids. It was suggested that larger GO provided a wider hydrophobic region with stronger phospholipid extraction activity, whereas smaller GO showed an enhanced cleavage effect against *S. mutans*." "This may be due to the blade effect caused by the sharp edges" (Ng & Shamsi, 2022). "Research revealed that having electrolytes in the GO suspension inhibits the activity of sharp edges by interacting with the ions, resulting in aggregate formation." It was reported that the accumulation of GO boosts the surface energy, thus hindering the cutting effect of the sharp edges. This research also showed that the ideal concentration for the effectiveness of sharp edges is under 6 μg/mL when GO is dissolved in any solution. Yet, with increased concentration, antibacterial action is solely noted in GO, which is dispersed in water. This was additionally supported by detecting cellular debris and alterations in bacterial morphology on the AFM (Palmieri et al., 2017). "One study also suggested that electron transfer from the membrane of *E. coli* to the oxygen functional groups of GO coated on metals can potentially lead to a loss of membrane integrity. Another study also demonstrated that the availability of peaks or sharp edges could promote a charge transfer, thereby leading to cell membrane destruction. This was evident by the increase in the bacteria's uptake of PI, a membrane-impermeable dye." The concentration of CBNs can also influence their anti-microbial activity (Ng & Shamsi, 2022). A higher concentration of GO was demonstrated to cause intracellular leakage. GO also seemed to enhance the membrane peroxidase activity, which enhanced penetration into *E. coli* in a concentration and time-dependent manner (Nanda et al., 2016). Another study demonstrated that 50–200 μg/mL of GO could cause membrane perturbation of *Pseudomonas syringae* and *Xanthomonas campestris pv. Undulosa* (Chen et al., 2014). Studies have also reported that the peptidoglycan layer is one of the differentiating factors in the efficacy of GO. Due to thinner peptidoglycan layer, gram-negative bacteria are more prone to lacerations and piercing of phospholipid membrane by GO (Ng & Shamsi, 2022).

11.3.2 Generation of Reactive Oxygen Species

Mitochondria are the major cellular organelles responsible for energy production and metabolic regulation. In particular, they are involved in various redox reactions through the mitochondrial electron transport chain (ETC), where purification occurs, generating the most reactive oxygen species (ROS). ROS regulates cell survival, death, differentiation, and cell signaling. CBN is known to generate ROS when antioxidants in bacterial cells are depleted, causing oxidative stress. Sometimes, this can act as a double-edged sword, causing oxidative stress in bacteria or cytotoxic stress in humans when ingested (Cao et al., 2025). In bacteria, "ROS are formed as a result of aerobic respiration, and their production is balanced by antioxidant mechanisms for cell protection, but excessive ROS formation with additional damage leads to the oxidation of biomolecules (lipid components of cell membranes, DNA, proteins), which eventually causes serious damage to the cell". The main regulators of oxidative protection in *E. coli* are the oxyR regulon, which responds to increased intracellular hydrogen peroxide concentration, and the soxRS regulon (soxS gene), whose expression is activated by superoxide anion. Metal NPs are well-known oxidants, but carbon nanoparticles can also induce oxidative stress. The first mention of oxidative stress as a possible mechanism for the antibacterial action of CNTs was found in the study of Kang et al. (2007). It was discovered that multiple genes within the soxRS and oxyR regulons are activated following the exposure of cells to SWCNTs and multi-walled CNTs (MWCNT). Martin et al. demonstrated that a 24-h exposure to 100 mg/L of SWCNTs and MWCNTs resulted in the generation of ROS and heightened oxidative stress in the pathogenic yeast *Candida albicans* and the bacteria *Staphylococcus aureus* and *Pseudomonas aeruginosa* (Martel Martín et al., 2021). In the presence of oxygen and under light, carboxylated SWCNTs produce singlet oxygen, superoxide, and hydroxyl radicals. In this scenario, the functionalized nanotubes may serve as electron donors or be involved in electron transfer from alternative donors utilizing NADH. Oxidative stress may arise from the characteristics of the carbon NPs and their capacity to promote ROS production inside the cell. In the initial scenario, the cause of oxidative stress is the transition metal contaminants utilized as catalysts in the fabrication of non-metallic nanoparticles, such as SWCNTs. In the second scenario, the comparatively stable free radical intermediates found on the particle surface or the redox groups are created due to the functionalization of the NPs. With the rise in the share of SWCNTs showing metallic characteristics, there was also a corresponding increase in the oxidation of glutathione, a compound that regulates the cell's redox state (Vecitis et al., 2010). Maksimova et al. demonstrated notable variations in the gene expression of soxS when subjected to CNTs, while no expression was observed for the gene oxyR. It is established that SWCNTs cause greater harm to bacterial cells compared to MWCNTs (Maksimova et al., 2023). It has been proposed that CNTs do not directly induce oxidative stress. Instead, it arises from damage to the cell membrane and uncoupling the respiratory chain, resulting in an overproduction of free radicals within the cell. Thus, the study's results confirm that oxidative stress is one of the main ways NPs affect microbial cells, and the formation of oxidative stress has been observed under the influence of some carbon NPs and metals. Studies have recently shown that GO also exhibits oxidative stress action. It has previously been shown that *E. coli* can experience oxidative stress when coated with smaller GO nanosheets on a nitrocellulose filter surface. This might be because more oxygen could adsorb onto the GO nanosheets because of the increased defect concentrations on the carbon structure. This is evidenced by the greater ratio of the

D band to the G band in Raman spectroscopy. Glutathione oxidation and this ratio are highly correlated ($R^2 = 0.96$). In addition, it has been reported that the amount of oxygen-containing functional groups present on GO nanosheets does not affect their oxidation ability. Another study reported that ROS are formed when oxygen is adsorbed on defective sites, which ultimately oxidizes glutathione. Another study also identified defects that lead to the formation of hydroxyl radicals. In Raman studies, an increase in the D band indicates defects. According to one theory, carbonyl groups in the peptide bonds of bacterial cell membranes are attacked by hydroxyl radicals. Defects and tighter bacterial trapping effects have been reported to enhance GO–bacteria interactions, promote charge transfer, and increase oxidative stress (Ng & Shamsi, 2022). A different study suggested the oxidative stress-inducing effect as an antibacterial mechanism for immobilized flat GO on the PET substrate. The immobilized GO prevents it from penetrating or trapping the *E. coli*. Consequently, this research indicated that oxidative stress resulted from electron transfer through the direct interaction of the bacteria with the GO surface. Similarly, another research demonstrated that the availability of the GO basal plane is crucial for its antibacterial effectiveness. "However, a simulation study indicated that additional oxidation in the oxygen-containing functional groups might enhance the pathways of the oxidation reaction" (Ng & Shamsi, 2022).

11.3.3 Biofilm Disruption

Bacteria are encased in aggregates, creating biofilms by producing an extracellular polymeric matrix (EPM). The extracellular polymeric matrix consists of extracellular polysaccharides (EPS), extracellular DNA, and various macromolecular complexes, including proteins, lipids, flagella, and pili (Maksimova & Zorina, 2024). The process of biofilm development by various microbes takes place in four stages: attachment, microcolony formation, maturation of the biofilm, and dispersion (Cui et al., 2022; Kumar et al., 2023). The creation of an extracellular polymer matrix is essential. This matrix shields biofilm cells against the impacts of unfavorable environmental conditions (Maksimova & Zorina, 2024). Therefore, comprehending the mechanism underlying the anti-biofilm activity of CBNs is crucial. The engagement of CBN with biofilms initiates during the EPS phase, creating a thick aqueous solution that hinders the movement of nanomaterials and directly impacts cells. Besides surface chemistry, the dimensions of the nanomaterial play a crucial role in facilitating interactions with biofilm components for penetrating the biofilm matrix. Forming biofilms is viewed as a way for microorganisms to adapt to stress. This procedure is intricate, particularly at the expression level, and necessitates numerous factors.

The development of bacterial biofilms relies on two key regulatory elements: quorum sensing (QS) and the second messenger bis-(3′-5′)-cyclic dimeric guanosine monophosphate (c-di-GMP), which enhances the synthesis of extracellular polysaccharides. Starvation leads bacterial cells to reduce levels of c-di-GMP as phosphodiesterase is activated, aiding in removing biofilm cells. The sigma subunit

RpoS of RNA polymerase is recognized for its significant role in the formation of biofilms. The activation of the rpoS gene during the biofilm formation of *E. coli* was discovered to promote biofilm development by enhancing the expression of genes related to adhesion and stress response while reducing the expression of genes linked to flagellar production and energy metabolism. The incorporation of carboxylated CNTs in the environment was found to elevate rpoS expression in *E. coli*, which grew larger. In response to oxidative stress, the bacteria elevated the expression of genes linked to EPS secretion, biofilm development, and antibiotic resistance (Maksimova & Zorina, 2024).

Recently, the anti-biofilm potential of GO has also been demonstrated (Ng & Shamsi, 2022). Ricci et al. pointed out the inhibitory effect of GO on *S. aureus* and *E. coli* biofilms at concentrations of 10 and 100 μg/mL, respectively. Although no cytotoxic effect was observed at 100 μg/mL, the changes in gene expression suggest dysregulation of OC, COLI, and RUNX2 genes, which affect key genes in the bone repair pathway (Ricci et al., 2017). Recently, carbon quantum dots (CQDs) have also been shown to be effective. CQDs with a size of less than 10 nm can be easily modified by adding functional groups to enhance their ability to generate ROS via a photodynamic mechanism. Thus, they are effective against bacterial biofilms. In one study, Wang et al. developed mixed CQD and PLGA nanomaterials to encapsulate various antibiotics. Namely, Azithromycin and tobramycin combine chemotherapy and photothermal effects to remove biofilms. Another study by Lin et al. showed that CQDs synthesized from *Lactobacillus plantarum* (LP) inhibited *E. coli* biofilm formation without cytotoxic effects (Xie et al., 2025).

11.4 Multi-Drug Resistance and Carbon Nanoparticles

A significant danger facing the healthcare sector today is the increase in bacteria resistant to antibiotics. Lethal multidrug-resistant (MDR) infectious bacteria are increasingly emerging, becoming increasingly prevalent, and presenting a significant risk to human health. Multi-drug resistance (MDR) refers to microorganisms' lack of sensitivity or resistance to one or more administered antibiotics. WHO studies indicate that *E. coli* shows resistance to cephalosporins and fluoroquinolones, *Shigella* to fluoroquinolones, *Neisseria gonorrhoeae* to cephalosporins, and *Mycobacterium tuberculosis* to rifampicin, isoniazid, and fluoroquinolones, which are typical bacteria responsible for infections. Previously, these antibiotic-resistant bacterial strains were uncommon and primarily associated with hospital-acquired infections, but they are now increasingly widespread. This issue affects both gram-positive and gram-negative bacteria. Nonetheless, anti-microbial resistance is not restricted to bacteria. Certain nosocomial fungal infections, including *Candida albicans* and *Candida auris*, have shown multi-drug resistance. Researchers theorize that acquiring plasmids via horizontal gene transfer from the environment is responsible for antibiotic resistance. Certain bacterial species have been noted to possess a range of distinct defense strategies to prevent the detrimental impacts of antibiotics, such as efflux pumps, decreased permeability of the LPS layer, secretion of degrading enzymes, and movement of targets (Bharadwaj et al., 2022).

11.4.1 Mechanism of MDR

11.4.1.1 Destruction of the Antibiotic Molecule

The main resistance method to β-lactams involves the enzyme β-lactamase, which degrades these compounds. “This enzyme cleaves the amide bond of the β-lactam ring, making the antibiotic ineffective. β-lactamases were initially identified in the 1940s, just prior to the commercialization of penicillin, yet evidence suggests they have been around for millions of years. After extensive penicillin use, infections resulting from antibiotic-resistant *Staphylococcus aureus* have gained clinical importance. The resistance mechanism was found to be a penicillinase encoded by a plasmid, which was easily transferred among *S. aureus* strains, leading to a swift spread of the resistance trait. To tackle this issue, novel β-lactam compounds, including ampicillin, were created with expanded activity and decreased vulnerability to penicillinase. In the 1960s, researchers found a new plasmid-encoded β-lactamase capable of hydrolyzing ampicillin in Gram-negative bacteria.” This enzyme was named TEM-1 in honor of a patient named Temoneira. Since then, the emergence of enzymes capable of degrading every new molecule that has come to the market has invariably coincided with the birth of the next β-lactam. This process vividly illustrates how bacteria have evolved adaptively under the influence of antibiotics, as shown in Fig. 11.4.

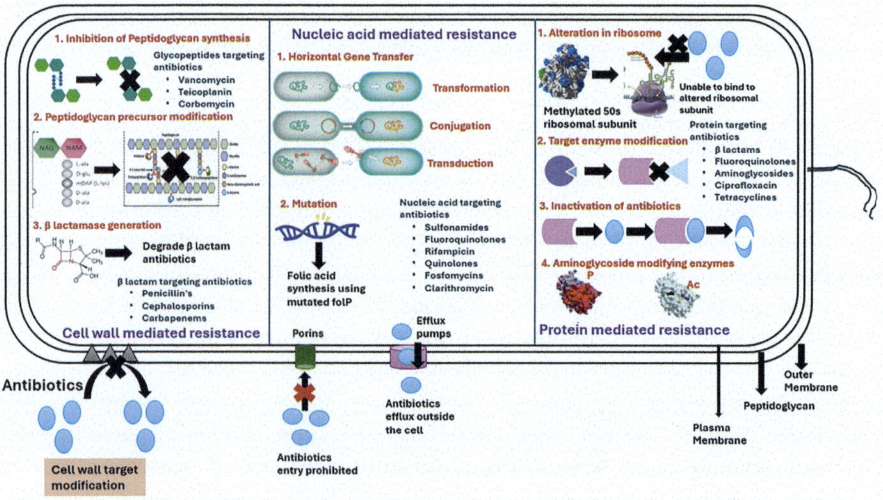

Fig. 11.4 Diverse multi-drug resistance mechanism of bacteria

11.4.1.2 Decreased Permeability

Numerous antibiotics utilized in clinical settings have their targets located inside the cell or, for Gram-negative bacteria, within the cytoplasmic membrane (inner membrane). Consequently, chemicals must traverse the cytoplasm and/or outer membrane to demonstrate their anti-microbial action. Bacteria have evolved protective strategies that limit the absorption of anti-microbial substances and block antibiotics from accessing their intracellular or periplasmic targets. Due to the reasons above, these mechanisms are especially critical for Gram-negative bacteria as they restrict the ingress of chemicals from their surroundings. Indeed, the outer membrane serves as the initial barrier against the infiltration of numerous harmful substances, such as anti-microbial agents. Alterations primarily influence hydrophilic substances in the permeability of the outer membrane, such as β-lactams, tetracyclines, and certain fluoroquinolones, since they frequently utilize porins (water-filled diffusion channels) to navigate this barrier. The glycopeptide antibiotic vancomycin is ineffective against Gram-negative pathogens because it cannot cross the outer membrane, providing an excellent example of how effective this natural barrier is. "Similarly, *Pseudomonas* and *A. baumannii* have an inherently low resistance to β-lactams compared to *Enterobacteriaceae*, which may be partly due to reduced abundance and/or differential expression of porins" (Munita & Arias, 2016).

11.5 Biocompatibility and Toxicity of Carbon Nanomaterials

As discussed in the above section, "carbon nanomaterials have unique physicochemical properties. Studies have shown that hybridization state, isotope, and size can affect toxicity, accumulation, and dispersibility. However, if these resources are used extensively, they will inevitably enter the marine environment and the atmosphere, causing environmental problems. Studies have shown they can harm human cells if ingested in large quantities. The widespread use of nanomaterials raises health concerns. Nanomaterials can generally cause cytotoxicity in various ways, including generating reactive oxygen species (ROS), DNA damage, lysosomal damage, mitochondrial dysfunction leading to immune responses, and cell death or necrosis. Oxidative stress is a key mechanism underlying nanomaterial-induced toxicity due to increased reactive chemical species (Kyriakides et al., 2021; Makhado et al., 2024). Significant work has been undertaken to assess the possible toxicity of carbon nanomaterials in vitro and in vivo. Numerous researchers have acknowledged the varied toxic properties of carbon nanoparticles, which are influenced by several factors, such as the physicochemical characteristics of the particular CBN, the targeted cell type, and the particle dispersion method. Recorded negative effects have encompassed cell death by generating reactive oxygen species, DNA injury, lysosomal impairment, mitochondrial failure, and apoptosis or necrosis. Although carbon nanomaterials are biologically toxic, they have been extensively researched for their effects on the immune system, such as pulmonary macrophage activation and inflammatory responses. The involvement of CBN in activating or inhibiting immune cells has also been studied" (Sarkar et al., 2021; Yuan et al., 2019).

The majority of research indicates that natural fullerenes are harmful. Natural fullerenes possess strong lipophilicity and can produce ROS, leading to oxidative stress. Incorporating hydrophilic groups into fullerenes has been proposed to increase their solubility and decrease ROS production, thus enhancing their biocompatibility. The harmfulness of nanotubes primarily relies on their dimensions, surface properties, and the existence of undesired impurities. Although CNTs in their natural state are considered harmful to the biological environment, graphene and GO have demonstrated negative, pro-inflammatory and positive, anti-inflammatory characteristics, respectively.

Furthermore, research findings indicate that nanodiamonds possess outstanding biocompatibility and minimal toxicity (Jayaprakash et al., 2024). The cytotoxicity of CBN is one of the main reasons its industrial potential has not been fully exploited, especially in biomedical applications. However, functionalizing CBN by adding various organic or inorganic groups can potentially improve the biocompatibility of CNPs. Adding dextran, polyethylene glycol (PEG), or bovine serum albumin (BSA) can improve the toxicity profile of CBN, such as GO. GO DEX surface coating increased HeLa cell viability by 80% at concentrations close to 450 μg/mL. This is thought to be due to the adsorption of proteins on GO, which weakens the interaction. Functionalization with BSA did not result in a significant decrease in cell proliferation. In contrast, GO-PEG conjugates showed good biocompatibility with A549 cells without cytotoxicity up to 100 μg/mL. Another study showed that GO-containing gallic acid exposure to zebrafish embryos for 72 h improved the toxicity profile without compromising the antibacterial properties (Ng & Shamsi, 2022). "Among the various CBNs, CDs are considered most biocompatible and have been shown to possess low toxicity, which qualifies them to be used in bioimaging and drug delivery. The ester and ether groups on the surface of these particles are hydrophilic and biocompatible, making them ideal for use in any biomedical application (Jayaprakash et al., 2024). Thus, the biocompatibility of CBNs can be boosted to an inordinate scope by innumerable surface modifications, permitting a reduction in their intrinsic toxicity and an enhancement in their coupling with biological systems."

11.6 Applications of Carbon Nanomaterials as Anti-microbial Agents

Biomedical materials and devices are a key emphasis of carbon nanomaterial research. Over the years, a variety of applications for carbon nanotubes have been anticipated, including but not limited to biosensors, drug and vaccine delivery vehicles, and new biomaterials, as noted earlier. They have also been employed as nanofillers in polymeric materials to surge mechanical features and form extremely anisotropic nanocomposites suggestively. They can also be utilized to engender electrically conductive polymers and tissue engineering assemblies that are proficient in providing delimited electrical stimulus. While most clinical studies have aimed to comprehend the impact of CBNs on malignant cells, very few have been focused on studying anti-microbial effects and other disorders. Studies have used CBNs to guide the surgery of different cancers, viz. papillary thyroid (NCT02724176), lymph node biopsy for cervical cancer (NCT05167149, NCT03778268, and NCT04951245), endometrial cancer (NCT03778255) and lymph node tracing for gastrectomy (NCT05229874 and NCT04759820), colorectal cancer (NCT06783985) and rectal carcinoma post-neoadjuvant radiochemotherapy (NCT03550001), axillary metastatic lymph node resection in breast cancer (NCT04744506, NCT04482803, and NCT05939830), and laparoscopic colorectal cancer surgery (NCT03350945). Recently, a clinical trial sponsored by Bin He at Xinhua Hospital, Shanghai Jiao Tong University of Medicine aimed at evaluating the sensitivity, specificity, and effectiveness of GQDs combined with silicon (Si) nanowire photoelectrochemical immunosensor in the early detection of myocardial infraction on 100 participants (NCT04390490).

Additionally, 2C Tech Corp. demonstrated the potential of intravitreal QDs in advanced retinitis pigmentosa (NCT05841862). Another study used graphene oxide for the surgical management of periodontal intrabody in dogs (NCT05341245). Studies have also been designed to explore the efficacy of fullerenes in preventing radiation-induced dermatitis in patients with head and neck malignancy (NCT06484166). Although various carbon-based nanomaterials (CBNs) have shown promising antimicrobial properties, there remains a significant gap in advancing these findings to clinical trial stages. However, in the context of anti-microbial activity, we suggest two main applications of carbon nanomaterials—wound healing and management and medical devices.

11.6.1 Wound Healing and Management

Wound management intends to deliver an apt setting for the wound to heal without breaking down the skin (Chauhan et al., 2023). Burn wound infections are prompted by diverse pathogenic microbes, including bacteria, fungi, and viruses (Sangnim et al., 2024). For active wound supervision, it is imperative to comprehend the extent, deepness, harshness, and site of the wound. Carbon nanomaterials have been recommended to provide a minor advantage over other conventional approaches for treating persistent wound infections. "Carbon nanomaterials have anti-microbial, antioxidant, and anti-inflammatory properties, as recommended in the previous section, and a high surface area and surface-to-volume ratio. These properties are necessary for effective wound healing (Chauhan et al., 2023; Sangnim et al., 2024). This enables safe and effective inclusion of CBNs into biomaterials (Sangnim et al., 2024). CNTs have been suggested to disrupt the integrity of bacterial cell membranes, especially when combined with silver nanoparticles (AgNPs), nanocomposites, etc. SWCNTs and MWCNTs have been shown to disrupt membrane integrity by generating ROS and heavy metal residues through their metabolic activity and shape. Research has indicated that CNT composites prevent the proliferation of many harmful bacteria in vitro. CNTs have been demonstrated to boost the anti-microbial effectiveness of specific compounds. Research indicates that CNTs prevent the *A. baumannii* strain from being resistant to multiple medications."

Another study reported that the antibacterial effect was enhanced when CNTs were combined with AgNPs and titanium dioxide nanoparticles (TiO2NP). A similar report was also made for hydroxyl-functionalized single-walled nanotubes from *Paracoccus denitrificans* (Nandhini et al., 2024). MWCNTs have been shown to control infection and accelerate the regenerative process of fibroblast proliferation and re-epithelialization, which has been shown to be effective not only in burns but also in diabetic wounds that are difficult to heal. Khalid et al. observed that diabetic wounds closed and re-epithelialized more quickly, in addition to the formation of healthy granulation tissue. This process was enhanced by reducing the levels of substances that trigger inflammation like elevated IL-1a, TNF-α, and VEGF (Khalid et al., 2022). In another study, Du et al. confirmed the antibacterial characteristics of graphene. They successfully reduced the risk of wound infection caused by *Escherichia coli* and *Staphylococcus aureus* by employing a nitrocellulose membrane infused with nanoporous graphene. This potentially accelerated wound closure and recovery (Du et al., 2021). Similarly, Gupta et al. noted that a nanofibrous film with GO, silver, and magnesium demonstrated accelerated wound contraction and re-epithelialization (Gupta et al., 2022). In 2014, a research team from Biomedical Carbon Technology Co., Ltd. in Taiwan began a clinical trial to investigate the therapeutic effects of silver-impregnated activated carbon fiber dressings on deep skin burns (NCT01598493). Their study found that carbon fibers have a large surface area and high porosity, effectively absorbing toxins. They also found that far-infrared rays emitted by carbon fibers promote blood circulation and promote tissue regeneration. They also found that silver nanoparticles exhibit antibacterial activity by penetrating cell walls and denaturing essential proteins. The study included approximately 30 patients. These results may suggest the development of innovative solutions for improving wound dressings. Several studies have focused on combining nanomaterials with various materials to improve the biocompatibility of medical devices such as stents, cardiovascular grafts, and prostheses. These modifications have been reported to change the physicochemical properties of biomaterials, improve mechanical strength and hemocompatibility, and reduce antithrombotic properties. In addition, conjugation with nanomaterials has been suggested to help promote re-epithelialization and regeneration, strengthen stents against restenosis, and reduce the risk of infection and biofilm formation.

One widely used material for manufacturing cardiovascular devices, namely stents, is SAE 316 L stainless steel. However, this material is prone to thrombosis and restenosis. Studies have shown that coating 316 L stents with a graphene layer can improve their physical and chemical properties. These graphene-coated stents reduce toxicity to endothelial cells and help cells adhere, proliferate, and metabolize. Graphene also effectively inhibits thrombus formation by minimizing platelet adhesion and aiding in the regeneration of blood vessel walls. CNP has also been used to make heart valves. Hastalex, a material developed from functionalized graphene oxide and poly(carbonate-urea) urethane, showed superior mechanical properties, hemocompatibility, and calcification resistance as evaluated by GOTE-TEX. Replacing the valve with polyurethane or heparin nanotubes significantly reduced clotting, further improving performance. In addition, cardiac stents, grafts, orthopedic replacements, or implants can also be coated with carbon nanoparticles. In addition to proliferation, it can also correct regeneration and osteoblast adhesion. It also protects the implant from damage for years after surgery. Interestingly, GO combined with poly(vinylidene fluoride) has also been demonstrated to have the potential to be used as a polymer nanogenerator for self-pacing (Holmannova et al., 2022). Azimi et al. entrenched this pacemaker in dogs and found that PNG harvested 0.487 μJ from every heartbeat. This value is apparently more than the threshold energy required to stimulate and sustain the human heart (Azimi et al., 2021). The research also showed the promise of nanodiamond-like CNPs for use in medicine. It can undoubtedly lower the likelihood of thrombosis and prevent the development of fibrin deposits. CNPs might also serve as a component in a material designed to repair damaged teeth. Incorporating nanodiamonds into the resin enhanced the quality of dental fillings by boosting the flexural strength, elastic modulus, and surface roughness. Nanodiamonds integrated into the resin have been suggested to inhibit the adhesion of *Candida albicans*, which is crucial for preventing denture-related stomatitis (Holmannova et al., 2022).

As mentioned earlier, CBN has a wide range of applications, from bioimaging to drug delivery to potential anti-microbial agents. However, the toxicity and biocompatibility of CBN remain serious concerns. Regulatory frameworks for approval and monitoring should be developed to ensure the safety and efficacy of these nanomaterials in anti-microbial testing and research. Appropriate disposal systems should be developed to minimize potential biological hazards. Efforts are underway to actively monitor and assess the safety of CBN in various countries. In 2011, the European Commission released its initial recommendations regarding the definition of nanomaterials. In the EU, nanomaterials fall under the substance definitions of REACH and CLP. EU institutions, such as the Organization for Economic Co-operation and Development (OECD), are creating approaches for assessing and testing nanomaterials. The OECD has a distinct recommendation that offers direction on nanomaterial evaluation, encompassing factors for cytotoxicity assessment. In the United States, the rules governing nanomaterials are viewed as analogous to those for other chemicals and products overseen by the FDA. Likewise, India has established guidelines for the safe utilization and disposal of nanomaterials created by the Centre for Knowledge Management of Nanoscience and Technology (CKMNT) within the Department of Science and Technology (DST). The International Organization for Standardization (ISO) has a technical committee focused on nanotechnology (ISO/TC229) that is engaged in creating standards for the field of nanotechnology. These ethical guidelines might encompass monitoring cytotoxicity testing techniques to ensure safety in consumer goods and medical uses (Mondal et al., 2024).

11.7 Challenges and Future Considerations

Although carbon nanomaterials are potentially anti-microbial agents, they are prone to multiple challenges for safe and effective use (Mondal et al., 2024). Firstly, CNPs can contribute to anti-microbial resistance over time, like all other anti-microbial agents and antibiotics. An investigation is required to comprehend the factors of resistance and the approaches to alleviating this threat. Secondly, understanding the long-term biocompatibility and possible toxicity of anti-microbial CNPs is essential for assessing their effects on human health and the environment. The harmfulness of CNPs can be linked to their physicochemical characteristics, such as size, shape, charge, functionalization, and corona development. Research studies and epidemiological analyses have demonstrated that CNPs exhibit cytotoxic and genotoxic effects on common tissue types and biological systems. There are also reports indicating that they can cause negative effects on the kidneys, immune system, skin, and eyes. The probable reason for the damage is thought to be the interaction of CNPs with the cell membrane, nucleus, or mitochondria, which is linked to heightened production of reactive oxygen species (ROS) and the activation of signaling pathways that lead to inflammation, along with elevated production of pro-inflammatory cytokines, such as TNF-α and IL-6. Inflammation combined with ROS and the direct engagement of CNPs with the nucleus leads to DNA damage and interferes with DNA-repair mechanisms. This frequently results in the buildup of mutations and the development of malignancy (Holmannova et al., 2022). Multiple clinical studies are being undertaken to explore the toxic effects of CBNs when exposed while breathing. This may lead to cardiovascular damage. NCT03659864, initiated by the University of Edinburgh, aims to explore the effects of environmental as well as synthetic carbonaceous nanoparticles (viz. GO, diesel exhaust nanoparticles, carbon black nanoparticles, etc.) on lung, blood vessels, blood clotting, and levels of eicosanoids in blood and urine. Thirdly, understanding the environmental effects of anti-microbial nanoparticles discharged into the environment, whether by design or accident, raises worries about their influence on ecosystems, aquatic species, and soil vitality. Examining their destiny, movement, and environmental impacts is crucial. Another important avenue to consider when considering nanoparticles as anti-microbial agents is their effect on beneficial microbiota, especially considering their therapeutic potential. To ensure the biocompatibility and toxicity of carbon nanomaterials, it is essential to optimizing the formulation by altering the size, shape, surface charge, etc. In addition, future studies should aim to understand the effectiveness of these nanomaterials through clinical and preclinical studies to ascertain their claim.

11.8 Conclusions

Anti-microbial resistance (AMR) is swiftly growing due to the overuse of antibiotics. This has given rise to a global health crisis. New and innovative solutions with high specificity are being understood to target AMR. New found interest in nanomaterials, due to their smaller size, higher surface area, and greater surface area to volume ratio, provides a unique alternative to traditional anti-microbial agents. Among the variety of nanomaterials, carbon-based nanomaterials like nanotubes have been considered most effective in inhibiting microbial growth due to their diverse physicochemical properties. This has been facilitated through ROS generation, cell membrane disruption, altering gene expression, and biofilm disruption. Despite their high potential as anti-microbial agents, carbon nanomaterials lack specificity. They have been suggested to cause cytotoxicity in humans. The bioavailability and toxicity of CBNs are still a matter of concern. We recommend optimizing CBN structure and size, in addition to the use of adjuvants, to increase their bioavailability and reduce toxicity. This would also reduce environmental pollution considerably. Despite their visible toxicity and ill effects, carbon nanomaterials present themselves as viable and effective candidates as anti-microbial agents if used effectively. More preclinical and clinical evidence is required to validate their role as anti-microbial agents for wound healing and management.

References

Abo-Neima, S. E., Motaweh, H. A., & Elsehly, E. M. (2020). Anti-microbial activity of functionalised carbon nanotubes against pathogenic microorganisms. *IET Nanobiotechnology, 14*(6), 457–464. https://doi.org/10.1049/iet-nbt.2019.0342

Ajith, M. P., Pardhiya, S., & Rajamani, P. (2022). Carbon dots: An excellent fluorescent probe for contaminant sensing and remediation. *Small, 18*(15), 2105579. https://doi.org/10.1002/smll.202105579

Al-Jumaili, A., Alancherry, S., Bazaka, K., & Jacob, M. V. (2017). Review on the anti-microbial properties of carbon nanostructures. *Materials, 10*(9), 9. https://doi.org/10.3390/ma10091066

Anand, A., Unnikrishnan, B., Wei, S.-C., Chou, C. P., Zhang, L.-Z., & Huang, C.-C. (2019). Graphene oxide and carbon dots as broad-spectrum anti-microbial agents—A minireview. *Nanoscale Horizons, 4*(1), 117–137. https://doi.org/10.1039/C8NH00174J

Arias, L. R., & Yang, L. (2009). Inactivation of bacterial pathogens by carbon nanotubes in suspensions. *Langmuir, 25*(5), 3003–3012. https://doi.org/10.1021/la802769m

Ayanda, O. S., Mmuoegbulam, A. O., Okezie, O., Durumin Iya, N. I., Mohammed, S. E., James, P. H., Muhammad, A. B., Unimke, A. A., Alim, S. A., Yahaya, S. M., Ojo, A., Adaramoye, T. O., Ekundayo, S. K., Abdullahi, A., & Badamasi, H. (2024). Recent progress in carbon-based nanomaterials: Critical review. *Journal of Nanoparticle Research, 26*(5), 106. https://doi.org/10.1007/s11051-024-06006-2

Azimi, S., Golabchi, A., Nekookar, A., Rabbani, S., Amiri, M. H., Asadi, K., & Abolhasani, M. M. (2021). Self-powered cardiac pacemaker by piezoelectric polymer nanogenerator implant. *Nano Energy, 83*, 105781. https://doi.org/10.1016/j.nanoen.2021.105781

Azizi-Lalabadi, M., Hashemi, H., Feng, J., & Jafari, S. M. (2020). Carbon nanomaterials against pathogens; the anti-microbial activity of carbon nanotubes, graphene/graphene oxide, fullerenes, and their nanocomposites. *Advances in Colloid and Interface Science, 284*(102), 250. https://doi.org/10.1016/j.cis.2020.102250

Berry, T. D., Filley, T. R., Clavijo, A. P., Bischoff Gray, M., & Turco, R. (2017). Degradation and microbial uptake of C_{60} fullerols in contrasting agricultural soils. *Environmental Science & Technology, 51*(3), 1387–1394. https://doi.org/10.1021/acs.est.6b04637

Bharadwaj, A., Rastogi, A., Pandey, S., Gupta, S., & Sohal, J. S. (2022). Multidrug-resistant bacteria: Their mechanism of action and prophylaxis. *BioMed Research International, 2022*, 1–17. https://doi.org/10.1155/2022/5419874

Bolshakova, O., Lebedev, V., Mikhailova, E., Zherebyateva, O., Aznabaeva, L., Burdakov, V., Kulvelis, Y., Yevlampieva, N., Mironov, A., Miroshnichenko, I., & Sarantseva, S. (2023). Fullerenes on a nanodiamond platform demonstrate antibacterial activity with low cytotoxicity. *Pharmaceutics, 15*(7), 7. https://doi.org/10.3390/pharmaceutics15071984

Cao, J., Wu, Q., Liu, X., Zhu, X., Huang, C., Wang, X., & Song, Y. (2025). Mechanistic insight on nanomaterial-induced reactive oxygen species formation. *Journal of Environmental Sciences, 151*, 200–210. https://doi.org/10.1016/j.jes.2024.03.009

Chauhan, N., Saxena, K., & Jain, U. (2023). Carbon-based nanomaterials in wound care management: A new and pristine strategy. *Biomedical Materials & Devices, 1*(1), 108–121. https://doi.org/10.1007/s44174-022-00030-3

Chausali, N., Saxena, J., & Prasad, R. (2021). Nanobiochar and biochar based nanocomposites: Advances and applications. *Journal of Agriculture and Food Research, 5*, 100191. https://doi.org/10.1016/j.jafr.2021.100191

Chen, J., Peng, H., Wang, X., Shao, F., Yuan, Z., & Han, H. (2014). Graphene oxide exhibits broad-spectrum anti-microbial activity against bacterial phytopathogens and fungal conidia by intertwining and membrane perturbation. *Nanoscale, 6*(3), 1879–1889. https://doi.org/10.1039/C3NR04941H

Coville, N. J., Mhlanga, S. D., Nxumalo, E. N., & Shaikjee, A. (2011). A review of shaped carbon nanomaterials: Review article. *South African Journal of Science, 107*(3), 1–15. https://doi.org/10.10520/EJC97121

Cui, F., Li, T., Wang, D., Yi, S., Li, J., & Li, X. (2022). Recent advances in carbon-based nanomaterials for combating bacterial biofilm-associated infections. *Journal of Hazardous Materials, 431*(128), 597. https://doi.org/10.1016/j.jhazmat.2022.128597

Du, S., Liu, B., Li, Z., Tan, H., Qi, W., Liu, T., Qiang, S., Zhang, T., Song, F., Chen, X., Chen, J., Qiu, H., & Wu, W. (2021). A nanoporous graphene/nitrocellulose membrane beneficial to wound healing. *ACS Applied Bio Materials, 4*(5), 4522–4531. https://doi.org/10.1021/acsabm.1c00261

Georgakilas, V., Perman, J. A., Tucek, J., & Zboril, R. (2015, May 27). *Broad family of carbon nanoallotropes: Classification, chemistry, and applications of fullerenes, carbon dots, nanotubes, graphene, nanodiamonds, and combined superstructures* (world) [Review-article]. ACS Publications; American Chemical Society. https://doi.org/10.1021/cr500304f

Gilsanz-Muñoz, M. F., Martínez-Martínez, M., Pérez-Piñeiro, J., Roldán, M., Arce, M. P., Blasco, R., Rico-San Román, L., Esperón-Fajardo, F., Cerpa-Naranjo, A., & Martín-Maldonado, B. (2024). Assessing the anti-microbial efficacy of graphene oxide and its PEGylated derivative against Staphylococcus aureus. *Science, 6*(4), 4. https://doi.org/10.3390/sci6040066

Gupta, S., Prasad, P., Roy, A., Alam, M. M., Ahmed, I., & Bit, A. (2022). Metallic ion-based graphene oxide functionalized silk fibroin-based dressing promotes wound healing via improved bactericidal outcomes and faster re-epithelization. *Biomedical Materials, 17*(3), 035010. https://doi.org/10.1088/1748-605X/ac64dd

Haung, C.-F., Chan, Y.-H., Chen, L.-K., Liu, C.-M., Huang, W.-C., Ou, S.-F., Ou, K.-L., & Wang, D.-Y. (2013). Preparation, characterization, and properties of anticoagulation and antibacterial films of carbon-based nanowires fabricated on surfaces of Ti implants. *Journal of The Electrochemical Society, 160*(6), H392. https://doi.org/10.1149/2.109306jes

Ho, C. S., Wong, C. T. H., Aung, T. T., Lakshminarayanan, R., Mehta, J. S., Rauz, S., McNally, A., Kintses, B., Peacock, S. J., de la Fuente-Nunez, C., Hancock, R. E. W., & Ting, D. S. J. (2024). Anti-microbial resistance: A concise update. *The Lancet Microbe*. https://doi.org/10.1016/j.lanmic.2024.07.010

Holmannova, D., Borsky, P., Svadlakova, T., Borska, L., & Fiala, Z. (2022). Carbon nanoparticles and their biomedical applications. *Applied Sciences, 12*(15), 15. https://doi.org/10.3390/app12157865

Ibrahim, A. A., Khan, T., Nowlin, K., Averitt, J., Pathiraja, G., LaJeunesse, D., Obare, S. O., & Dellinger, A. L. (2024). A rapid one-step synthesis of silver and copper coordinated chlorine functionalized fullerene nanoparticles with enhanced antibacterial activity. *Nanoscale Advances, 6*(23), 5833–5852. https://doi.org/10.1039/D4NA00732H

Jayaprakash, N., Elumalai, K., Manickam, S., Bakthavatchalam, G., & Tamilselvan, P. (2024). Carbon nanomaterials: Revolutionizing biomedical applications with promising potential. *Nano Materials Science*. https://doi.org/10.1016/j.nanoms.2024.11.004

Johnston, H. J., Hutchison, G. R., Christensen, F. M., Peters, S., Hankin, S., Aschberger, K., & Stone, V. (2010). A critical review of the biological mechanisms underlying the in vivo and in vitro toxicity of carbon nanotubes: The contribution of physico-chemical characteristics. *Nanotoxicology, 4*(2), 207–246. https://doi.org/10.3109/17435390903569639

Kang, S., Pinault, M., Pfefferle, L. D., & Elimelech, M. (2007). Single-walled carbon nanotubes exhibit strong anti-microbial activity. *Langmuir: The ACS Journal of Surfaces and Colloids, 23*(17), 8670–8673. https://doi.org/10.1021/la701067r

Kang, S., Mauter, M. S., & Elimelech, M. (2008). Physicochemical determinants of multiwalled carbon nanotube bacterial cytotoxicity. *Environmental Science & Technology, 42*(19), 7528–7534. https://doi.org/10.1021/es8010173

Khalid, A., Madni, A., Raza, B., ul Islam, M., Hassan, A., Ahmad, F., Ali, H., Khan, T., & Wahid, F. (2022). Multi-walled carbon nanotubes functionalized bacterial cellulose as an efficient healing material for diabetic wounds. *International Journal of Biological Macromolecules, 203*, 256–267. https://doi.org/10.1016/j.ijbiomac.2022.01.146

Kumar, L., Bisen, M., Harjai, K., Chhibber, S., Azizov, S., Lalhlenmawia, H., & Kumar, D. (2023). Advances in nanotechnology for biofilm inhibition. *ACS Omega, 8*(24), 21391–21409. https://doi.org/10.1021/acsomega.3c02239

Kyriakides, T. R., Raj, A., Tseng, T. H., Xiao, H., Nguyen, R., Mohammed, F. S., Halder, S., Xu, M., Wu, M. J., Bao, S., & Sheu, W. C. (2021). Biocompatibility of nanomaterials and their immunological properties. *Biomedical Materials (Bristol, England), 16*(4). https://doi.org/10.1088/1748-605X/abe5fa

Li, P., Sun, L., Xue, S., Qu, D., An, L., Wang, X., & Sun, Z. (2022). Recent advances of carbon dots as new anti-microbial agents. *SmartMat, 3*(2), 226–248. https://doi.org/10.1002/smm2.1131

Liu, S., Zeng, T. H., Hofmann, M., Burcombe, E., Wei, J., Jiang, R., Kong, J., & Chen, Y. (2011). Antibacterial activity of graphite, graphite oxide, graphene oxide, and reduced graphene oxide: Membrane and oxidative stress. *ACS Nano, 5*(9), 6971–6980. https://doi.org/10.1021/nn202451x

Liu, D., Mao, Y., & Ding, L. (2018). Carbon nanotubes as anti-microbial agents for water disinfection and pathogen control. *Journal of Water and Health, 16*(2), 171–180. https://doi.org/10.2166/wh.2018.228

Maduraiveeran, G., & Jin, W. (2021). Carbon nanomaterials: Synthesis, properties and applications in electrochemical sensors and energy conversion systems. *Materials Science and Engineering: B, 272*, 115341. https://doi.org/10.1016/j.mseb.2021.115341

Makhado, B. P., Oladipo, A. O., Gumbi, N. N., De Kock, L. A., Andraos, C., Gulumian, M., & Nxumalo, E. N. (2024). Unravelling the toxicity of carbon nanomaterials—From cellular interactions to mechanistic understanding. Toxicology in Vitro, 100, 105898. https://doi.org/10.1016/j.tiv.2024.105898

Maksimova, Y. G., & Zorina, A. S. (2024). Antibiofilm and probiofilm effects of nanomaterials on microorganisms (review). *Applied Biochemistry and Microbiology, 60*(1), 1–16. https://doi.org/10.1134/S0003683824010125

Maksimova, Y., Zorina, A., & Nesterova, L. (2023). Oxidative stress response and E. coli biofilm formation under the effect of pristine and modified carbon nanotubes. *Microorganisms, 11*(5), 5. https://doi.org/10.3390/microorganisms11051221

Maleki Dizaj, S., Mennati, A., Jafari, S., Khezri, K., & Adibkia, K. (2015). Anti-microbial activity of carbon-based nanoparticles [Text/html]. *Advanced Pharmaceutical Bulletin; eISSN 2251-7308*. https://doi.org/10.5681/APB.2015.003

Martel Martín, S., Barros, R., Domi, B., Rumbo, C., Poddighe, M., Aparicio, S., Suarez-Diez, M., & Tamayo-Ramos, J. A. (2021). Low toxicological impact of commercial pristine multi-walled carbon nanotubes on the yeast Saccharomyces cerevisiae. *Nanomaterials, 11*(9), 9. https://doi.org/10.3390/nano11092272

Mekuye, B., & Abera, B. (2023). Nanomaterials: An overview of synthesis, classification, characterization, and applications. *Nano Select, 4*(8), 486–501. https://doi.org/10.1002/nano.202300038

Minsu, P., Hyewon, Y., & Seokwoo, J. (2020). 7—Graphene-based quantum dot emitters for light-emitting diodes. In T.-W. Lee (Ed.), *Graphene for flexible lighting and displays* (pp. 117–150). Woodhead Publishing. https://doi.org/10.1016/B978-0-08-102482-9.00007-1

Mondal, S. K., Chakraborty, S., Manna, S., & Mandal, S. M. (2024). Anti-microbial nanoparticles: Current landscape and future challenges. *RSC Pharmaceutics, 1*(3), 388–402. https://doi.org/10.1039/D4PM00032C

Munita, J. M., & Arias, C. A. (2016). Mechanisms of antibiotic resistance. *Microbiology Spectrum, 4*(2), 4.2.15. https://doi.org/10.1128/microbiolspec.VMBF-0016-2015

Nanda, S. S., Yi, D. K., & Kim, K. (2016). Study of antibacterial mechanism of graphene oxide using Raman spectroscopy. *Scientific Reports, 6*(1), 28443. https://doi.org/10.1038/srep28443

Nandhini, J., Karthikeyan, E., & Rajeshkumar, S. (2024). Nanomaterials for wound healing: Current status and futuristic frontier. *Biomedical Technology, 6*, 26–45. https://doi.org/10.1016/j.bmt.2023.10.001

Ng, I. M. J., & Shamsi, S. (2022). Graphene oxide (GO): A promising nanomaterial against infectious diseases caused by multidrug-resistant bacteria. *International Journal of Molecular Sciences, 23*(16), 9096. https://doi.org/10.3390/ijms23169096

Nishshankage, K., Fernandez, A. B., Pallewatta, S., Buddhinie, P. K. C., & Vithanage, M. (2024). Current trends in anti-microbial activities of carbon nanostructures: Potentiality and status of nanobiochar in comparison to carbon dots. *Biochar, 6*(1), 1–20. https://doi.org/10.1007/s42773-023-00282-2

Nurtay, L., Benassi, E., Nazir, F., Dastan, D., Utupova, A., Dautov, A., Dukenbayev, K., Xie, Y., Pham, T. T., & Fan, H. (2023). Novel carbon nanozymes with enhanced phosphatase-like catalytic activity for anti-microbial applications. *Discover Nano, 18*(1), 76. https://doi.org/10.1186/s11671-023-03856-y

Palmieri, V., Bugli, F., Lauriola, M. C., Cacaci, M., Torelli, R., Ciasca, G., Conti, C., Sanguinetti, M., Papi, M., & De Spirito, M. (2017). Bacteria meet graphene: Modulation of graphene oxide nanosheet interaction with human pathogens for effective anti-microbial therapy. *ACS Biomaterials Science & Engineering, 3*(4), 619–627. https://doi.org/10.1021/acsbiomaterials.6b00812

Panáček, D., Belza, J., Hochvaldová, L., Baďura, Z., Zoppellaro, G., Šrejber, M., Malina, T., Šedajová, V., Paloncýová, M., Langer, R., Zdražil, L., Zeng, J., Li, L., Zhao, E., Chen, Z., Xiong, Z., Li, R., Panáček, A., Večeřová, R., … Zbořil, R. (2024). Single atom engineered antibiotics overcome bacterial resistance. *Advanced Materials, 36*(50), 2410652. https://doi.org/10.1002/adma.202410652

Parambil, A. M., Prasad, A., Tomar, A. K., Ghosh, I., & Rajamani, P. (2024). Biogenic carbon dots: A novel mechanistic approach to combat multidrug-resistant critical pathogens on the global priority list. *Journal of Materials Chemistry B, 12*(1), 202–221. https://doi.org/10.1039/D3TB02374E

Prestinaci, F., Pezzotti, P., & Pantosti, A. (2015). Anti-microbial resistance: A global multifaceted phenomenon. *Pathogens and Global Health, 109*(7), 309. https://doi.org/10.1179/2047773215Y.0000000030

Rao, N., Singh, R., & Bashambu, L. (2021). Carbon-based nanomaterials: Synthesis and prospective applications. *Materials Today: Proceedings, 44*, 608–614. https://doi.org/10.1016/j.matpr.2020.10.593

Ricci, R., Leite, N. C. S., da-Silva, N. S., Pacheco-Soares, C., Canevari, R. A., Marciano, F. R., Webster, T. J., & Lobo, A. O. (2017). Graphene oxide nanoribbons as nanomaterial for bone regeneration: Effects on cytotoxicity, gene expression and bactericidal effect. *Materials Science and Engineering: C, 78*, 341–348. https://doi.org/10.1016/j.msec.2017.03.278

Saleemi, M. A., Kong, Y. L., Yong, P. V. C., & Wong, E. H. (2021). An overview of anti-microbial properties of carbon nanotubes-based nanocomposites. *Advanced Pharmaceutical Bulletin, 12*(3), 449. https://doi.org/10.34172/apb.2022.049

Salesa, B., Martí, M., Frígols, B., & Serrano-Aroca, Á. (2019). Carbon nanofibers in pure form and in calcium alginate composites films: New cost-effective antibacterial biomaterials against the life-threatening multidrug-resistant Staphylococcus epidermidis. *Polymers, 11*(3), 453. https://doi.org/10.3390/polym11030453

Sangnim, T., Puri, V., Dheer, D., Venkatesh, D. N., Huanbutta, K., & Sharma, A. (2024). Nanomaterials in the wound healing process: New insights and advancements. *Pharmaceutics, 16*(3), 300. https://doi.org/10.3390/pharmaceutics16030300

Sarkar, P., Ghosal, K., Chakraborty, D., & Sarkar, K. (2021). Chapter 20—Biocompatibility and biomedical applications of various carbon-based materials. In S. Thomas, C. Sarathchandran, S. A. Ilangovan, & J. C. Moreno-Piraján (Eds.), *Handbook of carbon-based nanomaterials* (pp. 829–875). Elsevier. https://doi.org/10.1016/B978-0-12-821996-6.00015-4

Snow, S. D., Lee, J., & Kim, J.-H. (2012). Photochemical and photophysical properties of sequentially functionalized fullerenes in the aqueous phase. *Environmental Science & Technology, 46*(24), 13227–13234. https://doi.org/10.1021/es303237v

Varghese, M., & Balachandran, M. (2021). Antibacterial efficiency of carbon dots against Gram-positive and Gram-negative bacteria: A review. *Journal of Environmental Chemical Engineering, 9*(6), 106821. https://doi.org/10.1016/j.jece.2021.106821

Vecitis, C. D., Zodrow, K. R., Kang, S., & Elimelech, M. (2010). Electronic-structure-dependent bacterial cytotoxicity of single-walled carbon nanotubes. *ACS Nano, 4*(9), 5471–5479. https://doi.org/10.1021/nn101558x

Xie, Y., Liu, H., Teng, Z., Ma, J., & Liu, G. (2025). Nanomaterial-enabled anti-biofilm strategies: New opportunities for treatment of bacterial infections. *Nanoscale*. https://doi.org/10.1039/D4NR04774E

Xin, Q., Shah, H., Nawaz, A., Xie, W., Akram, M. Z., Batool, A., Tian, L., Jan, S. U., Boddula, R., Guo, B., Liu, Q., & Gong, J. R. (2019). Antibacterial carbon-based nanomaterials. *Advanced Materials (Deerfield Beach, Fla.), 31*(45), e1804838. https://doi.org/10.1002/adma.201804838

Yang, Y., Shi, Q., Feng, J., Shu, X., & Feng, J. (2014). Preparation and antibacterial properties of an activated carbon sphere–quaternary phosphonium salt composite. *RSC Advances, 4*(92), 50708–50712. https://doi.org/10.1039/C4RA07282K

Yuan, X., Zhang, X., Sun, L., Wei, Y., & Wei, X. (2019). Cellular toxicity and immunological effects of carbon-based nanomaterials. *Particle and Fibre Toxicology, 16*(1), 18. https://doi.org/10.1186/s12989-019-0299-z

Chapter 12
Case Studies: Carbon in Sustainable Environmental Detection and Remediation

K. G. Kaushani, S. H. Kelaniyagama, T. N. Dharmapriya, and Gayan Priyadarshana

Abstract The increasing concerns over environmental pollution and climate change have driven global efforts to develop sustainable detection and remediation techniques. Carbon-based materials are essential in tackling these challenges due to their distinctive characteristics, including high surface area, adjustable porosity, electrical conductivity, and chemical stability. This chapter explores real-world case studies demonstrating the application of carbon-based materials, both bulk and nano-carbon, in addressing environmental challenges. Bulk carbon materials, including activated carbon, biochar, carbon black, and carbon fiber, find extensive application in water purification, air filtration, soil remediation, and photocatalysis, showcasing their cost-effectiveness in effectively removing various contaminants from air, water, and soil. Meanwhile, nano-carbon materials, including carbon nanotubes, graphene, activated carbon nanoparticles, carbon quantum dots, and nanodiamonds, offer enhanced capabilities in sensing hazardous pollutants, adsorbing toxic gases, removing heavy metals, and degrading chemical and biological contaminants. Carbon-based sensors, known for their high sensitivity and selectivity, are increasingly used in real-time monitoring of pollutants like heavy metals, volatile organic compounds, and greenhouse gases. Case studies illustrate these sensors' advanced solutions for early detection and rapid response. Innovative remediation technologies that leverage carbon materials' adsorption, catalysis, and filtration properties are also explored. Examples from water treatment plants, air purification systems, and soil remediation projects emphasize their cost-effectiveness, recyclability, and scalability. The effectiveness and sustainability of these approaches are emphasized concerning the circular economy by producing carbon materials from renewable resources like agricultural waste and biomass. It

K. G. Kaushani · S. H. Kelaniyagama · G. Priyadarshana (✉)
Department of Materials and Mechanical Technology, Faculty of Technology, University of Sri Jayewardenepura, Nugegoda, Sri Lanka
e-mail: gayanp@sjp.ac.lk

T. N. Dharmapriya
Department of Environmental Technology, Faculty of Technology, University of Colombo, Colombo, Sri Lanka

A. M. Parambil et al. (eds.), *Carbon*,
https://doi.org/10.1007/978-3-031-90613-8_12

highlights the growing significance of low-cost, eco-friendly carbon sources in environmental applications. Furthermore, it discusses challenges, including the regeneration of spent materials, large-scale implementation, and the need for environmental impact assessments. A comprehensive analysis of bulk and nano-carbon materials highlights their relative strengths, challenges, and future potential for advancing sustainable environmental detection and remediation.

Keywords Carbon · Bulk carbon · Environmental detection · Nano-carbon · Remediation

Abbreviations

AC	Activated carbon
CNS	Carbon nanostructure
CNT	Carbon nanotube
CO	Carbon monoxide
CO_2	Carbon dioxide
CQD	Carbon quantum dots
GHG	Greenhouse gas
GO	Graphene oxide
HRGO	Hydrothermally reduced graphene oxide
MOF	Metal-organic framework
MWCNT	Multi-walled carbon nanotube
NM	Nanomaterial
NO_x	Nitrogen oxides
QF	Quartz fiber
SO_2	Sulfur dioxide
SWCNT	Single-walled carbon nanotube
VOC	Volatile organic compounds

12.1 Introduction

Environmental pollution, driven by urbanization, industrial expansion, and advancements in the transportation sector, has caused substantial damage to ecosystems and threatened human health. The release of harmful pollutants into the atmosphere from diverse human activities, including metal mining and smelting, agriculture, industrial processes, and vehicle emissions, has emerged as a significant environmental concern. These pollutants include pesticides, heavy metals, dyes, greenhouse gases (GHGs), industrial waste, sewage, endocrine-disrupting chemicals, and volatile organic compounds (VOCs) (Gopinath et al., 2021). These pollutants have contributed to environmental issues, such as acid rain, global warming, contamination of water bodies, and the depletion of the ozone layer (Brião et al., 2020).

Finding efficient ways to remove these pollutants from different environmental media is crucial in addressing the urgent problem of environmental pollution. A range of strategies has been explored to address the removal of heavy metals, dyes, and emerging pollutants from aqueous systems, including adsorption, coagulation, filtration, and precipitation (Wan et al., 2016; Ahamad et al., 2019). Among these approaches, adsorption stands out as the most economical and environmentally friendly method, showing greater effectiveness in eliminating pollutants from water bodies. Unlike other methods, which often entail significant operational costs, generate sludge, and require substantial capital investments, adsorption presents a viable alternative that aligns with environmental preservation goals (Gopinath et al., 2021). The usage of a variety of materials to improve the adsorption process has increased recently due to developments in passivation technology.

Inorganic compounds, such as dolomite, kaolinite, layered double hydroxides, and clay minerals, have been used to increase the effectiveness of pollutant removal. Additionally, biomaterials like chitosan, lignin, and microbial biomass have gained popularity as effective passivation due to their environmental friendliness and abundance (Yadav et al., 2021; Lico et al., 2019; Wang et al., 2020).

Carbon-based materials have gained significant attention from researchers due to their classification as environmentally friendly and recyclable additives. These materials encompass a variety of substances, including carbon nanotubes (CNTs), carbon quantum dots (CQDs), carbon black, activated carbon (AC), graphene, and graphene oxide (GO) (Gopinath et al., 2021; Wang et al., 2022a, 2022b). The unique properties of carbon-based materials make them particularly effective in addressing environmental challenges. Notably, biochar, a form of carbon derived from biomass, has been extensively employed in various applications, including soil fertilization, carbon sequestration, nutrient retention, and heavy metal immobilization (Huang et al., 2019; Liu et al., 2020a, 2020b). Additionally, AC is the most commonly used carbon adsorbent for water remediation (Yadav et al., 2021).

The advantages of carbon-based materials extend beyond their effectiveness as adsorbents. Carbon nanostructures (CNSs) are notable for their exceptional chemical, electrical, thermal, and mechanical properties, making them invaluable across various sectors, including biomedicine, electronics, energy storage, sensors, and environmental remediation (Hussain, 2015). CNSs offer a high cation exchange capacity, elevated pH levels, and a porous structure rich in organic functional groups. These characteristics contribute to their large specific surface area, excellent thermal stability, a variety of pore sizes, high biocompatibility, and reduced mass transfer resistance. As a result, these CNSs are effectively employed in the adsorption of contaminants across various applications, such as heavy metal immobilization through soil treatment (Wang et al., 2020), reducing global warming through gas purification (Zhang et al., 2022), water remediation, anaerobic digestion, catalysis, and composting (Gopinath et al., 2021; El-Khouly & Fathy, 2021).

Integrating carbon-based materials into environmental monitoring and cleanup efforts presents an innovative and effective approach to addressing the pollution crisis. The diverse functionalities of these materials enable them to selectively adsorb and immobilize a wide range of pollutants, enhancing the overall effectiveness of remediation solutions. Additionally, their environmentally friendly nature and recyclability align with sustainable development goals by fostering a circular economy that minimizes waste and maximizes resource utilization.

This chapter thoroughly analyzes how carbon-based materials may be used to address environmental issues, particularly focusing on their applications in pollution detection and remediation. Through a series of case studies, we will assess the effectiveness of various carbon-based bulk and NMs (nanomaterials) in practical scenarios, highlighting their transformative potential for environmental remediation and monitoring efforts. Each case study will delve into the specific methodologies employed, the outcomes achieved, and the broader implications of these findings for future research and applications in environmental science.

12.2 Specific Examples of Bulk Carbon Materials in Environmental Remediation

Carbonaceous materials, particularly AC and biochar, have become popular as green materials for environmental insulation and restoration owing to their distinct physicochemical properties. These materials have large surface areas, high porosity, and strong adsorption capacities, enhancing their efficiency in multiple environmental uses. Ranging from water treatment and air filtration to bulk carbon materials and soil purification, these materials are adaptable and conform to the current trend of sustainability. Specific examples of the uses of these bulk carbon materials in various environmental remediation contexts are given below.

12.2.1 Activated Carbon in Water Purification

One of the materials most frequently used in water purification is AC. Its large surface area and diffusion of a porous structure allow a wide range of pollutants, including heavy metallic elements, pesticides, drugs, and organics, to be absorbed (Lundquist et al., 2019). This makes it effective in refining potable water, treating industrial effluents, and treating municipal water supply systems.

Deng et al. (2021), in a remarkable investigation, reported the potential use of coconut shell-based AC to remove heavy metals like lead (Pb) and cadmium (Cd) from polluted water. The coconut shell AC's large surface area and active functional groups, which may interact complexly and electrostatically with the metal ions, are responsible for the observed removal rates of over 90% for certain metals. This research focuses on AC's contribution to resolving heavy metal contamination, among the notable environmental challenges many countries face worldwide.

Additionally, AC is highly effective at eliminating new pollutants from water sources, such as pharmaceuticals and endocrine disruptors. The utilization of AC in the Sheng et al. (2016) research focused on eliminating pharmaceutical waste from wastewater treatment plants, namely the residues of ibuprofen, carbamazepine, and diclofenac. Due to its significant efficiency in the adsorption process, AC could remove as much as 90% of the contaminants in the first few hours of treatment. The work has shown how efficiently AC helps in phosphorus removal from aqueous liquors, even considering the growing concentration of such contaminants in water supplies around the world.

12.2.2 Biochar in Water Purification

Biochar, made by pyrolysis, a processing method that utilizes heat to convert organic matter in an oxygen-free atmosphere, has proved very affordable. AC, which is advanced and normally not easily accessible in rural or economically poor places such as developing countries, has been substituted by biochar hay for some time. Besides, it is often made from agricultural wastes such as rice husks, which makes it ideal for water treatment.

A conspicuous one was done by Zhiyuan Liu et al. (2020b) to assess the biochar's potential to remediate nitrate and phosphate lithophytic pollution from agricultural runoff. Biochar created from rice husks could remove nitrate ions from water at a level of up to 75% and phosphorus at 85% from water bodies. High concentrations of such nutrients lead to eutrophication in the water bodies, depleting oxygen levels and killing the animals. Therefore, biochar can serve as a remedy for reducing nutrient pollution, which in turn helps conserve water bodies.

Biochar has been studied for its potential to remove toxic metals from water. A study by Zhang et al. (2018) investigated removing arsenic from aquifers using pine wood and straw biochar. The study showed that biochar's surface characteristics, including surface microporosity and functional groups, led up to 95% adsorption of the organic compound arsenic. Because of this, biochar is a viable and affordable option for treating arsenic pollution of groundwater, a widespread problem worldwide, particularly in South Asia. Apart from recovering metals and nutrients, biochar also serves the purpose of adsorbing organic pollutants. For instance, biochar has been utilized to capture polycyclic aromatic hydrocarbons and other persistent organic pollutants from polluted water. The capability of biochar to adsorb such toxic materials on its surface and immobilize them through bonding with its functional groups within its mesoporous structure has made it a new concern in water treatment.

12.2.3 Activated Carbon in Air Filtration

In the field of air filtration, AC is mostly used in the elimination of VOCs, particulate matter, and gases, such as sulfur dioxide (SO_2), nitrogen oxides (NO_x), and carbon monoxide (CO). It is an essential component of air-cleaning systems used in both homes and factories because of its large surface area and capacity to adsorb gaseous pollutants. Chen and Yuan (2021) conducted research on the utilization of AC in the treatment of gases containing VOCs emitted from industrial exhaust systems. AC filters efficiently eradicated up to 98% of VOCs comprising pathological compounds, such as benzene, toluene, and xylene. This ability to efficiently prevent the discharge of hazardous chemical gases and vapors into the atmosphere explains the importance of AC in the fight against air pollution from various industries.

The use of AC extends even further in indoor air filtration systems, especially in spaces with high levels of all carbon compounds, including those released by construction materials, cleaning agents, and upholstery. Park and Hwang (2014) investigated using AC air purifiers to remove formaldehyde, a common VOC. Among the key findings was that using AC filters reduced the indoor air concentration of formaldehyde by as much as 88%, thus improving indoor air quality by lowering the risk of exposure to harmful substances. This shows the use of AC in air pollution, such as how it can reduce outdoor pollution, which is highly common when cities are overpopulated, and indoor pollution is inside buildings with several occupancy levels.

Apart from the VOCs, AC has also been useful as an air-cleaning agent, deodorizing the air and eliminating other gaseous pollutants. It uses its physical adsorption properties to obtain diverse contaminants from factory smoke to cooking odors, making it useful in any environment where air quality needs improvement.

12.2.4 Biochar in Soil Remediation

Because biochar improves soil fertility, immobilizes heavy metals, and increases carbon capture, its usage as a soil amendment has produced a favorable reaction for environmental sustainability. Because of its porous nature and surface-active functional groups, such waste can be adsorbed from water bodies and soil to minimize leaching to groundwater and plant uptake.

Biochar produced from sewage sludge used to clean soils with metal pollutants was studied in a key publication in the *Journal of Hazardous Materials* by Huang et al. (2019). The study indicated that biochar effectively reduces the bioavailability of the three metals by up to 70%, thereby limiting their uptake by plants and minimizing leaching into the surrounding soil. This protects the food chain from metal pollution and helps rehabilitate the polluted soil for crop production.

Aside from this, biochar has been a main subject of research in carbon sequestration and soil management for climate change mitigation. The research by Lehmann et al. (2024) demonstrated that biochar utilization in combination with common agricultural soils could increase soil organic carbon by up to 25–30%. It was also reported that biochar enhanced the physicochemical characteristics of the soil, like structure, water-holding capacity, and other nutrients, making it a useful input in agricultural practices. By improving soil quality and storing carbon, biochar helps not only rehabilitate certain sites but also combat global warming.

By successfully lowering the bioavailability of Cd in polluted soil, nanobiochar enhanced microbial diversity and plant development. In *Brassica chinensis* L., experiments showed reduced absorption of Cd and increased microbial biomass, with higher abundances of *Actinobacteria* and *Bacteroidetes*. This demonstrates the potential of nanobiochar for remediating soil (Liu et al., 2020a, 2020b).

Biochar has also been employed in the remediation of soils impacted by organic contaminants, like pesticides, gasoline, and other aliphatic hydrocarbons. Even with added nitrogen, its high surface area and functional groups allow it to capture and fix such contaminants, thereby inhibiting the dispersion of such contaminants in the environment. For this reason, biochar can be used for any kind of soil pollution due to production, agriculture, or cataclysms.

12.3 Case Studies of Carbon Nanomaterials in Environmental Applications

NMs are essential for the detection and control of pollutants through processes, such as adsorption, degradation, and isolation. Adsorption by NMs, nanocatalysts' degradation, and nanofillers' isolation or filtration are key mechanisms used to monitor and manage pollutants, even at low environmental concentrations (Gabal et al., 2019). Because of their large specific surface areas, conductivity, tunable surface chemistry, and numerous adsorption sites, functionalized carbon-based NMs can form potent complexes with organic molecules and heavy metals on a solid surface, making them effective adsorbents for the removal of pollutants from soil, wastewater, and air (Liu et al., 2022).

12.3.1 Pollutant Detection

Carbon-based NMs, functioning as nanosensors, can identify contaminants by detecting surface markers or amplifying the analytical signal. Their unique properties, such as large surface area, composition, electrical, magnetic, mechanical, and optical characteristics, make these nanosensors highly effective for sensitively detecting even trace amounts of contaminants, offering a more reliable approach than traditional methods (Yadav et al., 2021).

CNTs can significantly enhance pollution detection sensitivity and specificity, especially when functionalized. Non-functionalized CNTs often lack the precision required for efficient pollutant detection. However, their capacity to capture target pollutants can be greatly improved through covalent or non-covalent functionalization. Functionalization introduces active sites that specifically bind with contaminants, enhancing the electrochemical properties of CNTs and making them more effective for pollutant detection (Schroeder et al., 2019).

Covalent functionalization entails affixing functional groups to their surface to increase CNTs'stability and robustness under a variety of environmental circumstances. For example, single-walled carbon nanotubes (SWCNTs) functionalized with hexafluoroisopropanol (HFIP) have been used to detect contaminants like dimethyl methylphosphonate (DMPP). Covalent modifications improve pollutant detection by facilitating efficient chemical reactions, such as hydrogen bonding, thereby increasing the sensitivity and effectiveness of the detection process (Yadav et al., 2021).

Researchers have explored different functional groups, such as thiourea, urea, and squaramide, to functionalize SWCNTs for developing chemo-resistive sensors that can detect explosive vapors, including nitromethane and cyclohexanone, at low concentrations, as well as VOCs. These functionalized SWCNTs offer high sensitivity, repeatability, and stability in water and air environments, making them suitable for practical applications (Schroeder et al., 2019).

Functionalizing CNTs with metal oxides, such as tin oxide, through sol-gel techniques has enabled the detection of gaseous pollutants like nitrogen oxides (NO and NO_2), ethanol, and acetylene at high temperatures (Cargnello et al., 2012). In addition to detecting gaseous pollutants, CNTs can be utilized to identify organic compounds sensitively. The application of MWCNTs functionalized with silica networks and gold nanoparticles has broadened the capabilities of CNT-based sensors, allowing for the sensitive detection of compounds like dopamine and ascorbic acid (Schroeder et al., 2019). Another significant application of CNTs is in the identification of explosive chemicals. For instance, CNTs functionalized with pyrene units and amino groups have successfully detected trinitrotoluene (TNT), highlighting their effectiveness in material identification. Additionally, CNTs modified with cyclodextrin and boronic acid have been employed to selectively capture glucose and various chemical intermediates. These applications illustrate how functionalized CNTs can selectively bind to target contaminants, thereby enhancing the sensitivity and specificity of their detection capabilities (Wei et al., 2014). In addition, non-covalent interactions, particularly hydrophobic interactions with polymeric surfactants, can greatly enhance the adsorption capacity of CNTs. This capability enables CNTs to function as effective platforms for the adsorption of targeted conjugates, thereby facilitating the efficient detection of a wide range of environmental contaminants, such as biotin and staphylococcal protein A. Consequently, this characteristic broadens the scope of CNT applications in environmental monitoring and diagnostics, advancing sophisticated sensors that offer increased stability and sensitivity (Schroeder et al., 2019).

One of the critical pilot studies investigated SWCNTs as a high-sensitivity gas sensor for detecting nitrogen dioxide and ammonia in the ambient atmosphere. The sensing mechanism for these sensors involves a change in electrical resistance upon exposure to target gases by using the interaction between pollutant molecules and the SWCNT surface. Results showed detection limits as low as 10 ppb for NO_2 and 25 ppb for NH_3, demonstrating the unparalleled sensitivity and selectivity of the material (Zhang et al., 2022). This pilot-scale implementation demonstrates the potential of CNT-based devices in real-time air quality monitoring.

In one more pilot study combining MWCNTs with conducting polymers, the application of hybrid sensors was evaluated to detect VOCs in polluted air. The hybrid material increased sensor response due to the synergistic effect of CNTs and the polymer, showing a high selectivity toward toluene and benzene. This study provided the framework for preparing low-cost, scalable sensors for urban pollution monitoring (Cho et al., 2017). For the detection of 4-nitrophenol (4-NP) in polluted water, a nitrogen-doped hierarchical porous carbon (NPC)-based sensor electrode modified with chitosan (CTS) on an indium-tin oxide (ITO) electrode (CTS/NPC/ITO) showed excellent sensitivity and selectivity. The electrode demonstrated exceptional stability, repeatability, and real-sample application (Hu et al., 2018).

12.3.2 Remediation of Toxic Gases

The adsorption of combustion gases and by-product vapors is essential for companies since it entails the capture of harmful gases such as SO_2 and NO_x, along with greenhouse gases like carbon dioxide (CO_2) and methane. This process is crucial in reducing environmental pollution and curbing GHG emissions (Gopinath et al., 2021). The use of CNT-supported air filters to efficiently remove air pollutants demonstrates one of the main environmental applications of CNSs in air purification. Because of their high-energy binding sites, CNTs serve as highly effective sorbents, making them particularly well-suited for improving air purification, especially in removing hazardous air pollutants (El-Khouly & Fathy, 2021). Activated biochar compounds and various carbon-based NMs were used to investigate the adsorption-based sequestration of gases (Asgarpour Khansary et al., 2020).

CNTs are being explored for their potential use in capturing GHGs as part of global warming mitigation efforts, with promising results. CNTs may be employed as nano-adsorbents to separate different airborne contaminants because of their improved qualities and high affinity for selective contaminant adsorption. These materials possess a large surface area, increased pore diameter, and pore volume, contributing to their effectiveness (Saleem et al., 2022). The surface characteristics, resembling densely packed structures, increase the number of reactive sites that can bind to specific airborne contaminants (Gopinath et al., 2021). Functionalizing the nano-adsorbents by introducing specific groups can further enhance the active sites, potentially improving their adsorptive capacity (Baby et al., 2019). Methane and CO_2 emissions from automobile exhaust and industrial use chimneys are reduced by using CNTs. Additionally, they can capture GHGs generated by coal mining operations and power plants (Saleem et al., 2022). Research has been done on CNTs with single and multiple walls, alone and in conjunction with chemical substances like polyethylenimine (PEI) (Keller et al., 2018). Zhu et al. (2019) studied the integration of metal-organic frameworks (MOFs) with CNTs and GO for the removal of VOCs from the environment, aiming to enhance environmental remediation efforts (Zhu et al., 2019). The findings revealed that incorporating nanocarbons into the MOF matrix can significantly improve the mechanical, electrical, and chemical durability and thermal stability of the original MOFs, creating a synergistic effect that enhances both materials.

CNT filters have recently been utilized in air purification processes, such as particle removal, for filtering the inlet airflow of gas turbines in industrial plants (Gabal et al., 2019). High-efficiency particulate air (HEPA) filters and AC filter systems effectively address indoor particulate matter (PM) and ozone pollution. Polymer nanofibers have become the most commonly used nanofibers in air filtration systems due to their efficiency (Bratovcic, 2019; Li et al., 2014; Parham et al., 2013; Park & Hwang, 2014). A study by Li et al. (2014) discusses an aerosol-based method for producing CNT/quartz fiber (QF) filters with gradient nanostructures for large-scale manufacturing. Incorporating CNTs into the QF filter significantly enhanced the filter's performance in air purification.

In contrast to other carbon-based NMs, GO is very successful in reactively adsorbing basic gases because of its layered structure and noticeable acidity. Sui and Han (2015) studied the CO_2 adsorption capacity of hydrothermally reduced graphene oxide (HRGO) and found it to be 2.4 mmol g^{-1} at 1.0 bar and 273 K. The hydrothermal method created a 3D porous network in HRGO with a high surface area and pore volume. Compared to samples synthesized at 80 and 120 °C, HRGO demonstrated a higher adsorption capability at 100 °C. The study highlighted the role of polar groups and hydrogen bonding between CO_2 and oxygen-containing functional groups in enhancing adsorption. Several studies have explored hybrid designs that combine metal-oxide nanoparticles with graphene or its derivatives. These frameworks have improved gas sensing abilities, especially regarding sensitivity and selectivity at ambient temperatures (Gupta Chatterjee et al., 2015).

Filters in air particle monitoring systems were modified using carbon NMs, such as reduced GO, graphitic carbon nitride nanosheets, and CNTs. CNT-modified filters demonstrated greater particulate matter adsorption efficiency because of their wide surface area and pore volume. They were also able to detect heavy metals like Cd, providing improved monitoring capabilities (Attia et al., 2024). The application of several carbon-based NMs for the adsorption of harmful gases is demonstrated in case studies in Table 12.1. These examples demonstrate the potential of various materials in environmental remediation by highlighting their efficacy in absorbing pollutants, including CO_2, SO_2, and NO_x.

Table 12.1 Adsorption of toxic gases from the environment utilizing carbon-based nanomaterials

Nanoadsorptive materials	Types of nanoparticles	Surface area (m^2/g)	Conditions for adsorption	Target pollutant gases adsorbed	Adsorption capacity	Source
Carbon nanotubes (CNTs)	Transition metal-encapsulated nitrogen-doped carbon nanotubes (M@NCNTs; M = Fe/Co/Ni)	392.3 m^2/g for NCNT; 196.8 m^2/g for Fe/Fe_3C encapsulated NCNT; 249.6 m^2/g for Co-encapsulated NCNT; and 274.0 m^2/g for Ni-encapsulated NCNT	Temperature: 25–100 °C; pressure: 2–20 bar	CO_2	1.1–14.0 mmol/g for Co; 1.0–13.2 mmol/g for Ni; 1.5–15.4 mmol/g for Fe/Fe_3C; and 0.6–7.4 mmol/g for NCNT	Ghosh and Ramaprabhu (2019)
	CNT hollow fiber-silica composite immobilized with 20 wt% PEI	283.2	Temperature: 80 °C; pressure: 0.15 bar; flow rate: 150 mL/min; duration: 45 min	CO_2	1.92 mmol/g	Keller et al. (2019)
	Single-walled CNTs	–	Temperature: 27 °C; pressure: 100–500 kPa	N_2, SO_2	17.5–34 mmol/g (SO_2); 0.4–2.3 mmol/g (N_2)	Hu et al. (2019)
	Double-walled CNTs	–	Temperature: 30 °C; pressure: 0–3 bar	SO_2, N_2, H_2S, CO_2	5.5 mmol/g (SO_2); 0.42 mmol/g (N_2); 0.22 mmol/g (H_2S); 2.54 mmol/g (CO_2)	Majumdar et al. (2018)
	Highly porous multi-walled CNTs impregnated with 20 wt% PEI	34.7	Temperature: 30 °C; pressure: Ambient; flow rate: 150 mL/min	CO_2	93.3 mg/g	Keller et al. (2018)
	CNTs on a quartz filter		Room temperature; pressure: 10^{-5} Pa	VOCs		

(continued)

Table 12.1 (continued)

Nanoadsorptive materials	Types of nanoparticles	Surface area (m^2/g)	Conditions for adsorption	Target pollutant gases adsorbed	Adsorption capacity	Source
Graphene	Pillared graphene structures	–	Pressure: 35 bar	CO_2, H_2, CH_4	58.9 wt% (CO_2); 0.40 wt% (H_2); 22.3 wt% (CH_4)	Pedrielli et al. (2018)
Carbon quantum dots (CQDs)	CQDs-modified FeOOH nanocomposites	127.2	Room temperature, flow rate: 3 L/min, duration: 36 s, gas concentration: 50 ppm	NO	22 wt%	Huang et al. (2018)

12.3.3 Remediation of Soil and Water

Wastewater treatment from both the industrial and urban environments has attracted a lot of interest. As the population grows rapidly and various environmental pollutants increase, ensuring access to clean water has become more difficult, turning water scarcity into a pressing issue that demands urgent action. Contaminants such as dissolved or suspended sediments, synthetic or natural debris, and toxic heavy metals pose a significant threat to both the future of humanity and aquatic ecosystems. Human health and societal progress are closely tied to the availability of clean water. Consequently, treating wastewater streams before they are reintroduced into the environment has become an urgent priority to ensure the safe recirculation of clean water (El-Khouly & Fathy, 2021). Various strategies for environmental remediation are classified into physical, chemical, biological, and advanced oxidation processes (Gopinath et al., 2021). Among these methods, adsorption systems are widely used for their simplicity, ease of operation, and high efficiency in removing contaminants from water sources and soil. The potential of CNTs and graphene-based compounds as efficient adsorption materials for the treatment of wastewater, able to remove a variety of contaminants, has been emphasized by recent studies (Baby et al., 2019; Chung et al., 2021; Myung et al., 2019). ACs have long been used in gas adsorption and wastewater treatment due to their large surface area and well-defined porous structure, making them highly effective for these applications (Gopinath et al., 2021). However, the sorption capacity of AC sorbents is limited when dealing with traditional contaminants like dyes and large-molecular-size oil molecules. This hinders their effectiveness in wastewater treatment, posing challenges for researchers aiming to remediate surface and groundwater. CNSs have proven to perform exceptionally well in a range of industrial and environmental applications, such as fuel storage, photocatalysis, separation, purification, and regeneration. Their popularity as adsorbents has grown due to their rapid equilibrium times, ease of regeneration, and ability to functionalize their surfaces with different groups, enhancing their adsorption capacity (Ali et al., 2019).

12.3.3.1 Adsorption of Heavy Metal Ions

The surface characteristics of the NMs and their functional groups determine the main methods by which CNSs adsorb metal ions from aqueous solutions. These mechanisms include physical adsorption, precipitation, sorption/reduction, ion exchange, and electrostatic interactions. Together, these processes facilitate the effective uptake of metal ions by CNSs (Xu et al., 2018).

The ability of carbon-based NMs to adsorb heavy metals is often influenced by the interaction between metal ions and surface oxygen-containing groups. Because of their manufacturing method, GOs are naturally rich in oxygen-containing functional groups, unlike CNTs, which usually need acid or oxidation treatments to improve their surface with such groups. This inherent functionalization allows GOs to be effective adsorbents for heavy metals without additional treatments (Liu et al., 2022; Manna & Retna Raj, 2018).

The large surface area, ease of surface functionalization, porous structure, and nanoscale dimensions of single-walled carbon nanotubes (SWCNTs) make them well-suited for controlling environmental pollution. These features make SWCNTs highly effective for applications in water purification, offering significant potential for removing pollutants (Baby et al., 2019). Alijani and Shariatinia (2018) developed a SWCNT-based nanocomposite that incorporates magnetite cobalt sulfide. The resulting nanocomposites demonstrated exceptional efficacy in mercury removal, achieving an adsorption rate of over 99.56% within 7 min. In contrast, SWCNTs alone could adsorb only 45.39% of the mercury. Zazouli et al. (2014) developed SWCNT nanocomposites functionalized with L-cysteine and utilized them for mercury removal from water. The study found that the SWCNT-cysteine nanocomposites achieved a mercury adsorption efficiency of 95%. Additionally, a SWCNT-polysulfone nanocomposite membrane was developed by Gupta et al. (2015) to remove heavy metals, with the incorporation of SWCNTs, resulting in a smoother surface and reduced pore size. This enhanced membrane achieved high rejection rates, eliminating 94.2% of Pb^{2+}, 87.6% of As^{3+}, and 96.8% of Cr^{6+}. In comparison, a membrane without SWCNTs only rejected 30.3%, 28.5%, and 28.3% of the ions Cr^{6+}, As^{3+}, and Pb^{2+}, respectively. These findings demonstrate how the addition of SWCNTs significantly improved the membrane's efficiency.

Additionally, multi-walled carbon nanotubes (MWCNTs) have been used extensively in water treatment, particularly for the adsorption of heavy metal ions via chemical interactions with their functional groups. Research has demonstrated that oxidized MWCNTs have a high capacity for adsorption and are effective in eliminating Cr^{6+}, Pb^{2+}, and Cd^{2+} ions from water (Baby et al., 2019). MWCNT composites containing compounds like iron (III), zinc, and aluminum oxide are synthesized through co-precipitation. These composites have proven effective in removing arsenic, copper, nickel, chromium, and lead ions. Temperature and pH levels are two variables that affect these nanocomposites'adsorption effectiveness; adjusting these may improve the sorption process (Gusain et al., 2019). Elsehly et al. (2016) utilized commercial MWCNTs to eliminate iron and manganese from aqueous solutions. At concentrations of 50 ppm, the MWCNTs were able to achieve removal efficiencies of 71.5% for iron and 52% for manganese.

Because graphene-based materials have special physicochemical qualities, such as a huge surface area, exceptional electrical conductivity, high thermal mobility, remarkable mechanical strength, and adjustable surface chemistry, their usage in water treatment is growing in popularity (Tabish et al., 2018; Mohan et al., 2018). Tabish et al. (2018) synthesized porous graphene, which is used as an adsorbent to remove heavy metal ions and other impurities from water. Their research demonstrated that this porous graphene material achieved an 80% effectiveness in removing As^{3+} from aqueous solutions. Additionally, the regeneration and recycling of the material were shown to maintain its water treatment capabilities. Guo et al. (2018) developed a partially reduced GO nanocomposite using an in situ co-precipitation technique with Fe_3O_4 to remove Pb^{2+} ions from water. The efficiency of the suggested nanocomposite in eliminating Pb^{2+} ions from aqueous solutions was demonstrated by its adsorption capacity of 373.14 mg/g. Zhang et al. (2018) modified reduced graphene oxide with 4-sulfophenylazo (rGOs) and used it to eliminate various heavy metal ions from aqueous solutions. The modified material demonstrated impressive adsorption capacities, with 689 mg/g values for Pb^{2+}, 59 mg/g for Cu^{2+}, 66 mg/g for Ni^{2+}, 267 mg/g for Cd^{2+}, and 191 mg/g for Cr^{3+}, respectively.

Graphene-based sensors have been successfully implemented in industrial settings to monitor wastewater effluents for hazardous pollutants. For example, a water treatment plant in Singapore deployed graphene oxide (GO) nanosheets integrated into an electrochemical sensing platform. The system demonstrated high efficiency in detecting heavy metals, such as arsenic and chromium, at concentrations below regulatory limits. The sensors provided real-time data, enabling timely interventions and compliance with environmental standards (Rodríguez et al., 2020). This case study illustrates the practical deployment of carbon nanomaterials in addressing global water pollution challenges.

A review of the use of CQDs in the adsorption of radionuclides, including Sr(II), Eu(III), and U(VI), demonstrated their effectiveness in removing these radionuclides from solutions (Wang et al., 2022a, 2022b). Furthermore, selective fluorescence detection of Hg^{2+} (83 nM) and sulfide (11 nM) with low cytotoxicity was made possible by nitrogen- and sulfur-codoped carbon dots (N, S-CDs) with 16.1% quantum yield. These nanoparticles were effectively used in environmental monitoring (Wu & Tong, 2019). Wang et al. (2017) utilized an in situ microplasma-assisted technique to create the CQDs, which were subsequently incorporated into functionalized ordered mesoporous silica ($SBA\text{-}NH_2$), as demonstrated in Fig. 12.1a. This composite maintained the enormous specific surface area of the mesoporous silica while maintaining the fluorescence capabilities of the CQDs. As a result, the composite can simultaneously monitor the adsorption process and exhibit higher effectiveness in uranium adsorption.

A patented methodology, US 10,376,845 B2 (2019), describes the preparation of fluorescent carbon quantum dot sensors for detecting heavy metal ions in water, such as lead (Pb^{2+}) and mercury (Hg^{2+}). When these ions are present, CQDs exhibit substantial fluorescence quenching, which makes it possible to detect them quickly and precisely. The novelty incorporates CQDs into portable devices for a low-cost, environmentally friendly approach toward monitoring water quality. The commercial potential of this patent lies in its applicability to large-scale environmental surveillance.

12.3.3.2 Adsorption of Organic and Inorganic Pollutants

The surface functional groups of various carbon-based NMs differ, indicating that the interactions between organic and inorganic contaminants and these materials can vary. Nevertheless, the primary mechanisms to remove organic and inorganic contaminants typically include electrostatic interactions, π–π interactions, physical

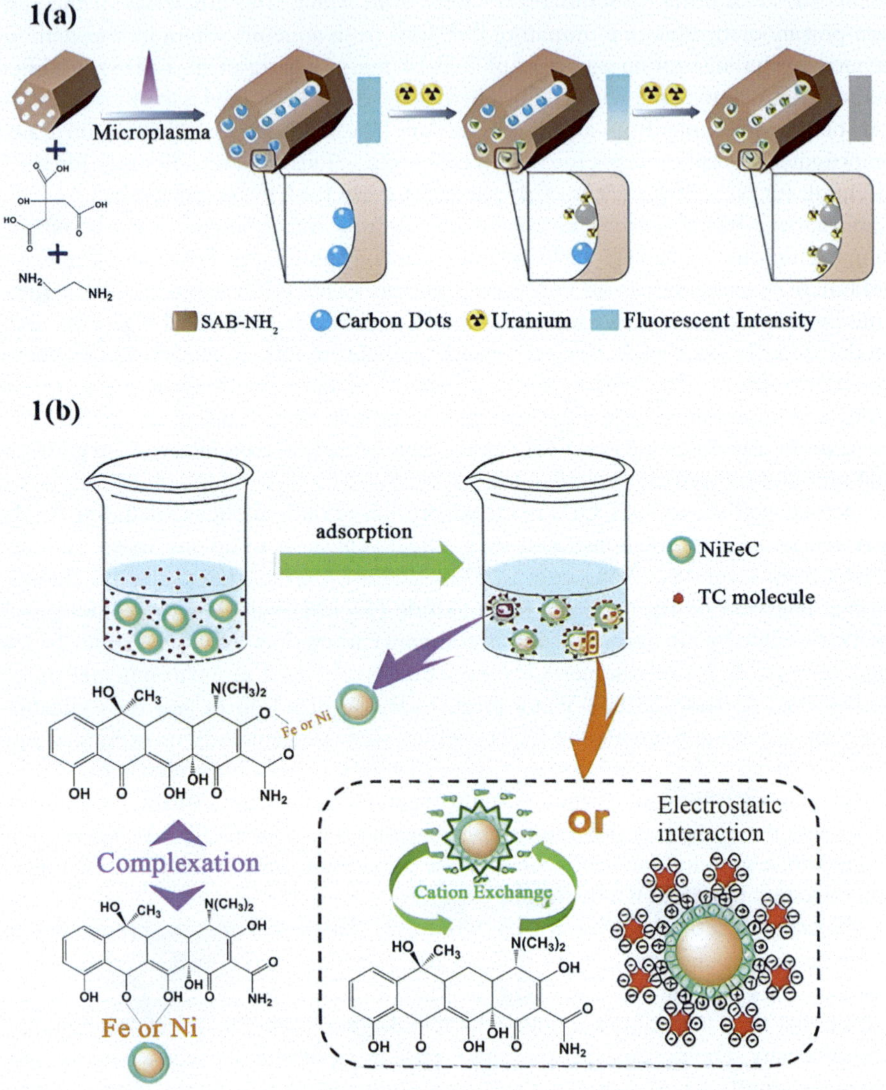

Fig. 12.1 Illustration of (**a**) adsorption and monitoring of uranium with CQDs/SBA-NH_2 nanocomposites. (Reproduced with permission from Wang et al. Copyright 2017, American Chemical Society.) (**b**) Adsorption of TC on $NiFe_2O_4$/CQDs nanocomposites. (Reproduced with permission from Liu et al. Copyright 2017, Elsevier)

adsorption, Fenton/Fenton-like processes, and photocatalytic degradation (Huang et al., 2019).

Jeong et al. (2017) employed various carbon-based NMs, including expanded graphite (EG), SWCNTs, and MWCNTs, to treat wastewater. According to the findings, MWCNTs had a noticeably greater adsorption capability than EG. Chen and

Yuan (2021) used three distinct types of CNSs: SWCNTs, oxidized SWCNTs, and non-porous graphite powder to investigate the removal of six compounds from aqueous solutions: thiophene, pyrimidine, benzene, aniline, 2-aminopyrimidine, and 4,6-diaminopyrimidine. Their study revealed significant variations in the affinities of the different organic compounds for the three adsorbents. Another study investigated the adsorption of fluoride onto SWCNTs in various aqueous media (Balarak et al., 2016). The research examined several factors affecting fluoride removal, including fluoride concentration in the solution, contact duration, dosage, temperature, pH, and agitation speed. The findings indicated that at 298 K and a pH 6.0, the maximum fluoride adsorption capacity reached approximately 60 mg/g. Additionally, the study suggested that fluoride anions could adsorb to SWCNTs naturally and through an endothermic process. Recent studies have demonstrated that CNTs and their functionalized derivatives outperform commercial ACs and clays as adsorbents for organic dyes (Rajabi et al., 2017).

The elimination of certain organic dyes, such as malachite gray, rhodamine B, and acriflavine, has been studied using GO foam. At 310 K, the obtained GO foam demonstrated impressive sorption capabilities, achieving values of 460.0 mg/g for acriflavine, 351.4 mg/g for rhodamine B, and 235.4 mg/g for malachite gray (Jayanthi et al., 2016). Yu et al. (2017) recently investigated tetracycline's sorption onto a magnetic graphene oxide sponge (MGOS, FeO_4/GO). At 473 mg/g, the Fe_3O_4/GO sponge demonstrated an impressive tetracycline adsorption capacity, around 50% exceeding that of GO alone.

Pesticides are among the most hazardous organic pollutants in aquatic environments due to their potential to cause cancer in both humans and animals. The potential for utilizing graphene-based materials as effective adsorbents for cleaning up pesticide-contaminated wastewater is significant. Tang et al. (2013) used graphene and TiO_2 as a new magnetic photocatalyst to extract the common herbicide 2,4-dichlorophenoxyacetic acid (2,4-D) from water under simulated solar light irradiation. They revealed a high adsorption capacity for organic pollutants and excellent visible spectrum light absorption. This combination achieved nearly 100% photocatalytic efficiency, surpassing the 33% efficiency observed with commercial P25.

Furthermore, the CQDs/$ZnFe_2O_4$ nanocomposite exhibits strong adsorption potential and can be effectively used to remove methyl orange (MO) (Shi et al., 2018). Additionally, to remove tetracycline (TC), a solvothermal one-pot method was employed to synthesize the $NiFe_2O_4$/CQDs nanocomposite, as demonstrated in Fig. 12.1b. With the $NiFeO_4$/CQDs nanocomposites, this method greatly increased the optimum capacity for adsorption, which was 591.72 mg/g. These results demonstrate the promising potential of $NiFeO_4$/CQDs nanocomposites for TC removal from wastewater (Liu et al., 2017). Table 12.2 summarizes case studies involving various carbon-based NMs utilized in water treatment processes.

Table 12.2 Summary of case studies highlighting the use of various carbon-based nanomaterials in water treatment processes

Nanoadsorptive materials	Types of nanoparticles	Target pollutant	Surface area (m^2/g)	Conditions for adsorption	Adsorption capacity	Source
Single-walled carbon nanotubes (SWCNTs)	Iron(II) chloride tetrahydrate/iron(III) chloride hexahydrate/ammonia	Xylene	187.34	Temperature: 25 °C; pH: 8.0	50 mg/g	Pourzamani et al. (2017)
	Iron-SWCNTs	Arsenic(V)	–	Temperature: 25 ± 1 °C; duration: 0–24 h; pH: 4.0; dosage of adsorbent: 0.4 g/L	49.65 mg/g	Ma et al. (2017)
	CNT-powdered activated carbon composite	Cadmium(II)		Room temperature; duration: 40 min; pH: 7.0	98.35%	AlSaadi et al. (2016)
	Potassium permanganate-CNTs	Mercury(II)	199.4	Room temperature; duration: 28 min; pH: 5.5; dosage of adsorbent: 5 mg	187.0 mg/g	AlOmar et al. (2017)
Multi-walled carbon nanotubes (MWCNTs)	Iron-MWCNTs	Arsenite(III); arsenic(V)	–	Temperature: 25–45 °C; duration: 60–90 min; pH: 7.0; dosage of adsorbent: 0.2 g/L	210.0–220.0 mg/g	Alijani and Shariatinia (2017)
	Iron(III) oxide-MWCNTs	Toluene; paraxylene	216	Temperature: 25 °C; pH: 7.0	381.18 mg/g (toluene); 458.52 mg/g (paraxylene)	Abbas et al. (2017)
	MWCNT-carboxylic acid cysteamine composite	Amido black 10B	–	Temperature: 25 °C; dosage of adsorbent: 0.5 mg; pH: 6.0	131 mg/g	Sadegh et al. (2016)
	Magnetic ammonia-functionalized MWCNTs	Methylene blue	202.4	Temperature: 25–65 °C; duration: 1–120 min; pH: 2.0–10.0; dosage of adsorbent: 0.8 g/L	178.57 mg/g	Ahamad et al. (2019)
	MWCNTs	Unleaded gasoline	108.78	Room temperature; duration: 5–120 min; dosage of adsorbent: 0.2–0.8 g	40–90%	Lico et al. (2019)

Graphene	GO-hydrated manganese oxide nanocomposite	Lead(II)	383.9	Temperature: 25 °C; pH: 5.0; duration: 24 h; dosage of adsorbent: 0.01 g	553.6 mg/g	Wan et al. (2016)
	GO	Methyl orange (MO); rhodamine blue (BR 12)	133	Temperature: 20–35 °C; duration: 10–120 min; pH: 3.0–9.0; dosage of adsorbent: 20 mg	16.8 mg/g for MO; 63.7 for RB 12	Robati et al. (2016)
	Magnetic freeze-dried GO/ferric oxide nanocomposite	Rhodamine blue	720.09	Temperature: 25 °C; duration: 24 h; pH: 7.0; dosage of adsorbent: 0.007 g	126.6 mg/g	Faghihi et al. (2016)
	Graphene hydrogel	Ciprofloxacin	–	Temperature: 75 °C; duration: 6 h	348 mg/g	Yiran Sun et al. (2019)
	Graphene nanoplatelets (GNP) and graphene magnetite (GM)	Emulsified oil	120–150	Temperature: 25–50 °C; duration: 0–120 min; pH: 2.0–12.0; dosage of adsorbent: 1.0–7.0 g/L (GM)	100 mg/g for GNP; 85 mg/g for GM	Abou Chacra et al. (2018)
	GO coated with highly stable lanthanum-substituted manganese ferrites (LMF)	Perfluorooctanoic acid	–	Room temperature; duration: 24 h; dosage of adsorbent: 1.0 g/L	1.61 mg/g	Elanchezhiyan et al. (2020)
Carbon quantum dots (CQDs)	CQDs-decorated Bi_2WO_6 nanocomposite	VOCs	–	Under UV irradiation and visible light	Enhanced photocatalytic oxidation activity for acetone and toluene	Qian et al. (2016)
	ZnS/CQDs nanocomposite	Alizarin red S	–	Duration: 250 min, under visible light	89%	Kaur et al. (2016)

12.4 Challenges, Insights, and Scaling Potential of Carbon-Based Materials

12.4.1 Lessons Learned

Carbonaceous materials, particularly AC and biochar, have demonstrated a high adsorption capability on a wide range of contaminants because of their high surface area and considerable porosity. These materials are used for environmental remediation. Extensive studies have been carried out to determine how effective AC is in removing heavy metals, organic impurities, and pharmaceuticals and their residues from various sources of water. According to reviews, AC is perfect for use in any water purification system since it can remove more than 90% of the toxic pollutants (Pb and Cd) in polluted wastewater (Deng et al., 2021). Alternatively, biochar is a cheaper substitute for AC. Biochar is a biomass product obtained from the pyrolysis of agricultural waste, and this material has been highly successful in decontaminating soil and water. It has been shown that biochar improves soil quality, sequesters heavy metals, and enables plant nutrient retention, thus contributing to sustainable land use (Sun et al., 2021).

The use of bulk carbon in different facets of ecological improvement has enhanced the ecological conditions and provided affordable means, particularly in third-world countries. Biomass is present in abundant quantities that are helpful for the generation of biochar, and this is usable by communities that have no means for the adoption of high-level filtration systems. Besides, biochar's high carbon replacement potential has also become a critical weapon in reducing CO_2 levels in the atmosphere, aiding in climate change control (Lehmann et al., 2024).

Among the prominent types of nano-carbon materials are CNTs, graphene, and CQDs, which have improved environmental detection and remediation. These materials are nanoscale in structure, having a very high surface area and variable surface chemistry; hence, a higher efficiency in the adsorption of trace pollutants and detection of hazardous substances is possible. For example, due to their good adsorption capacity, CNTs can effectively absorb VOCs and heavy metals (El-Khouly & Fathy, 2021). Graphene, a nano-carbon material, is widely recognized for its strength and ability to adsorb and break down contaminants at the molecular level. Its sensors can detect trace pollutants, such as heavy metals and VOCs, enhancing environmental monitoring. Similarly, CQDs can degrade organic pollutants in water, offering benefits for water treatment. The advanced properties of nano-carbon materials give them an advantage over bulk materials in environmental cleanup. However, challenges related to their complex fabrication and cost-effective production limit their broader application.

Carbon materials from renewable natural resources, like biochar from agricultural residues (e.g., rice husks and wood chips), show great potential for cost-effective environmental management. Biochar offers an affordable method for water treatment and soil reclamation, making it particularly beneficial for developing countries with limited access to expensive filtration systems. Additionally, producing biochar from waste materials is environmentally friendly, contributing to sustainable waste management strategies for cleaner resource utilization (Lehmann et al., 2024).

In agricultural practices, biochar serves as a valuable soil amendment, particularly in water management, by integrating active and passive resource management approaches. Farmers can convert crop residues into biochar to enhance soil quality while reducing the need for chemical fertilizers and minimizing pollution. This dual benefit has led to the growing adoption of biochar as a key component in promoting sustainable agriculture and land management practices (Sun et al., 2021). Additionally, biochar acts as a stable carbon sink, enhancing carbon sequestration and mitigating global warming by lowering CO_2 levels in the atmosphere (Lehmann et al., 2024). It also improves soil structure, moisture retention, and fertility, increasing crop productivity. Encouraging the broader use of biochar in agriculture could yield ecological and economic advantages (Zhang et al., 2018).

12.4.2 Challenges Encountered

When it comes to utilizing carbonaceous materials, particularly AC, a significant challenge arises regarding the regeneration of spent materials. After their initial use, AC can become saturated with contaminants, leading to diminishing returns in pollutant adsorption. Additionally, the current dimensions, methods, and potential production rates of nano-carbon materials cannot be further enhanced. This limitation challenges the broader application of these materials in environmental management, especially in developing regions where such solutions are most needed (Asgarpour Khansary et al., 2020). One of the most critical considerations in applying carbonaceous adsorbents, particularly AC, is their regeneration potential. Over time, AC can become saturated with impurities, rendering it less effective for contaminant adsorption. While some methods, such as thermal treatment and chemical washing, can help restore a portion of the material's adsorption capacity, these processes are often energy-intensive and costly (Sheng et al., 2016).

Additionally, these regeneration techniques may result in the release of secondary pollutants, exacerbating existing environmental issues. Therefore, addressing the challenges of longevity and cost-effectiveness in developing carbon-based materials is essential. One potential solution for enhancing the regeneration of AC is the incorporation of biomass, such as biochar, which may facilitate more sustainable and efficient recovery methods (Huang et al., 2019). Despite their remarkable pollution detection and destruction capabilities, the massive synthesis of nano-carbon materials, like graphene and nanotubes, is still a major issue. The complex, interdependent processes involved in their production require capital-intensive equipment and specialized expertise, limiting their widespread use in environmental management, especially in developing regions (Asgarpour Khansary et al., 2020). Therefore, efficient regeneration methods are essential to enhance the durability and cost-effectiveness of carbonaceous materials.

Nanotechnology's concerns for human health and the environment must be fully addressed before it can be extensively used. Research on NMs' use in environmental cleaning is still in its infancy, and it is still uncertain how they will affect ecosystems and the well-being of people in the long run (Saleem et al., 2022). Comprehensive environmental assessments are crucial before scaling up carbon-based materials, particularly nano-carbon products, as these processes pose potential risks (Ghosh & Ramaprabhu, 2019). For instance, water treatment processes may inadvertently introduce NMs into aquatic systems, adversely affecting sensitive organisms due to their small size and high reactivity (Zhu et al., 2019). While carbon-based materials demonstrate sustainability and efficacy, ensuring they do not create new environmental issues is vital for future applications. Continued research is necessary to understand their environmental behavior and to establish safe disposal and recycling methods.

12.4.3 Potential for Scaling to Broader Applications

There is great potential in exploring how carbon-based materials can be integrated into existing environmental systems. For instance, AC filters can be employed in municipal drinking water treatment plants, which enhance the quality of the water and extract impurities from it. In the same way, pollution induced by construction activities could be monitored using nano-carbon sensors integrated into the air pollution monitoring systems (El-Khouly & Fathy, 2021).

Scaling the use of carbon-based materials can also be done more efficiently without incurring unnecessary costs, as this type of infrastructure can be deployed using available facilities. This strategy then further lessens the need for completely different infrastructure, which allows for savings in cost and complications caused by attempts to disperse these technologies to larger spheres (Gupta et al., 2015).

The transition to carbon-based materials will likely face troubles unless the industry, research agencies, and governmental organizations unite. The collaboration of scientists, engineers, and manufacturers is important and will speed up the pace of innovations concerning these materials' synthesis, production, and application. Government help regarding the financing and regulations controlling the adoption of carbon-based technologies in the broader use of these materials will be one of the additional aids (Baby et al., 2019). Partnering with industry and academia will be equally important in overcoming the challenges of scaling nano-carbon materials. There will be an adequate need for research on such more economical and cost-effective synthesis techniques to bring down the costs of production of these materials to wider applications (Asgarpour Khansary et al., 2020).

The capacity to scale carbon-based solutions will also be important in emerging nations where it is becoming more difficult to find clean soil, water, and air. Biochar especially has great potential for environmental cleanup as a low-cost and locally produced material. There are favorable circumstances for developing countries to gain from low-cost bio-resources due to the introduction of biochar for agricultural and water treatment activities instead of the procurement of expensive biochar (Yuqing Sun et al., 2021). The production of biochar using biomass available in the locality, such as agricultural residues, enhances the chances of the local communities dealing with the problem of environmental pollution and, at the same time, generating income. Governments and international bodies must endorse the promotion of the usage of biochar in developing areas via education, training, and monetary support (Huang et al., 2019).

12.5 Comparative Analysis of Bulk and Nano-Carbon Materials in Environmental Applications

Bulk carbon materials and carbon nanostructures show the particular benefits and disadvantages pertaining to the use of environmental detection and remediation applications. From this perspective, a comparison was performed concerning different properties: surface characteristics, porosity, optical properties, adsorption efficiencies, and sustainability (Table 12.3).

Table 12.3 Comparison of bulk and nano-carbon materials in environmental detection and remediation

Attribute	Bulk carbon materials (e.g., AC, biochar)	Nano-carbon materials (e.g., CNTs, graphene)
Surface area	Moderate (e.g., activated carbon: ~1000 to 1500 m^2/g)	Very high (e.g., graphene: ~2600 m^2/g; CNTs: ~100 to 300 m^2/g for MWCNTs)
Porosity	High, primarily micropores and mesopores	Tailored porosity with high accessibility to active sites
Surface chemistry	Less diverse functional groups; dependent on precursor and activation methods	High functional group diversity; functionalization possible to enhance reactivity
Adsorption efficiency	Moderate; effective for general contaminants (e.g., organic pollutants, metals)	High; superior for specific pollutants due to higher binding capacity
Optical properties	Limited or negligible	Unique optical properties such as fluorescence, tunable bandgaps
Cost	Low, inexpensive, and readily available	High; expensive due to sophisticated production methods
Sustainability	Renewable sources (e.g., biochar from biomass); carbon sequestration potential	Less sustainable; may involve toxic precursors and sophisticated production methods
Environmental impact	Generally low; biochar is eco-friendly and improves soil health	Potentially high if nanomaterials leach into ecosystems; long-term effects not fully understood
Applications	Large-scale, cost-effective uses (e.g., municipal water treatment, industrial air filters)	Specialized uses in advanced applications (e.g., detection of trace contaminants, remediation of specific sites)
Ease of manufacturing	Simple and scalable processes (e.g., pyrolysis for biochar, activation for AC)	Complex requires advanced techniques like chemical vapor deposition, exfoliation, or arc discharge
Durability	Relatively durable under various environmental conditions	May degrade under specific conditions (e.g., oxidation, aggregation)

CNSs, like CNTs and graphene, have exceptional adsorption behavior due to their high surface area and rich surface chemistry, which might build strong interactions with pollutants. However, bulk carbon materials—like biochar and activated carbon—are more affordable, simpler to produce, and more sustainable, so they are chosen for widespread use in industrial air filtration and municipal water treatment systems (Saleem et al., 2022). While CNSs are more effective in contaminant removal, their high production cost and possible environmental impact hinder their extensive use. Bulk carbon materials, particularly biochar, provide additional ecological benefits, including carbon sequestration, toward supporting climate change mitigation efforts. However, the long-term environmental risks of nanocarbons, especially their possible impacts on aquatic ecosystems and soil health, need further investigation (Ghosh & Ramaprabhu, 2019).

12.6 Concluding Remarks

Using carbon-based materials in environmental detection and remediation has shown immense potential for addressing pressing environmental issues, including water contamination, soil degradation, and air pollution. As research progresses and innovations emerge, several key insights are shaping the future of these technologies. These insights focus on integrating carbon-based materials into the circular economy, advancing CNSs, expanding applications to new environmental challenges, and fostering collaboration among key stakeholders.

12.6.1 Integration of Carbon-Based Technologies into the Circular Economy

One of the most promising developments for carbon-based materials is their incorporation into the circular economy, which aims at waste reduction and waste utilization in such a way as to maximize resource efficiency. This fit-for-purpose economic model allows for synthesizing carbon-based materials such as biochar and AC from agricultural waste. Carbon-based technologies, transforming waste into useful materials, provide a more environmentally friendly approach than conventional pollution management systems. The case of biochar application, for instance, illustrates how mangrove habitats and other carbon-based resources can be utilized in a forward engineering manner without being detrimental to the environment. Biochar production is done through pyrolysis, where escaping organics from organic wastes are converted into a stable, carbon-rich material. It has tended to absorb many pollutants in both soil and water and thus helps in carbon sequestration, addressing climate change.

More work is needed to consider the life cycle of such materials, including their production, use, disposal, and, most importantly, regeneration, before carbon-based technologies can be considered to have achieved circular economy status. The need to find environmentally sound approaches for the further utilization of exhausted carbon materials, such as AC used for filtering, is crucially important. This also involves systems that help regenerate such materials by low-energy input processes or processes that generate low or no secondary waste. Hence, it follows that life cycle evaluation would be an asset in orienting future developments of carbon technologies.

12.6.2 Development of Advanced Carbon Nanomaterials

Sustainability CNSs such as CNTs, graphene, and CQDs are envisaged to be very constructive in the future in the detection and remediation of the environment. These materials are appropriate for the elimination of low-concentration contaminants from the environment because of their features, which include a large surface area, variable surface chemistry, and strong electrical conductivity. It has been demonstrated that CNTs and graphene are highly efficient at eliminating organic pollutants and hazardous metals from water. In contrast, CQDs have been effective in detecting and removing pollutants.

However, the CNSs have some disadvantages; for example, the upscaling of nano-carbon materials and their economic feasibility are critical hurdles. These materials are usually fabricated using complicated and costly processes. This means that the technology cannot be applied in beyond-scale environmental applications. To achieve this, synthesis methods that employ fewer resources and are more effective must be researched. Further nanotechnology and materials science development will be important for solving this problem. For example, the future production of such NMs utilizing environmentally oriented synthesis methods will mitigate the negative environmental impacts of the material's synthesis and allow the aforementioned methods to be implemented for more extensive purposes.

In addition, with the rising application of CNSs, there is a need to investigate how these materials affect the environment or health over time. One such concern is the threat that NMs pose to the environment and human civilization if used improperly due to their microscopic dimension and activity level. This will entail conducting a reasonably detailed study of potential risk factors and taking additional precautions related to the use of nano-carbon materials in the environment. This also involves the creation of policies and frameworks that control the practice application of NMs.

12.6.3 Expansion of Carbon-Based Technologies to New Environmental Applications

The capabilities of carbon-based technologies are not restricted to only water and air purification systems. As these technologies continue to improve, they are expected to emerge as well in solving additional environmental problems. One such problem is microplastic contamination, which is ubiquitous in oceans and freshwater bodies. In that direction, nanostructured carbon-based materials may be developed to extract microplastics from water.

The issue of pharmaceutical waste is another environmental issue that carbon-based technologies can help address. Several substances from this family are not effectively destroyed in the sewage treatment cycle and thus remain in the water bodies, which may negatively influence the biological organisms inhabiting these waters. Elderly citizens prevent this from occurring by installing special carbon-based filtration systems that will adsorb the residues of the drug from water-based solutions.

Besides these services, CNSs can also play a role in reducing GHG emissions. For instance, carbon capture and storage (CCS) could be enhanced by employing CNSs, which can act as more effective adsorbents for CO_2 and other greenhouse gases (GHGs), thereby contributing to the reduction of emissions from fossil fuel sources. In addition, carbon sensors based on real-time monitoring of the environment should provide data on the concentration of pollution and aid environmental management on the part of politicians or agencies involved in environmental protection. These sensors can monitor air and water pollution status and soil quality, thus allowing timely control measures against environmental hazards.

References

Abbas, A., Abussaud, B. A., Ihsanullah, Al-Baghli, N. A. H., & Redhwi, H. H. (2017). Adsorption of toluene and paraxylene from aqueous solution using pure and iron oxide impregnated carbon nanotubes: Kinetics and isotherms study. *Bioinorganic Chemistry and Applications, 2017*, 18–22. https://doi.org/10.1155/2017/2853925

Abou Chacra, L., Sabri, M. A., Ibrahim, T. H., Khamis, M. I., Hamdan, N. M., Al-Asheh, S., Alrefai, M., & Fernandez, C. (2018). Application of graphene nanoplatelets and graphene magnetite for the removal of emulsified oil from produced water. *Journal of Environmental Chemical Engineering, 6*(2), 3018–3033. https://doi.org/10.1016/j.jece.2018.04.060

Ahamad, T., Naushad, M., Eldesoky, G. E., Al-Saeedi, S. I., Nafady, A., Al-Kadhi, N. S., Al-Muhtaseb, A.'a. H., Khan, A. A., & Khan, A. (2019). Effective and fast adsorptive removal of toxic cationic dye (MB) from aqueous medium using amino-functionalized magnetic multiwall carbon nanotubes. *Journal of Molecular Liquids, 282*, 154–161. https://doi.org/10.1016/j.molliq.2019.02.128

Ali, I., Basheer, A. A., Mbianda, X. Y., Burakov, A., Galunin, E., Burakova, I., Mkrtchyan, E., Tkachev, A., & Grachev, V. (2019, March). Graphene based adsorbents for remediation of noxious pollutants from wastewater. *Environment International, 127*, 160–180. https://doi.org/10.1016/j.envint.2019.03.029

Alijani, H., & Shariatinia, Z. (2017). Effective aqueous arsenic removal using zero valent iron doped MWCNT synthesized by in situ CVD method using natural α-Fe_2O_3 as a precursor. *Chemosphere, 171*, 502–511. https://doi.org/10.1016/j.chemosphere.2016.12.106

Alijani, H., & Shariatinia, Z. (2018). Synthesis of high growth rate SWCNTs and their magnetite cobalt sulfide nanohybrid as super-adsorbent for mercury removal. *Chemical Engineering Research and Design, 129*, 132–149.

AlOmar, M. K., Alsaadi, M. A., Hayyan, M., Akib, S., Ibrahim, M., & Hashim, M. A. (2017). Allyl triphenyl phosphonium bromide based DES-functionalized carbon nanotubes for the removal of mercury from water. *Chemosphere, 167*, 44–52. https://doi.org/10.1016/j.chemosphere.2016.09.133

AlSaadi, M. A., Al Mamun, A., Alam, M. Z., Amosa, M. K., & Atieh, M. A. (2016). Removal of cadmium from water by CNT-PAC composite: Effect of functionalization. *Nano, 11*(1), 1650011. https://doi.org/10.1142/S1793292016500119

Asgarpour Khansary, M., Aroon, M. A., & Shirazian, S. (2020). Physical adsorption of CO_2 in biomass at atmospheric pressure and ambient temperature. *Environmental Chemistry Letters, 18*(4), 1423–1431. https://doi.org/10.1007/s10311-020-01011-y

Attia, Y. A., Ezet, A. E., Saeed, S., & Galmed, A. H. (2024). Nano carbon-modified air purification filters for removal and detection of particulate matters from ambient air. *Scientific Reports, 14*(1), 621. https://doi.org/10.1038/s41598-023-50902-x

Baby, R., Saifullah, B., & Hussein, M. Z. (2019). Carbon nanomaterials for the treatment of heavy metal-contaminated water and environmental remediation. *Nanoscale Research Letters, 14*(1), 341. https://doi.org/10.1186/s11671-019-3167-8

Balarak, D., Mahdavi, Y., Bazrafshan, E., Mahvi, A. H., & Esfandyari, Y. (2016). Adsorption of fluoride from aqueous solutions by carbon nanotubes: Determination of equilibrium, kinetic, and thermodynamic parameters. *Fluoride, 49*(1), 71.

Bratovcic, A. (2019). Different applications of nanomaterials and their impact on the environment. *International Journal of Material Science and Engineering, 5*(1), 1–7. https://doi.org/10.14445/23948884/ijmse-v5i1p101

Brião, G., de Vargas, J., de Andrade, R., da Silva, M. G. C., & Vieira, M. G. A. (2020). Removal of toxic metals from water using chitosan-based magnetic adsorbents. A review. *Environmental Chemistry Letters, 18*(4), 1145–1168. https://doi.org/10.1007/s10311-020-01003-y

Cargnello, M., Grzelczak, M., Rodríguez-González, B., Syrgiannis, Z., Bakhmutsky, K., La Parola, V., Liz-Marzán, L. M., Gorte, R. J., Prato, M., & Fornasiero, P. (2012). Multi-walled carbon nanotubes drive the activity of metal@ oxide core–shell catalysts in modular nanocomposites. *Journal of the American Chemical Society, 134*(28), 11760–11766.

Chen, G., & Yuan, H. (2021). Effects of carbon nanomaterials on the migration and fate of organic pollutants in the ecological environment. *Ferroelectrics, 570*(1), 206–217. https://doi.org/10.1080/00150193.2020.1762424

Cho, Y., Shin, N., Kim, D., Park, J. Y., & Hong, S. (2017). Nanoscale hybrid systems based on carbon nanotubes for biological sensing and control. *Bioscience Reports, 37*(2), BSR20160330. https://doi.org/10.1042/BSR20160330

Chung, J. H., Hasyimah, N., & Hussein, N. (2021). Application of carbon nanotubes (CNTs) for remediation of emerging pollutants—A review. *Tropical Aquatic and Soil Pollution, 2*(1), 13–26. https://doi.org/10.53623/tasp.v2i1.27

Deng, Z., Sun, S., Li, H., Pan, D., Patil, R. R., Guo, Z., & Seok, I. (2021). Modification of coconut shell-based activated carbon and purification of wastewater. *Advanced Composites and Hybrid Materials, 4*(1), 65–73. https://doi.org/10.1007/s42114-021-00205-4

Elanchezhiyan, S. S. D., Prabhu, S. M., Kim, Y., & Park, C. M. (2020). Lanthanum-substituted bimetallic magnetic materials assembled carboxylate-rich graphene oxide nanohybrids as highly efficient adsorbent for perfluorooctanoic acid adsorption from aqueous solutions. *Applied Surface Science, 509*, 144716. https://doi.org/10.1016/j.apsusc.2019.144716

El-Khouly, S. M., & Fathy, N. A. (2021). A review on nano-carbon materials for pollution remediation. *Egyptian Journal of Chemistry, 64*(12), 6929–6951. https://doi.org/10.21608/EJCHEM.2021.80926.4007

Elsehly, E. M. I., Chechenin, N. G., Bukunov, K. A., Makunin, A. V., Priselkova, A. B., Vorobyeva, E. A., & Motaweh, H. A. (2016). Removal of iron and manganese from aqueous solutions using carbon nanotube filters. *Water Science and Technology: Water Supply, 16*(2), 347–353. https://doi.org/10.2166/ws.2015.143

Faghihi, A., Vakili, M. H., Hosseinzadeh, G., Farhadian, M., & Jafari, Z. (2016). Synthesis and application of recyclable magnetic freeze-dried graphene oxide nanocomposite as a high capacity adsorbent for cationic dye adsorption. *Desalination and Water Treatment, 57*(47), 22655–22670. https://doi.org/10.1080/19443994.2015.1134357

Gabal, E., Chatterjee, S., Ahmed, F. K., & Abd-Elsalam, K. A. (2019). Carbon nanomaterial applications in air pollution remediation. In *Carbon nanomaterials for agri-food and environmental applications*. Elsevier. https://doi.org/10.1016/B978-0-12-819786-8.00007-4

Ghosh, S., & Ramaprabhu, S. (2019). Green synthesis of transition metal nanocrystals encapsulated into nitrogen-doped carbon nanotubes for efficient carbon dioxide capture. *Carbon, 141*, 692–703. https://doi.org/10.1016/j.carbon.2018.09.083

Gopinath, K. P., Vo, D. V. N., Prakash, D. G., Joseph, A. A., Viswanathan, S., & Arun, J. (2021). Environmental applications of carbon-based materials: A review. *Environmental Chemistry Letters, 19*(1), 557–582. https://doi.org/10.1007/s10311-020-01084-9

Guo, T., Bulin, C., Li, B., Zhao, Z., Yu, H., Sun, H., Ge, X., Xing, R., & Zhang, B. (2018). Efficient removal of aqueous Pb(II) using partially reduced graphene oxide-Fe_3O_4. *Adsorption Science and Technology, 36*(3–4), 1031–1048. https://doi.org/10.1177/0263617417744402

Gupta Chatterjee, S., Chatterjee, S., Ray, A. K., & Chakraborty, A. K. (2015). Graphene-metal oxide nanohybrids for toxic gas sensor: A review. *Sensors and Actuators, B: Chemical, 221*(2), 1170–1181. https://doi.org/10.1016/j.snb.2015.07.070

Gupta, S., Bhatiya, D., & Murthy, C. N. (2015). Metal removal studies by composite membrane of polysulfone and functionalized single-walled carbon nanotubes. *Separation Science and Technology (Philadelphia), 50*(3), 421–429. https://doi.org/10.1080/01496395.2014.973516

Gusain, R., Kumar, N., Fosso-Kankeu, E., & Ray, S. S. (2019). Efficient removal of Pb(II) and Cd(II) from industrial mine water by a hierarchical MoS2/SH-MWCNT nanocomposite. *ACS Omega, 4*(9), 13922–13935. https://doi.org/10.1021/acsomega.9b01603

Hu, L., Peng, F., Xia, D., He, H., He, C., Fang, Z., Yang, J., Tian, S., Sharma, V. K., & Shu, D. (2018). Carbohydrates-derived nitrogen-doped hierarchical porous carbon for ultrasensitive detection of 4-nitrophenol. *ACS Sustainable Chemistry & Engineering, 6*(12), 17391–17401. https://doi.org/10.1021/acssuschemeng.8b05169

Hu, Z., Xie, H., Wang, Q., & Chen, S. (2019). Adsorption and diffusion of sulfur dioxide and nitrogen in single-wall carbon nanotubes. *Journal of Molecular Graphics and Modelling, 88*, 62–70. https://doi.org/10.1016/j.jmgm.2019.01.003

Huang, Y., Gao, Y., Zhang, Q., Zhang, Y., Cao, J. J., Ho, W., & Lee, S. C. (2018). Biocompatible FeOOH-carbon quantum dots nanocomposites for gaseous NOx removal under visible light: Improved charge separation and high selectivity. *Journal of Hazardous Materials., 354*, 54. https://doi.org/10.1016/j.jhazmat.2018.04.071

Huang, Q., Song, S., Chen, Z., Hu, B., Chen, J., & Wang, X. (2019). Biochar-based materials and their applications in removal of organic contaminants from wastewater: State-of-the-art review. *Biochar, 1*(1), 45–73. https://doi.org/10.1007/s42773-019-00006-5

Hussain, C. M. (2015). *Carbon nanomaterials as adsorbents for environmental analysis* (Vol. 6). Wiley Blackwell. https://doi.org/10.1002/9781118845530.ch14

Jayanthi, S., Eswar, N. K., Singh, S. A., Chatterjee, K., Madras, G., & Sood, A. K. (2016). Macroporous three-dimensional graphene oxide foams for dye adsorption and antibacterial applications. *RSC Advances, 6*(2), 1231–1242. https://doi.org/10.1039/c5ra19925e

Jeong, K., Kim, D. G., & Ko, S. O. (2017). Adsorption characteristics of effluent organic matter and natural organic matter by carbon based nanomaterials. *KSCE Journal of Civil Engineering, 21*(1), 119–126. https://doi.org/10.1007/s12205-016-0421-9

Kaur, S., Sharma, S., & Kansal, S. K. (2016). Synthesis of ZnS/CQDs nanocomposite and its application as a photocatalyst for the degradation of an anionic dye, ARS. *Superlattices and Microstructures, 98*, 86–95. https://doi.org/10.1016/j.spmi.2016.08.011

Keller, L., Ohs, B., Lenhart, J., Abduly, L., Blanke, P., & Wessling, M. (2018). High capacity polyethylenimine impregnated microtubes made of carbon nanotubes for CO_2 capture. *Carbon, 126*, 338–345. https://doi.org/10.1016/j.carbon.2017.10.023

Keller, L., Ohs, B., Abduly, L., & Wessling, M. (2019). Carbon nanotube silica composite hollow fibers impregnated with polyethylenimine for CO_2 capture. *Chemical Engineering Journal, 359*, 476–484. https://doi.org/10.1016/j.cej.2018.11.100

Lehmann, J., & Joseph, S. (Eds.). (2024). *Biochar for environmental management: Science, technology and implementation*. Taylor & Francis.

Li, P., Wang, C., Li, Z., Zong, Y., Zhang, Y., Yang, X., Li, S., & Wei, F. (2014). Hierarchical carbon-nanotube/quartz-fiber films with gradient nanostructures for high efficiency and long service life air filters. *RSC Advances, 4*(96), 54115–54121. https://doi.org/10.1039/c4ra08746a

Lico, D., Vuono, D., Siciliano, C., Nagy, J. B., & De Luca, P. (2019, September 2017). Removal of unleaded gasoline from water by multi-walled carbon nanotubes. *Journal of Environmental Management, 237*, 636–643. https://doi.org/10.1016/j.jenvman.2019.02.062.

Liu, F., Zhang, W., Chen, W., Wang, J., Yang, Q., Zhu, W., & Wang, J. (2017). One-pot synthesis of $NiFe_2O_4$ integrated with EDTA-derived carbon dots for enhanced removal of tetracycline. *Chemical Engineering Journal, 310*, 187–196. https://doi.org/10.1016/j.cej.2016.10.116

Liu, W., Li, Y., Feng, Y., Qiao, J., Zhao, H., Xie, J., Fang, Y., Shen, S., & Liang, S. (2020a). The effectiveness of nanobiochar for reducing phytotoxicity and improving soil remediation in cadmium-contaminated soil. *Scientific Reports, 10*(1), 858. https://doi.org/10.1038/s41598-020-57954-3

Liu, Z., Sun, Y., Xu, X., Qu, J., & Qu, B. (2020b). Adsorption of Hg(II) in an aqueous solution by activated carbon prepared from rice husk using KOH activation. *ACS Omega, 5*(45), 29231–29242. https://doi.org/10.1021/acsomega.0c03992

Liu, Z., Ling, Q., Cai, Y., Xu, L., Su, J., Yu, K., Wu, X., Xu, J., Hu, B., & Wang, X. (2022). Synthesis of carbon-based nanomaterials and their application in pollution management. *Nanoscale Advances, 4*(5), 1246–1262. https://doi.org/10.1039/d1na00843a

Lundquist, N. A., Sweetman, M. J., Scroggie, K. R., Worthington, M. J. H., Esdaile, L. J., Alboaiji, S. F. K., Plush, S. E., Hayball, J. D., & Chalker, J. M. (2019). Polymer supported carbon for safe and effective remediation of PFOA-and PFOS-contaminated water. *ACS Sustainable Chemistry & Engineering, 7*(13), 11044–11049. https://doi.org/10.1021/acssuschemeng.9b01793

Ma, Q., Yu, Y., Sindoro, M., Fane, A. G., Wang, R., & Zhang, H. (2017). Carbon-based functional materials derived from waste for water remediation and energy storage. *Advanced Materials, 29*(13). https://doi.org/10.1002/adma.201605361

Majumdar, S., Maurya, M., & Singh, J. K. (2018). Adsorptive separation of CO_2 from multicomponent mixtures of flue gas in carbon nanotube arrays: A grand canonical Monte Carlo study. *Energy and Fuels, 32*(5), 6090–6097. https://doi.org/10.1021/acs.energyfuels.8b00649

Manna, B., & Retna Raj, C. (2018). Nanostructured sulfur-doped porous reduced graphene oxide for the ultrasensitive electrochemical detection and efficient removal of Hg (II). *ACS Sustainable Chemistry & Engineering, 6*(5), 6175–6182. https://doi.org/10.1021/acssuschemeng.7b04884

Mohan, V. B., Lau, K. T., Hui, D., & Bhattacharyya, D. (2018, January). Graphene-based materials and their composites: A review on production, applications and product limitations. *Composites Part B: Engineering, 142*, 200–220. https://doi.org/10.1016/j.compositesb.2018.01.013

Myung, Y., Jung, S., Tung, T. T., Tripathi, K. M., & Kim, T. Y. (2019). Graphene-based aerogels derived from biomass for energy storage and environmental remediation. *ACS Sustainable Chemistry & Engineering, 7*(4), 3772–3782. https://doi.org/10.1021/acssuschemeng.8b04202

Parham, H., Bates, S., Xia, Y., & Zhu, Y. (2013). A highly efficient and versatile carbon nanotube/ceramic composite filter. *Carbon, 54*, 215–223. https://doi.org/10.1016/j.carbon.2012.11.032

Park, K. T., & Hwang, J. (2014). Filtration and inactivation of aerosolized bacteriophage MS2 by a CNT air filter fabricated using electro-aerodynamic deposition. *Carbon, 75*, 401–410. https://doi.org/10.1016/j.carbon.2014.04.019

Pedrielli, A., Taioli, S., Garberoglio, G., & Pugno, N. M. (2018). Gas adsorption and dynamics in pillared graphene frameworks. *Microporous and Mesoporous Materials, 257*, 222–231. https://doi.org/10.1016/j.micromeso.2017.08.034

Pourzamani, H., Parastar, S., & Hashemi, M. (2017, July). The elimination of xylene from aqueous solutions using single wall carbon nanotube and magnetic nanoparticle hybrid adsorbent. *Process Safety and Environmental Protection, 109*, 688–696. https://doi.org/10.1016/j.psep.2017.05.010

Qian, X., Yue, D., Tian, Z., Reng, M., Zhu, Y., Kan, M., Zhang, T., & Zhao, Y. (2016). Carbon quantum dots decorated Bi_2WO_6 nanocomposite with enhanced photocatalytic oxidation activity for VOCs. *Applied Catalysis B: Environmental, 193*, 16–21. https://doi.org/10.1016/j.apcatb.2016.04.009

Rajabi, M., Mahanpoor, K., & Moradi, O. (2017). Removal of dye molecules from aqueous solution by carbon nanotubes and carbon nanotube functional groups: Critical review. *RSC Advances, 7*(74), 47083–47090. https://doi.org/10.1039/c7ra09377b

Robati, D., Mirza, B., Rajabi, M., Moradi, O., Tyagi, I., Agarwal, S., & Gupta, V. K. (2016). Removal of hazardous dyes-BR 12 and methyl orange using graphene oxide as an adsorbent from aqueous phase. *Chemical Engineering Journal, 284*, 687–697. https://doi.org/10.1016/j.cej.2015.08.131

Rodríguez, C., Tapia, C., Leiva-Aravena, E., & Leiva, E. (2020). Graphene oxide–ZnO nanocomposites for removal of aluminum and copper ions from acid mine drainage wastewater. *International Journal of Environmental Research and Public Health, 17*(18), 6911. https://doi.org/10.3390/ijerph17186911

Sadegh, H., Zare, K., Maazinejad, B., Shahryari-Ghoshekandi, R., Tyagi, I., Agarwal, S., & Gupta, V. K. (2016, March). Synthesis of MWCNT-COOH-cysteamine composite and its application for dye removal. *Journal of Molecular Liquids, 215*, 221–228. https://doi.org/10.1016/j.molliq.2015.12.042

Saleem, H., Zaidi, S. J., Ismail, A. F., & Goh, P. S. (2022). Advances of nanomaterials for air pollution remediation and their impacts on the environment. *Chemosphere, 287*(P2), 132083. https://doi.org/10.1016/j.chemosphere.2021.132083

Schroeder, V., Savagatrup, S., He, M., Lin, S., & Swager, T. M. (2019). Carbon nanotube chemical sensors. Review-article. *Chemical Reviews, 119*(1), 599–663. https://doi.org/10.1021/acs.chemrev.8b00340

Sheng, C., Agwu Nnanna, A. G., Liu, Y., & Vargo, J. D. (2016). Removal of trace pharmaceuticals from water using coagulation and powdered activated carbon as pretreatment to ultrafiltration membrane system. *Science of the Total Environment, 550*, 1075–1083. https://doi.org/10.1016/j.scitotenv.2016.01.179

Shi, W., Guo, F., Wang, H., Liu, C., Fu, Y., Yuan, S., Huang, H., Liu, Y., & Kang, Z. (2018). Carbon dots decorated magnetic $ZnFe_2O_4$ nanoparticles with enhanced adsorption capacity for the removal of dye from aqueous solution. *Applied Surface Science, 433*, 790–797. https://doi.org/10.1016/j.apsusc.2017.10.099

Sui, Z. Y., & Han, B. H. (2015). Effect of surface chemistry and textural properties on carbon dioxide uptake in hydrothermally reduced graphene oxide. *Carbon, 82*(C), 590–598. https://doi.org/10.1016/j.carbon.2014.11.014

Sun, Y., Yang, Y., Yang, M., Yu, F., & Ma, J. (2019). Response surface methodological evaluation and optimization for adsorption removal of ciprofloxacin onto graphene hydrogel. *Journal of Molecular Liquids, 284*, 124–130. https://doi.org/10.1016/j.molliq.2019.03.118

Sun, Y., Xiong, X., He, M., Xu, Z., Hou, D., Zhang, W., Ok, Y. S., Rinklebe, J., Wang, L., & Tsang, D. C. W. (2021, April). Roles of biochar-derived dissolved organic matter in soil amendment and environmental remediation: A critical review. *Chemical Engineering Journal, 424*, 130387. https://doi.org/10.1016/j.cej.2021.130387

Tabish, T. A., Memon, F. A., Gomez, D. E., Horsell, D. W., & Zhang, S. (2018). A facile synthesis of porous graphene for efficient water and wastewater treatment. *Scientific Reports, 8*(1), 1–14. https://doi.org/10.1038/s41598-018-19978-8

Tang, Y., Zhang, G., Liu, C., Luo, S., & Xu, X. (2013). Magnetic TiO_2-graphene composite as a high-performance and recyclable platform for efficient photocatalytic removal of herbicides from water. *Journal of Hazardous Materials, 252–253*, 115–122. https://doi.org/10.1016/j.jhazmat.2013.02.053

US Patent No. 10,376,845 B2. (2019). *Fluorescent carbon quantum dot-based sensors for heavy metal detection.*

Wan, S., He, F., Wu, J., Wan, W., Gu, Y., & Gao, B. (2016). Rapid and highly selective removal of lead from water using graphene oxide-hydrated manganese oxide nanocomposites. *Journal of Hazardous Materials, 314*, 32–40. https://doi.org/10.1016/j.jhazmat.2016.04.014

Wang, Z., Xu, C., Lu, Y., Wu, F., Ye, G., Wei, G., Sun, T., & Chen, J. (2017). Visualization of adsorption: Luminescent mesoporous silica-carbon dots composite for rapid and selective removal of U(VI) and in situ monitoring the adsorption behavior. *ACS Applied Materials and Interfaces, 9*(8), 7392–7398. https://doi.org/10.1021/acsami.6b13427

Wang, Y., Tsang, Y. F., Wang, H., Sun, Y., Song, Y., Pan, X., & Luo, S. (2020). Effective stabilization of arsenic in contaminated soils with biogenic manganese oxide (BMO) materials. *Environmental Pollution, 258*, 113481. https://doi.org/10.1016/j.envpol.2019.113481

Wang, H., Xu, Y., Xu, D., Chen, L., Qiu, X., & Zhu, Y. (2022a). Graphitic carbon nitride for photoelectrochemical detection of environmental pollutants. *ACS ES&T Engineering, 2*(2), 140–157. https://doi.org/10.1021/acsestengg.1c00337

Wang, Z., Zhang, L., Zhang, K., Lu, Y., Chen, J., Wang, S., Hu, B., & Wang, X. (2022b). Application of carbon dots and their composite materials for the detection and removal of radioactive ions: A review. *Chemosphere, 287*, 132313.

Wei, L., Lu, D., Wang, J., Wei, H., Zhao, J., Geng, H., & Zhang, Y. (2014). Highly sensitive detection of trinitrotoluene in water by chemiresistive sensor based on noncovalently amino functionalized single-walled carbon nanotube. *Sensors and Actuators, B: Chemical, 190*, 529–534. https://doi.org/10.1016/j.snb.2013.09.017

Wu, H., & Tong, C. (2019). Nitrogen-and sulfur-codoped carbon dots for highly selective and sensitive fluorescent detection of Hg^{2+} ions and sulfide in environmental water samples. *Journal of Agricultural and Food Chemistry, 67*(10), 2794–2800. https://doi.org/10.1021/acs.jafc.8b07176

Xu, J., Cao, Z., Zhang, Y., Yuan, Z., Lou, Z., Xu, X., & Wang, X. (2018). A review of functionalized carbon nanotubes and graphene for heavy metal adsorption from water: Preparation, application, and mechanism. *Chemosphere, 195*, 351–364. https://doi.org/10.1016/j.chemosphere.2017.12.061

Yadav, N., Garg, V. K., Chhillar, A. K., & Rana, J. S. (2021, May). Detection and remediation of pollutants to maintain Ecosustainability employing nanotechnology: A review. *Chemosphere, 280*, 130792. https://doi.org/10.1016/j.chemosphere.2021.130792

Yu, B., Bai, Y., Ming, Z., Yang, H., Chen, L., Hu, X., Feng, S., & Yang, S. T. (2017). Adsorption behaviors of tetracycline on magnetic graphene oxide sponge. *Materials Chemistry and Physics, 198*, 283–290. https://doi.org/10.1016/j.matchemphys.2017.05.042

Zazouli, M. A., Yousefi, Z. A., Cherati, J. Y., Tabarinia, H., Tabarinia, F., & Adergani, B. A. (2014). Evaluation of L-cysteine functionalized single-walled carbon nanotubes on mercury removal from aqueous solutions. *Journal of Mazandaran University of Medical Sciences, 24*(111), 10–21.

Zhang, C. Z., Chen, B., Bai, Y., & Xie, J. (2018). A new functionalized reduced graphene oxide adsorbent for removing heavy metal ions in water via coordination and ion exchange. *Separation Science and Technology (Philadelphia), 53*(18), 2896–2905. https://doi.org/10.1080/01496395.2018.1497655

Zhang, L., Lei, Y., He, P., Wu, H., Guo, L., & Wei, G. (2022). Carbon material-based aerogels for gas adsorption: Fabrication, structure design, functional tailoring, and applications. *Nanomaterials, 12*(18). https://doi.org/10.3390/nano12183172

Zhu, L., Meng, L., Shi, J., Li, J., Zhang, X., & Feng, M. (2019, November 2018). Metal-organic frameworks/carbon-based materials for environmental remediation: A state-of-the-art mini-review. *Journal of Environmental Management, 232*, 964–977. https://doi.org/10.1016/j.jenvman.2018.12.004

Chapter 13
Challenges in Implementing Carbon-Based Remediation

Ajith Manayil Parambil and Eepsita Priyadarshini

Abstract The unique physicochemical properties of carbon-based materials have made them a promising substrate in environmental remediation. Because of their high surface area, porosity, strong adsorption, and catalytic potential are being widely investigated for remediation purposes alone or in combination with other materials. Purposes. However, practically implementing them in real-time applications faces several challenges that have hindered their widespread applications. In this chapter, we have discussed the key challenges, including technical limitations (synthesis and optimization of carbon-based materials), scalability issues such as cost-effectiveness in large-scale production, and regulatory and safety issues related to real-time deployment. In addition, the challenges in site-specific implementation, long-term persistence, and socio-economic barriers to specific public perception and stakeholder engagement are also elaborately discussed. Finally, we have discussed future directions to circumvent these challenges and advancements in designing materials, sustainable production, and deployment strategies. Addressing these challenges and employing a multidisciplinary approach would provide a comprehensive understanding of the challenges faced in carbon-based remediation technologies.

Keywords Carbon-based materials · Environmental remediation · Large-scale applications · Contaminants · Regulatory frameworks

A. M. Parambil (✉)
Nanotechnology Centre, Centre for Energy and Environmental Technologies, VŠB–Technical University of Ostrava, Ostrava-Poruba, Czech Republic

E. Priyadarshini (✉)
School of Biosciences and Technology, Galgotias University,
Greater Noida, Uttar Pradesh, India
e-mail: eepsita@galgotiasuniversity.edu.in

A. M. Parambil et al. (eds.), *Carbon*,
https://doi.org/10.1007/978-3-031-90613-8_13

13.1 Introduction

Accelerating urbanization and modernization has resulted in an enhancement in environmental pollution. The untreated discharge of toxic pollutants (antibiotics, heavy metals, insecticides, pesticides, and endocrine disruptors) is a growing environmental concern and is also affecting human health (Gopinath et al., 2020; Wang et al., 2022a, 2022b). It is, therefore, imperative to treat/remove such pollutants from wastewater, and researchers are making a continuous effort to develop cost-effective and efficient materials capable of removing pollutants from wastewater. The fascinating properties of carbon-based materials have made them an excellent substrate in environmental remediation. High surface area, porosity, ease of functionalization, and tunable properties are a few of the properties to name (Liu et al., 2022; Priya et al., 2024; Piumie et al., 2018). Activated carbon, carbon nanotubes, graphene, and biochar have significant applications in contaminants'adsorption, degradation, and immobilization (Priya et al., 2024). For instance, Rivas-Sanchez et al. (2022) reported the application of alginate-functionalized carbon-based nanocomposites for bioremediation. Scientific reports collected over the last decade (2012–2021) have shown a considerable increase in the use of carbon-based nanomaterials for the removal of pollutants, such as hydrogen sulfide, heavy metals, organic dyes, and pesticides with average adsorption and degradation efficiency of >80% and >98%, respectively (Gopinath et al., 2020). The principal reason behind this high removal percentage is the advantages of high specific surface areas and ample adsorption sites. This allows the formation of strong complexes between carbon-based nanomaterials and organic pollutants or heavy metals, facilitating the effective removal of the latter from wastewater (Chen & Yuan, 2021; Cheng et al., 2021). Precipitation, sorption, ion exchange, electrostatic interaction, and adsorption are the principal interaction mechanisms that allow these contaminates' uptake and removal (Xu et al., 2018). The feasibility and potential of these carbon-based nanomaterials have urged scientists, industrialists, and policymakers to seek sustainable and effective solutions for emerging environmental challenges.

While carbon-based materials offer numerous advantages in environmental remediation, their practical implementation is restricted due to several limitations. A plethora of technical, economic, regulatory, and social factors affect their real-time applications. High production cost, optimization during scaling-up, stability, toxicity, and harmful intermediates are some technical and scalability challenges that circumvent their applications. In addition, real-world applications suffer from unpredictable results and reproducibility compared to laboratory studies. This chapter provides a theoretical understanding of the challenges encountered when using carbon-based nanomaterials for real-world applications. Outlining these challenges will provide a framework and opportunities for collaborations among scientists, engineers, policymakers, and communities. This would pave the way for responsible innovations, improvement, and deployment of carbon-based remediation technologies, contributing to global sustainability goals.

13.2 Challenges in the Application of Carbon-Based Nanomaterial for Remediation

13.2.1 Technical Challenges

While carbon-based materials have shown considerable efficacy in environmental remediation, several technical challenges limit their applications in real-world scenarios. The complexity in design, optimization of material, environmental variability, stability, durability, and material regeneration are some of the common technical challenges. Formulation of contaminate-specific carbon-based material is crucial. However, it is equally challenging due to the complexity and variability. While lab-scale synthesis of carbon materials allows precise control over size, properties, surface area, and functional groups, these are not directly translated to real-time field applications. Specifically, the efficiency of adsorption and removal decreases compared to that obtained at a laboratory scale. For example, activated carbon and biochar had shown effective adsorption of heavy metals and organic pollutants under laboratory conditions, while a decrease in efficiency was observed when a mixed-contaminant system was used mainly because of competition. Functionalization techniques such as amination or oxidation enhance the specificity and removal efficiency (Zhang et al., 2023). In addition, techniques such as alkaline treatment, acidic treatment, microbial immobilization, impregnation technique, and biological modifications are used to enhance the adsorption efficiency of activated carbon. However, these methods suffer from a lack of scalability. Besides, the preparation of hybrid carbon materials, for example, graphene with metal-organic frameworks, has shown improved adsorption efficiency but simultaneously introduces complexity of synthesis along with stability and characterization issues (Pet et al., 2024; Li et al., 2022). In addition, combining hybrid materials requires extensive chemical and thermal treatments to introduce functionality, which adds to costs and the generation of by-products, increasing sustainability concerns.

Environmental conditions and variability are other major issues regarding real-scenario applications. pH, temperature, salinity, and the presence of other similar ions are some factors that restrict the potential for remediation by carbon-based materials. For example, adsorption efficiency varies with pH alteration; materials optimized for adsorption under acidic conditions may behave poorly under neutral or alkaline conditions (Dai et al., 2020). Similarly, salinity causes a reduction in the adsorption efficacy of carbon materials for heavy metals because of competition among ions (Nguyen et al., 2023). In addition, natural organic matter (NOM) gets deposited on active sites of carbon surface, thereby decreasing the efficiency of adsorption. While surface coatings have shown promise in resisting fouling, their long-term performance and efficacy are uncertain (Wang et al., 2022a, 2022b).

Another significant drawback is the limited reusability after being exposed to harsh environmental conditions. Exposing the carbon-based materials to repeated adsorption–desorption cycles results in material degradation, thereby reducing the efficacy with time. Further, the regeneration of carbon material becomes an

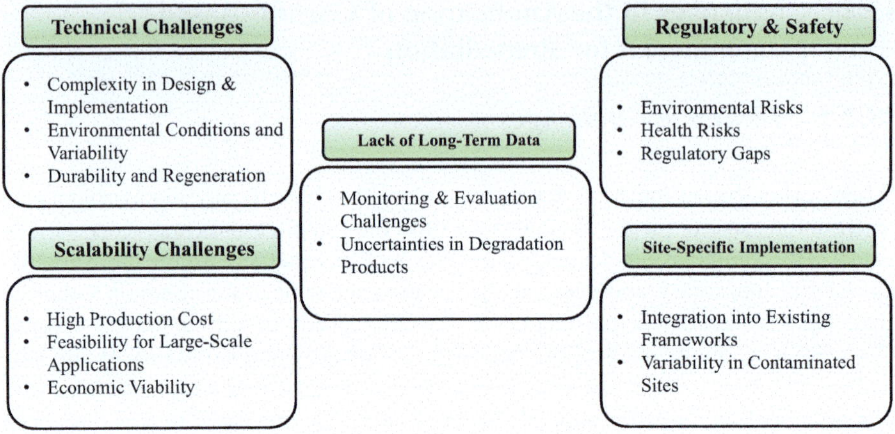

Fig. 13.1 The various challenges encountered in the utilization of carbon-based materials for environmental applications

energy-intensive process with high-temperature and chemical treatments that refute the applications in environmental remediation (Kumar et al., 2023). To alleviate these challenges, microwave-assisted regeneration and anti-fouling coatings are being used; however, these are limited to specific materials and have durability issues (Singh et al., 2022; Patel et al., 2023). The various challenges encountered in using carbon-based materials for environmental applications are shown in Fig. 13.1.

13.2.1.1 Scalability Challenges

Scaling-up refers to the field-scale implementation of laboratory research and is associated with several interrelated challenges, including high production cost, logistic issues, lack of standardization, and economic viability. These challenges have seriously restricted the widespread adoption of carbon-based materials in environmental remediation applications.

The generation of carbon-based materials with specific functionalization and effective remediation strategies requires energy-intensive and resource-demanding methods. Chemical vapor deposition (CVD) and thermal activation are a few functionalization strategies. However, they have high production costs, making large-scale production economically impractical. Generation of graphene and carbon nanotubes also suffers from the same issue. Researchers are exploring cost-effective alternatives such as activated carbon and other carbon-based materials from agriculture and food waste to circumvent this. However, these materials lack precision and specificity regarding contaminate adsorption (Ahmed et al., 2023).

Logistics is another challenge regarding large-scale deployment for real-time application of carbon dots. Managing logistic hurdles such as transport, storage, and integration into prevailing remediation systems are major issues. It becomes tedious and difficult to transport large amounts of materials to remote areas besides being costly and environmentally unsustainable. Additionally, storing bulk materials is difficult and requires specialized facilities to prevent contamination of engineered nanoscale materials. The third challenge is the uniform application of carbon-based materials across contaminated sites. Inconsistent synthesis, characterization, and properties of carbon-based materials are additional bottlenecks when scaling up. Alterations in functional group density, variation in pore structure, changes in the density of functional groups on the surface of carbon material, and pore variations introduce unpredictable results. Universal protocol standardization is difficult as real-time implementation is subjected to seasonal variations in environmental conditions. In addition, applying the same material at different locations is not possible as the performance varies, necessitating regulatory approvals.

13.2.1.2 Regulatory and Safety Concerns

Carbon-based nanomaterials have been considerably used in real-time applications. However, prior to their use, proper approval and compliance with environmental laws are necessary. The production, application, and disposal of carbon-based materials should be per protocols specified by government agencies and regulatory bodies. These frameworks consider the potential risks of by-products produced by carbon materials, their harmful effect on the environment, and their persistence in soil and water bodies. Engineered carbon-based materials (graphene, fullerenes, and carbon nanotubes) have altered properties that can interact unpredictably with the ecosystem. Therefore, their uncontrolled release into the environment can result in bioaccumulation, be toxic to aquatic organisms, and affect humans by entering the food web. The introduction of 1.00 g/L of carbon nanotube to the aquatic system was reported to result in growth inhibition and mortality in invertebrates. The type of CNT, precleaning protocol used, and material dispersion were a few factors that affected the extent of toxicity. Microscopic examination of test organisms (*Hyalella Azteca, Chironomus dilutus,* and *Lumbriculus variegatus*) showed accumulation of CNTs in the gut as well as outer surface (Mwangi et al., 2012). The generation of reactive oxygen species, oxidative stress, inflammation followed by cellular damage, and activation of the stress response pathway are possible ways CNTs induce toxicity (Punt et al., 2022). Accumulation and subsequent toxicity of graphene oxide sheets were also reported in test organisms, such as *Chlorella vulgaris, Artemia salina*, and zebrafish. Biodistribution analysis revealed the penetration of nano-sized GO into the intestinal wall and interaction with vital organs, leading to increased mortality and impairment in reproductive function (Hashemi et al., 2024).

Besides, exposure to engineered carbon materials via production, handling, or application has potential health effects on humans. Being nanoscale, they can easily penetrate biological barriers. For example, multi-walled carbon nanotubes (MWCNTs) induce tumor growth and DNA damage in healthy cells (Sargent et al., 2014). MWCNTs are known for pulmonary, hematologic, and cardiovascular toxicity (Nasirzadeh et al., 2020). Likewise, graphene nanoparticles increase the generation of oxygen free radicals, known to damage DNA and proteins, inducing apoptosis and necrosis (Seabra et al., 2014; Sanchez et al., 2012). Carbon-based nanomaterials have some activation of pulmonary macrophage and inflammation. In addition, single-walled carbon nanotubes (SWCNTs) are reported to induce acute and chronic pulmonary pathologies systematic damage in blood chemistry, liver, and cardiovascular system (Erdely et al., 2009; Ajith and Paulraj, 2022). They can also cross the blood–brain barrier and enter the brain (Yang et al., 2007).

Rapid advancements in technological interventions for carbon-based materials have outpaced the development of regulatory frameworks. The current regulatory framework is inadequate to account for the unique properties of carbon nanomaterials, resulting in inconsistency in their evaluation and approval. Insufficient testing protocols, lack of standardization or variability in synthesis and characterization, and global variability in regulatory frameworks are the key challenges in regulatory landscapes regarding nanomaterials. The International Organization for Standardization (ISO) is a non-governmental international organization involved in producing international standards in nanotechnologies. However, it is complicated as there are chances of duplication when similar work is documented by individuals and companies or regulatory agencies.

13.2.1.3 Long-Term Data Deficiency

While carbon-based materials have been significantly used in remediation, their behavior and interactions with ecosystems over an extended time period are still speculative and insufficiently understood. This increases concerns about its toxicological impact, safety, sustainability, and unintentional consequences.

The environmental persistence of carbon-based materials is a key uncertainty and depends largely on the starting material. Engineered nanomaterials, in particular carbon nanotubes (CNTs), graphene oxide, and activated carbon, are often synthesized for stability and high durability and therefore have prolonged persistence in the environment. This thus increases their chances of accumulation in soil, water, and sediments. Additionally, the natural breakdown of carbon-based materials involves interactions with microbial communities. While microbes degrade some carbon materials, studies report that some carbon nanomaterials resist microbial decomposition, increasing concerns related to their long-term persistence. Apart from this, another major concern is the bioaccumulation of carbon-based materials and their transfer through food chain, affecting higher tropic levels. A study reported the agglomeration of SWCNT in green algae, decreasing its growth and chlorophyll production along with reduced cell viability. Further, these SWCNTs were transferred from primary producers to filter-feeding consumers, suggesting their entry into the food web (Al-Shaeri et al., 2022). These studies suggest an interaction between carbon-based materials and biological systems, the interaction mechanism being very complex as several factors such as size, shape, surface functionalization, and environmental conditions such as pH and salinity affect it.

As already discussed, carbon nanomaterials can induce oxidative stress, membrane damage, and changes in aquatic species; however, the long-term interaction is not clearly understood. MWCNTs at 1 mg/g are known to reduce the microbial biomass in soil and alter the bacterial composition after 1 year of accumulation and interaction as analyzed via PCR sequencing (Ge et al., 2016). Likewise, graphene in soil reduced the activity of bacteria associated with soil ammoxidation, thereby affecting the soil nitrogen cycle (Ren et al., 2015). Hao et al. (2018) reported a decrease in the density of soil microorganisms with an increase in graphene oxide concentration (50–500 mg/kg). Graphene is known to induce microbial toxicity via oxidative stress. It attacks cell membranes, lipids, and proteins, ultimately resulting in microbial cell imbalance. However, it is difficult to determine the long-term impact of graphene or other carbon-based materials on the ecosystem due to the lack of a dedicated monitoring program. Most studies focus on analyzing the immediate efficacy of these materials on environmental contaminants, often overlooking their fate after the remediation process. It is indispensable to monitor the fate, transport, and transformation of these materials in the diverse ecosystem to understand long-term safety and sustainability. Besides, there are chances that these carbon-based materials can undergo physical, chemical, or biological transformations in the environment, thereby generating secondary products that can be pollutants with unknown or harmful effects.

13.2.1.4 Site-Specific Implementation Challenges

The ultimate application of carbon-based materials for remediation is their site-specific implementation. This requires careful analysis of several factors at the contamination site to optimize the remediation efforts and minimize environmental and economic costs.

Analysis of contaminated sites involves understanding the site geochemistry (soil, water, and sediment composition). Additionally, factors such as pH, redox potential, organic matter content, and ionic strength are crucial for a comprehensive assessment. Anegbe et al. (2024) stated that various environmental factors affect the adsorption efficiency of graphene oxide for emerging contaminants, such as pharmaceuticals, pesticides, and personal care products. Other than factors, profiling of the contaminant and hydrogeological assessment are also necessary. Contaminant profiling involves identifying the type, concentration, and distribution of contaminants such as heavy metals, organic pollutants, and PFAs so that the carbon material can be tailored accordingly. In addition, the water flow pattern analysis, permeability, and aquifer characteristics of the contamination site are also necessary (Vasilachi et al., 2021; Esteban-Arranz et al., 2018). The lab-scale synthesized carbon materials would not necessarily perform well under real-time conditions. For instance, carbon dots may be efficient for the remediation of heavy metals, while graphene may be suitable for organic contaminants. Therefore, depending on the environmental parameters, selection, optimization, and surface functionalization of carbon-based materials can be optimized to enhance the sensitivity and limit of detection.

Customization of these materials per site-specific parameters would improve the process of remediation (Liu & Kujawinski, 2015; Ersan et al., 2017). Further, functionalizing carbon materials with molecules such as amine or carboxyl groups will help remove contaminants more efficiently. The presence of functional groups facilitates interaction with contaminants via hydrogen bonding, π–π stacking, or other van der Waals or covalent interactions. For instance, the presence of hydroxyl and carboxyl groups (oxygen-rich functional groups) enhances the adsorption of polar contaminants. In addition, the interaction of graphene oxide with contaminants depends on the surface charge present on the nanomaterial surface (Yang et al., 2019).

The next step after optimization and analysis of the contaminant site is the deployment of carbon-based materials at the site. In situ and ex situ application strategies are used for this. In situ involves directly applying carbon-based materials at the site of contamination by mixing them with soil or treating groundwater. This cost-effective approach with minimal disruption is suitable when widespread contamination occurs in aquifers or soils. Conversely, ex situ application involves excavating or external treatment of contaminated soil or water. This is an expensive method but provides a controlled environment and hence is beneficial when it comes to complex mixtures of contaminants. Another approach that is used is the involvement of a permeable reactive barrier involving the use of reactive medium as activated carbon in the flow path of contaminated groundwater for the capture and degradation of pollutants. While deployment is easy, the implementation of sensors and analytical tools to monitor the performance of remediation material in real-time is critical. Conducting small-scale pilot studies before full-scale implementation ensures effectiveness under real-time conditions. To circumvent these issues, scientists are using predictive modeling and geographic information system to understand contaminant behavior and optimize remediation plans.

13.3 Future Direction and Potential Solutions

Carbon-based materials have emerged as a promising material for the treatment and removal of contaminates. As the use of these materials is still gaining prominence, addressing the challenges and exploring innovative approaches will shape their future applications. Future directions to address these challenges involve a combination of technological advancements, regulatory adjustments, and sustainability-driven practices. Some key directions and potential solutions to overcome current limitations have been discussed below.

13.3.1 Advancing the Synthesis and Design of Materials

Researchers should focus on specific functionalization of carbon-based materials so that a selective contaminant is detected with utmost selectivity and sensitivity. Further, synthesis should focus on the generation of hybrid materials, such as conjugates with metal oxides or polymers that enhance multifunctionality. Hybrids of carbon materials and polymers improve the adsorption and catalytic activity along with environmental stability compared to single components (Navrotskaya et al., 2020). The focus should also be on the generation of biodegradable and cost-effective carbon materials from agricultural waste, which would provide a sustainable alternative to engineered nanomaterial (Yaah et al., 2021).

13.3.2 Scaling-Up Production

Environmental bioremediation involves large-scale synthesis of materials, which requires optimization of several parameters to ensure similar properties of materials. Chemical vapor deposition and hydrothermal synthesis are a few methods that facilitate the cost-effective fabrication of high-quality carbon materials and allow efficient large-scale production. Besides, researchers can focus on using low-cost waste materials or a green synthesis approach using plant or agricultural products that would make the production process economically viable and environmentally friendly (Padhan et al., 2025; Vijeata et al., 2024).

13.3.3 Collaborative and Interdisciplinary Research

A collaborative research approach between material scientists, environmental engineers, ecologists, and policymakers would be beneficial in bridging the knowledge gaps and ensuring successful implementation. In addition, a public–private agency partnership would help fund the large-scale implementation and commercialization of carbon-based remediation technology.

13.3.4 Use of Enhanced Approach for Field Applications

Improvements in deployment methods, such as the use of nanocarriers that would allow controlled release, would significantly improve the efficiency of carbon material in situ. In addition, as already discussed, using permeable reactive barriers along with carbon materials would be a promising strategy. Integrating carbon-based materials with microbial remediation approach would enhance the degradation process as it would use a synergistic approach of materials and microbes. Furthermore, the integration of real-time monitoring along with remediation strategies would optimize the removal of contaminants even under varying environmental conditions.

13.3.5 Safety and Regulatory Concerns

Further research should focus on the generation of non-toxic carbon-based materials and evaluation of their long-term environmental and ecological impact. For this, researchers should focus on understanding the degradation pathway, the toxic possible effects on in vivo systems, persistence, and bioaccumulation risks. Based on these, clear guidelines for safe use and disposal of these carbon-based materials need to be framed by government and regulatory bodies. The major focus of these frameworks should be safe use with minimal exposure risk to humans and biological systems along with strategies for appropriate disposal.

13.3.6 Implementation of AI and Simulation Studies

Machine learning along with AI-driven approaches can help optimize the synthesis process of carbon materials by assisting in the identification of appropriate materials using data from a large dataset. Additionally, advanced modeling tools and simulation techniques assist in the prediction of carbon-based materials under various environmental conditions, favoring the design of an effective remediation strategy.

13.4 Conclusion

Using carbon-based materials has provided enormous promise in environmental remediation to address pollution related to air, water, and soil. This chapter highlights the advantages of improving carbon material design, providing cost-effective and sustainable techniques and the necessity to address the regulatory gaps for ensuring real-time carbon-based material implementation. Additionally, understanding the long-term impact of these materials on the environment would help implement site-specific strategies and foster collaboration between interdisciplinary researchers and stakeholders, which are crucial steps in comprehending the full potential of carbon-based remediation. We postulate that adopting a multidisciplinary approach integrating scientific innovations with societal and environmental responsibility would greatly assist in the transition of carbon-based materials from lab-scale to real-time applications.

References

Ahmed, M. B., et al. (2023). Biochar from agricultural residues for environmental remediation: A cost-benefit analysis. *Environmental Research, 225*, 115495.

Ajith, M.P. and Paulraj, R., 2022. Nanoparticles and their role in environmental decontamination technologies. In Nanomaterials for Advanced Technologies (pp. 127–140). Singapore: Springer Nature Singapore. https://doi.org/10.1007/978-981-19-1384-6_7

Al-Shaeri, M., Paterson, L., Stobie, M., Cyphus, P., & Hartl, M. G. J. (2022). Trophic transfer of single-walled carbon nanotubes at the base of the food chain and toxicological response. *Nanomaterials, 12*(24), 4363. https://doi.org/10.3390/nano12244363

Anegbe, B., Ifijen, I. H., Maliki, M., et al. (2024). Graphene oxide synthesis and applications in emerging contaminant removal: A comprehensive review. *Environmental Sciences Europe, 36*, 15. https://doi.org/10.1186/s12302-023-00814-4

Chen, G. B., & Yuan, H. Y. (2021). Effects of carbon nanomaterials on the migration and fate of organic pollutants in the ecological environment. *Ferroelectrics, 570*(1), 206–217.

Cheng, G., Zhang, A., Zhao, Z., Chai, Z., Hu, B., Han, B., Ai, Y., & Wang, X. (2021). Extremely stable amidoxime functionalized covalent organic frameworks for uranium extraction from seawater with high efficiency and selectivity. *Science Bulletin, 66*, 1996–2001.

Dai, Y., Wang, W., Lu, L., Yan, L., & Yu, D. (2020). Utilization of biochar for the removal of nitrogen and phosphorus. *Journal of Cleaner Production, 257*, 120573.

Erdely, A., et al. (2009). Cross-talk between lung and systemic circulation during carbon nanotube respiratory exposure: Potential biomarkers. *Nano Letters, 9*(1), 36–43. https://doi.org/10.1021/nl801828z

Ersan, G., Apul, O. G., Perreault, F., & Karanfil, T. (2017). Adsorption of organic contaminants by graphene nanosheets: A review. *Water Research, 126*(1), 385–398.

Esteban-Arranz, A., Compte-Tordesillas, D., Muñoz-Andrés, V., Pérez-Cadenas, M., & Guerrero-Ruiz, A. (2018). Effect of surface, structural and textural properties of graphenic materials over cooperative and synergetic adsorptions of two chloroaromatic compounds from aqueous solution. *Catalysis Today, 301*, 104–111. https://doi.org/10.1016/j.cattod.2017.03.048

Ge, Y., Priester, J. H., Mortimer, M., Chang, C. H., Ji, Z., Schimel, J. P., & Holden, P. A. (2016). Long-term effects of multi-walled carbon nanotubes and graphene on microbial communities in dry soil. *Environmental Science & Technology, 50*, 3965–3974.

Gopinath, K. P., Vo, D. V. N., Prakash, D. G., Joseph, A. A., Viswanathan, S., & Arun, J. (2020). Environmental applications of carbon-based materials: A review. *Environmental Chemistry Letters, 19*(1), 557–582.

Hao, Y., Ma, C., Zhang, Z., Song, Y., Cao, W., Guo, J., Zhou, G., Rui, Y., Liu, L., & Xing, B. (2018). Carbon nanomaterials alter plant physiology and soil bacterial community composition in a rice-soil-bacterial ecosystem. *Environmental Pollution, 232*, 123–136.

Hashemi, E., et al. (2024). Impacts of graphene oxide contamination on a food web: Threats to somatic and reproductive health of organisms. *Ecotoxicology and Environmental Safety, 285*, 117032. https://doi.org/10.1016/j.ecoenv.2024.117032

Kumar, A., et al. (2023). Thermal regeneration of activated carbon: Balancing efficacy and environmental impact. *Journal of Environmental Management, 342*, 118663.

Li, Y., et al. (2022). Hybrid carbon nanomaterials for targeted contaminant removal: Challenges and perspectives. *Journal of Cleaner Production, 371*, 133416.

Liu, Y., & Kujawinski, E. B. (2015). Chemical composition and potential environmental impacts of water-soluble polar crude oil components inferred from ESI FT-ICR MS. *PLoS One, 10*(9), e0136376. https://doi.org/10.1371/journal.pone.0136376

Liu, Z., Ling, Q., Cai, Y., Xu, L., Su, J., Yu, K., Wu, X., Xu, J., Hu, B., & Wang, X. (2022). Synthesis of carbon-based nanomaterials and their application in pollution management. *Nanoscale Advances, 4*(5), 1246–1262. https://doi.org/10.1039/D1NA00843A

Mwangi, J. N., Wang, N., Ingersoll, C. G., Hardesty, D. K., Brunson, E. L., Li, H., & Deng, B. (2012). Toxicity of carbon nanotubes to freshwater aquatic invertebrates. *Environmental Toxicology and Chemistry, 31*(8), 1823–1830. https://doi.org/10.1002/etc.1888

Nasirzadeh, N., Mohammadian, Y., & Fakhri, Y. (2020). Concentration and cancer risk assessment of asbestos in Middle East countries: A systematic review-meta-analysis. *International Journal of Environmental Analytical Chemistry, 7*, 1–15.

Navrotskaya, A. G., Aleksandrova, D. D., Krivoshapkina, E. F., Sillanpää, M., & Krivoshapkin, P. V. (2020). Hybrid materials based on carbon nanotubes and nanofibers for environmental applications. *Frontiers in Chemistry, 8*, 546. https://doi.org/10.3389/fchem.2020.00546

Nguyen, T. H., et al. (2023). Salinity and its impact on the adsorption performance of carbon-based materials. *Water Research, 231*, 119615.

Padhan, S. R., et al. (2025). Synthesis of carbon nanomaterials from agro-industrial wastes and their extensive applications. In N. Talreja, D. Chauhan, & M. Ashfaq (Eds.), *Waste-derived carbon nanostructures*. Springer. https://doi.org/10.1007/978-3-031-75247-6_3

Patel, H., et al. (2023). Anti-fouling strategies for carbon-based materials: A review of techniques and applications. *Advances in Colloid and Interface Science, 313*, 102817.

Pet, I., Sanad, M. N., Farouz, M., ElFaham, M. M., El Hussein, A., El Sadek, M. S. A., Althobiti, R. A., & Ioanid, A. (2024). Recent developments in the implementation of activated carbon as heavy metal removal management. *Water Conservation Science and Engineering, 9*, 62. https://doi.org/10.1007/s41101-024-00287-3

Piumie, R. P., Power, A., Chandra, S., & Chapman, J. (2018). Graphene, electrospun membranes, and granular activated carbon for eliminating heavy metals, pesticides, and bacteria in water and wastewater treatment processes. *Analyst, 143*(23), 5629–5645.

Priya, A. K., Muruganandam, M., & Suresh, S. (2024). Bio-derived carbon-based materials for sustainable environmental remediation and wastewater treatment. *Chemosphere, 362*, 142731. https://doi.org/10.1016/j.chemosphere.2024.142731

Punt, A., Louisse, J., Pinckaers, N., Fabian, E., & van Ravenzwaay, B. (2022). Predictive performance of next generation physiologically based kinetic (PBK) model predictions in rats based on in vitro and in silico input data. *Toxicological Sciences, 186*, 18–28. https://doi.org/10.1093/toxsci/kfab150

Ren, W., Ren, G., Teng, Y., Li, Z., & Li, L. (2015). Time-dependent effect of graphene on the structure, abundance, and function of the soil bacterial community. *Journal of Hazardous Materials, 297*, 286–294.

Rivas-Sanchez, A., Cruz-Cruz, A., Gallareta-Olivares, G., González-González, R. B., Parra-Saldívar, R., & Iqbal, H. M. N. (2022). Carbon-based nanocomposite materials with multifunctional attributes for environmental remediation of emerging pollutants. *Chemosphere, 303*(1), 135054. https://doi.org/10.1016/j.chemosphere.2022.135054

Sanchez, V. C., Jachak, A., Hurt, R. H., & Kane, A. B. (2012). Biological interactions of graphene-family nanomaterials: An interdisciplinary review. *Chemical Research in Toxicology, 25*(1), 15–36.

Sargent, L. M., Porter, D. W., Staska, L. M., Hubbs, A. F., Lowry, D. T., Battelli, L., et al. (2014). Promotion of lung adenocarcinoma following inhalation exposure to multi-walled carbon nanotubes. *Particle and Fibre Toxicology, 11*(3), 3.

Seabra, A. B., Paula, A. J., de Lima, R., Alves, O. L., & Durán, N. (2014). Nanotoxicity of graphene and graphene oxide. *Chemical Research in Toxicology, 27*(2), 159–168.

Singh, R., et al. (2022). Microwave-assisted regeneration of carbon-based adsorbents: Current status and future directions. *Renewable and Sustainable Energy Reviews, 169*, 112976.

Vasilachi, I. C., Asiminicesei, D. M., Fertu, D. I., & Gavrilescu, M. (2021). Occurrence and fate of emerging pollutants in water environment and options for their removal. *Water, 13*, 181. https://doi.org/10.3390/w13020181

Vijeata, A., Chaudhary, G. R., Chaudhary, S., Ibrahim, A. A., & Umar, A. (2024). Recent advancements and prospects in carbon-based nanomaterials derived from biomass for environmental remediation applications. *Chemosphere, 357*, 141935. https://doi.org/10.1016/j.chemosphere.2024.141935

Wang, J. Q., Zhang, S., Cao, H., Ma, J. Z., Huang, L. T. Y., Yu, S. J., Ma, X. Y., Song, G., Qiu, M. Q., & Wang, X. X. (2022a). Water purification and environmental remediation applications of carbonaceous nanofiber-based materials. *Journal of Cleaner Production, 331*, 130023.

Wang, Z., et al. (2022b). Anti-fouling strategies for carbon adsorbents in natural water matrices. *Science of the Total Environment, 850*, 158038.

Xu, J., Cao, Z., Zhang, Y., Yuan, Z., Lou, Z., Xu, X., & Wang, X. (2018). A review of functionalized carbon nanotubes and graphene for heavy metal adsorption from water: Preparation, application, and mechanism. *Chemosphere, 195*, 351–364.

Yaah, V. B. K., Ojala, S., Khallok, H., Laitinen, T., & de Oliveira, S. B. (2021). Hybrid carbon materials: Synthesis, characterization, and application in the removal of pharmaceuticals from water. *Journal of Water Process Engineering, 43*, 102279. https://doi.org/10.1016/j.jwpe.2021.102279

Yang, S.-T., et al. (2007). Biodistribution of pristine single-walled carbon nanotubes in vivo. *The Journal of Physical Chemistry C, 111*(48), 17761–17764.

Yang, X., Wan, Y., Zheng, Y., He, F., Yu, Z., Huang, J., Wang, H., Ok, Y. S., Jiang, Y., & Gao, B. (2019). Surface functional groups of carbon-based adsorbents and their roles in the removal of heavy metals from aqueous solutions: A critical review. *Chemical Engineering Journal, 366*, 608–621. https://doi.org/10.1016/j.cej.2019.02.119

Zhang, W., et al. (2023). Advanced functionalization of carbon-based adsorbents for environmental applications. *Environmental Science & Technology, 57*(3), 1450–1462.

Index

A. M. Parambil et al. (eds.), *Carbon*,
https://doi.org/10.1007/978-3-031-90613-8

The manufacturer's authorised representative in the EU is Springer Nature Customer Service Centre GmbH, Europaplatz 3, 69115 Heidelberg, Germany. If you have any concerns regarding our products, please contact ProductSafety@springernature.com

Printed and bound by CPI Group (UK) Ltd, Croydon, CR0 4YY
30/09/2026
02225233-0008